Bacterial and Fungal Diseases of Plants in the Tropics

Bacterial and Fungal Diseases of Plants in the Tropics

George F. Weber

University of Florida Press
Gainesville • 1973

632.3
Web

A University of Florida Press Book

Library of Congress Cataloging in Publication Data

Weber, George Frederick, 1894-
Bacterial and fungal diseases of plants in the tropics.

Includes bibliographies.
1. Plant diseases—Tropics. 2. Bacteria, Phytopathogenic—Tropics. 3. Fungi, Phytopathogenic—Tropics. I. Title.
SB605.T7W4 632'.3'0913 78-137853
ISBN 0-8130-0320-2
2. Florida- Plant diseases
3. Bacterial diseases of plants.
4. Fungous diseases of plants.

COPYRIGHT © 1973 BY THE STATE OF FLORIDA
BOARD OF TRUSTEES OF THE INTERNAL
IMPROVEMENT TRUST FUND

All rights reserved

PRINTED IN FLORIDA, U.S.A.

9/85

To Kate C. Weber

Contents

Foreword

ONE OF THE continuing problems facing agricultural development in tropical and semitropical areas, which are frequently the underdeveloped areas of the world, is the identification and treatment of bacterial and fungal diseases of plants. The need for a textbook which clearly describes the symptoms and the causal agents of specific plant diseases has long been evident, for use abroad as well as in United States universities in training students from tropical and semitropical areas.

It has been pointed out that training in plant pathology in the United States focuses primarily on basic principles of the discipline. This, perhaps, is as it should be. On the other hand those who will be working in tropical countries have a need for more descriptive and applied phases of the subject.

Out of a background of forty-two years of successful teaching and research, covering work in both basic and applied phases of plant pathology, Dr. George F. Weber sets his hand to meeting the needs of the student and the working plant pathologist in the tropical and semitropical areas of the world. He has carefully selected the more important diseases of over one-hundred economic plants and described their symptoms. In addition, symptoms and parasites are illustrated with numerous photographs. The less important organisms frequently associated with cultivated plant diseases are also listed, although not treated in detail.

Confronted with an immediate disease problem, one might wish for more detail on control methods, but Dr. Weber wisely treats of such methods sparingly, knowing full well of ever changing recommendations, the diversity of conditions under which plants are grown, and the continuous development of resistant plants through selection and hybridization.

It has been my privilege to have known Dr. Weber since 1934. He has been an exceedingly productive investigator and teacher in dealing with basic principles as well as coming to grips with the applied and practical aspects of plant disease identification and control. He has constantly attempted to make his teaching meaningful and realistic. Certainly this volume reflects that approach in meeting the needs of those who are serving and will serve in tropical and semitropical areas of the world.

J. Wayne Reitz
President Emeritus
University of Florida

Preface

NO CLASSIFICATION is perfect: nature recognizes no very marked divisions. As Professor Massee, of Kew, used to say 'why make a fence! Some rooster is sure to get on top of it with his head on one side and his tail on the other!' And often the higher realms of perfection are of little practical use. It is vastly more important to help twenty students to a better knowledge of a group of plants than to tickle the fancy and win the praise of one who no longer needs help. Anyone leaving the beaten track is subject to criticism, when he should get only sympathy." William A. Murrill. 1917. *American Journal of Botany* 4(6):315.

The purpose of this volume is to bring to teachers, students, agricultural agents, extension advisers, growers, crop production managers, farmers, and landowners a guide to the identification of plant diseases through the use of diagnostic symptoms and the characteristics of the causal parasite.

The data have been accumulated during the past half century by the writer as a student, research specialist, teacher, student counselor in the Agency for International Development, consultant, editor, and the author of more than a hundred publications. Additional data have been supplied by scores of students from tropical countries who have attended the College of Agriculture, University of Florida, and its several branch stations in order to pursue advanced studies in the area of plant pathology. Discussions concerning their local plant disease problems have been of considerable importance in relation to the selections included here.

The book is arranged alphabetically by common host name. For each host there is a list of specific diseases and causal agents, followed by a description of symptoms and the essential characteristics and etiology of the parasite. The diseases listed are arranged by the phylogenic relations of the parasites. Each section is concluded with a list of references and a list of additional fungi that may be infrequently associated with the host, but nevertheless cause severe damage.

Rhizoctonia occurs in the literature frequently. During the hundred years since it was first described, the name of the fungus has been changed to *Corticium.* Through recent studies, this genus has been subdivided into several new genera including: *Botryobasidium*, containing *B. microsclerotia, B. rolfsii,* and *B. salmonicolor*; *Ceratobasidium* with a single species, *C. stevensii*; and *Thanatephorus* with two species, *T. cucumeris* and *T. sasakii. Sclerotium,* also an imperfect fungus with worldwide distribution, has been studied, and the spore forms discovered place it in *Macrophomina phaseolina,* with *Sclerotium bataticola* as a synonym, *Stromatinia cepivorum* with *Sclerotium cepivorum* as a synonym, and *Leptosphaeria salvinii* with *Sclerotium oryzae* as a synonym. In addition to *Rhizoctonia, Corticium,* and *Sclerotium,* the abovementioned binomials will be used in the text.

The nomenclature of host plants and authorities used follows very closely that presented by Dr. L. H. Bailey in *Manual of Cultivated Plants* and Dr. J. K. Small in *Manual of the Southeastern Flora.* Other references dealing with tropical plants have been consulted, particularly Dr. H. F. MacMillan's *Tropical Planting and Gardening.*

The binomials of parasitic organisms have been obtained from the current literature in general and in most instances coincide with the contents of the *Index of Plant Diseases in the United States,* United States Department of Agriculture Handbook Number 165. Authorities for binomials of cryptogams have been derived from this handbook and from L. Roger's list in *Encyclopedie Mycologique* 19:2923–32. There has not always been complete agreement among literature references, and it is hoped that the choices have not been unreasonable and that should mistakes occur they will be forgiven. They are entirely the responsibility of the writer.

This book is constructed primarily as a source of authoritative, accurate, condensed, diagnostic information acquired from and supported by existing cited literature and the author's researches and observations. Excellent monographs and books dealing exhaustively with diseases of specific hosts or groups of hosts are plentiful and usually well distributed. A bibliography of the contributions in subjects related to plant diseases is supplied and is supplemented by a list of useful reference journals. The illustrations, mostly originals by the author, are credited otherwise.

The initial idea for this work and the encouragement necessary for its continuation came from the author's association with foreign students, and the project has become increasingly necessary and real as a contribution to them.

Acknowledgments

THE LIST OF COLLEAGUES and friends to whom my indebtedness is extended is very long indeed. I cannot sufficiently thank them for their help which has so greatly enhanced whatever value this volume may have in lightening the burden of those responsible for leading and guiding those who are uninformed but eager to learn.

The writer is permanently grateful to the following members of the Department of Plant Pathology, University of Florida, who have read the manuscript in its entirety or in part, for their generosity in the allotment of their valuable time: Drs. A. A. Cook, Phares Decker, T. E. Freeman, H. H. Luke, C. R. Miller, H. N. Miller, D. E. Purcifull, D. A. Roberts, R. E. Stall, F. W. Zettler, and Chairman, L. H. Purdy. I cannot sufficiently thank them for their abundant help. They have been painstakingly critical in regard to good usage of technical material and have offered many valuable suggestions applicable to content, arrangement, organization, and manner of presentation.

The contributions of the following people, through correspondence, by individual contact in the United States or abroad, and as students attending the university while undergoing special training or studying for advanced degrees, have been freely given and have been carefully considered in relation to the selection of many of the hosts and diseases presented herein. They have been important participants, probably unknowingly, in the accumulation of the contents of this volume and it is hoped that they may experience some pleasure in their endeavors. To them I am most cordially extending thanks and appreciation in acknowledging my obligation. Brazil: J. S. Aroeira, A. C. Batista, E. France, F. Galli, D. C. Giocometti, Ear Kimmal, L. G. Lordehlo. Cambodia: Sam E. Jalcofy. Colombia: José V. Arboleda. Costa Rica: C. E. Fernandez, Ed. Jimenez, Manuel Jimenez. Cuba: Hera Acuna, Julio Capo-Mass. Ecuador: Mario Jalil. Ethiopia: Amare Aetakum. El Salvador: Gilberto R. Aguila, Oscar Ancalmo, Carlos Buryos, Aristides Escobar, Karl Flores, José E. Funes, Bernardo Patino, Tomas Villanova. Greece: P. I. Constantinou, C. Catseurbos. Guatemala: Mario Fernandez. Honduras: Emilio Coto, Miguel A. Elvir, Jorge Maradiaga, Zuniga Ridoniel, Bernardo Roehrs. India: Sheth Anilkumat, R. Radhikrishna, G. N. Safaua, P. P. Ninjappa. Indonesia: Faisal Aman, Ishak Mohamad, Ida M. Oka. Iran: Mohammod Akhauzodegan, Hojjute S. Moosa, Aziz Shiralipour. Malayasia: Abraham Yusof. Mexico: Dagoberto Aguilar, Badillo Pacheco. Nicaragua: Vernon D. Bent, José C. Camales. Pakistan: Abdul G. Kausar. Panama: Alberto Broce. Peru: Paul H. Figueroa, Rafael Franciosi. Philippines: Juan T. Carlos, Tricita Hidalgo, Miouisio Minora, Faustino T. Orillo. Puerto Rico: Marciso Almeyda. Taiwan: Sing Ching Chen, Y. J. Chin, Ching-Chen Chow, Chalo Tsing, Chin-Chyu Tu. Thailand: Saksiri Kirtpredi, Sasipalin Pisit, Sowart Ratanaivorabhan, Sompark Siddhipongse, Sataeoth Pramaun, Phongsayam Sausar. Turkey: Ismail Baykal. Uruguay: Oscar Abaracon. Venezuela: R. E. Pontis. Viet Nam: Ha Thu Nguyen, Van Hank Nguyen.

Among the numerous citizens of foreign countries who have kindly furnished information during the past several years concerning the diseases of paramount importance in their respective regions, the following should be mentioned especially and to each and every one of them I express my sincere thanks: Leopoldo Abrego, El Salvador; Dr. Enrique Ampuero, Ecuador; Basil Anastasiadis, Greece; Robert Armour, Honduras; Dr. M. M. Ben Halim, Lybia; Taye Beyuneh, Ethiopia; Dr. José Calvo, El Salvador; Dr. Winit Changsri, Thailand; Dr. G. G. Divinagracia, Philippines; Sok Doeung, Cambodia; Adrian Fajardo, Peru; Juan Ferrer, Panama; José Gonçalves, Brazil; Dr. Luis Gonzales, Costa Rica; Dr. Fereidoon Hashewi, Iran; Sastra A. Hidir, Indonesia; C. F. Loh, Malaysia; Dr. Kishwar Maur, India; Dr. I. Malca, Panama; Dr. Simon Malo, Ecuador; Dr. A. S. Muller, Honduras; Dr. R. M. Natour, Israel; Son Hoang Nguyen, Viet Nam; Limhuot Nong, Cambodia; Faisal Osman, Sudan; Edwardo Porros, Nicaragua; Dr. J. Enrique Rivera, Mexico; Abraham Ziver, Chile.

Additional assistance through consultation, correspondence, and conversation is cheerfully and happily acknowledged to Drs. C. H. Blazquez, J. F. Darby, R. A. Conover, E. P. DuCharme, J. W. Kimbrough, R. D. Magie, J. P. Jones, R. R. Kincaid, Hugh Popenoe, and E. T. York, all of the Institute of Food and Agricultural Sciences of the University of Florida, and to Dr. L. F. Haines of the University College of the University of Florida.

The writer is most grateful for the help and assistance rendered by Lulu Mae Marshall, chief office administrator, and Shirley Strouse of her staff for materials used, typing of the manuscript, letter writing, and general cooperation. Again may I express my appreciation to these people who have contributed so generously. The undertaking probably would not have been completed without their devoted help.

Introduction

SINCE THE BEGINNING of civilization the tiller of the soil has been robbed of some of the fruits of his labours by plant diseases." C. H. Gadd (Ceylon).

The low elevations in the tropics are generally considered to be wet and warm, conditions conducive to luxuriant plant growth. The highlands provide ecological conditions in relation to altitude that correspond to latitude in other crop producing areas of the world. These environmental factors are favorable to the development of plant diseases. To improve and increase the food derived from primitive agriculture, a knowledge of plant diseases and some idea of how to combat them must be supplied.

The production of food and fiber is most essential for survival. Loss of these products attributed to pathogenic organisms occurs wherever plants are grown. Ecological factors in relation to plant development are highly variable and must be considered when selecting types and quantity of plants to grow. As these conditions vary around the world, crop plants likewise vary. Causes of some losses are associated with fungi. Therefore, it is most desirable to prevent such disease-producing plants from destroying those under cultivation. This, then, becomes a problem of plant disease control. Some suggestions are implied here, although specific formulas and directions are intentionally avoided. Effective methods and means for crop plant preservation in one area may not be most efficient in another.

The application of disease control in the early stages of experimentation usually centers around the recognition of the symptoms and their association with various phases of the life cycle of the organisms present. This information is pertinent to proper diagnosis, supported by Koch's Postulates. When these facts are known there is established a factual basis for methods of control. If this information is not available, control may not be successful. On this premise, emphasis is directed toward an accurate understanding of plant disease symptoms and detailed knowledge of the life cycles, etiology, and morphology of fungi. The most urgent and pressing problems which the inhabitants of many tropical lands must face is how to close the gap between production and consumption of food. Investigations on the scale of scientific research into the cause of plant diseases are still largely in the hands of very few. With political freedom and its associated responsibilities, there develops a change from being dependent on outsiders to show the way to carrying the burden themselves. Increased food production requires more cultivated lands, fertile and supplied with economic plants and readily available water. These basic requirements must of necessity be correlated with a knowledge of cultural methods and the use and storage of yields. Even when the prospects appear to be very good, plant diseases have been known to be destructive and cause famine. Information from throughout the world must be accumulated, compiled, and supplied to struggling populations so that they may help themselves to a better existence.

The environment surrounding growing plants is most important since the plant is stationary and must survive or perish under existing or developing conditions supplied naturally or artificially. Meteorological factors are mostly included under the headings of temperature, light, air, humidity, topography, and mechanical forces such as wind damage, hail, and flooding. These are highly variable and as a group are exceedingly influential in plant distribution and survival and are often directly related to the prevalence of disease. Other contributing influences are associated with the soil as the medium in which the plant grows. Again, there is a certain overlapping of temperature and moisture, but added to these are the importance of soil texture, reaction, compactness, toxicity, deficiencies, slope, origin, and age. The biological aspects naturally introduce most of the diseases of plants such as those caused by animals, nematodes, insects, parasitic plants, viruses, bacteria, and fungi. It is to these that more detail is given in order to aid in their prevention, control, and elimination.

In a large number of specific areas in certain states and countries, laws, rules, and regulations are in force for the specific purpose of excluding by the existence of a quarantine the transportation of herbage of any kind into designated places. These become operational and are executed through inspection by competent inspectors with the authority of refusal of movement or confiscation and destruction. The first operation of control is *exclusion* by any means possible of any disease that may threaten the productive life of a plant and its fruit. This is brought about by the formation of a barrier that is sufficient to prevent the ingress of outsiders. Physical structures such as covered houses, flowing water, open ditches, and windbreaks have been used. The entire universal application of quarantine and regulations between states, countries, and continents is promulgated and enforced as a means of plant disease control.

The process of freeing an area of contaminated and diseased plants and reducing the population of soil inhabitors suggests the possibilities of roguing and destruction of plants or plant parts that might spread diseases. This operation should be followed by rotation and fallow as a means of reducing susceptible hosts or alternate hosts of specific parasites. There are certain soil treatments that should be used in reducing populations of detrimental organisms. Applications of heat at specific temperatures and gas at lethal concentrations associated with antibiotics are usually successful in at least commercial control.

Control of disease-producing, soil-inhabiting parasites associated with soil treatments is the protection given to seeds to rid them of contaminants by aging them sufficiently so that many associated organisms fail to survive. Seed are protected by applying certain fungicides and coating them completely prior to admitting them to the growing medium. There are ways of protecting plants to a certain extent by raising barriers around them, producing trellises, raising them above soil surface, alternating with a crop-free season, supplying a trap crop, trenching to prevent root intermingling, high budding of perennials, and off-season or out-of-season planting of a susceptible crop, resulting in disease escape. There are many examples illustrating the successful use and application of each of these methods.

If, however, none of the precautionary applications is sufficient to prevent invasion of the protected area, then it becomes necessary to use *eradication* for plant disease control. This implies the destruction of parasites and the diseases they cause. Certain forms of seed and soil treatments and fumigation are generally applied. Often plant parts are removed and frequently whole plants extirpated. There are many control formulae applied to entire plants and to their products in a wide range of operations. The success obtained is usually correlated with strict adherence to methods of operation. When operations associated with the parasites have not been sufficient to prevent crop losses it becomes apparent that *protection* of plants in their healthy condition is necessary.

Fungicides of various compositions, applied as dust or as sprays, are in most instances intended to function as protectants. They form a coating or layer of material that is toxic to the spores of fungi, and, if properly formulated and carefully applied at the proper time, are nondetrimental to the tenderest plant parts. To be most effective applications to growing plants must be made at 7 to 10 day intervals for continuous complete coverage. Protection is gained by coating seed with a fungicide that is retained through the germination period. Spreaders and stickers are useful on many occasions to assure complete and prolonged benefits.

Frequently, eradicant fungicides are profitably used in proper disinfection and sterilization of soil and growth media. They are used effectively as sprays in more concentrated form as applicants to plants that are deciduous and dormant.

Chemotherapy is also a promising treatment for plants, both as a protectant and eradicant. It offers much for future consideration but is not extensively applied at the present time. Systemic fungicides also require increased attention.

Most of these operations are usually considered temporary and effective for only a short time during growing season and must of necessity be repeated. This form of control is adequately demonstrated by the application of fungicidal sprays and dusts to fields, groves, and orchards. There are certain chemical solutions used in disinfecting washes sufficient to prevent invasion through natural openings or through mechanical injuries. Certain plants are protected from disease organisms, because of natural spaces through which the parasite does not survive. However, there is always the failure of commercial control, and then it becomes necessary to further manipulate the cultivated plant.

The lengthy duration of corrective methods is associated with resistance of the plant to pathogens. The discovery of resistance associated with natural chimeras and sports as plant variations in nature have been successful. Plant cultivators must be continually alert in order to recognize the external manifestations associated with resistance. They are seldom detected unless one is looking for such occurrences. Actually, however, the plant breeder, knowledgable in genetics and the manipulation of chromosomes, makes possible new segregates that may contain the desirable characteristics and be resistant. They perfect the selection of hybrids that survive successive inoculations by pathogens that cause disease. Resistance is not a guarantee of permanent disease control; it is, however, a means of producing profitable crop plants and may lead to the discovery of immunity at any time as a final successful way of growing plants in the presence of virile parasites. This is possible when *resistance* to disease is recognized in plants. There has been a great deal of searching for plants that develop in the presence of virulent disease-producing parasites. The variation in plants which continues to take place in nature offers the keen observer an opportunity to utilize their inherited tendencies.

Abaca, *Musa textilis* Nees

Moko wilt, *Pseudomonas solanacearum* E. F. Smith
Dry sheath rot, *Marasmius semiustus* Berkeley & Curtis
Anthracnose, *Gloeosporium musarum* Cooke & Massee
Stem rot, *Helminthosporium torulosum* (Sydow) Ashby
Stalk rot, *Fusarium moniliforme* Sheldon var. *subglutinans* Wollenweber
Wilt, *Fusarium oxysporum* Schlechtendahl f. 3. Wollenweber
Vascular disease, *Fusarium oxysporum* f. sp. *cubense* (E. F. Smith) Wollenweber
Other fungi associated with abaca

Moko wilt, *Pseudomonas solanacearum* E. F. Sm.

Symptoms

Early symptoms are rusty brown linear streaks along the veins on some or all of the leaves. The disease spreads laterally from vein to vein. Severely diseased leaves show parallel streaks from midrib to blade margins. Necrotic streaking is accompanied by considerable yellowing and is followed by browning, drying, and death of the leaf. The leaf petiole may collapse at the base or farther up the midrib and hang down. Often the central leaves show the streaks, or they may wilt abruptly. The entire plant may wilt with no streak apparent. Cross sections of the pseudostem and rhizome reveal a brown to almost black discoloration of the vascular tissue. Sections of the rhizome show discolored spots where the infected vascular tissue connects with the young suckers, which then become inoculated.

Etiology

The organism is a short, gram-negative rod, single or in pairs, motile by a single polar flagellum, and measures $1–5 \times 0.5\ \mu$. Culture colonies are opalescent, dirty white becoming brown, small, irregular, smooth, and shiny. Optimum temperature, 36–37C.

References

4, 7.

Dry sheath rot, *Marasmius semiustus* Berk. & Curt.

Symptoms

The disease spreads in the pseudostem but usually is found first on the roots and corms. The leaf sheaths become watery and turn brown. As the fungus develops, it grows into the next interior leaf sheath, causing the invaded sheaths to stick together because of an abundance of white mycelium that continues to penetrate deeper. As the outer sheath becomes more completely invaded, it turns brown and dies but remains in place. Under favorable moisture and temperature conditions, mushroomlike fruiting bodies appear on the dead tissue.

Etiology

The mycelium penetrates inward and usually extends to the innermost part of the pseudostem; sometimes the flowering stalk is invaded. Rhizomorphlike mats develop on the outer dead parts and produce sporophores on the accumulated debris at ground level. The caps of these mushrooms are brown to yellow and 5–15 mm in diameter. The gills on the lower side of the pileus are wide-spaced and nearly white; the stipe is white, attached eccentrically, and 7–9 mm high. The basidiospores are white or hyaline, oval, 1-celled, and $7–9 \times 5–6\ \mu$.

References

9, 10, 12.

Anthracnose, *Gloeosporium musarum* Cke. & Mass.

Symptoms

All aboveground parts of the plant may be affected by the disease. Discolored spots appear most frequently on the leaf petiole and sheath in the pseudostem. The lesions are dark-colored, elongate, slightly sunken, more or less dry, and sometimes expose the vascular fibers. Spots on the leaf blade are brown to bleached gray. They are circular at first, but later become elongate in the direction of the veins.

Etiology

The mycelium is hyaline, septate, branched, mostly intercellular, and causes the tissue to become dark-colored. The acervuli form under the cuticle and emerge by its rupture. The acervuli are pink during wet, humid weather because of the abundance of conidia; in dry weather they are mostly black. They are 150–400 μ in diameter and are crowded with numerous conidiophores, 30–40 μ long. The conidia are hyaline, 1-celled, with 2 vacuoles, 10–15 μ, and usually have rounded ends and parallel sides.

Reference

1.

Stem rot, *Helminthosporium torulosum* (Syd.) Ashby

Symptoms

Small, brown spots with dark centers surrounded by a lighter border are the early indications of the disease. As they increase in size and number, they coalesce into a larger spot that becomes sunken and forms a scarlike injury. The center of the oval, elongate, dark brown to black area becomes covered with the mycelium of the fungus. These sunken, cankerlike areas continue to develop. When several are present, the stem may be more or less girdled and weakened and may lean over.

Etiology

Mycelium develops over the surfaces of the exposed diseased areas and supports conidiophores that are brownish, 48–170 μ high, and 10–15 μ in diameter. The conidia are hyaline becoming smoky olive, crescent-shaped or curved, 6–14-septate, distinctly nodulose, and 25–59 × 8–12 μ. The ascospore stage is ***Deightoniella torulosa*** (Sydow) Ellis.

References

1, 6, 12.

Stalk rot, *Fusarium moniliforme* Sheldon var. *subglutinans* Wr.

Symptoms

The conspicuous manifestation of the disease is the dying and rotting of the heart leaves. This continues until the downward advancement of the disease reaches the growing point, there killing the plant and causing decay of the rhizome. There is no wilting of the foliage to any extent before the plant is killed.

Etiology

The mycelium is septate and branched; it is more or less colorless, but may be highly colorful, pink to blue and purple, under certain conditions and is variable as to color in culture. The microconidia are 2–5-septate; macroconidia measure 29–61 × 3–5 μ. No chlamydospores are formed.

Reference

8.

Wilt, *Fusarium oxysporum* Schlecht. f. 3. Wr.

Symptoms

The first indication of the disease is inward curling near the tips of the lower leaves followed by a noticeable stunting. Then drooping and wilt begin at the tips and proceed toward the petiole, resulting in yellow to brown discolorations of the blade. More of the older leaves become yellow as new leaves appear. A red to violet color is present in the vascular tissue or corms. Such parts and accompanying roots rapidly decay, and the plant dies.

Etiology

The mycelium is hyaline, white en masse, cottony, branched, and somewhat constricted at the septa. The conidiophores are hyaline, septate, and branched in whorls. Microconidia are hyaline, 1–2-celled, mostly nonseptate, and average 8–9 × 3–4 × 3–5 μ. Chlamydospores are produced abundantly in culture and measure 5–7 × 3–6 μ.

Reference

3.

Vascular disease, *Fusarium oxysporum* f. sp. *cubense* (E. F. Sm.) Wr.

Symptoms

The sheaths composing the pseudostem may show some external discoloration, pointing to internal symptoms of the disease. Leaves sometimes show streaks of rusty-colored tissue following the veins from the midrib to the leaf margins. Often the leaves seem to be bunched, as caused by a shortening of the internodes; thus, the plant appears stunted. In many cases there is only slight internal vascular discoloration. Some longitudinal splitting of the sheaths or leaf bases occurs. New stems may grow out and away from diseased plants. During the decorticating process, diseased plants exhibit greater difficulty in the separation of the fibers from the vascular tissue, and these tissues possess decreased tensile strength. Diseased fibers are distinctly fuzzy. Laboratory studies and cross inoculations have shown that there is some doubt as to the similarity of this vascular disease and Panama disease of banana. These experiments show also that there is a wide variation in resistance to this disease among abaca selections.

Etiology

The mycelium is hyaline, white when dense and felted, and in culture may become bluish or rose red, depending on the medium. It is branched, septate, and produces sporodochia. The microconidia are produced in heads abstricted from tips of secondary branches and enclosed in mucilage. They are hyaline, oval, 1-celled, and $5–8 \times 2–3$ μ. The macroconidia are formed on short conidiophores clustered in sporodochia. They are hyaline, mildy bent, mostly septate, while some are up to 5-septate, and $20–30 \times 4–5$ μ. Chlamydospores are oval, thick-walled, granular, and 9×7 μ.

References

11, 12.

Other fungi associated with abaca

Botryodiplodia theobromae Patouillard
Cercospora musae Zimmermann
Erwinia carotovora (Jones) Holland
Macrophoma musae (Cooke) Berlese & Voglino
Macrophomina phaseolina (Tassi) Goidanich
Phytophthora parasitica Dastur
Pythium butleri Subramaniam
Sclerotium rolfsii Saccardo
Thielaviopsis paradoxa (De Seynes) Hoehnel
Ustilaginoidella musaeperdae Essed

References: abaca

1. Agati, J. A. 1925. The anthracnose of abaca, or Manila hemp. Philippine Agriculturist 13:337–44.
2. Agati, J. A., et al. 1934. Further studies on the stem rot of abaca in the Philippines. Philippine J. Agr. 5:191–211.
3. Castillo, B. F., and M. S. Celino. 1940. Wilt disease of abaca, or Manila hemp (Musa textilis Nees). Philippine Agriculturist 29:65–85.
4. Elliott, C. 1951. Manual of bacterial plant pathogens. 2d ed. Chronica Botanica: Waltham, Mass., p. 139.
5. Loegering, W. Q. 1953. Marasmius stenophyllus on abaca (Musa textilis) in Central America. Phytopathology 43:479 (Abstr.).
6. Mendiola-Ela, V., and M. O. San Juan. 1954. Leaf spot and stem rot of abaca. Philippine Agriculturist 38:251–71.
7. Palo, M. A., and M. R. Calinisan. 1939. The bacterial wilt of the abaca (Manila hemp) plant in Davao: 1. Nature of the disease and pathogenicity tests. Philippine J. Agr. 10:373–95.
8. Ramos, M. M. 1933. Mechanical injuries to roots and corms of abaca in relation to heart-rot disease. Philippine Agriculturist 22:322–37.
9. Ramos, M. M. 1941. Dry sheath rot of abaca caused by Marasmius and suggestions for its control. Philippine J. Agr. 21:31–39.
10. Robinson, B. B., and F. L. Johnson. 1953. Abaca—a cordage fiber. U.S. Dept. Agr. Monograph 21:39.
11. Waite, B. H. 1954. Vascular disease of abaca or Manila hemp in Central America. Plant Disease Reptr. 38:575–78.
12. Wardlaw, C. W. 1961. Diseases of the banana and of the Manila hemp plant. 2d ed. John Wiley & Sons: N.Y., pp. 307–12.

Acacia, *Acacia eburnea* Willdenow

Rust, *Ravenelia esculenta* (Barclay) Narasimhan & Thirumalachar
Other fungi associated with acacia

Rust, *Ravenelia esculenta* (Barcl.) Naras. & Thirum.

Symptoms

The axillary buds become hypertrophied, elongated, and turn orange. Malformed branches and flower heads develop similarly. The diseased flower heads produce pods that are 10–15 times normal size. The hypertrophied parts are tender and free from fibers. The malformed shoots present many irregular and curious shapes. All the infected parts show 10–15 per cent sugars and other nutritional components. Infected parts are collected and used as food.

Etiology

The dormant mycelium, as an obligate parasite, is perennial in the host tissue and invades new growth, developing pycnia and aecia. Aeciospores are ovate to cuboid, verrucose, and $30–40 \times 16–22$ μ. Uredia and telia form small, inconspicuous leaf spots. Uredospores are brown, ovate, verrucose, and $18–25 \times 15–18$ μ. The dark brown telia are formed in old uredia later in the season. The teliospores are pedicellate with a head including 4–8 spores. The pedicels are compound and up to 15 μ long. The teliospores are brown, 1-celled, smooth, and $20–25 \times 18–20$ μ. Basidiospores produced on the 4-celled basidium are globular, measuring 7×10 μ.

References

1, 2.

Other fungi associated with acacia

Ravenelia acaciae-arabicae Mundkur & Thirumalachar
Ravenelia angustissima (Miller) Kuntze
Ravenelia hieronymi Spegazzini
Ravenelia papillosa Spegazzini
Ravenelia roemerianae Long
Ravenelia thornberiana Long
Ravenelia versatilis (Peck) Dietel

References: acacia

1. Narasimhan, M. J., and M. J. Thirumalachar. 1961. Ravenelia esculenta, an edible rust fungus. Phytopathol. Z. 41:97–102.
2. Thirumalachar, M. J. 1941. Hapalophragnium ponderosum Syd., on Acacia leucophlaea Willd. J. Indian Botan. Soc. 20:293–98.

Agave, *Agave sisalana* Perrine

Zebra leaf spot, *Phytophthora arecae* (Coleman) Pethybridge
Leaf spot, *Nectriella miltina* (Montagne) Saccardo
Sooty mold, *Dimerosporium agavectova* Patouillard & Harsot
Leaf spot, *Dothiorella sisalanae* Roger
Black leaf spot, *Diplodia natalensis* Evans
Anthracnose, *Colletotrichum agaves* Cavara
Other fungi associated with agave

Zebra leaf spot, *Phytophthora arecae* (Col.) Pethyb.

Symptoms

The disease begins as small leaf lesions that rapidly enlarge, developing alternate concentric rings or zones of dark purple and green with pale greenish yellow margins. The centers gradually darken, and sometimes a sticky exudate appears. The tissue becomes dry and rough, with gray-yellow-white rings. A bole rot and a spike rot often follow in association with the secondary invasion of cut-leaf scars and a soft basal rot of the spike. The group of symptoms is attributed to the causal fungus and secondary invaders such as *Aspergillus niger* and *Mycothecium verrucaria.*

Etiology

The mycelium is hyaline, nonseptate, branched, and produces haustoria. The hyaline chlamydospores are spherical and 22–50 μ in diameter. Sporangia are hyaline, oval to lemon-shaped, papillate, and 32×58 μ. Zoospores are hyaline, bicilliate and 8×15 μ; 15–30 are produced per sporangium. Oospores are hyaline, spherical, smooth, and 18–41 μ in diameter.

References

1, 4, 10.

Leaf spot, *Nectriella miltina* (Mont.) Sacc.

Symptoms

Large brown to black spots on the leaves are more or less circular to oval and may be separate or coalescing. Spots appear on both leaf surfaces, and the tissue is softened sufficiently to free the vascular fibers, which become discolored.

Etiology

The stromata in pink or red groups are small, about 2 mm in diameter. They are formed by the dark mycelium which is branched, septate, and produces a large number of hyaline, 1-celled, cylindrical, straight or curved conidia, measuring 16–26 × 4.5–7 μ. The perithecia are reddish brown and form in the same stromata at a later time. The asci contain 8 hyaline, ovoid, very small ascospores, measuring 3–6 × 2–3 μ. The vegetative stage is *Tubercularia agaves*.

Reference

9.

Sooty mold, *Dimerosporium agavectova* Pat. & Har.

Symptoms

The leaf spots are generally circular, superficial, up to 30 mm in diameter, usually nonconfluent and dark, with clumps of black mycelium, 4–5 μ in diameter, radiating from a center. Young plants are sometimes killed.

Etiology

The perithecia are hemispherical and up to 90 μ high; they produce 8-spored asci, lack paraphyses, and are 36–40 × 25 μ. The ascospores are pale brown, 2-celled, straight or curved, more pointed at one end, and 26 × 8 μ.

Reference

9.

Leaf spot, *Dothiorella sisalanae* Roger

Symptoms

Spots on the leaves, usually on both surfaces, are black, round to elongate, and subepidermal becoming erumpent.

Etiology

The stromata are about twice as long as wide, 400–800 μ long and up to 500 μ high. The pycnidia are mostly scattered and sunken in the stromata; they are black, oval to subglobular, and up to 200 μ in diameter. The pycnidiospores are hyaline, ovoid, 1-celled, with straight sides and rounded ends, and measure 10–12 × 3–6 μ.

Reference

8.

Black leaf spot, *Diplodia natalensis* Evans

Symptoms

The earliest indications of the disease are large, black, circular to slightly elongate, sunken spots on the older leaves. The disease spots develop slowly, as do the leaves, so that a year or more passes before the spots become conspicuous on the harvested leaves. The black spots cause the fibers to be black and reduce their tensile strength and commercial value.

Etiology

The fungus mycelium is dark, septate, branched, and rather plentiful. It accumulates subepidermally where the black, globose, ostiolate pycnidia develop, usually in old lesions that may be on discarded leaves. The pycnidiospores, sometimes exuded in tendrils, are white, oval, 1-celled, and soon mature to dark, striate, 2-celled conidia, measuring $24–30 \times 12–15\ \mu$. The sexual stage is *Physalospora rhodina.*

References

5, 6.

Anthracnose, *Colletotrichum agaves* Cav.

Symptoms

Light brown, somewhat depressed leaf spots up to 2 cm in diameter develop on the leaves. They are at first surrounded by a pale green halo that later becomes black. The marginal tissues usually become brown and later change to white as concentric circles are formed by reddish brown stromatic masses of mycelium.

Etiology

The hyaline conidiophores are upright, measuring $30–40 \times 3–4\ \mu$. The setae are brown, sparingly branched, up to 100 μ high, and about 5 μ wide. The conidia are hyaline, pinkish, 1-celled, with straight sides and rounded ends, and measure $28–35 \times 6–7\ \mu$. The ascospore stage is *Glomerella cingulata.*

References

2, 3, 7.

Other fungi associated with agave

Botryodiplodia theobromae Patouillard
Coniothyrium concentricum (Desmazieres) Saccardo
Coryneum congoense Torrey
Cucurbitaria agaves Sydow & Butler

Diplodia agaves Moesz & Göllner
Hendersonia agaves Maublanc
Leptosphaeria agaves Sydow & Butler
Macrophoma brevipes (Penzig & Saccardo) Berlese & Voglino
Macrosporium lanceolatum Massee
Marssonina agaves Earle
Melanconium americanum Peck & G. W. Clinton
Metasphaeria polygonati Saccardo & Fautrey
Mycosphaerella agavis Massalongo
Nectria bouanseana Saccardo
Phoma agaves Saccardo
Phyllosticta agaves Maublanc
Pleospora bataanensis Petrak
Plowrightia agaves Maublanc
Septoria megaspora Spegazzini
Stagonospora gigantea Heald & Wolf
Tubercularia agaves Patouillard

References: agave

1. Ashby, S. F. 1929. Strains and taxonomy of Phytophthora palmivora Butler (P. faberi Maubl.). Trans. Brit. Mycol. Soc. 14:18–38.
2. Barthelet, J. 1946. L'Anthracnose des agaves. Ann. Epiphyties 8:111–20 (1942). (R.A.M. 25:394.)
3. Butler, E. J. 1918. Fungi and diseases in plants. Thacker: Calcutta, India, pp. 374–76.
4. Clinton, P. K. S., and W. T. H. Peregrine. 1963. The zebra complex of sisal hybrid no. 11648. E. African Forestry J. 29: 110–13.
5. Crandall, B. S., L. Abrego, and B. Patino. 1954. Mechanics of the control of henequen black leaf spot in El Salvador. Plant Disease Reptr. 38:380–83.
6. Crane, J. C., and F. L. Wellman. 1951. Edad del henequen en relacion con los caracteristicas de la fibra. Intern. Inst. Agr. Sci. (Turrialba) 1:74–77.
7. Morstatt, H. 1930. Blattkrankheiten der sisalagave. Der Trapenpflenzen 33:307–12. (R.A.M. 10:31–32.)
8. Roger, L. 1938. Quelques champignons exotiques nouveaux ou peu connus.—III. Bull. Soc. Mycol. France 54:48–54.
9. Roger, L. 1953. Phytopathologie des pays chauds. Encyclopedie Mycologique 18: 1274, 1465–67.
10. Tucker, C. M. 1931. Taxonomy of the genus Phytophthora de Bary. Missouri Univ. Agr. Expt. Sta. Res. Bull. 153:1–197.

Allspice, *Pimenta dioica* (Linnaeus) Merrill

Rust, *Puccinia psidii* Winter

Rust, *Puccinia psidii* Wint.

Symptoms

Infection spots on young foliage, the florescence, and succulent young twigs appear first as mere pinpoints. They enlarge up to 1 cm in diameter and an abundance of light yellow powdery spores is produced on the surface of these spots. They are produced mostly on the lower leaf surface.

Etiology

Light infections result from scattered sori. When severe, however, the spots may coalesce and the entire leaf may be involved; the leaf becomes brown and is shed. Premature shedding results in the formation of new foliage which in turn becomes diseased. Plants are weakened and finally die. Nondiseased foliage becomes immune to infection after about a month. Only the uredinial and telial spore stages are known in Jamaica.

Reference

1.

Reference: allspice

1. MacLachlan, J. D. 1938. A rust of the pimento tree in Jamaica, B.W.I. Phytopathology 28:157–70.

Annona, *Annona* spp.

Fruit rot, *Phytophthora parasitica* Dastur var. *macrospora* Ashby

Fruit rot, *Phytophthora parasitica* Dast. var. *macrospora* Ashby

Symptoms

The earliest symptoms appear as water-soaked areas that enlarge and cover the entire fruit, which gradually becomes brown, shriveled, dry, hard, and drops from the plant. All stages of the fruit are affected, but maturity determines the severity of the disease; all diseased fruits, however, are a total loss. No other parts of the plant are reported affected. The disease is most prevalent during humid and rainy weather. *Annona* species affected include: *A. cherimola* Miller, cherimoya; *A. diversifolia* Linnaeus, custard apple; *A. muricata* Linnaeus, soursop; *A. palustris* Linnaeus, alligator apple; *A. reticulata* Linnaeus, custard apple; and *A. squamosa* Linnaeus, sugar apple.

Etiology

The fungus produces abundant sporangia and chlamydospores; the sporangia are $40 \times 19\ \mu$. The oospores in culture are $21\ \mu$ in diameter.

Reference

1.

Reference: annona

1. Rao, V. G., M. K. Desai, and N. B. Kulkarni. 1962. A new Phytophthora fruit rot of Annona squamosa from India. Plant Disease Reptr. 46:874–76.

Avocado, *Persea americana* Miller

Fruit spot, *Pseudomonas syringae* van Hall
Crown gall, *Agrobacterium tumefaciens* (E. F. Smith & Townsend) Conn

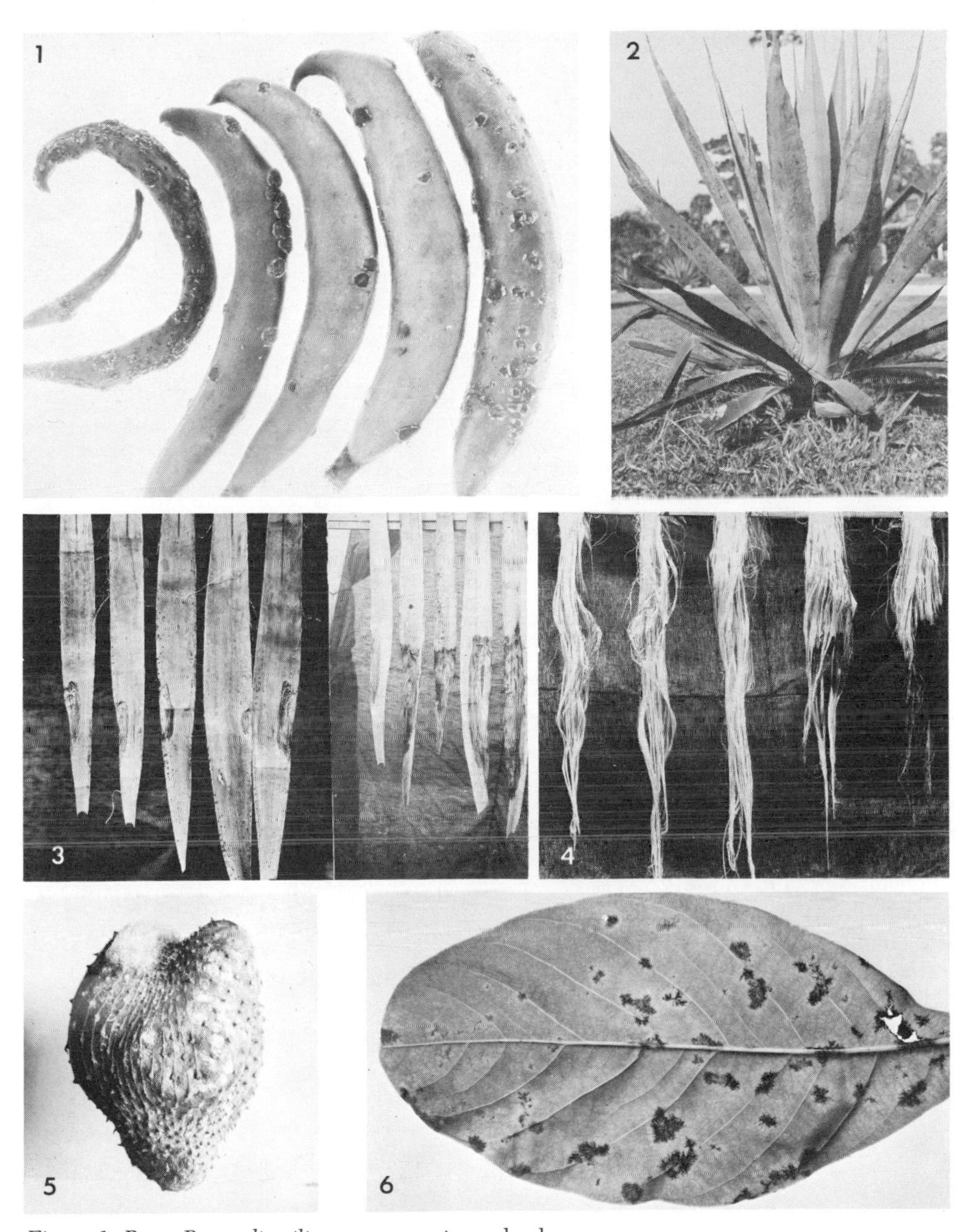

Figure 1. Rust, *Ravenelia siliquae*, on acacia seedpods.

Figure 2. Black leaf spot, *Diplodia natalensis*, on agave. (Photograph by L. Abrego.)

Figure 3. Black leaf spot, *Diplodia natalensis*, on foliage of henequen, showing stages of disease development. (Photograph by L. Abrego.)

Figure 4. Black leaf spot, *Diplodia natalensis*, on henequen, showing fiber deterioration. Three at left healthy. (Photograph by L. Abrego.)

Figure 5. Anthracnose, *Colletotrichum gloeosporioides*, on annona. (Photograph by H. G. McMillan.)

Figure 6. Powdery mildew on annona. (Photograph by H. G. McMillan.)

Root rot, *Phytophthora cinnamomi* Rands
Collar rot, *Phytophthora cactorum* (Lebert & Cohn) Schroeter
Soft rot, *Rhizopus nigricans* Ehrenberg
Leaf spot, *Phyllachora gratissima* Rehm
Mushroom root rot, *Clitocybe tabescens* (Scopoli ex Fries) Bresadola
Shoestring root rot, *Armillaria mellea* Vahl ex Fries
Root rot, *Rhizoctonia solani* Kuehn
Root rot, *Rosellinia bunodes* (Berkeley & Broome) Saccardo
Root rot, *Rosellinia necatrix* (Hartig) Berkeley
Surface rot, *Dothiorella gregaria* Saccardo
Fruit rot, *Phomopsis perseae* Zerova
Stem-end rot, *Diplodia natalensis* P. Evans
Leaf spot, *Pestalotia adusta* Ellis & Everhart
Anthracnose, *Colletotrichum gloeosporioides* Penzig
Rusty blight, *Colletotrichum nigrum* Ellis & Halsted
Scab, *Sphaceloma perseae* Jenkins
Powdery mildew, *Oidium* sp.
Leaf spot, *Cercospora purpurea* Cooke
Smudgy spot, *Helminthosporium* sp.
Pink rot, *Cephalothecium roseum* Corda
Wilt, *Verticillium albo-atrum* Reinke & Berthold
Algal spot, *Cephaleuros virescens* Kunze
Other fungi associated with avocado

Fruit spot, *Pseudomonas syringae* van Hall

Symptoms

Blemishes on fruit, stem, and leaves usually develop into more mature spots. On the skin of the fruit they are brown to black, circular to irregular, 3–6 mm in diameter, characterized by cracks and crevices, and often surround the fruit. The colored areas often appear raised and their surfaces smooth and hard.

Etiology

The organism is a gram-negative rod, single or in chains, forms capsules, is motile by 1 to several polar flagella, and measures $0.75–1.5 \times 1.5–3\ \mu$. Agar colonies are white with transparent margins, round, smooth, and convex. A green fluorescent pigment forms in agar culture. Optimum temperature 28–30C.

References

1, 6, 23.

Crown gall, *Agrobacterium tumefaciens* (E. F. Sm. & Towns.) Conn

Symptoms

Typical cultures were obtained from a rough, cracked, swollen gall taken from a Fuerte avocado tree. Inoculations were made on avocado seedlings and several other plants from this tissue. All developed the small swellings and galls that are typical of this disease.

Etiology

The organism is a gram-negative rod, produces capsules, is motile by 1–4 flagella, and measures 0.4–0.8 × 1–3 μ. Agar colonies are white, translucent, small, circular, and glistening. Optimum temperature, 25–28C.

References

6, 16.

Root rot, *Phytophthora cinnamomi* Rands

Symptoms

Foliage of infected trees is often smaller than that of healthy trees and may be a yellow green color. Wilting sometimes occurs, and a certain amount of leaf shedding takes place. New growth is meager, and some dieback appears in the upper part of the tree. There is a reduced set of fruit, and these may be definitely smaller than normal. Nursery and young grove trees show the effects of the disease and are more frequently killed than older trees. The feeder roots are readily killed and become blackened; they are scarcely present in old, diseased trees.

Etiology

In artificial culture, the fungus develops hyaline hyphae that appear almost white in rapidly growing colonies and produce a sort of wavelike rhythm as they expand. Hyphae range from 4.7 to 6.2 μ in diameter. Chlamydosporelike bodies are spherical, measuring 29 μ. Sporangia develop on long sporangiophores which measure 55 × 38 μ. Sporangial germination produces small, hyaline zoospores, especially in running water. Considerable movement takes place just before an opening in the apex develops through which the zoospores escape. Zoospores are active for a short period, then come to rest and germinate by producing mycelium.

References

4, 25, 28, 29.

Collar rot, *Phytophthora cactorum* (Leb. & Cohn) Schroet.

Symptoms

Under highly favorable conditions, the fungus will persist almost indefinitely around avocado trees. The disease affects the wood and may progress extensively before any surface evidence appears. Narrow streaks develop in the cambium up the main stem but do not spread extensively laterally before surface evidence of a disease is present. The root system is often blackened. The larger roots are directly reduced in function, and some or most of the feeder roots are killed. The trunk is seldom completely girdled.

Etiology

The mycelium is hyaline, nonseptate, branched, and intercellular. The sporangia are pale yellow, almost spherical to ovoid and elongate, granular, with prominent papillae, and measure 20–46 × 18–35 μ. They may appear singly but are usually in groups; they have short pedicels and germinate by producing about 20 biflagellate zoospores. These swarm spores measure up to 10 μ in diameter. The chlamydospores are yellowish, spherical, and 20–40 μ in diameter, averaging 35 μ. The oogonia are yellowish, spherical, and smooth; the oospores are very similar, but thick-walled.

References

11, 24.

Soft rot, *Rhizopus nigricans* Ehr.

Symptoms

An affected fruit decays rapidly, and, if the air is dry, only very dense, short, dark, feltlike mold appears in spots on the surface, usually occupying depressions or protected places. If the atmosphere is moist very long, gray mold appears. The decay appears only on softened fruit at the stem end or on previously wounded areas. Under favorable conditions, the decay may advance in soft flesh as much as an inch in 2 or 3 days.

Etiology

The aerial mycelium is faintly colored, rhizoids are numerous, and sporangiophores are erect and nonseptate, arising in clusters along the stolons. The sporangia are globose, and black at maturity with dark spores that measure 11–14 μ. Zygospores are black, warty, and 150–200 μ.

Reference

11.

Leaf spot, *Phyllachora gratissima* Rehm

Symptoms

On the upper leaf surface, this fungus produces black crustations, in the form of a stroma, surrounded by lesser manifestations producing secondary circles as brown areas that are yellowish on the opposite side. The disease is not generally important.

Etiology

Clypei are black, oval to elliptical or indefinite, and amphigenous; asci are cylindrical, 90–130 μ long, and contain 8 hyaline, uniseriate, 1-celled ascospores, measuring $10\text{–}18 \times 7\text{–}10$ μ.

Reference

20.

Mushroom root rot, *Clitocybe tabescens* (Scop. ex Fr.) Bres.

Symptoms

Diseased trees may show a slow type of decline in which growth stops, and the foliage complement becomes thin and yellowish before shedding, dieback, and eventual death. However, there may occur a rapid killing of the tree, with all foliage intact, resulting from the girdling of the main trunk at the crown or near ground level.

Etiology

The mycelium is white and forms flat, sheetlike accumulations in the cortex, cambium, and down to the wood. The mushroom type of sporophores are produced aerially around the crown of the plant in clusters. They are tan to honey-colored with an umbrellalike cap on a central stipe. There is no annulus or cup. Spores are white, oval, 1-celled, and 8–10 × 3–6 μ.

Reference

19.

Shoestring root rot, *Armillaria mellea* Vahl ex Fr.

Symptoms

This root rot and crown canker cause a slow decline of the aerial parts of the tree or, in most instances, quickly kill the infested plant. The deterioration of the aerial parts, manifested by some loss of foliage and twig dieback, results from root decay that has not been too severe. The rapid killing occurs when a crown canker quickly girdles the main trunk.

Etiology

The fungus produces mycelium that survives in the soil and penetrates the host plant directly or through wounds. The fungus invades the cortical areas to the wood, where it produces white mycelium and black shoestringlike rhizomorphs that invade the woody parts and penetrate the surrounding soil. White spores are produced on honey-colored, gilled, mushroomlike sporophores. The 4–5 in. caps are umbrella-shaped, centrally stipitate, with an annulus, and are 4–8 in. tall. The hyaline basidiospores are elliptical and 9 × 6 μ.

References

5, 11, 18.

Root rot, *Rhizoctonia solani* Kuehn

Symptoms

The manifestations of the disease are slightly variable, depending on whether the cotyledons, embryo, radicle, or rootlets are attacked. The cotyledons or embryo are destroyed by the rotting of the young shoots or radicle. If germination takes place before serious decay sets in, young plants may develop to various stages but usually succumb due to a lack of roots. Root rot in seedling stages is often serious and results in girdling and killing of the young plant. The older the plant when root rot develops, the less the chances are that it will be serious. The greatest loss is in the nursery, where seed are placed for germination and where seedlings are lined out after removal from the seed germination area.

Etiology

The tan to brown mycelium is branched and septate. There are no conidiophores or conidia produced. Brown, irregularly shaped sclerotia are frequently found on plant parts.

Reference

15.

Root rot, *Rosellinia bunodes* (Berk. & Br.) Sacc.

Symptoms

The external aerial manifestations of the disease take the form of a plant declining because of malnutrition; this is indicated by lack of growth, shedding of foliage, and some dieback in the smaller twigs. The roots have been largely invaded by the white mycelium, killing the cortical tissue down to the wood. They become brown or black with irregular, intermittent nodules. The root system is rapidly killed, and the plant eventually dies. The sporulating structures appear on the roots and crown.

Etiology

The mycelium forms in dark rhizomorphic strands in the wood but is usually white where exposed. The conidial form develops in small, brown columns up to 3 mm high and 50 μ in diameter. Composed of fascicles of filaments, brown, and septate, the conidia are 1-celled, elongate, and $4–6 \times 2$ μ. The black, globose perithecia form in encrusted stroma, usually at the crown. The asci are long and slender, averaging $300 \times 12–15$ μ. The ascospores are dark brown, narrow, $80–110 \times 7–12$ μ, with fine, drawn-out, hairlike extremities about 25 μ long. The conidial stage is *Graphium* sp.

Reference

20.

Root rot, *Rosellinia necatrix* (Hartig) Berk.

Symptoms

This fungus causes a root rot of a number of perennial plants and can be used to inoculate avocado seedlings, although not isolated from avocado. Under favorable conditions the inoculated plants are killed in less than 2 months. The fungus invades the root system and surrounding soil. Wilting is the first symptom. The cotyledons and roots are invaded by the mycelium, and white mycelial plates are produced in the cortex.

Etiology

The fungus produces conidia on the surface of invaded cortical tissue, and this vegetative stage is known as *Dematophora necatrix* Hartig. The perithecial

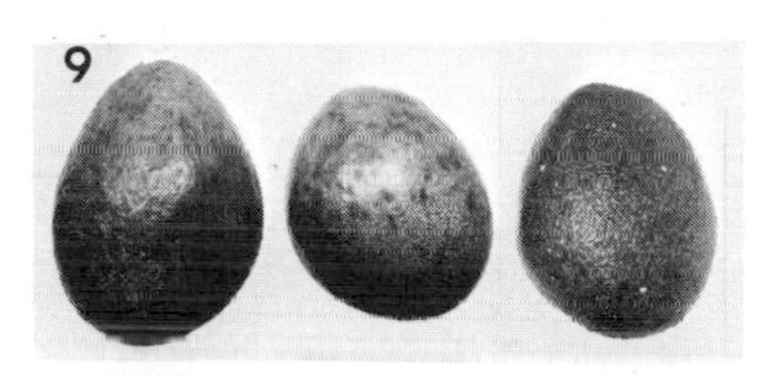

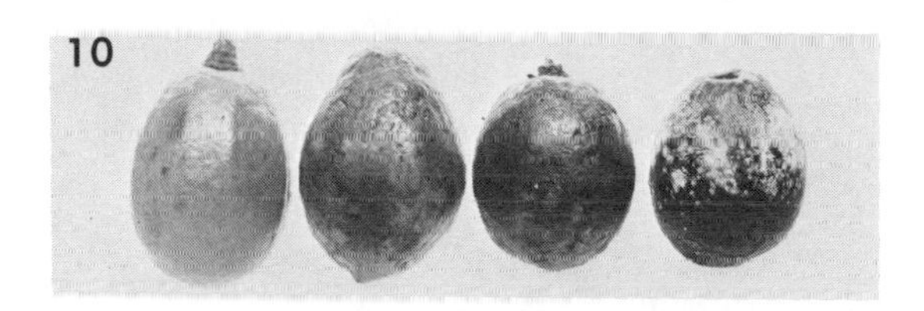

Figure 7. Root rot, *Phytophthora cinnamomi,* causing decline of avocado trees.

Figure 8. Shoestring root rot, *Armillaria mellea,* on avocado. (Photograph by G. A. Zentmyer.)

Figure 9. Left: Early symptoms of surface rot, *Dothiorella gregaria,* on avocados. Right: Healthy avocado. (Photograph by G. A. Zentmyer.)

Figure 10. Anthracnose, *Colletotrichum gloeosporioides,* on avocado. (Photograph by G. A. Zentmyer.)

Figure 11. Scab, *Sphaceloma perseae,* on foliage and stems of avocado.

Figure 12. Wilt, *Verticillium* sp., on avocado. (Photograph by G. A. Zentmyer.)

stage, found on the roots of diseased plants, is characterized by the production of black, superficial, subglobose, somewhat flattened, and ostiolate fruiting structures. These structures produce an abundance of ascospores which are dark, elongate, pointed at each end, 1-celled, and 31–48 × 5–8 μ.

References

10, 17.

Surface rot, *Dothiorella gregaria* Sacc.

Symptoms

This rot appears on the fruit when it begins to soften as it matures. The disease begins at the stem end as a small ring of affected tissue, dark brown and firm in texture. It spreads toward the blossom end, finally involving the entire fruit. Except for discoloration, very little external change is noted, and the surface remains smooth and firm. Later, a surface growth of short, feltlike fungus mass appears on the fruit, varying in color and amount. Some of the fungus is at first white and gradually darkens. Black pycnidia, which often characterize the disease, and perithecia appear on the fruit and on marginal burned leaves.

Etiology

The pycnidia develop in the rind of the fruit, among the aerial mycelium, and on the margins of browning leaves and twigs on a leathery stroma; they are erumpent, and about 200 μ in diameter. The spores are hyaline, fusoid, continuous, and 18–31 × 4–8 μ. Perithecia are immersed in elongate stroma, and are cone-shaped, with papillate ostioles about the same size as pycnidia and are often found among them. Paraphyses are filiform, asci are clavate, and ascospores are hyaline, fusoid, continuous, and 16–23 × 5–7 μ. The ascospore stage is *Botryosphaeria ribis.*

References

8, 12, 22.

Fruit rot, *Phomopsis perseae* Zerova

Symptoms

This rot is a discoloration initiated around the collar of fruit detached from the peduncle; it frequently spreads and destroys the entire fruit. The invaded tissue is blackish externally and lighter internally. The flesh becomes softened but usually does not cause a wet, leaking decay.

Etiology

The fungus produces black pycnidia mostly on the woody cortical tissue of the twigs, surviving structures of the floral panicle, and fruit. The conidia are hyaline, oval, 1-celled, and 5–8 × 3–4 μ. They are occasionally accompanied by hyaline, long, filiform, nonseptate stylospores, measuring 20–30 × 2–3 μ.

Reference

24.

Stem-end rot, *Diplodia natalensis* P. Evans

Symptoms

After removal of the fruit from the pedicel, the wound becomes infected by the fungus, and a dark, slate-colored decay begins and advances with increasing vigor, eventually involving the entire fruit. The internal discoloration may be slightly in advance of the external symptom. The injured branches or twigs may show cankerlike disease symptoms.

Etiology

The fungus produces pale to black, septate, branched mycelium on host tissue, and under certain conditions pycnidial pustules develop under the cuticle of the rind of the fruit. The pustules are submerged, later erumpent, black, oval, ostiolate, and produce an abundance of dark, oval, 2-celled, faintly striate spores, measuring $24–30 \times 12–15$ μ. The ascospore stage is *Physalospora rhodina.*

References

20, 24.

Leaf spot, *Pestalotia adusta* Ell. & Ev.

Symptoms

Brown spots or dead areas appear on opposite sides of the foliage, mostly on older, declining parts. They are irregular in shape, usually more or less circular, and up to 6 mm in diameter.

Etiology

Black, small, subepidermal, erumpent, more or less spherical pustules, 70–120 μ in diameter, may appear on both surfaces, but are most often found above. Conidia are hyaline, oblong or fusoid, erect, 5-celled, and $16–20 \times 5–7$ μ; the basal cell has a short pedicel.

Reference

9.

Anthracnose, *Colletotrichum gloeosporioides* Penz.

Symptoms

Definite spots scattered over the fruit surface are mostly black to dark brown, circular, and 0.5–2 cm in diameter; the centers may be slightly sunken. Under moist conditions, numerous small, waxy, flesh-colored spore masses may be seen breaking through the epidermis. Surfaces of larger spots frequently appear zonated, cracked, or fissured. The fungus, a wound infector, is a weak parasite and rarely becomes destructive except to tissue of low vitality. It is also saprophytic and can exist and propagate readily on dead twigs, leaves, and fruit of various plants.

Etiology

The acervuli measure 90–250 μ in diameter. They appear as blisterlike specks in the diseased tissue, becoming erumpent as the epidermis breaks irregularly. Setae are fuliginous, septate, up to 100 μ high, but frequently absent. Stroma are composed of fungus cells and the conidiophores. Conidia are hyaline, 1-celled, cylindrical with rounded ends, and 18–25×4–5 μ. The ascospore stage is *Glomerella cingulata.*

References

7, 11, 20, 21.

Rusty blight, *Colletotrichum nigrum* Ell. & Halst.

Symptoms

This fungus, about which there exists some confusion at present, is thought to be the cause of a kind of rusty discoloration on the exterior area of growing fruit, foliage, and young branches. In general, the discolored area may be circular or rectangular, and a fine sandpaper roughness may appear on external fruit areas. No internal decay has been noted. This disease is generally of minor importance.

Etiology

Numerous acervuli appear on indefinite, depressed areas; they are subepidermal, becoming erumpent, and 80–120 μ in diameter. The areas are darkened by the brown setae that are up to 100 μ high. The conidia are hyaline, cylindrical with rounded ends, and 18–25×3–6 μ. The ascospore stage is probably a *Glomerella* sp.

Reference

24.

Scab, *Sphaceloma perseae* Jenkins

Symptoms

Definite spots or patches on leaves, young shoots, and fruit may be scattered or numerous, and severe attacks may cause leaf curling, twig distortion, and malformed fruit. Spots are purplish brown to dark brown. The centers of older leaf spots are light brown, dry, dead, more or less spongy tissue. On fruit, spots may be small and scattered or may combine to form irregular areas of hard, glazed tissue involving most of the surface. They are dark brown, and, as the fruit increase in size, the surfaces of such areas become marked with fissures or cracks. The fungus is most active in cool, moist weather. As soon as the leaf tissue hardens, the fungus is unable to cause further infection. This critical period for leaf infection is probably a month or less in duration and is much shorter than that for fruit.

Etiology

Hyaline hyphae develop acervuli or sporodochia that are olive to dark, scattered, and up to 70 μ long. Wedge-shaped conidiospores are 12 μ high. One to several conidia are produced; they are hyaline, oval to oblong, continuous, 1-celled or in a chain, and 5–8 $\times$ 3–4 μ.

References

13, 21, 24.

Powdery mildew, *Oidium* sp.

Symptoms

The affected terminal leaves of tender shoots show a dark green discoloration on the upper surfaces along the midrib. The white, powdery fungus growth covers and is confined to the undersurfaces of infected leaves; the spots are rarely conspicuous on the upper side except in faint outline. The infected tissue has a purplish cast.

Etiology

Mycelium is white, septate, and superficial. Conidiophores are sparse, erect, and septate. Conidia are hyaline, 1-celled, oval, catenulate, and 40 $\times$ 15–20 μ.

References

21, 24.

Leaf spot, *Cercospora purpurea* Cke.

Symptoms

Small, brown to dark brown, angular spots develop on the leaves. These first appear on the undersurface as purplish, slightly raised, blisterlike swellings not visible from the upper side. Later, the affected area turns brown, the tissues collapse, and the spots have a sunken appearance. On fruit, small, greenish white dots appear which later develop into slightly sunken, irregular surface blotches. The spots enlarge and develop into circular areas of hard, brown, dead tissue, the surfaces of which become cracked or fissured. Spots coalesce to form irregular areas of dead rind tissue. Young fruit and those that are nearly mature are practically immune to the fungus, but during the period between, infections may readily occur.

Etiology

Dark-colored stroma, up to 100 μ in diameter, appear on the brown spots; fascicles are pale to olivaceous. Conidiophores are multiseptate, straight, or undulate, develop mostly on the lower leaf surface, and measure 20–200 $\times$ 3–5 μ. Conidia, produced amphigenously, are obclavate, colored, rounded at the base with acute tips, up to 9-septate, straight or curved, and 20–100×2–5 μ. The ascospore stage is probably *Mycosphaerella perseae.*

References

3, 21, 24.

Smudgy spot, *Helminthosporium* sp.

Symptoms

Spots on the cortex of green twigs and on petioles or leaves are not sharply outlined. The size varies from very small to those that are confluent into a nearly continuous area. Leaf petioles are more frequently spotted than leaves and fruit. The spots are entirely superficial.

Etiology

The organism has not been described or given a specific name.

Reference

11.

Pink rot, *Cephalothecium roseum* Cda.

Symptoms

The disease is recognized by the dark, sunken spots on the surface of fruit that are dying back and in cankers and wounds. The fungus is considered a weak parasite favored in its development by declining host tissues and wounds.

Etiology

The mycelium of the fungus is white and under favorable conditions develops an abundance of pink spores over the exposed surfaces. The conidiophores are hyaline, erect, simple, and continuous; the conidia are light rose, oblong to ovate, and constricted at the septum.

Reference

26.

Wilt, *Verticillium albo-atrum* Reinke & Berth.

Symptoms

There is usually a wilting of the foliage on certain branches of the tree, but occasionally the foliage of an entire tree collapses, and death follows rapidly. The leaves turn brown and may remain on the tree for extended periods. Examination of the wood in diseased trees reveals dark brown streaks in the xylem tissue, varying from a few faint brown lines to abundant discoloration.

Etiology

The mycelium is hyaline becoming dark, septate, and branched. The conidiophores are elongate, narrow, septate, tapering, and branched in whorls. Conidia are abscised at the tips in chains. They are hyaline, oval, 1-celled, and 4–6×2 μ.

References

14, 27.

Algal spot, *Cephaleuros virescens* Kunze

Symptoms

The leaf-spotting organism is one of the green algae often encountered as an epiphyte, but it frequently assumes a more or less parasitic role in which the thallus penetrates the leaf and extends beneath the cuticle. Usually the leaf spots are scattered and seldom exceed 1/2 cm in diameter. They are circular, slightly raised, often appearing on opposite surfaces, and become hairy and tinged with a reddish color.

Etiology

The alga is green, elongating radially by dichotomous branches. Single stalks arise vertically and develop into sporangiophores about 1 mm high and 50 μ in diameter, swollen into a vesicle at the tip and supporting up to 6 sporangia, usually 3–4, on curved pedicels. The sporangia are oval, about 35 μ in diameter, and germinate by producing orange zoospores which come to rest after an active period and germinate by producing new thalli.

References

2, 20.

Other fungi associated with avocado

Asterina delitescens Ellis & G. Martin
Coccoidella scutula (Berkeley & Curtis) Hoehnel
Helminthosporium fumosum Ellis & G. Martin
Irene perseae (F. L. Stevens) Toro
Irenopsis martiniana (Gaillard) F. L. Stevens
Meliola amphitricha Fries
Penicillium digitatum (Fries) Saccardo
Pestalotia versicolor Spegazzini
Phyllosticta micropunctate Cooke
Physalospora fusca N. E. Stevens
Pythium debaryanum Hesse
Sclerotinia sclerotiorum (Libert) De Bary
Sphaerostilbe repens Berkeley & Broome
Sporotrichum citri Butler
Trachysphaera fructigena Tabor & Bunting

References: avocado

1. Bryan, M. K. 1928. Lilac blight in the United States. J. Agr. Res. 36:225–35.
2. Butler, E. J. 1918. Fungi and diseases in plants. Thacker: Calcutta, India, pp. 412–22.
3. Chupp, C. 1953. A monograph of the fungus genus Cercospora. Published by the author: Ithaca, N.Y., pp. 55–56.
4. Crandall, B. S. 1948. Phytophthora cinnamomi root rot of avocados under

tropical conditions. Phytopathology 38: 123–30.
5. Darley, E. F., and G. A. Zentmyer. 1957. Oak root fungus on avocados. Calif. Avocado Soc. Yearbook 41:80–81.
6. Elliott, C. 1951. Manual of bacterial plant pathogens. 2d ed. Chronica Botanica: Waltham, Mass., pp. 5–12, 88–93.
7. Fawcett, H. S. 1936. Citrus diseases and their control. 2d ed. McGraw-Hill: N.Y., pp. 301–8.
8. Grossenbacher, J. G., and B. M. Duggar. 1911. A contribution to the life history, parasitism and biology of Botryosphaeria ribis. N.Y. State Agr. Expt. Sta. Tech. Bull. 18:114–88.
9. Guba, E. F. 1961. Monograph of Monochaetia and Pestalotia. Harvard University Press: Cambridge, Mass., pp. 117–21.
10. Hansen, H. N., H. E. Thomas, and H. Earl Thomas. 1937. The connection between Dematophora necatrix and Rosellinia necatrix. Hilgardia 10:561–64.
11. Horne, W. T. 1934. Avocado diseases in California. Calif. Univ. Agr. Expt. Sta. Bull. 585:3–72.
12. Horne, W. T., and D. F. Palmer. 1935. The control of dothiorella rot on Avocado fruits. Calif. Univ. Agr. Expt. Sta. Bull. 594:3–16.
13. Jenkins, A. E. 1934. Sphaceloma perseae, the cause of avocado scab. J. Agr. Res. 49:859–69.
14. Marlatt, R. B., and S. Goldweber. 1969. Verticillium wilt of avocado (Persea americana) in Florida. Plant Disease Reptr. 53:583–84.
15. Mircetich, S. M., and G. A. Zentmyer. 1964. Rhizoctonia seed and root rot of avocado. Phytopathology 54:211–13.
16. Munnecke, D. E., and G. A. Zentmyer. 1964. Crown gall on avocado. Phytopathology 54:1302–3.
17. Raabe, R. D., and G. A. Zentmyer. 1955. Susceptibility of avocados to dematophora root rot. Plant Disease Reptr. 39: 509–10.
18. Rhoads, A. S. 1945. A comparative study of two closely related root-rot fungi, Clitocybe tabescens and Armillaria mellea. Mycologia 37:741–66.
19. Rhoads, A. S. 1956. The occurrence and destructiveness of Clitocybe root rot of woody plants in Florida. Lloydia 19: 193–239.
20. Roger, L. 1953. Phytopathologie des pays chauds. Encyclopedie Mycologique. 18:1286–97, 1376–81, 1404–16, 1445–46; 19:2260–67.
21. Ruehle, G. D. 1958. The Florida avocado industry. Florida Univ. Agr. Expt. Sta. Bull. 602:74–79, 81–82.
22. Smith, C. O. 1934. Inoculations showing the wide host range of Botryosphaeria ribis. J. Agr. Res. 49:467–76.
23. Smith, C. O., and H. S. Fawcett. 1930. A comparative study of the citrus blast bacterium and some other allied organisms. J. Agr. Res. 41:233–46.
24. Stevens, H. E., and R. B. Piper. 1941. Avocado diseases in Florida. U.S. Dept. Agr. Circ. 582:1–45.
25. Tucker, C. M. 1929. Avocado root diseases. Porto Rico Agr. Expt. Sta. Ann. Rept. 1928:29–32.
26. Yale, J. W., Jr., and G. R. Johnstone. 1951. The occurrence and effect of Cephalothecium on avocado. Colorado Avocado Soc. Yearbook for 1957, pp. 179–85.
27. Zentmyer, G. A. 1949. Verticillium wilt of avocado. Phytopathology 39:677–82.
28. Zentmyer, G. A., and A. O. Paulus. 1957. Phytophthora avocado root rot. Calif. Univ. Agr. Expt. Sta. Circ. 465:1–15.
29. Zentmyer, G. A., A. O. Paulus, and R. M. Burns. 1962. Avocado root rot. Calif. Univ. Agr. Expt. Sta. Circ. 511:1–18.

Banana, *Musa* spp.

Head rot, *Erwinia musa* Warren
Rhizome rot, *Erwinia carotovora* (Jones) Holland
Moko disease, *Pseudomonas solanacearum* E. F. Smith
Javanese wilt, *Pseudomonas musae* Gäumann
Blood disease, *Xanthomonas celebensis* (Gäumann) Dowson
Bacterial leaf spot, *Xanthomonas musicola* Rangel & Rangel
Bacterial wilt, *Xanthomonas musacearum* Yirgou & Bradbury; of ensete, *Ensete ventricosum* Welwitsch & Cheesman
Whiskers, *Rhizopus nigricans* Ehrenberg
Cigar end, *Trachysphaera fructigena* Tabor & Bunting
Fruit rot, *Sclerotinia sclerotiorum* (Libert) De Bary
Sigatoka, *Mycosphaerella musicola* Leach
Bonnygate disease, *Calostilbe striispora* (Ellis & Everhart) Seaver

Rust, *Uromyces musae* P. Henning
Soil rot, *Thanatephorus cucumeris* (Frank) Donk
Pseudostem rot, *Botryobasidium rolfsii* (Saccardo) Venkatarayan
Shoestring root rot, *Armillaria mellea* Vahl ex Fries
Mushroom root rot, *Clitocybe tabescens* (Scopoli ex Fries) Bresadola
Stem rot, *Marasmius semiustus* Berkeley & Curtis
Stone fungus, *Laccocephalum basilapiloides* McAlpine & Tipper
Fruit rot, *Botryodiplodia theobromae* Patouillard
Freckle, *Macrophoma musae* (Cooke) Berlese & Voglino
Charcoal rot, *Macrophomina phaseolina* (Tassi) Goidanich
Fruit rot, *Dothiorella gregaria* Saccardo
Tip rot, *Hendersonula toruloides* Nattrass
Anthracnose, *Gloeosporium musarum* Cooke & Massee
Leaf lesions, *Pestalotia leprogena* Spegazzini
Leaf spot, *Pestalotia palmarum* Cooke
Tip-end rot, *Botrytis cinerea* Persoon ex Fries
Sour rot, *Oospora rosea-flavida* Saccardo
Pitting, *Piricularia grisea* (Cooke) Saccardo
Cigar end, *Verticillium theobromae* (Turconi) Mason & Hughes
Leaf speckle, *Cloridium musae* Stahel
Black tip, *Deightoniella torulosa* (Sydow) Ellis
Leaf spot, *Cordana musae* (Zimmermann) Hoehnel
Zonate eyespot, *Drechslera gigantea* (Heald & Wolf) Ito
Brown spot, *Cercospora hayi* Calpouzos
Squirter, *Nigrospora sphaerica* (Saccardo) Mason
Blackhead, *Thielaviopsis paradoxa* (De Seynes) Hoehnel
Panama wilt, *Fusarium oxysporum* f. sp. *cubense* (E. F. Smith) Snyder & Hansen
Heart-leaf rot, *Fusarium moniliforme* Sheldon var. *subglutinans* Wollenweber
Diamond spot, *Fusarium roseum* (Link) Snyder & Hansen
Other fungi associated with banana

Head rot, *Erwinia musa* Warren

Symptoms

The failure of newly planted rhizomes to sprout, stunting and poor growth of young plants, and the unstable condition of bearing plants that are subject to falling over or being pushed over are external appearances that generally characterize diseased plants. Areas of brown, water-soaked tissue appear on the surface and advance inward from the rhizome surface. The areas become variously pocketed internally, the pockets surrounded by dark borders. Young plants show yellowing and stunting, and mature plants exhibit instability and readily fall over, often splitting the rhizome, half attached to the pseudostem. Young suckers are subject to infection although some may escape. Roots are not directly attacked.

Etiology

The organism is a gram-negative rod, produces no capsules, is mostly single or in chains, is motile by 2–5 peritrichous flagella, and measures $0.9–2 \times 0.4–0.7$ μ. Agar colonies are white, iridescent, round, smooth, convex, entire, slightly rough, punctiform, and produce pectic enzymes.

References

43, 54.

Rhizome rot, *Erwinia carotovora* (Jones) Holland

Symptoms

General aspects of the disease are rotting of newly planted rhizomes, stunting and poor growth of young plants, and tipping over of bearing plants. The rhizome becomes diseased in local areas, developing soft, spongy, yellow to dark brown spots, which leave cavities as they enlarge. Young plants become yellow, stunted, and are easily pushed over. Mature plants seldom show any external symptoms; they carry a full complement of foliage and also a marketable bunch of fruit, except possibly the fingers, which may be undersized.

Etiology

The organism is a gram-negative rod, motile by 2–6 peritrichous flagella, and $0.6–0.9 \times 1.5–2\ \mu$. Agar colonies are grayish white, circular, and glistening. The classification of this organism may be controversial. Optimum temperature, 27C.

References

9, 52.

Moko disease, *Pseudomonas solanacearum* E. F. Sm.

Symptoms

External symptoms on Cavendish, plantain, ensete, and abaca include a yellowish tinge of lower leaves which extends to others, except unrolled heart leaf. Coloration tends to be evenly distributed over laminae and more conspicuous on one side of the midrib than on the other. In later stages all leaves hang down, and the plant rots and falls to the ground. The disease inhibits further maturation of half-grown fruit. Symptoms are seldom seen on vigorously growing young plants. Internal symptoms include discoloration of all vascular strands from pale yellow to dark brown; this is evident with the appearance of soft petioles in fleshy rhizomes. Opaque, dirty white, bacterial exudate appears upon cutting of the rhizome or trunk. Gros Michel does not exhibit symptoms on leaves other than the unfolding central leaf. Infection is localized only in the pseudostem, and symptoms are very pronounced on fingers, which turn yellow and finally blacken and rot.

Etiology

The organism is a short, gram-negative rod, often in pairs, motile by a single polar flagellum, and measuring $1.5 \times 0.5\ \mu$. Culture colonies are dirty white, small, and smooth. They are black when cultured on sterile potato blocks. Optimum temperature, 35–37C.

References

4, 5, 11, 35, 54, 55.

Javanese wilt, *Pseudomonas musae* Gäum.

Symptoms

No external symptoms are evident in 90 per cent of the cases. Yellow discoloration and dead strands are found in the rhizome. Symptoms consist of longitudinal splitting of the outer leaf sheath, premature breaking down and wilting of leaves, and red coloration of exuding cell sap. In severe infection, extension of discoloration progresses upward to the pseudostem. Symptoms may also be caused by a mixture of moko and panama disease organisms.

Etiology

The organism is gram-positive, oval to pear-shaped, 1–3-flagellate, and measures $0.8 \times 1.2\ \mu$. Agar colonies are light yellow, round, and smooth; margins are entire to undulate.

References

9, 12.

Blood disease, *Xanthomonas celebensis* (Gäum.) Dows.

Symptoms

Discoloration of one of the younger leaves and outward spreading to other young leaves is the developmental symptom, followed by spreading of the light, yellow brown stripes from the midrib. It is commonly found with a single discolored leaf. The yellowing of the leaf crown is complete after only 1 week. Fruit hangs down, becomes yellow, and decays. The disease attacks primarily the vascular bundles, causing yellow striping and drooping of leaves. Blood-colored slimy exudate appears upon cutting of the stem. Discolored vascular strands are chiefly localized in the central part of the stem. Discoloration of cell walls extends to parenchyma. Damaged tissues are watery in consistency, smell sour, and develop a yellow or brown discoloration of the bundles of the fruit skins.

Etiology

The organism is a gram-negative rod, with a single polar flagellum, and measures $0.9 \times 1.5\ \mu$. Agar colonies are grayish yellow.

Reference

9.

Bacterial leaf spot, *Xanthomonas musicola* Rangel & Rangel

Symptoms

Chlorotic linear streaks appear along the veins of the blade. No other parts are affected. The streaks may coalesce, forming irregular brownish patches which later become black with a yellowish halo. Large blotches of the infected tissue may roll and dry, appearing scorched. Plants may be stunted and the production of fruit reduced.

Etiology

The organism is a gram-negative rod, single or in chains, with a single polar flagellum, and measuring 0.6–0.9 × 1.2–2.1 μ. On agar, colonies are yellow, glistening, shiny, and filiform.

Reference

30.

Bacterial wilt, *Xanthomonas musacearum* Yirgou & Bradbury, of ensete, *Ensete ventricosum* Welwitsch & Cheesman

Symptoms

A droopy or limp condition first affects the heart leaf or one of the young leaves. The folds of these leaves show grayish brown decaying areas that exude a slimy secretion. Later, the leaves break at the union with the petioles. Gradually, all leaves wilt, break over, and wither. Stem and rhizome sections reveal a discolored vascular system and often pockets of bacteria. The plants finally die, collapse, and are consumed in a soft, slimy decay.

Etiology

The organism is a gram-negative rod, single or in chains, motile by a single polar flagellum, and measuring 1.8–2 × 0.7–0.9 μ. Agar colonies are light yellow, convex, and mucoid to butyrous. A yellow fluorescent pigment is produced in the culture medium. Optimum temperature, 25–28C.

Reference

55.

Whiskers, *Rhizopus nigricans* Ehr.

Symptoms

Beginning as a soft, wet rot, the disease progresses rapidly in the fleshy tissues. After 4–5 days, these tissues become soft, stringy, and leak upon breaking of the skin. Infected areas turn brown, later followed by the appearance of coarse, whiskery, extramatrical growth.

Etiology

It is a heterothallic fungus with siphonaceous, coenocytic mycelium, stolons, and rhizoids. The erect, aseptate sporangiophores bear globose, dark-colored sporangia, 100–150 μ in diameter. The airborne spores are brown, ovoid, subglobose, with walls marked externally with longitudinal ridges, and measure 11 × 14 μ. It is usually considered a wound parasite. The fungus produces black, sexual, warty zygospores, measuring 150–200 μ.

Reference

15.

Cigar end, *Trachysphaera fructigena* Tabor & Bunt.

Symptoms

The first evidence is a slight olive brown, water-soaked discoloration, at the blossom end of the fingers. This rapidly darkens. The disease develops to the young fruit and becomes less progressive as the fruit matures. It then causes blackening of the skin, shrinkage, and corrugated folds that advance from the stigma end of the finger.

Etiology

The siphonaceous hyphae collect in the subepidermal areas and rupture the epidermis, producing an abundance of conidiophores. The conidiophores are hyaline, elongate, myceliumlike, branched, and produce terminal or successive indeterminate whorls of conidia. The conidia are hyaline, spherical, echinulate, averaging 35 μ in diameter, and terminating in pedicels that are up to 30 μ long. The fungus produces oospores within the host tissue that are spherical to elongate, irregular in size and shape, with a bumpy surface, and measure 40×24 μ. Conidia germinate directly.

References

6, 47.

Fruit rot, *Sclerotinia sclerotiorum* (Lib.) d By.

Symptoms

Infection begins at the perianth, extends along the finger, and has a brownish, water-soaked appearance, which later turns dark brown. Affected inner tissue appears reddish. Central coloration is seen in advance of the penetrating hyphae. Sclerotia in various shapes and sizes are found mostly on the outside of the fruit, but may also be found internally.

Etiology

Sclerotia are a few millimeters to 3 cm long. They are oval to flat, depending on expansion space when formed, and black, with a white interior. Asci are cylindrical and $130–135 \times 10$ μ. Spores are ellipsoidal, minutely guttulate, and $9–13 \times 4–6.5$ μ. Apothecia are scattered, cupulate, and 4–8 mm or more in diameter; stems are slender, and paraphyses are clavate. Mycelium penetrates either wounded or weak, uninjured tissue. The fungus grows at temperatures lower than those used for banana storage.

References

34, 54.

Sigatoka, *Mycosphaerella musicola* Leach

Symptoms

The first indication of disease is the appearance of light brownish green, indistinct linear markings parallel to the veins. Occasionally, markings become dry, and muddy black, later turning to dirty white or grayish in the center, surrounded by a narrow, well-defined, dark brown margin. Numerous black specks are present in the center. Frequently, spots coalesce, forming large dead patches. Destruction of the leaf blade is followed by rotting of the petiole and midrib by secondary fungi and bacteria. Immature fruit bunches fail to ripen.

Etiology

The fungus forms conidia soon after the spots turn dark brown. The conidia are usually produced in great abundance in sporodochia on the upper surface of the leaf. Conidia are narrow, elongated, 0–4-septate, and measure 23–66 × 2–5 μ. The conidia are formed on the ends of dusky, brown, upright hyphae projecting in clusters from the leaf surface. The perithecia are blackish, scattered, submerged becoming erumpent, with protruding ostioles. Asci are clavate and measure 28–36 × 8–10 μ. Ascospores are hyaline, 2-celled, unequal, and 14–18 × 3–4 μ. The imperfect stage is *Cercospora musae.*

References

7, 17, 38, 42, 45, 54.

Bonnygate disease, *Calostilbe striispora* (Ell. & Ev.) Seaver

Symptoms

Outer leaves frequently show a narrow zone of pale brown, dry tissue, extending along the blade margin and bordered on the inner side by a narrow, bright yellow band. In younger, inner leaves, marginal bands are usually broader, dry, and yellow, resulting in slow growth and failure of bunch production. The trunk is easily broken over at bulb level. The lower part of the trunk, just above the bulb, turns black. The outer cortical region is bright yellow, orange, or red. Infection appears as a rounded, water-soaked area, bounded by a narrow red line.

Etiology

The fungus is a wound parasite. The vegetative stage appears as an erect, hairlike structure composed of slender stalks, 90–120 μ, arising from the orange, roundish cushion. The conidia are bluntly spindle-shaped, 3-septate, with 2 larger central cells, and measuring 42–48 μ. Paraphyses are hyaline, numerous, septate, and 150–180 μ; the perithecia often develop from the same cushion as conidiophores, arising just beneath the surface of the cushion. The perithecia are pear-shaped, covered with yellow, spiral, septate, rough hairs, with paraphyses, and measure 850–1000 × 450–600 μ. The asci are 250–300 × 20–24 μ, and ascospores are brown, 2-celled, fusiform, slightly constricted, longitudinally striate, and 40–45 × 9.5–11.5 μ. The vegetative stage is *Calostilbella calostilbe.*

References

27, 34, 54.

Rust, *Uromyces musae* P. Henn.

Symptoms

The fungus produces brown to black, circular to lenticel-shaped, erumpent uredosori, up to 240 μ in diameter, on the lower surface of the leaves.

Etiology

The uredospores are brown, subglobose to ellipsoidal, echinulate, and 18–24 × 16–20 μ. The telia are black, circular to oblong, and produce teliospores measuring 22–30 × 15–21 μ on persistent pedicels.

Reference

28.

Soil rot, *Thanatephorus cucumeris* (Frank) Donk

Symptoms

This is a root disease that reduces the water supply. It is characterized by yellowing and wilting of leaves just before fruit bearing. Yellowing and drying extends from margin to midrib, wilting from oldest to youngest leaves. Young fruit turn black, failing to ripen. Roots are blackened superficially or killed.

Etiology

Hyphae are colorless, turning brown at maturity, septate, branched at a right angle, constricted, and 7–10 μ in diameter. Mycelium may develop above ground level and grow upward on the outside of old pseudostem structures. In this location small, brown, irregularly shaped sclerotia may appear.

Reference

54.

Pseudostem rot, *Botryobasidium rolfsii* (Sacc.) Venkat.

Symptoms

The fungus, which is soil inhabiting, produces white mycelium that attacks the outer leaf sheaths. The resulting decay rapidly devitalizes the outer layer and continues to penetrate the successive ones, spreading laterally. As the foliage is cut off by the girdling effect, the plant shows less vigor. The disease frequently causes a rapid collapse of the plant and results in its falling over. The fungus produces brown, small, spherical, seedlike sclerotia on the outer surface of the leaf sheath, often with an abundance of white mycelium.

Etiology

The hyphae are white en masse, individually hyaline, branched, and septate. The sclerotia are superficial, white at first, brown at maturity, spherical, and 1–3 mm in diameter. Basidia are hyaline, clavate to elongate, and 16 × 6 μ. The 4 sterigmata are subulate and up to 4 μ long. The basidiospores are hyaline, oval, apiculate, flat on the interior side, and 6 × 4 μ.

References

33, 54.

Shoestring root rot, *Armillaria mellea* Vahl ex Fr.

Symptoms

Diseased plants cease to grow, pseudostems are short and slender, and foliage appears off-color, becomes yellow, and droops. The plants may be killed, showing symptoms of plants girdled at the crown or soil line. The rhizomes become filled with whitish mycelium and rhizomorphs. This fungus produces black, shoestringlike rhizomorphs.

Etiology

The spore-producing sporophore is a honey-colored mushroom, 5–10 cm in diameter, and up to 10 cm tall. It is centrally stipitate, with radiating gills, and a more or less distinct annulus surrounding the stipe. The basidiospores, produced in abundance, are hyaline, white in quantity, and 8–9 $\times$ 5–6 μ.

References

14, 54.

Mushroom root rot, *Clitocybe tabescens* (Scop. ex Fr.) Bres.

Symptoms

Yellowing and wilting of the foliage indicate a trunk girdling effect and result in the death of the plant. Examination of the roots and stem at the soil line reveals a wet, brownish discoloration with sheets of mycelium and a development of rhizomorphs between the leaf petiole bases and within the pseudostem. The true stem also contains the fungus, and the secondary rhizomes or suckers are invaded. The presence of mycelial mats or shields, originating near the crown of the plant, is usually a reliable diagnostic sign.

Etiology

The fruiting of the fungus occurs after the disease is well advanced. The sporophores are honey-colored mushrooms with cap, gills, and a central stipe; they lack an annulus, and are produced in small clumps. The sporophore may be up to 14 cm tall and the cap up to 10 cm across. The basidiospores are white and 8–10 $\times$ 4–6 μ. The mycelium mats and rhizomorphs are white.

References

31, 32.

Stem rot, *Marasmius semiustus* Berk. & Curt.

Symptoms

Older leaves become functionless from fungal activity in the root and on the basal region of the pseudostem; they turn brown and when removed show layers and patches of white mycelium on and between the dead leaf sheaths and also oval, brown patches at various points, penetrating to the center of the pseudostem. At the base of the trunk the infected outer leaf sheaths are moist and soft, with velvety white mycelium and a characteristic mushroom odor. Infection begins at the soil level.

Etiology

The fungus hyphae are abundant, and thickened strands may seal one sheath to the next. The mushrooms have brownish or salmon yellow caps on slender, globose stipes. The pileus is 5–15 mm across and may be central or eccentric. Spores are white and papillate, measuring 7–8 $\times$ 5–6 μ. The stipe is 7–9 mm high.

References

27, 33, 54.

Stone fungus, *Laccocephalum basilapiloides* McAlp. & Tipp.

Symptoms

External indications are similar to those of other diseases in which roots fail to function properly or are damaged. Soils are compact in irregularly shaped masses of an almost stonelike density.

Etiology

The mycelium is concentrated in certain portions of soil, enclosing root fibers and preventing normal functioning. These clumps or masses of soil have a mottled appearance upon being broken open. No fruiting structures have been described and no spores are known.

Reference

54.

Fruit rot, *Botryodiplodia theobromae* Pat.

Symptoms

The fungus is a wound parasite, formed as a stalk rot, finger dropping, tip rot, and fruit blemish in the field, and a transit and storage rot. The decay spreads uniformly along the fruit, causing a progressive brownish black discoloration of the skin. With the approach of maturity, entire fingers are affected with a soft, pulpy, and finally semiliquid decay. Later, the skin becomes black, soft, wrinkled, and encrusted with pycnidia. Generally, a characteristic grayish green mycelium appears as a dense weft over the skin. A downward rotting of the main stalk of the whole bunch may occur.

Etiology

The mycelium varies from hyaline to olive black, is abundant, coarse, and produces large fructifications consisting of stroma of matted hyphae, with a few or many imbedded pycnidia. On fruit, the pycnidia are spherical to flask-shaped, separate or grouped. Pycnidia have short, straight necks and ostioles. Conidia, borne on short conidiophores, are accompanied by paraphyses. Conidia are brown to olive black, 1-septate, striated, and 26 $\times$ 15 μ. The ascospore stage is *Physalospora rhodina.*

Figure 13. Moko, *Pseudomonas solanacearum,* on banana. (Photograph by L. Abrego.)
Figure 14. Moko, *Pseudomonas solanacearum,* on banana. Internal symptoms. (Photograph by L. Abrego.)
Figure 15. Sigatoka, *Mycosphaerella musicola,* on banana. (Photograph by L. Abrego.)
Figure 16. Sigatoka, *Cercospora musae* (the imperfect stage), on banana. (Photograph by E. Ampuero.)
Figure 17. Mushroom root rot, *Clitocybe tabescens,* on banana stem. (Photograph by A. S. Rhoads.)
Figure 18. Anthracnose, *Gloeosporium* sp., on banana.

Figure 19. Blackhead, *Thielaviopsis paradoxa,* on banana. (Photograph by A. P. Martinez.)
Figure 20. Blackhead, *Thielaviopsis paradoxa,* on banana.
Figure 21. Panama wilt, *Fusarium oxysporum* f. sp. *cubense,* on banana.
Figure 22. Cross section of banana stem showing dark *Fusarium*-invaded areas. (Photograph by A. S. Rhoads.)
Figure 23. Left: Cross section of banana pseudostem darkened by *Fusarium.* Right: Healthy pseudostem. (Photograph by A. S. Rhoads.)
Figure 24. Left: Pseudostem darkened by *Fusarium.* Longitudinal section. Right: Healthy banana pseudostem.

References

10, 34, 50, 54.

Freckle, *Macrophoma musae* (Cke.) Berl. & Vogl.

Symptoms

There is characteristic freckling of leaves and both green and mature fruit with numerous spots that are about 1 mm in diameter. Infection occurs only on the surface, making fruit unsightly and unsuitable for exportation. It is also found on the leaves, as numerous minute, grayish or brown to dark brown raised spots with black dots in the center. Spots become more or less round, with indefinite borders. Lesions are less raised on leaves and more abundant on the upper surface. Infected fruit is rough to the touch, due to projecting pycnidia. Black, infected fruit and leaves turn light brown later.

Etiology

Spots occur with clusters of shining black, subconical, partly imbedded, beakless, ostiolate pycnidia. Conidia measure 8.5–20 × 6–12 μ, are densely granular, 1-celled, oval or irregularly shaped, with a thick, hyaline envelope and frequently a short, hyaline appendage.

Reference

54.

Charcoal rot, *Macrophomina phaseolina* (Tassi) Goid.

Symptoms

The fungus is frequently found on roots and other parts of the plant that contact the soil. Infection of aerial parts occurs through wounds. Fruit decay occurs after harvest. The stalk or cushion is darkened by a firm decay. Small, hard, black sclerotia imbedded in the darkened tissue are diagnostic of the disease.

Etiology

In culture the mycelium is dark, flat, septate, branched, and forms abundant, smooth sclerotia that are 50–100 μ in diameter. Pycnidia are black, immersed becoming erumpent, globose, 100–200 μ, and contain conidiophores 10–15 μ high. Conidia are hyaline, elliptical to oval, 1-celled, continuous, and 16 × 32 μ. The vegetative stage is *Sclerotium bataticola*.

Reference

1.

Fruit rot, *Dothiorella gregaria* Sacc.

Symptoms

The finger decay begins at the blossom end and extends toward the peduncle, producing a black area that is definitely demarked from the uninvaded tissue by a narrow, brown, water-soaked zone. On the older decayed areas, black pycnidia protrude through the epidermis and erupt when mature, producing a white, powdery mass of pycnidiospores. This is a disease of immature fruit.

Etiology

In culture the mycelium is grayish, gradually becoming dark. The pycnidia, at first submerged in the outer layers of the peel, become erumpent; they are black, globose, with a papillate ostiole, and measure 141×242 μ. The pycnidiospores are hyaline, fusoid to ovoid, 1-celled, and 16×5 μ.

Reference

54.

Tip rot, *Hendersonula toruloides* Nattrass

Symptoms

A tip rot has appeared in the West Indies and Southeast Asia. The disease involves the young fruit from the blossom end, causing blackening and shriveling. During rainy seasons, the fruit decay rapidly and become mummified if not destroyed by insects.

Etiology

Pycnidia develop on the blackened peel, producing pycnidiospores which are 1–5-septate, often brown, and 17×5 mm.

Reference

25.

Anthracnose, *Gloeosporium musarum* Cke. & Mass.

Symptoms

The early stage of the disease appears in the field on flowers, skin, and distal ends of hands. Small, black spots increase in size, become sunken, coalesce, and form large spots. In severe infection, the whole fruit is covered with dark blemishes with characteristic bright salmon-colored pustules. Occasionally no form of spotting is visible on the banana before harvest, and the development of the common brown spotting begins during the ripening period. The skin spotting begins with brownish, small, circular, sunken spots. They enlarge rapidly and coalesce when numerous, occupying the entire surface. Penetration to flesh through the skin soon follows, and the interior becomes soft and dark-colored. No matter what time the inoculation occurs, the most striking aspects of the disease appear with the approach of maturity.

Etiology

The mycelium is colorless. Acervuli appear on the brown, circular, subepidermal lesions in close, compact masses. They rupture the epidermis and produce an abundance of pink spore masses. Conidia are oval or elongate, with a clear spot at the center. The fungus infects the fruit in the field and produces appressoria that survive but remain dormant until favorable environmental conditions arise; then the pegs become active and the mycelium generated from them penetrates the skin tissue by way of middle lamella, pushing through the cell wall. The hyaline conidia measure 12–15 × 5–6 μ. The ascigerous form is *Glomerella cingulata.*

References

16, 34, 39, 53.

Leaf lesions, *Pestalotia leprogena* Speg.

Symptoms

The fungus has frequently been associated with typical spots or pits on fruit and as a main stalk rot. Spots on the leaves are light tan, orbicular, and up to 5 mm in diameter. On fruit, the brown, circinate spots are leathery, often involving most of the fruit.

Etiology

Fruiting structures are black, erumpent, and 72–145 μ wide. Conidia are 5-celled, more or less elongate-elliptical to fusiform or curved, and 18–23 μ. The 2 end cells are hyaline, the 3 central cells colored. Three slender setulae occur on the terminal cell, and a single pedicel occurs on the basal cell.

References

13, 23.

Leaf spot, *Pestalotia palmarum* Cke.

Symptoms

Initially, the foliage lesions on *Musa* spp. and hybrid seedlings are small, up to 6 mm long, and narrowly linear, limited by the veins. They usually expand up to 4 × 0.9 cm and become a light yellow brown.

Etiology

The fungus produces numerous black pustules on killed host tissue that give rise to an abundance of 5-celled, elongate lenticular to clavate conidia, measuring 16–22 × 5–7 μ. The 3 center cells are dark-colored, the end cells hyaline. The terminal cell shows 3 setulae, which are up to 16 μ long; the basal cell has a short pedicel.

References

13, 48.

Tip-end rot, *Botrytis cinerea* Pers. ex Fr.

Symptoms

The disease occurs during wet, cool months. Infection begins with a dark brown discoloration at the floral or distal end of the finger. Internal brown rotting develops, delimited by a clear, watery margin extending along the inner pulp more rapidly than along the peel. On the surface there appears a fairly dense, brown mycelium with short, erect, branched conidiophores, bearing characteristic grayish brown masses of conidia.

Etiology

The mycelium is hyaline, rapid growing, branched, septate, producing dense, gray tufts turning brown, and composed of conidiophores which are erect, branched, and more or less dendritic. Conidia are subglobose to oval, 10–12 μ in diameter, and on certain hosts produce black sclerotia, irregular in size and shape.

Reference

54.

Sour rot, *Oospora rosca-flavida* Sacc.

Symptoms

The disease is found on leaves, but is characterized by a soft, wet, sour, dark-colored, leaking appearance on mature fruit. As a wound invader, it forms a thin, compact, water-soaked, often somewhat wrinkled layer over the affected surface.

Etiology

Conidia are hyaline, catenulate, oval, 1-celled, and $8–11 \times 2.5–3$ μ; the fertile hyphae are erect, measuring $40–45 \times 4.4–5$ μ.

References

10, 54.

Pitting, *Piricularia grisea* (Cke.) Sacc.

Symptoms

Finger dropping is due to the small, reddish pits of the finger stalks and cushions of the proximal or uppermost hands. Infection is initiated on immature fruit in the field and slowly develops during storage. Pits or spots, 4–8 mm in diameter, are localized on the exposed cushions and finger stalks of the two bottom hands. In addition, a slight pitting of similar origin may be present on one side of the main stalk.

Etiology

Circular, water-soaked spots appear on the leaf and bracts and with age develop a brown, moist decay. Conidiophores emerge in clusters of 2–5 from stomata. They are simple or sparingly branched, grayish, septate, and bear at their distal ends single, ovate, 2-septate conidia, measuring $24–29 \times 10–12$ μ.

References

2, 3, 18, 21, 54.

Cigar end, *Verticillium theobromae* (Turc.) Mason & Hughes

Symptoms

The disease essentially attacks immature fruit. Infection originates in the perianth and spreads slowly backward along the finger, causing blackening, shrinkage, and folding of skin tissues that are covered later by conidiophores and powdery gray conidia. Diseased ends resemble a burnt cigar tip. Presence of pathogens tends to make the perianth adhere persistently to the fingers. Internal dry rot of infected tissue is marked off sharply from healthy tissue. The disease does not progress in storage and does not cause premature ripening.

Etiology

The conidiophores are mostly solitary or loosely gregarious. They are pale yellow, septate, cylindrical, verticillately branched, basally thickened, tapering from below upwards, and measuring 100–400×4–6 μ. Conidia are hyaline, oblong or cylindrical, 4–6×2 μ, and borne in mucilaginous, globular, translucent heads.

References

18, 54.

Leaf speckle, *Cloridium musae* Stahel

Symptoms

The disease is characterized by spots and larger patches formed by a great number of isolated, minute, black or brown speckles. They are conspicuous on the upper surface but eventually cover the entire leaf. Hairlike structures protrude through stomata on the underside of the leaf.

Etiology

Hyphae of the fungus produce a fine rather loose network over the leaf surface and become attached at the stomata. The conidiophores are 100–200 μ long, arising from the mycelium on single speckles. Conidia are hyaline, oval, with a minute papilla at the point of attachment, and measure 5–8×2–4 μ.

Reference

41.

Black tip, *Deightoniella torulosa* (Syd.) Ell.

Symptoms

Black spot, pinhead, and trunk rot are other common names for this disease. It is first noticed on young, green fruit as minute, reddish brown spots with a slightly darker green, water-soaked, halo effect. The spots are confined almost entirely to the peel of the fruit and may develop in size up to 4 mm. A typical damping-off of seedlings has become a serious disease where hybrids are grown. The trunk rot originates on the outer sheath and gradually penetrates successively more inner sheaths until the pseudostem is so weakened that the plant falls over. Black spot is characterized by the pinpoint lesions on the principal veins of the leaf blade. The lesions enlarge and mature into brown, wedge-shaped, dead areas with yellow borders.

Etiology

Under favorable circumstances, the fungus sporulates abundantly on dying and dead leaf tissues and, consequently, inoculum is plentiful during the development of the fruit. The conidiophores are brown, of various lengths, usually somewhat elbowed, with swollen tips producing a single conidium, followed by another after the first is detached. The conidia are smoky olive, pear-shaped to elongate, with up to 12 septations; they measure $30–100 \times 12–16\ \mu$.

References

19, 44.

Leaf spot, *Cordana musae* (Zimm.) Hoehn.

Symptoms

The disease starts as pale brown, faintly concentric, oval spots and patches on leaf surfaces. The spots extend to larger areas and later develop into long strips of diseased tissue. Affected tissues are usually surrounded by a band of brilliant yellow or orange. Grayish brown spots appear on the underside of the leaf with less clearly defined zonation and border. Conidiophores are produced in large numbers on the underside and cause the smoky gray color.

Etiology

The fungus develops brown, septate, nodulose conidiophores, which measure $80–220\ \mu$, and bear 3 or more 2-celled, hyaline, obovate spores, measuring $12–21 \times 6–10\ \mu$.

References

34, 54.

Zonate eyespot, *Drechslera gigantea* (Heald & Wolf) Ito

Symptoms

The earliest manifestations are small, sunken, reddish brown to brown spots surrounded by a pale yellow border. They enlarge to elongate oval areas of up to 4 mm. The centers dry out and become gray with a pale yellow halo. Spots originate mostly on lower surfaces. They are usually scattered and may be found on the midrib and petiole. The disease is most common on young plants less than 3 m high.

Etiology

Limited amounts of brown, branched, septate mycelium appear on leaf spots that are of variable size. The conidiophores are usually long, up to $400\ \mu$, irregular, and up to 12 μ in diameter. The conidia are brown, 5-septate, long, cylindrical, and 300–315 × 15–21 μ.

References

8, 20, 21.

Brown spot, *Cercospora hayi* Calp.

Symptoms

Dark brown spots, 5-6 mm in diameter, occur on the rachis, the crescentic cushion, fingers, and pedicel. They have irregular margins surrounded by a water-soaked halo and are observed on fruit more than 50 days old. The spots do not enlarge with ripening.

Etiology

The mycelium is light-colored; it is not observed on spots, but is readily isolated from them. Conidia developed in culture on sterilized banana leaf tissue are hyaline, 5–45-septate, with acute tips, truncated bases, and measuring 90–150 × 3–4 μ.

Reference

16.

Squirter, *Nigrospora sphaerica* (Sacc.) Mason

Symptoms

This disease of banana is not apparent at the transport stage. Internal symptoms are darkening of the core center accompanied by darkening and decomposition of the pulp. Later, there is softening, with the flesh transformed into a mushy liquid that squirts out at the stalk end or at the side upon application of pressure. Earliest indication of squirter is recognized in the ripening room, during coloration of skin from green to faint yellow, and not in the field.

Etiology

In culture, hyphae are first hyaline, the submerged ones quickly turning brown, multiseptate, and contain a large amount of oil. White, cottony, strands of mycelium may develop. Conidia are sparse, black, borne on the tip of a more or less myceliumlike conidiophore, subspherical, and measure 15–18 μ. The fungus is a wound parasite.

References

36, 37, 54.

Blackhead, *Thielaviopsis paradoxa* (De Seyn.) Hoehn.

Symptoms

Often this fungus is observed as a wound parasite in transit and storage of many plants. It destroys the main stock, detached hands, and individual fingers, but it also causes a disease of underground parts. Characteristic brown or dark water-soaked patches extend from the surface into the cortex and also into the central tissues. Leaves are yellowish. The disease progresses slowly, weakens the plant, and lowers the quality of the bunches. Blackhead applies to any disease of the bulb showing superficial or internal brown or dark water-soaked patches.

Etiology

The mycelium is hyaline to gray, septate, branched, and abundant; it produces large oval, smoky brown or black spores, measuring 14–17 $\times$ 11 μ, usually in the form of curved chains at the end of hyaline, narrow, branched, and septate hyphae. In moist atmosphere, delicate, gray, superficial mycelium appears on the cut surface of infected tissue with characteristic sporulation. The fungus is not able to penetrate sound surfaces. It produces a characteristic pineapple odor. Hyaline endoconidiophores become septate, each cell rounding out to form a conidium, often formed in chains. Each conidium, formed within the fertile hyphae, is hyaline, rectangular, and 8–12 $\times$ 3.5–5 μ. The perfect stage is *Ceratocystis paradoxa*.

References

34, 54.

Panama wilt, *Fusarium oxysporum* f. sp. *cubense* (E. F. Sm.) Snyd. & Hans.

Symptoms

Infection may occur at any stage from very young suckers to fully grown plants. On Gros Michel and other varieties, infection starts as yellowing of lower and outer leaf blades, beginning along the margin and progressing inward toward the midrib. Coloration is in sharp contrast to the green healthy leaves. Affected leaves soon wilt and hang pendant until finally only the innermost leaf stands erect from the top of the plant; this leaf later withers followed by death of the entire plant. In addition, the leaf sheath splits longitudinally just above the soil level. A decided dwarfing and stunting of plants may occur. A transverse cut through an infected rhizome exposes a dark red to reddish brown and finally purple or black discoloration of diseased vascular strands. Roots growing from diseased rhizomes are blackened and decayed. The disease generally spreads through infected suckers used as planting stock.

Etiology

Sporodochia appear on the surface of petioles and leaves of infected plants and also emerge through stomata from a globose mass of pseudoparenchymatous tissue. Conidiophores emerge on fruiting bodies in all directions, verticillately branched with 2 or occasionally 3 one-celled branches in a whorl. Conidia are hyaline and 1-septate; microconidia, produced in great abundance, are ovate or elongate, and 5–7 $\times$ 2.5–3 μ. Macroconidia are hyaline, sickle-shaped, pedicellate, 3-septate, sometimes 4–5, and measure 22–36 $\times$ 4–5 μ. The fungus is soil-borne and infection always proceeds from root and rhizome upward.

References

27, 46, 49, 54.

Heart-leaf rot, *Fusarium moniliforme* Sheldon var. *subglutinans* Wr.

Symptoms

Disease symptoms are evident at distal ends of older leaves prior to shooting of the bunch. Also, infection of the young, unfolded central leaf by severe tip rot may occur, accompanied by a pronounced blackening of tissue that extends backwards along the unrolled heart leaf into the pseudostem. Finally the leaves constituting the crown tend to bend over at the junction with the pseudostem, and soon the whole trunk can readily be pushed over. Usually, older leaves are subject to a localized tip rot, causing the distal end, on which the spores of the fungus are abundantly developed, to fray and blacken.

Etiology

Microconidia are not borne in chains; they are unicellular and measure $5–16 \times 2–3.5$ μ. Macroconidia are slender, with a slightly constricted top and a pedicellate base, mostly 3-septate, measuring $25–38 \times 2.75–3.5$ μ. Chlamydospores are absent, sclerotia are dark blue, and mycelium is vinaceous to violet and spreads out or is erumpent. A perfect stage is *Gibberella moniliforme*.

References

29, 51, 54.

Diamond spot, *Fusarium roseum* (Lk.) Snyd. & Hans.

Symptoms

The first symptom of diamond spot is a yellow, slightly raised, inconspicuous blemish, 3–5 mm in diameter, on the surface of the peel. A longitudinal crack develops. The spot enlarges to $1–3.5 \times 5–1.5$ cm, the tissue shrinks, the yellow color darkens, and the spot becomes black. Spots continue to develop after harvest.

Etiology

The mycelium is white to pink or salmon, septate, and branched. Conidiophores are perpendicular to the hyphae and 45–125 μ high. Penicilli are irregularly branched and up to 140 μ high. Conidia are produced on verticils that typically become aggregated into wet, gelatinous balls. Conidia are colorless, elliptical, apiculate, smooth, 1-celled or 1-septate, and $5–7 \times 3–4$ μ. The perithecia, not frequently produced, are violet to black and semistromatic in structure. There are no paraphyses. Ascospores are hyaline, fusoid to elliptical, typically 3-septate, and $25–70 \times 4–6$ μ. The perfect stage is *Gibberella roseum*.

Reference

3.

Other fungi associated with banana

Alternaria musae Bouriquet & Bataille
Anthostomella molleriana Winter
Aposphaeria musarum Spegazzini
Capnodium musae Viégas
Cercospora koepkei Krueger
Cladosporium musae Mason
Diaporthe musae Spegazzini
Diplodia creba Saccardo
Dothidella musae Hoehnel
Fusarium sambucinum Fuckel
Guignardia musae F. L. Stevens
Leptosphaeria musarum Saccardo & Berlese
Macrophoma ensetes Saccardo & Scalla
Macrosporium ensetes Thuemen
Mycosphaerella musae (Spegazzini) Sydow
Naucoria musarum Patouillard & Desmazieres
Phoma musicola Spegazzini
Phyllosticta musae-sapientii Fragoso & Ciferri
Ramichloridium musae Stahel
Rosellinia bunodes (Berkeley & Broome) Saccardo
Sphaeropsis paradisiaca Montagne
Sphaerostilbe musarum Ashby
Stagnospora musae (Cooke) Saccardo

References: banana

1. Ashby, S. F. 1927. Macrophomina phaseoli (Maubl.) comb. nov. the pycnidial stage of Rhizoctonia bataticola (Taub.) Butl. Trans. Brit. Mycol. Soc. 12:141–47.
2. Beraha, L. 1962. Pitting disease of banana on the market. Plant Disease Reptr. 46: 354–55. (Abstr.)
3. Berg, L. A. 1968. Diamond spot of bananas caused by Fusarium roseum 'Gibbosum.' Phytopathology 58:388–89.
4. Buddenhagen, I. W. 1960. Strains of Pseudomonas solanacearum in indigenous hosts in banana plantations of Costa Rica, and their relationship to bacterial wilt of bananas. Phytopathology 50:660–64.
5. Buddenhagen, I. W. 1961. Bacterial wilt of bananas: history and known distribution. Trop. Agr. (London) 38:107–21.
6. Bunting, R. H., and H. A. Nade. 1924. Gold Coast plant diseases. Waterlow & Sons: London, pp. 31–34.
7. Calpouzos, L., et al. 1961. Relation of petroleum oil composition to phytotoxicity and sigatoka disease control on banana leaves. Phytopathology 51:317–21.
8. Drechsler, C. 1928. Zonate eyespot of grasses caused by Helminthosporium giganteum. J. Agr. Res. 37:473–92.
9. Elliott, C. 1951. Manual of bacterial plant pathogens. 2d ed. Chronica Botanica: Waltham, Mass., pp. 39–44, 109, 167.
10. Fawcett, H. S. 1936. Citrus diseases and their control. 2d ed. McGraw-Hill: N.Y., pp. 406–9, 449–53.
11. French, E. R., and L. Sequeira. 1970. Strains of Pseudomonas solanacearum from Central and South America: a comparative study. Phytopathology 60: 506–12.
12. Gäumann, E. 1921. On a vascular bacterial disease of the banana in the Dutch

East Indies. Mededel. Inst. Plantereziekten 48:134.
13. Guba, E. F. 1961. Monograph of Monochaetia and Pestalotia. Harvard University Press: Cambridge, Mass., pp. 167–71, 213–14.
14. Hansford, C. G. 1945. Uganda plant diseases. E. African Agr. J. 10:147–51.
15. Heald, F. deF. 1933. Manual of plant diseases. 2d ed. McGraw-Hill: N.Y., pp. 492–501.
16. Kaiser, W. J., and F. L. Lukezic. 1965. Brown spot disease of banana fruit caused by Cercospora hayi. Phytopathology 55:977–80.
17. Leach, R. 1941. Banana leaf spot Mycosphaerella musicola Zimm. Trop. Agr. (Trinidad) 18:91–95.
18. Loesecke, W. H. von. 1941. Bananas. Interscience: N.Y., pp. 156, 158–59.
19. Meredith, D. S. 1961. Fruit-spot (speckle) of Jamaican bananas caused by Deightoniella torulosa (Syd.) Ellis. Trans. Brit. Mycol. Soc. 44:95–104, 265–84, 391–405, 487–92.
20. Meredith, D. S. 1963. 'Eyespot,' a foliar disease of bananas caused by Drechslera gigantea (Heald & Wolf) Ito. Ann. Appl. Biol. 51:29–40.
21. Meredith, D. S. 1963. Further observations on the zonate eyespot fungus, Drechslera gigantea, in Jamaica. Trans. Brit. Mycol. Soc. 46:201–7.
22. Meredith, D. S. 1963. Latent infections in Pyricularia grisea causing pitting disease of banana fruits in Costa Rica. Plant Disease Reptr. 47:766–68.
23. Meredith, D. S. 1963. Pestalotia leprogena on leaves of Musa spp. in Jamaica. Trans. Brit. Mycol. Soc. 46:537–40.
24. Meredith, D. S. 1963. Pyricularia grisea (Cooke) Sacc. causing pitting disease of bananas in Central America. Ann. Appl. Biol. 52:453–63.
25. Meredith, D. S. 1963. Tip rot of banana fruits in Jamaica. Trans. Brit. Mycol. Soc. 46:473–81.
26. Meredith D. S. 1970. Banana leafspot disease (Sigatoka) caused by Mycosphaerella musicola Leach. Commonwealth Mycological Institute. Pathological Paper 11.
27. Nowell, W. 1924. Diseases of cropplants in the Lesser Antilles. West India Committee: London, pp. 241–47, 252–54.
28. Ocfemia, G. O. 1934. Two rusts hitherto unreported on economic hosts from the Philippine Islands. Philippine Agriculturist 23:880–85.
29. Ocfemia, G. O., and V. B. Mendiola. 1932. The fusarium associated with some field cases of heart rot of abaca. Philippine Agriculturist 21:296–308.
30. Rangaswami, G., and M. Rangarajan. 1965. A bacterial leaf spot disease of banana. Phytopathology 55:1035–36.
31. Rhoads, A. S. 1942. Notes on Clitocybe root rot of bananas and other plants in Florida. Phytopathology 32:487–96.
32. Rhoads, A. S. 1945. A comparative study of two closely related root-rot fungi, Clitocybe tabescens and Armillaria mellea. Mycologia 37:741–66.
33. Roger, L. 1951. Phytopathologie des pays chauds. Encyclopedie Mycologique 17:877, 952–77, 1080–84.
34. Roger, L. 1953. Phytopathologie des pays chauds. Encyclopedie Mycologique 18:1170–77, 1300–1308, 1488–91, 1758–71, 1804–19, 2007–8.
35. Sequeira, L. 1958. Bacterial wilt of bananas: dissemination of the pathogen and control of the disease. Phytopathology 48:64–69.
36. Simmonds, J. H. 1933. Squirter disease of bananas. Queensland Agr. J. 40:98–115.
37. Simmonds, J. H. 1938. Plant diseases and their control. David Whyte: Brisbane, Australia, pp. 21–22.
38. Simmonds, N. W. 1959. Bananas. Longmans: London, pp. 379–94.
39. Simmonds, J. H. 1963. Studies in the latent phase of Colletotrichum species causing ripe rots of tropical fruits. Queensland J. Agr. Sci. 20:373–424.
40. Snyder, W. C., and H. N. Hansen. 1945. The species concept in Fusarium with reference to discolor and other sections. Am. J. Botany 32:657–66.
41. Stahel, G. 1937. The banana leaf speckle in Surinam caused by Chloridium musae nov. spec. and another related banana disease. Trop. Agr. (Trinidad) 14:42–45.
42. Stahel, G. 1937. Notes on Cercospora leaf spot of bananas. Trop. Agr. (Trinidad) 14:257–64.
43. Stover, R. H. 1959. Bacterial rhizome rot of bananas. Phytopathology 49:290–92.
44. Stover, R. H. 1963. Leaf spot and damping-off of Musa seedlings caused by Deightoniella torulosa. Trop. Agr. (London) 40:9–14.
45. Stover, R. H. 1970. Leaf spot of bananas caused by Mycosphaerella musicola: role of conidia in epidemiology. Phytopathology 60:856–60.
46. Stover, R. H., and D. L. Richardson. 1968. 'Pelipita,' an ABB bluggoe-type

plantain resistant to bacterial and fusarial wilts. Plant Disease Reptr. 52: 901–3.
47. Tabor, R. J., and R. H. Bunting. 1923. On a disease of cocoa and coffee fruits caused by a fungus hitherto undescribed. Ann. Botany (London) 37: 153–57.
48. Vakili, N. G. 1963. A leaf spotting disease of Musa seedlings incited by Pestalotia palmarum Cooke. Plant Disease Reptr. 47:644–46.
49. Vakili, N. G. 1965. Fusarium wilt resistance in seedlings and mature plants of Musa spp. Phytopathology 55:135–40.
50. Voorhees, R. K. 1942. Life history and taxonomy of the fungus Physalospora rhodina. Florida Univ. Agr. Expt. Sta. Bull. 371:5–91.
51. Waite, B. H. 1956. Fusarium stalk rot of bananas in Central America. Plant Disease Reptr. 40:309–11.
52. Waldee, E. L. 1945. Comparative studies of some peritrichous phytopathogenic bacteria. Iowa State Coll. J. Sci. 19: 435–84.
53. Wardlaw, C. W. 1931. Notes on the parasitism of Gloeosporium musarum (Cooke & Massee). Trop. Agr. (Trinidad) 8:327–31.
54. Wardlaw, C. W. 1961. Banana diseases. Longmans: London.
55. Yirgou, D., and J. F. Bradbury. 1968. Bacterial wilt of enset (Ensete ventricosum) incited by Xanthomonas musacearum sp. n. Phytopathology 58:111–12.

Barbasco (cube), *Lonchocarpus utilis* Smith

Rust, *Dicheirinia archeri* Cummins
Target spot, *Thanatephorus cucumeris* (Frank) Donk
Leaf wilt, *Cephalosporium deformans* Crandall
Leaf spot, *Cercospora lonchocarpi* Stevenson

Rust, *Dicheirinia archeri* Cumm.

Symptoms

Early infection on leaves that have not become fully expanded causes distortion and malformation of the blades and areas between the veins. Blade infections show surface depressions on the top side and raised bumps on the bottom. The bumps develop into sori, which produce teliospores. There is very little killed tissue and only slight brown discoloration of areas immediately surrounding the infections. The spots rarely exceed 2–3 mm in diameter.

Etiology

The pycnidia and telia appear on the lower leaf surfaces, and the aecia and uredinia stages are lacking. Telia are chestnut brown, paraphysate, partially immersed, and 100–200 μ in diameter. Teliospores are subglobose, 19–38 $\times$ 32–40 μ, in 2-celled combination, each cell oblong to ellipsoidal, with tuberculate processes, and a hyaline, short pedicel.

References

2, 3.

Target spot, *Thanatephorus cucumeris* (Frank) Donk

Symptoms

The leaf spots are first evident by the appearance of wet, chlorotic, barely noticeable specks that rapidly develop into circular areas 1 mm or more in diameter surrounded by a yellow band. The spots enlarge, forming a series of concentric targetlike bands. The external hyphae grow out 1 mm or more in advance of the internal hyphae. The advance is correlated with night humidity and the band lines with daytime arrested growth. The spot expansion becomes static; a red brown border forms, and a silvery sheen appears on the lower surface.

Etiology

Brown, flat to oval vegetative sclerotia may form on killed, dry host tissue. Under humid conditions basidiospores, measuring 15–21×5–7 μ, are produced in abundance on lower stem and leaf tissues.

Reference

2.

Leaf wilt, *Cephalosporium deformans* Crandall

Symptoms

Small chlorotic spots, up to 2 mm in diameter, appear on the parenchyma tissue of the leaf blades before they reach full size. The spots are surrounded by a pinkish to pale purple halo and do not expand much more. They produce conidia in these lesions on the lower leaf surface. Sometimes the midrib or larger veins become infected and produce long, wet, discolored areas that often kill the terminal leaf tissue beyond.

Etiology

The mycelium is rosy white to dark, branched, sometimes tufted, and up to 3 μ in diameter. The conidiophores are hyaline, nonseptate, taper toward the tip, and measure 75×12 μ. Conidia are hyaline to rosy pink, elliptical, 1-celled, produced singly but often forming globose heads, and 8–17×5–7 μ.

Reference

2.

Leaf spot, *Cercospora lonchocarpi* Stevenson

Symptoms

This leaf spot is common wherever the host is grown and has caused considerable defoliation. The spots are often large, up to 20 mm in diameter, and yellow to tan surrounded by a narrow, reddish line and often a yellow halo.

Etiology

Stromata are mostly on the lower leaf surface. Fascicles are dense, and conidiophores are pale to dark brown, up to 3-septate, usually straight, and 10–45 ×2–4 μ. Conidia are yellow to olivaceous, cylindrical, septate, and 30–75×2–4 μ.

Reference

1.

References: barbasco

1. Chupp, C. 1953. A monograph of the fungus genus Cercospora. Published by the author: Ithaca, N.Y.
2. Crandall, B. S. 1950. Leaf diseases of Peruvian barbasco, Lonchocarpus utilis. Phytopathology 40:34–43.
3. Cummins, G. B. 1937. Descriptions of tropical rusts. Bull. Torrey Botan. Club 64:39–44.

Bean, *Phaseolus vulgaris* Linnaeus

Wilt, *Corynebacterium flaccumfaciens* (Hedges) Dowson
Halo blight, *Pseudomonas phaseolicola* (Burkholder) Dowson
*Brown spot, *Pseudomonas syringae* van Hall
Gall blight, *Pseudomonas viridiflava* (Burkholder) Clara
Bacterial blight, *Xanthomonas phaseoli* (E. F. Smith) Dowson
Blight, *Xanthomonas phaseoli* var. *fuscans* (Burkholder) Starr & Burkholder
Brown rot, *Pseudomonas solanacearum* E. F. Smith
Streak, *Bacillus lathyri* Manns & Taubenhaus
*Downy mildew, *Phytophthora phaseoli* Thaxter
Wilt, *Pythium butleri* Subramaniam
Root rot, *Pythium debaryanum* Hesse
Whiskers, *Rhizopus nigricans* Ehrenberg
*Yeast spot, *Nematospora coryli* Peglion
Powdery mildew, *Erysiphe polygoni* De Candolle
†Scab, *Elsinoë dolichi* Jenkins, Bitancourt, & Cheo
*Scab, *Elsinoë phaseoli* Jenkins
Cottony rot, *Sclerotinia sclerotiorum* (Libert) De Bary
*Pod blight, *Diaporthe phaseolorum* (Cooke & Ellis) Saccardo
Rust, *Uromyces phaseoli* var. *typica* Arthur & Cummins
Soil rot, *Thanatephorus cucumeris* (Frank) Donk
Southern blight, *Botryobasidium rolfsii* (Saccardo) Venkatarayan
Web blight, *Botryobasidium microsclerotia* (Matz) Venkatarayan
Leaf spot, *Phyllosticta phaseolina* Saccardo
Leaf spot, *Ascochyta phaseolorum* Saccardo
Anthracnose, *Colletotrichum lindemuthianum* (Saccardo & Magnus) Briosi & Cavara
*Stem anthracnose, *Colletotrichum truncatum* (Schweinitz) Andrus & W. D. Moore
Gray mold, *Botrytis cinerea* Persoon ex Fries
Ashy stem blight, *Macrophomina phaseolina* (Tassi) Goidanich
Leaf spot, *Alternaria brassicae* (Berkeley) Saccardo f. *phaseoli* Brunaud
Black root rot, *Thielaviopsis basicola* (Berkeley & Broome) Ferraris
Leaf spot, *Cercospora cruenta* Saccardo
Fusarium yellows, *Fusarium oxysporum* f. sp. *phaseoli* Kendrick & Snyder
Dry root rot, *Fusarium solani* (Martius) Appel & Wollenweber f. *phaseoli* (Burkholder) Snyder & Hansen
Texas root rot, *Phymatotrichum omnivorum* (Shear) Duggar
Angular leaf spot, *Phaeoisariopsis griseola* (Saccardo) Ferraris
Floury leaf spot, *Ramularia phaseolina* Petrak
Other fungi associated with bean

Wilt, *Corynebacterium flaccumfaciens* (Hedges) Dows.

Symptoms

This is a seed-borne disease and consequently kills seedlings and young plants. Surviving plants show stunting, top wilting, and leaf shedding; vascular tissue is often colored as well as one or both sutures. The spots on mature pods

*Diseases of lima bean.
†Diseases of hyacinth bean.

are greenish brown blotches rather than specific lesions or cankers. The use of uncontaminated seed is the principal means of control. There is no external sign of seed infection.

Etiology

The organism is a gram-negative rod, forms no capsules or chains, is motile by a single polar flagellum, and measures $0.6–3.0 \times 0.3–0.5\ \mu$. Agar colonies are yellow, smooth, wet, shiny, circular with entire margins, flat to slightly raised, semiopaque, and more or less viscid. Optimum temperature, 31C.

References

2, 6, 7.

Halo blight, *Pseudomonas phaseolicola* (Burkh.) Dows.

Symptoms

The distinguishing characteristic of the disease on bean plants is the reddish brown spots on the foliage, each surrounded by a wide halo, paler than the normal green. The areas become brown and die, and large portions of the leaf also die. On the pods the organism causes small, water-soaked spots that may be separate or may involve the suture. The spots later produce a white to milky exudate that often dries to a shiny, shieldlike crust. Diseased seed are often wrinkled and irregular in shape, and they are the principal means of dissemination of the organism.

Etiology

The organism is a gram-negative rod, single, in pairs, or in chains, forms capsules, is motile by a single flagellum, and measures $1.5–3.75 \times 0.7–1.5\ \mu$. Agar colonies are white to creamy with a bluish hue, circular, raised, thicker at the edges, and smooth or undulating. A green fluorescent pigment is produced in culture. Optimum temperature, 20-23C.

References

1, 6, 7.

Brown spot, *Pseudomonas syringae* van Hall

Symptoms

Although rarely found on bean, this disease is very common on a large number of hosts. On bean it produces a ring spot on the pods. On other parts of the plant it is similar to other bacterial diseases but is not seed-borne.

Etiology

The organism is a gram-negative rod, produces capsules, is motile by 1 to several polar flagella, and measures $0.75–1.5 \times 1.5–3\ \mu$. Agar colonies are white, transparent, circular, convex, and smooth or wrinkled. A green fluorescent pigment is produced in culture. Optimum temperature, 28–30C.

References

5, 7.

Gall blight, *Pseudomonas viridiflava* (Burkh.) Clara

Symptoms

This is a highly virulent, though rarely observed disease, that depends on wounds for ingress. A gall is produced on the stem where inoculated. Spots on green pods are reddish brown and limited in extent, forming lesions involving all local tissues. Reddish brown exudate may be evident on the pods.

Etiology

The organism is a gram-negative rod, single or in pairs, produces no capsules, is motile by 1 or 2 polar flagella, and measures 1.35–3.6 × 0.6–1.2 μ. Agar colonies are yellow to cream, glistening, and smooth with wavy edges. A green fluorescent pigment is produced in culture.

References

2, 7.

Bacterial blight, *Xanthomonas phaseoli* (E. F. Sm.) Dows.

Symptoms

The parasite is usually seed-borne and produces infection on cotyledons, the main stem, and young foliage. When not killed, the diseased seedling is a source of inoculum spread by rain and wind from the primary infections to adjacent plants. Mature plants show large, scaldedlike, brown spots on the foliage. Small water-soaked spots develop on the pods and cause rapid deterioration and loss of the pod. A yellow exudate is visible on lesions in early morning. The bacterium is worldwide in distribution.

Etiology

The organism is a gram-negative rod, single or in pairs, produces no capsules, has rounded ends, is motile by a single polar flagellum, and measures 0.3–3 × 0.4–0.8 μ. Agar colonies are yellowish, wet, glistening, circular, entire, and watery to butyrous.

References

2, 6, 7.

Blight, *Xanthomonas phaseoli* var. *fuscans* (Burkh.) Starr & Burkh.

Symptoms

This blight is not usually distinguishable from the common bacterial blight. Certain varieties susceptible to bacterial blight are resistant to this organism.

Etiology

The organism is a rod, single or in pairs, forms no capsules, is motile by a single polar flagellum, and measures 1.35–4×0.6–1.4 μ. Agar colonies are honey yellow, watery, butyrous, and entire to undulating. A brown pigment is produced in culture.

References

2, 7.

Brown rot, *Pseudomonas solanacearum* E. F. Sm.

Symptoms

This is a vascular disease and causes a distinct brown to slate discoloration of the root and stem tissues, resulting in a stunting of the plant followed by yellowing, wilting, leaf shedding, and death. There are no distinguishing spots or lesions formed on the foliage. The organism is soil-inhabiting and is not seed-borne. A whitish exudate may be visible on cut ends of diseased stems.

Etiology

The organism is a gram-negative rod, produces no capsules, is motile by a single polar flagellum, and measures 1.5–2×0.5 μ. Agar colonies are opalescent becoming brownish, wet, small, shiny, smooth, and irregular. Optimum temperature, 35–37C.

References

6, 7.

Streak, *Bacillus lathyri* Manns & Taub.

Symptoms

This bacterium is probably not a parasite on bean although reported as such.

Etiology

The organism is a gram-negative rod, produces no capsules or chains, is motile by numerous peritrichous flagella, and measures 0.7–1.5×0.6–0.9 μ. Agar colonies are yellow, glistening, smooth, raised, and grow rapidly. Optimum temperature, 28–30C.

Reference

7.

Downy mildew, *Phytophthora phaseoli* Thaxt.

Symptoms

Mildew is found on petioles, foliage, and flowers but is most conspicuous on the pods, where it produces a white, cottony mycelial growth over a portion or almost all of the outer surface. A blue to purple zone marks the border between healthy and diseased tissue. Pods are killed, become black and dry, and often remain attached.

Etiology

The mycelium is hyaline, branched, nonseptate, and produces absorbing haustoria. The conidiophores are branched, emerging from the host tissue with slightly swollen tips where single conidia are produced successively, producing swollen areas for each conidium or sporangium. The spores are hyaline, ovoid, give rise to about 15 biciliate zoospores, and measure $28–42 \times 17–27\ \mu$. Direct germination may occur. Oospores are hyaline to yellowish, spherical, smooth, and $26 \times 18\ \mu$.

References

5, 19.

Wilt, *Pythium butleri* Subr.

Symptoms

A soft, watery lesion forms on the stems at the soil line or slightly above. The fungus mycelium is white and may grow up the stem onto the leaves and pods. The stem cortex becomes softened and separates from the vascular tissue; the plant becomes chlorotic, wilts, and finally dies. This fungus also causes a soft, watery decay, an advanced stage of nesting in transit.

Etiology

The mycelium is hyaline, branched, nonseptate, and produces sporangia on myceliumlike sporangiophores. The sporangia may produce mycelium directly or zoospores in numbers.

References

9, 12.

Root rot, *Pythium debaryanum* Hesse

Symptoms

This disease develops on young seedlings, where it produces a typical damping-off in which the invaded tissues become soft but seldom discolored. Stem pithiness develops on older plants, and these plants usually wilt and die prematurely.

Etiology

The mycelium is plentiful in host plants. It is hyaline, branched, nonseptate, and about $5\ \mu$ in diameter. The sporangia, which are not abundant, are mostly terminal. The conidia are numerous, spherical, similar to sporangia, and germinate directly. Oospores are smooth, $10–18\ \mu$ in diameter, and germinate directly.

References

9, 19.

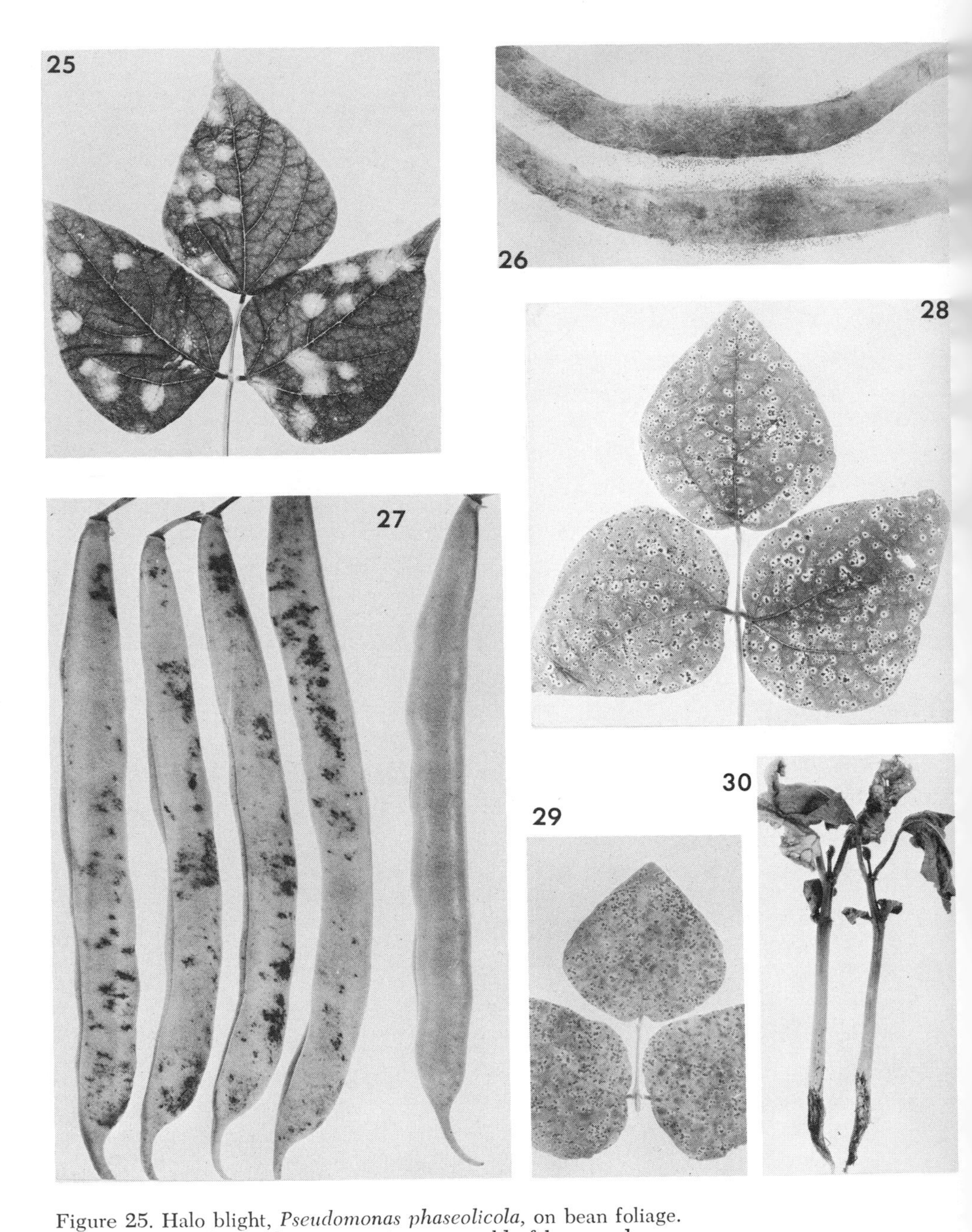

Figure 25. Halo blight, *Pseudomonas phaseolicola,* on bean foliage.
Figure 26. Whiskers, *Rhizopus nigricans,* a mold of bean pods.
Figure 27. Left: Powdery mildew, *Erysiphe polygoni,* on bean pods. Right: Healthy bean pod.
Figure 28. Rust, *Uromyces phaseoli* var. *typica.* Second and third stages on bean foliage.
Figure 29. Rust, *Uromyces phaseoli* var. *typica.* Second stage on bean foliage.
Figure 30. Soil rot, *Thanatephorus cucumeris,* showing stem girdling on bean.

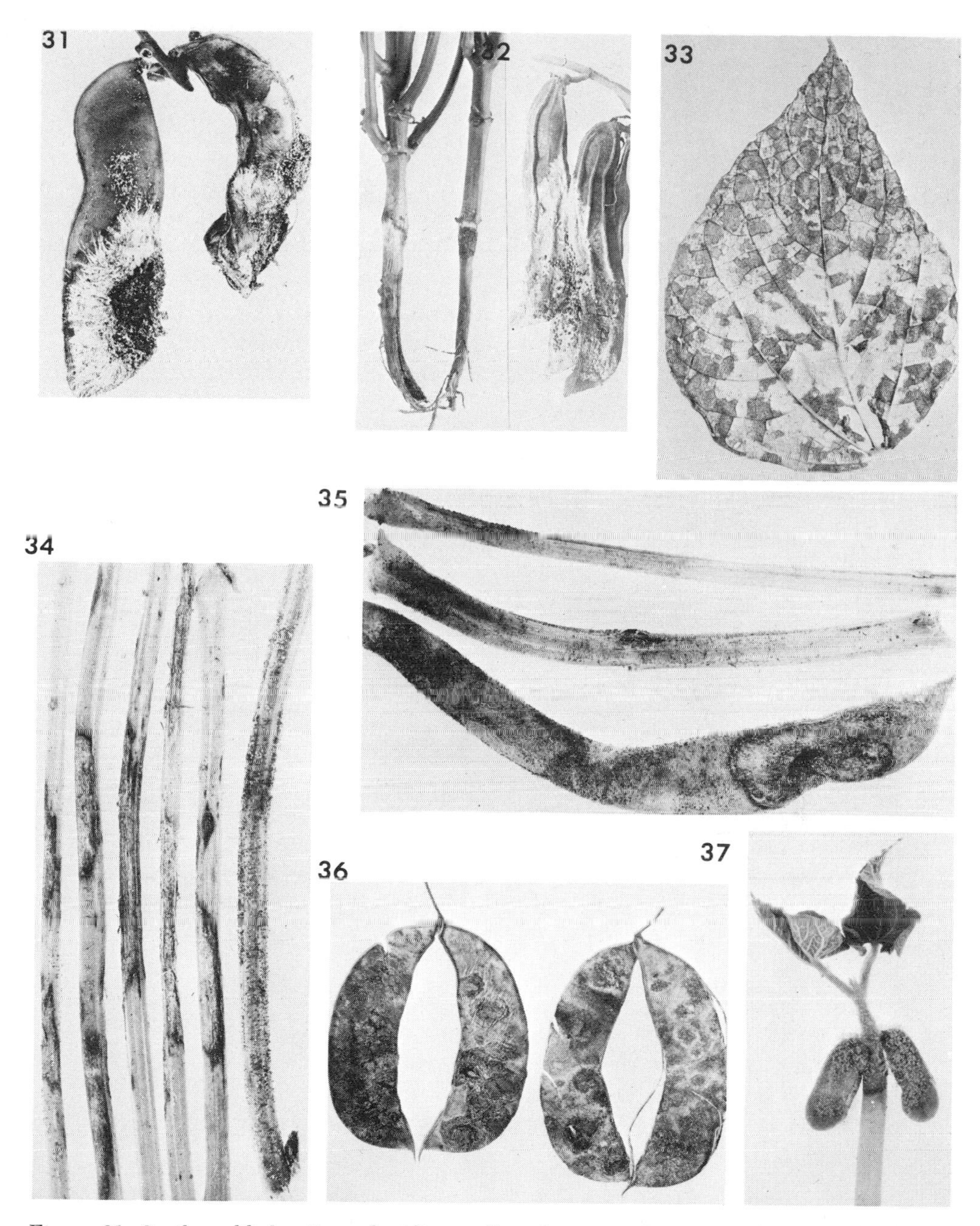

Figure 31. Southern blight, *Botryobasidium rolfsii,* showing soil contact infection on bean pods.
Figure 32. Southern blight, *Botryobasidium rolfsii,* on bean stems and velvet bean pods.
Figure 33. Angular leaf spot, *Phaeoisariopsis griseola,* on bean foliage.
Figure 34. Web blight, *Botryobasidium microsclerotia,* on bean petioles.
Figure 35. Web blight, *Botryobasidium microsclerotia,* on bean stems and pods.
Figure 36. Stem anthracnose, *Colletotrichum truncatum,* on lima bean pods, showing acervuli. (Photograph by J. F. Darby.)
Figure 37. Gray mold, *Botrytis cinerea,* on cotyledon and stem kills bean seedling.

Whiskers, *Rhizopus nigricans* Ehr.

Symptoms

This rot is the cause of rapid softening and decay of bean pods in packaging and in transit. It is only scatteringly found in the field on mechanically injured plants. It is generally recognized by the abundant mycelium and black-headed fruiting structures.

Etiology

The mycelium is hyaline, usually abundant, branched, and nonseptate. It produces stolons and conidiophores. The stolons advantageously spread the thallus, and the conidiophores or sporangiophores produce terminal clusters of conidia that measure 11–14 μ and are enclosed in a peridium. The zygospores are black, warty, spherical, and 150–200 μ. They are not frequently found.

Reference

9.

Yeast spot, *Nematospora coryli* Pegl.

Symptoms

Yeast spot has always been more or less associated with wounds caused by greenbug feeding punctures. The spots appearing on the young seed within the pod are at first irregular, slightly sunken, about 1 mm in diameter, and extend into the starchy part of the seed. The spot becomes tan to brown as the seed matures.

Etiology

The hyaline mycelium is scanty or nonexistent. The vegetative cells are yeastlike, spherical, and increase by budding. The asci are free-floating, cylindrical, with rounded ends, and measure 75–100 × 6–20 μ; they contain eight 2-celled ascospores which are in two groups of 4 at each extremity of the ascus. They measure 40–60 × 1–3 μ and germinate by producing buds, which enlarge, break apart, and become hyaline, spherical cells that continue the budding process. Agar colonies are glistening, entire, heaped, granular, and increase rapidly in size. As the development is slowed, the colonies spread out, and mycelial strands appear around their margins.

References

5, 9, 19, 21.

Powdery mildew, *Erysiphe polygoni* DC.

Symptoms

The disease is recognized by the white, powdery, superficial mycelium, conidiophores, and conidia growing on the foliage, pods, peduncles, petioles, and main stems of beans. It causes the foliage to become yellow and appear lifeless.

The leaves usually curl and eventually drop. Sometimes the black perithecial stage is present.

Etiology

The mycelium is septate, hyaline, white en masse, branched, and mostly superficial, supported by numerous epidermis-penetrating hyphae that harbor haustoria. The mycelium produces conidiophores that develop hyaline, terminal, 1-celled, ellipsoidal to slightly elongate conidia one after another, often in chains. The conidia measure 26–52×15–23 μ. The perithecia are black, spherical, appendaged, about 120 μ in diameter, and contain several asci, each with 2–4 ascospores that measure 24–28×11–13 μ.

References

5, 19.

Scab, *Elsinoë dolichi* Jenkins, Bitanc., & Cheo

Symptoms

Leaf spots tend to follow the venation and extend into the adjacent parenchyma. They develop shades of yellow, and the margins of the spots are slightly raised. Excessive vein infection may cause a general yellowing of the entire leaf. Spots on parenchyma leaf tissue may be circular and reach 4 mm in diameter. Stem lesions are brown to ash color with yellow to black borders. Pod lesions are circular, punctate, and about 5 mm in diameter. Lesions appear on both valves, the calyx, the pedicels, and along the suture; they are brown to purple black.

Etiology

The conidial stage consists of a hyaline to yellowish stromatic rind, raised, spreading, and disrupting the epidermis, and producing small, rounded surface cells about 30 μ thick. The conidiophores may be arranged in closely grouped fascicles about 20 μ high. Conidia are hyaline, spherical to elliptical, and 3–8×1–3 μ. The ascomata are more abundant on the leaf surface and originally infect the dark stromatic rind. They cover the lesion as dark, punctate bodies, measuring 100×60–300 μ. The asci are subglobose to ellipsoid and 20–32×15–22 μ. Ascospores are hyaline and up to 3-septate.

Reference

4.

Scab, *Elsinoë phaseoli* Jenkins

Symptoms

Scab occurs on the stem, leaf, and pods of lima beans, causing raised, wartlike protuberances that are buff-colored, ranging to red or brown, circular, elliptical to elongate, usually small but up to 1 cm in diameter. Spots appear on both leaf surfaces, and the pods are often malformed. Conidia are abundantly produced in dark-colored pycnidia in the brown spots.

Etiology

The mycelium is hyaline, scanty, submerged, and supports hyaline, short, stubby conidiophores that gradually darken and form compact sporodochia. The conidia are hyaline to pale-colored, and 10×4 μ. The asci, produced in erumpent ascomata, are about 30–40 μ in diameter. The ascospores are hyaline to pale-colored, oblong to elliptical, 3-septate, and $13–15 \times 5–6$ μ.

References

5, 9, 11, 19.

Cottony rot, *Sclerotinia sclerotiorum* (Lib.) d By.

Symptoms

Seedling infection in the field occurs during the cooler parts of the growing season in the form of a damping-off symptom. On older plants a stem-girdling canker develops at the soil line, eventually causing death. From the canker considerable aerial infection may develop in the form of a wilt of foliage and young branches. The fungus produces an abundance of white mycelium and many black, hard, superficial or within the stem or pod, loaflike or irregular sclerotia, measuring up to 1 cm in diameter. It is the most destructive bean disease in transit, causing nesting and a watery soft rot.

Etiology

The mycelium is hyaline, septate, branched, and under certain conditions develops short lateral branches that have sporiferous tips bearing microspores that are hyaline, spherical, and 2–4 μ in diameter. Sclerotia of various sizes and shapes are produced on host tissues. They are important in the spread of the fungus and its survival from season to season. Upon germination they produce apothecia that may be sessile or stipitate, depending on the location of the resting sclerotium. The apothecia are cup- to saucer-shaped, fleshy, and contain palisades of asci. Each ascus contains 8 uniseriately arranged ascospores that are hyaline, oval, 1-celled, and $11–16 \times 5–8$ μ.

References

5, 19.

Pod blight, *Diaporthe phaseolorum* (Cke. & Ell.) Sacc.

Symptoms

Blight is most important on the pods but usually appears first on the leaves as irregular-shaped brown spots with distinct borders. As the tissue is killed, numerous black fruiting bodies appear scattered or in zones. Pod infections develop as they mature, beginning anywhere, but usually involving the entire pod. As the pod becomes discolored, many black pycnidia appear. Their presence is diagnostic.

Etiology

The mycelium is hyaline, branched, septate, and develops the pycnidia that often appear in concentric rings in the spots, especially on the pods. The pycnidiospores are hyaline, 1-celled, oval, and 6–9×2–5 μ. Stylospores are hyaline, long, narrow, and bent. The perithecia are black and spherical, up to 300 μ in diameter. The asci are clavate, and the ascospores are hyaline, oblong, 1-septate, and 10–12×2–4 μ.

References

5, 19.

Rust, *Uromyces phaseoli* var. *typica* Arth. & Cumm.

Symptoms

Rust is an almost universal disease of beans to which all aboveground parts of the plant are susceptible. It usually appears first on the leaves in the form of light-colored, small, raised, blisterlike spots. The epidermis over these sori erupts and exposes the brown powderlike uredospores in large numbers. These spores reinfect the bean plant. The sori enlarge, often forming a secondary ring of sori around the original one. Highly susceptible plants may be killed from defoliation; others may linger longer but become unproductive.

Etiology

The mycelium is hyaline, branched, septate, and exists internally in the host tissue, producing sori of various kinds. The aecia and pycnia are in small groups and are rarely observed. The uredospores are brown, globoid to ellipsoidal, echinulated, thin-walled, and 20–23×16–23 μ. The teliospores are dark brown to almost black, globoid, 1-celled, thick-walled, with hyaline papilla, slightly verrucose, and measure 24–32×20–26 μ. The pedicel is fragile and usually shorter than the spore diameter.

References

5, 19.

Soil rot, *Thanatephorus cucumeris* (Frank) Donk

Symptoms

This rot causes a decay of seed as they germinate and a damping-off of seedlings. It causes reddish brown stem lesions below the soil surface that may be severe enough to girdle the stem, resulting in dead plants at various stages of maturity. Lesions are produced on pods that contact the soil. The fungus is soil inhabiting.

Etiology

The mycelium is light brown, septate, branched, and has no conidial spore. It survives between crops as mycelium and in the sclerotial stage. The sclerotia are brown, loaf-shaped to flat, and composed of compact fungus cells. They may vary in size from very small to 10 mm in diameter and germinate by producing mycelium. Under favorable conditions, the basidial stage can be seen superficially covering the plant stem with a delicate network of hyphal threads. The basidia are clustered, each producing 4 sterigmata and 4 basidiospores that are hyaline, elongate, and 7–13×4–7 μ. The sterile fungus is *Rhizoctonia solani.*

References

9, 17.

Southern blight, *Botryobasidium rolfsii* (Sacc.) Venkat.

Symptoms

The fungus develops more frequently in the warmer wet periods of the year. It attacks the plant at the soil line or slightly below, where a girdling occurs. The fungus produces white mycelium around the stem and in the surrounding soil. Brown, spherical sclerotia about 1 mm or less in diameter are usually present on killed plant stems.

Etiology

The fungus mycelium is hyaline to white in clumps. No conidia are produced. Brown, spherical, hard sclerotia are diagnostic. Basidia are rarely found. The basidiospores are hyaline, 1-celled, oval, and 5–10×3–6 μ. The sterile fungus is *Sclerotium rolfsii.*

References

17, 19.

Web blight, *Botryobasidium microsclerotia* (Matz) Venkat.

Symptoms

Leaf spots are brown, small, water-soaked, circular, and enlarge rapidly to irregular shapes up to 1 cm in size. The superficial mycelium grows up the stems, petioles, and peduncles onto the pods and leaves. The foliage is rapidly killed, and pod lesions become conspicuous as brown irregular lesions. All parts showing the disease produce an abundance of sclerotia.

Etiology

The fungus produces characteristic brown hyphae that are septate, branched, and silky. There are no vegetative conidia. The basidial stage develops on the leaf tissue, producing hyaline, oval, 1-celled basidiospores, measuring 5–6×11 μ. The small, brown sclerotia are diagnostic of the fungus. Most of them will pass through a 100 mesh screen. The sterile fungus is *Rhizoctonia microsclerotia.*

References

5, 15, 20.

Leaf spot, *Phyllosticta phaseolina* Sacc.

Symptoms

This is a very common disease wherever beans are grown and is associated mostly with mature and older leaves. The spots are up to 1 cm in diameter with brown centers and darker borders. Smaller spots sometimes appear on the pods. The leaf spots become speckled with small, black pycnidia.

Etiology

The mycelium is hyaline, branched, septate, and mostly submerged in host tissue. The pycnidia are black, mostly obovate, and up to 90 μ in diameter. Pycnidiospores are hyaline, oval to oblong, 1-celled, and 4–6×2–3 μ.

References

9, 19.

Leaf spot, *Ascochyta phaseolorum* Sacc.

Symptoms

This disease produces brown, circular, concentrically zoned areas speckled with minute, black, spore-producing pycnidia. When plentiful, the spots cause the leaves to drop. Lesions also appear on petioles, peduncles, and pods, and when on the main stem may cause a girdle, killing the plant.

Etiology

The mycelium is hyaline, branched, submerged, septate, and supports many brown to black pycnidia that measure 100–150 μ in diameter. The pycnidiospores are mostly 2-celled, sometimes 1- or more than 2-celled, and measure 18×5 μ or larger if more than 1-septate.

References

9, 22.

Anthracnose, *Colletotrichum lindemuthianum* (Sacc. & Magn.) Briosi & Cav.

Symptoms

Lesions are found on seed and all plant parts aboveground. The seed lesions are usually brown, rough, and slightly sunken. Stem, petiole, and leaf lesions appear as tan to brown areas that become sunken. The disease is manifested most distinctly on the pods, where more or less circular sunken lesions develop from mere specks up to 1 cm in diameter. The older spots on the pods usually produce many clusters of pink spores.

Etiology

The mycelium is hyaline to light-colored, branched, and septate. The accumulation of hyphae under the cuticle forms acervuli that rupture the epidermal covering. Short conidiophores support hyaline to pink, 1-celled, cylindrical conidia, measuring 13–22×4–6 μ. Setae are sometimes present. The ascospore stage of the fungus is *Glomerella cingulata*.

References

17, 19, 22.

Stem anthracnose, *Colletotrichum truncatum* (Schw.) Andrus & W. D. Moore

Symptoms

On the pods anthracnose forms circular to elongate, yellow pink spots that enlarge rapidly and become darker-colored and densely speckled with black acervuli. The stems, petioles, and foliage are also attacked. Girdling may occur if the lesion develops on the main stem at the soil line.

Etiology

The fungus hyphae are hyaline, branched, and septate, forming compact stromata supporting the acervuli which are often protruding and conspicuous. Setae are dark, prominent, and numerous. The conidia are hyaline to pink, 1-celled, sickle-shaped, and 18–30 × 3–4 μ.

References

5, 19.

Gray mold, *Botrytis cinerea* Pers. ex Fr.

Symptoms

Although not a serious disease, mold is often conspicuous on blossoms and young succulent tissues, causing them to become soft and wilt down. It is also found in transit on the pods, producing a mild form of nesting.

Etiology

The mycelium is light brown, variable in diameter, measuring up to 20 μ, branched, septate, and supports numerous branched conidiophores. The conidia are hyaline, 1-celled, oval, and 12–20 × 8–12 μ. Sclerotia are black, various shapes, and up to 4 mm in diameter.

References

9, 19.

Ashy stem blight, *Macrophomina phaseolina* (Tassi) Goid.

Symptoms

Blight appears as dark, sunken cankers on the stems of seedlings at the soil line. On older seedlings and maturing plants the stunted condition of the plant precedes the dropping of leaves followed by premature death.

Etiology

The disease is diagnosed by the black, very small, hard, imbedded, smooth, scattered sclerotia, measuring 50–150 μ in diameter. The pycnidia are observed less frequently than the sclerotia. They are black and submerged becoming erumpent. The conidia are 1-celled, fusiform, straight or slightly curved with pointed and blunt ends, respectively, and measure 15–30 × 5–8 μ.

References

5, 9, 19.

Leaf spot, *Alternaria brassicae* (Berk.) Sacc. f. *phaseoli* Brun.

Symptoms

Leaf spot has usually been considered a wound parasite and forms spots as a secondary invader. It forms circular, reddish brown, zonate spots with a darker brown border. Spores are produced abundantly, and they tend to darken the spots. This disease is usually found only on older foliage and not on stems or pods.

Etiology

The hyphae are brown, olivaceous, septate, branched, up to 8 μ wide, and produce erect, unbranched conidiophores. The conidia are borne singly or in chains of 2 or 3. They are smooth, long-beaked, obclavate, with 3–15 cross septa and several longitudinal septa, and measure 40–350 × 9–33 μ. There is much variation in spore characters.

References

5, 22.

Black root rot, *Thielaviopsis basicola* (Berk. & Br.) Ferr.

Symptoms

This fungus frequently causes a blackening of the roots and brown spots on the hypocotyl, more or less stunting the plant. It is not considered an important disease of beans.

Etiology

The mycelium is hyaline, septate, branched, and produces hyaline conidia endogenously on hyaline conidiophores. The conidia measure 8–23 × 3–5 μ. The chlamydospores are produced singly or in chains that break apart; they are dark brown, thick-walled, and 25–65 × 10–12 μ.

References

5, 9, 19.

Leaf spot, *Cercospora cruenta* Sacc.

Symptoms

Leaf spots up to 10 mm in diameter are generally prevalent on older leaves as reddish brown areas of variable size and shape, with darker borders. Spots are dry, crack, and portions fall out; some premature shedding of leaves may occur. The conidia develop on the spots from short, clustered conidiophores. Stems and pods usually escape with only minor spotting.

Etiology

The stromata are in stomatal openings mostly on the lower surface. Fascicles are dense, and conidiophores are pale olivaceous to brown, straight, geniculate, up to 2-septate, and up to 75 μ long. Conidia are pale brown, obclavate, curved, multiseptate, and 150–75 × 2–3 μ.

References

5, 9.

Fusarium yellows, *Fusarium oxysporum* f. sp. *phaseoli* Kendr. & Snyd.

Symptoms

This malady is often readily observed from some distance because of the greenish yellow to bright yellow leaves of diseased plants. The vascular tissue is invaded and discolored in stems, petioles, and peduncles. Some stunting occurs as lower leaves drop, but there is not a typical wilt.

Etiology

The mycelium is septate, branched, and hyaline at first becoming pale cream-colored and finally yellow. Chlamydospores are formed as the hyphae mature, and the microconidia are produced in abundance. The chlamydospores are hyaline, nonseptate, oval, and 6–15 × 2–4 μ. Macroconidia are rarely produced. They are elongate with rounded ends, curved, 2–3-septate, and 25–35 × 3–6 μ. There are no sclerotia or pionnotes.

References

9, 19.

Dry root rot, *Fusarium solani* (Mart.) Appel & Wr. f. *phaseoli* (Burkh.) Snyd. & Hans.

Symptoms

Dry rot first develops on the taproot in the seedling stage, producing a reddish pink color that increases in intensity and extends, involving larger portions of the main root which becomes brown. The smaller roots are killed, and the stem becomes hollow or pithy. New roots may develop near the soil line above the lesion. In severe cases the plant is nonproductive; in mild cases little damage results. The fungus is not seed-borne.

Etiology

The mycelium is hyaline, branched, septate, and is often evident around the stems of plants under favorable conditions. The microconidia are not frequently found although chlamydospores are usually plentiful, appearing intercalary or terminal, sometimes in chains, and measuring 8–11 μ in diameter. The macroconidia are hyaline, 3–4-septate, fusiform, and 44–50×4–6 μ. Pionnotes are conspicuous.

References

19, 22.

Texas root rot, *Phymatotrichum omnivorum* (Shear) Dug.

Symptoms

This is a root disease of beans resulting in stunting, a general unthrifty appearance of the plants, and often a sudden wilting. The fungus inhabits alkaline soil and damages the underground plant parts, producing a pinkish buff color of the roots.

Etiology

There are three stages of mycelium. The large-celled strands envelop roots in a brown, fuzzy growth. These strands are 20 μ in diameter. The acicular hyphae branch in pairs at right angles. White to buff-colored mats develop 4–8 in. across and 1/4 in. thick. They become powdery and produce smooth, ovate spores, measuring 6–8×5–6 μ. The final stage is the sclerotial formation of compact mycelial cells, dark-colored, firm, and resistant to deterioration.

References

9, 19.

Angular leaf spot, *Phaeoisariopsis griseola* (Sacc.) Ferr.

Symptoms

The earliest recognizable manifestations of the disease are somewhat smoky-dark areas on the foliage. As the disease develops, these areas become darker until they appear sooty. They are characteristically angular in outline. The darkness of the lesions results from the coremia and conidia produced mostly on the lower leaf surface. The spots on the upper surface of the leaves are brown. The disease also appears on stems, petioles, peduncles, and pods, but it is usually less prominent on these parts. Defoliation is often almost complete.

Etiology

The fungus produces stromata upon which columnar coremia develop. The coremia are black, dense, septate clusters of hyphal strands at the tips of which conidia are formed. The coremia vary in height up to 1 mm and when older may spread somewhat, forming a looser cluster that might suggest some *Cercospora* species. The conidia are lighter-colored than the coremial strands, cylindrical, slightly curved, and sometimes restricted. They measure 50–60× 7–8 μ and are 1–3-septate.

References

1, 10, 22.

Floury leaf spot, *Ramularia phaseolina* Petr.

Symptoms

The diagnostic sign of the disease is the white growth of conidiophores and conidia on the lower surface of infected leaves. Infection first appears on the older leaves and progresses upward. There is no definite demarcation of the diseased area, and it is recognized by the flourlike appearance of the fungus. The disease may appear any time after the cotyledon stage. Severe infection causes premature defoliation.

Etiology

The fungus produces fasciculate clusters of conidiophores that originally emerged from the stomata. The mycelium grows over the leaf surface and around trichomes of the host; when growth is luxuriant, it resembles a coremia and is easily distinguished by the white color of the fungus. The conidiophores are clustered, erect, often branched, septate, and $40–80 \times 3–8\ \mu$. Conidia are hyaline, oval to lemon-shaped, mostly nonseptate, and $7–18 \times 4–6\ \mu$.

References

3, 16.

Other fungi associated with bean

Achromobacter lipolyticum (Huss) Bergey et al.
Alternaria fasciculata (Cooke & Ellis) L. R. Jones & Grout
Aristastoma oeconomicum (Ellis & Tracy) Tehon
Ascochyta boltshauseri Saccardo
Cercospora canescens Ellis & G. Martin
Cercospora phaseoli Dearness & Bartholomew
Chaetoseptoria wellmanii Stevenson
Cladosporium herbarum Persoon ex Fries
Diplodia phaseolina Saccardo
Leptosphaeria phaseolorum Ellis & Everhart
Mycosphaera diffusa Cooke & Peck
Mycosphaerella cruenta (Saccardo) Latham
Phakopsora vignae (Bresadola) Arthur
Phomopsis phaseoli (Desmazieres) Grove
Phyllachora phaseoli (P. Henning) Theissen & Sydow
Phytophthora parasitica Dastur
Pullularia pullulans (De Bary) Berkhout
Pythium arrhenomanes Drechsler
Sclerophoma phaseoli Karakulin
Septoria phaseoli Maublanc
Stagonospora phaseoli Dearness
Stemphylium botryosum Wallroth

References: bean

1. Barros, O., et al. 1958. Angular leaf spot of bean in Colombia. Plant Disease Reptr. 42:420–24.
2. Burkholder, W. H. 1930. The bacterial diseases of the bean, a comparative study. N.Y. (Cornell) Agr. Expt. Sta. Mem. 127:1–88.
3. Cardona-Alvarez, C., and R. L. Skiles. 1958. Floury leaf spot (Mancha haricosa) of bean in Colombia. Plant Disease Reptr. 42:778–80.
4. Cheo, C. C., and A. E. Jenkins. 1945. Elsinoë and Sphaceloma diseases in Yunnan, China, particularly hyacinth bean scab and scab of castor bean. Phytopathology 35:339–52.
5. Chupp, C., and A. F. Sherf. 1960. Vegetable diseases and their control. Ronald Press: N.Y., pp. 104–65.
6. Dowson, W. J. 1949. Manual of bacterial plant diseases. Macmillan: N.Y., pp. 68–183.
7. Elliott, C. 1951. Manual of bacterial plant pathogens. 2d ed. Chronica Botanica: Waltham, Mass.
8. Green, D. E. 1946. Diseases of vegetables. Macmillan: London.
9. Harter, L. L., and W. J. Zaumeyer. 1944. A monographic study of bean diseases and methods for their control. U.S. Dept. Agr. Tech. Bull. 868:1–133.
10. Hocking, D. 1967. A new virulent form of Phaeoisariopsis griseola causing circular leaf spot of French beans. Plant Disease Reptr. 51:276–78.
11. Jenkins, A. E. 1931. Lima bean scab caused by Elsinoë. J. Agr. Res. 42:13–23.
12. Kendrick, J. B., Jr., and W. D. Wilbur. 1965. The relationship of population density of Pythium irregulare to preemergence death of lima bean seedlings. Phytopathology 55:1064. (Abstr.)
13. Lauritzen, J. I., L. L. Harter, and W. A. Whitney. 1933. Environmental factors in relation to snap-bean diseases occurring in shipment. Phytopathology 23:411–45.
14. Lefebvre, C. L., and J. A. Stevenson. 1945. The fungus causing zonate leaf-spot of cowpea. Mycologia 37:37–45.
15. Matz, J. 1917. A Rhizoctonia of the fig. Phytopathology 7:110–18.
16. Schieber, E. 1969. Ramularia leaf spot on beans in the highlands of Guatemala. Plant Disease Reptr. 53:415–17.
17. Townsend, G. R. 1939. Diseases of beans in southern Florida. Florida Univ. Agr. Expt. Sta. Bull. 336:1–59.
18. Townsend, G. R., and G. D. Ruehle. 1947. Diseases of beans in southern Florida. Florida Univ. Agr. Expt. Sta. Bull. (Rev.) 439:1–55.
19. Walker, J. C. 1952. Diseases of vegetable crops. McGraw-Hill: N.Y., pp. 10–56.
20. Weber, G. F. 1939. Web-blight, a disease of beans caused by Corticium microsclerotia. Phytopathology 29:559–75.
21. Wingard, S. A. 1922. Yeast-spot of lima beans. Phytopathology 12:525–32.
22. Zaumeyer, W. J., and H. R. Thomas. 1957. A monographic study of the bean diseases and methods for their control. U.S. Dept. Agr. Tech. Bull. 868:1–255.

Breadfruit, *Artocarpus* spp.

Soft fruit rot, *Rhizopus artocarpi* Raciborski
Fruit rot, *Physalospora rhodina* (Berkeley & Curtis) Cooke
Leaf blotch, *Phyllosticta artocarpicola* Batista
Anthracnose, *Colletotrichum artocarpi* Delacroix
Leaf spot, *Cercospora artocarpi* H. & P. Sydow
Other fungi associated with breadfruit

Soft fruit rot, *Rhizopus artocarpi* Rac.

Symptoms

The fungus produces a soft, almost superficial mycelium beginning near the stem end of the fruit or in the pollen-producing area. The gray, thin, external mycelium grows over the outside of the fruit and gradually darkens except at the growing margin. Finally, the entire inflorescence decays and is shed.

Etiology

The mycelium is hyaline, branched, nonseptate, and produces rhizoids. The sporangiophores are brownish, erect, and often branched. The sporangia are black, up to 190 μ in diameter, with a delicate covering. The columella is prominent. The conidia are dark, oval to oblong, 1-celled, and 12–16 μ in diameter.

References

2, 5.

Fruit rot, *Physalospora rhodina* (Berk. & Curt.) Cke.

Symptoms

The infected rind of the fruit becomes soft and dark brown in circular necrotic patches, beginning usually from a point of mechanical injury mostly resulting after the fruit are removed from the tree.

Etiology

Numerous pycnidia develop over the outer surface. The pycnidiospores are produced abundantly, often exuding in tendrils from the ostioles. Frequently, the spores are hyaline and 1-celled but later become darker and 2-celled, typical of *Diplodia natalensis*, the vegetative stage of the fungus.

Reference

4.

Leaf blotch, *Phyllosticta artocarpicola* Batista

Symptoms

A browning and drying begin at the tip of the leaf and extend downward and along the margins, producing large, necrotic areas surrounded by a light violet discoloration. Diseased leaves roll up and sometimes break off but usually remain attached. The yellow to tan areas become dotted with the fruiting structure of the fungus.

Etiology

Small, black, spherical, ostiolate, fruiting pycnidia appear in a more or less scattered pattern over the leaf blotches. They measure 112–144 × 96–120 μ. The pycnidiospores are produced in abundance and frequently are exuded from the ostioles in tendrils. They are hyaline, 1-celled, oval, and 5–7 × 4–5 μ.

Reference

1.

Anthracnose, *Colletotrichum artocarpi* Del.

Symptoms

Large, indistinct, yellow to brown spots appear on the leaves, probably secondary to earlier injury. The sporulating acervuli are at first subcuticular, later becoming superficial.

Etiology

The conidia are hyaline, oval to oblong, 1-celled, straight or curved, with 2 oil droplets, and 12–14×5–6 μ. The setae are brown, mostly straight, septate, and 100×4–5 μ. The perithecia are on the upper leaf surface, black, globose, and ostiolate. The asci are subclavate, and cylindric, and the ascospores are hyaline, 1-celled, oblong to oval, and 15×4 μ. The perfect stage has been designated as *Glomerella artocarpi.*

Reference

5.

Leaf spot, *Cercospora artocarpi* H. & P. Syd.

Symptoms

Yellow to dark brown leaf spots appear between the principal veins in an irregular fashion and along the margins, up to 20 mm in extent.

Etiology

The globular stromata appear on the upper leaf surfaces; they are brown to black and 20–60 μ in diameter. The fascicles are dense and spreading. The conidiophores are tan to dark brown, 3–7-septate, straight or curved, unbranched, and 40–125×3–5 μ. Conidia are pale brown, cylindric, 3–7-septate, and 30–75×3–5 μ.

Reference

3.

Other fungi associated with breadfruit

Botryobasidium salmonicolor (Berkeley & Broome) Venkatarayan
Botrytis artocarpi Viégas
Didymosphaeria oliveiana Dias et al.
Diplodia artocarpi Saccardo
Pestalotia funiera Desmazieres
Phomopsis artocarpi Sydow
Physopella artocarpi Arthur
Phytophthora coleocasiae Raciborski
Rosellinia bunodes (Berkeley & Broome) Saccardo
Sphaerostilbe repens Berkeley & Broome

References: breadfruit

1. Ananthanarayanan, S. 1964. A new leaf blotch of breadfruit from India. Plant Disease Reptr. 48:292–93.
2. Chowdhury, S. 1949. Some studies in the Rhizopus rot of jack fruit. J. Indian Botan. Soc. 28:42–50.
3. Chupp, C. 1953. A monograph of the fungus genus Cercospora. Published by

the author: Ithaca, N.Y., p. 392.
4. Rao, V. G. 1963. A new fruit rot of jack, Artocarpus integrifolia, from India. Plant Disease Reptr. 47:257–58.
5. Roger, L. 1951. Phytopathologie des pays chauds. Encyclopedie Mycologique 17: 593–95; 1953, 18:1427.

Broad bean, *Vicia faba* Linnaeus

Powdery mildew, *Erysiphe polygoni* De Candolle
Rust, *Uromyces fabae* (Persoon) De Bary
Blight, *Ascochyta fabae* Spegazzini
Red spot, *Botrytis fabae* Sardiña
Leaf spot, *Cercospora fabae* Fautrey
Foot rot, *Fusarium avenaceum* var. *fabae* Yu
Wilt, *Fusarium oxysporum* f. sp. *fabae* Yu & Fang
Root rot, *Fusarium solani* f. sp. *fabae* Yu & Fang
Other fungi associated with broad bean

Powdery mildew, *Erysiphe polygoni* DC.

Symptoms

White small, powdery spots appear on both surfaces of the leaves. The spots enlarge, often coalesce, and involve most of the leaf. The foliage infection causes malformations of young leaves. Stems and petioles show reddish brown lesions caused by the fungus. Young pods may be malformed, while old pods merely show the colored lesions.

Etiology

The mycelium produces haustoria, is superficial, hyaline, white en masse, branched, and septate. Conidiophores are septate and measure 20–68 × 6–9 μ. Conidia are hyaline with rounded ends, 1-celled, about twice as long as wide, and 38 × 20 μ. The perithecia are brown, scattered, globose, and average 100 μ in diameter. Appendages are brown, short, and undulating. There are 4–8 asci per perithecium. There are 3–8 hyaline ascospores per ascus; each measures 20–12 μ.

Reference

10.

Rust, *Uromyces fabae* (Pers.) d By.

Symptoms

Whitish spots, which later turn brown, appear on the foliage and on other aerial plant parts. The spots are usually scattered in initial infections and later may be plentiful, representing secondary infection. The disease is not usually serious but often causes some defoliation.

Etiology

The fungus produces pycnia, aecia, uredia, and telia on this host. The uredosori are most commonly observed and cause most of the damage, as the spores from these sori are capable of reinfecting the host. They appear on both leaf surfaces, petiole, and stem.

References

1, 5.

Blight, *Ascochyta fabae* Speg.

Symptoms

The fungus causes spots on leaves, stem, and pods. The spots vary in shape from circular to linear and in diameter from 2 to 22 mm. They develop light-colored centers and dark reddish borders. They may be zonate, and the centers frequently fall out. The size and color of the spots vary noticeably between dry and wet weather.

Etiology

Also variable with the weather is the arrangement of the pycnidia, scattered when dry, in circles when wet. Stem lesions are usually deeply sunken, and pod lesions reduce seed development. Pycnidiospores are hyaline, oblong, usually straight, 1–3-septate and 17.9×5.9 μ.

Reference

2.

Red spot, *Botrytis fabae* Sardiña

Symptoms

The disease is first manifested as minute red spots on the leaf blade. As the spots enlarge the centers become sunken and remain red, while the rest of the spot changes to brown with a colored margin. The spots, often numerous on a single leaf, are round to oval or oblong, about 1 mm in diameter, and may occasionally enlarge up to 5 mm but usually remain small. Spots may coalesce, and tips or leaf margins may fade and become papery and die. In moist weather these areas become overgrown with the fungus, giving the surface a gray appearance. Stem and petiole lesions are more or less elongate, sunken, with a deep red margin. Pods become infected, but incur little damage.

Etiology

The mycelium is intercellular, septate, branched, coarse, and develops rapidly. Sclerotia appear in culture but seldom in nature. They measure 0.5–3× 0.6–3.5 μ and are often variable in size and shape. Conidiophores are septate and twisted; they branch at the tips and then produce side branches upon which the conidia form in profuse clusters. Conidia are hyaline, subglobose to oval, and 16×13 μ. Microconidia are often produced in culture.

References

3, 4, 6, 9.

Leaf spot, *Cercospora fabae* Fautr.

Symptoms

The disease appears primarily on the foliage and sometimes on petioles and stems in the form of dark brown spots. These spots enlarge rapidly to about 7 mm, are often zonate, and develop a light gray color at the centers because of spore production. A slightly raised, deep red margin surrounds the spots. At times the centers of older, larger spots fall out.

Etiology

Spores may vary in size, quantity, and septation. Conidiophores are produced in clusters; there are usually about 6 in each cluster. The conidia average $64 \times 4.2\ \mu$ with an average of 9 septations. *Cercospora zonate* is considered a synonym.

Reference

7.

Foot rot, *Fusarium avenaceum* var. *fabae* Yu

Symptoms

The leaves are pale green to light yellow with black lesions on the margins. The top leaves may assume a more erect position and are more rigid than they are normally. Some twisting and curling may be manifested, followed by drying. The stems blacken and wither. Plants may wilt quickly or require 3 or 4 weeks. The root system is mostly diseased as a dry decay, while some vascular discoloration is evident in the lower stem. Inoculation experiments have produced the disease only on broad bean, of many crop plants.

Etiology

Mycelium is mostly white to rose-colored, septate, branched, and sometimes zonate in culture. Microconidia are ovoid to oblong, seldom septate, and $12 \times 3\ \mu$. Macroconidia form in sporodochia, are yellow to cinnamon, spindle-shaped to slightly curved, up to 6-septate, mostly 3-septate, and $34 \times 6\ \mu$. Chlamydospores are 1-celled, spherical to elongate, and about 10 μ in diameter.

Reference

8.

Wilt, *Fusarium oxysporum* f. sp. *fabae* Yu & Fang

Symptoms

The yellowing of the leaves is the most conspicuous symptom of the disease. The foliage of diseased plants becomes rigid, and the plants gradually succumb. Underground parts remain relatively normal except for some characteristic reddish brown discoloration of the vascular tissue. The fungus is very similar to the wilt disease of peas but is pathogenic only on broad bean.

Etiology

The mycelium is usually fluffy, dirty white, changing to purple, rose, or greenish slate to yellowish as it develops on various artificial media. It is septate, branched, and supports an abundance of microconidia which form on aerial mycelium. The microconidia are hyaline, ovoid, oblong to elliptical, 1-celled, and 7.1×3.3 μ. Macroconidia, sporodochia, and pionnotes are rarely produced. Macroconidia are hyaline, 3-septate, slightly curved, and 31×4 μ. Chlamydospores are dark brown, terminal or intercalary, spherical to obovate, smooth, and 6–9×5–8 μ.

Reference

13.

Root rot, *Fusarium solani* f. sp. *fabae* Yu & Fang

Symptoms

The fungus causes a typical root disease. The roots and lower stems become dark, dry, shriveled, and die. The leaves become yellow; black lesions eventually show along the margins and often involve the entire leaf. The fungus is very similar to *F. solani* but is pathogenic only on broad bean.

Etiology

The mycelium varies widely in color on various culture media. It is whitish when young, becoming buff, cinnamon, ivory, blue, rose, yellow, or any combination of colors. It is matted or cottony, producing branched conidiophores and ovoid, oblong, or short, rod-shaped microconidia which measure 6–7×2–3 μ. The macroconidia are scattered or in sporodochia or pionnotes. They are hyaline, spindle-shaped, slightly curved, slightly pedicellate, up to 6-septate, mostly 3-septate, and 30–39×4–6 μ. Chlamydospores are mostly 1-celled, spherical to oblong, and 10–11×10 μ.

Reference

13.

Other fungi associated with broad bean

Alternaria tenuis Nees
Ascochyta viciae Libert
Botryobasidium rolfsii (Saccardo) Venkatarayan
Colletotrichum viciae Dearness & Overholts
Fusarium culmorum (W. G. Smith) Saccardo
Phymatotrichum omnivorum (Shear) Duggar
Pythium debaryanum Hesse
Sclerotinia sclerotiorum (Libert) De Bary
Stemphylium botryosum Wallroth
Thanatephorus cucumeris (Frank) Donk
Thielaviopsis basicola (Berkeley & Broome) Ferraris

References: broad bean

1. Arthur, J. C., and G. B. Cummins. 1962. Manual of the rusts in United States and Canada. Hafner: N.Y., pp. 242–43.
2. Beaumont, A. 1950. On the Ascochyta spot disease of broad beans. Trans. Brit. Mycol. Soc. 33:345–49.
3. Butler, E. J., and S. G. Jones. 1949. Plant pathology. Macmillan: London, pp.

593–96.
4. Deverall, B. J., and R. K. S. Wood. 1961. Infection of bean plants (Vicia faba L.) with Botrytis cinerea and B. fabae. Ann. Appl. Biol. 49:461–72.
5. Kaiser, W. J., K. E. Mueller, and D. Danesh. 1967. An outbreak of broadbean diseases in Iran. Plant Disease Reptr. 51:595–99.
6. Wilson, A. R. 1937. The chocolate spot disease of beans (Vicia faba L.) caused by Botrytis cinerea Pers. Ann. Appl. Biol. 24:258–88.
7. Woodward, R. C. 1932. Cercospora fabae Fautrey, on field beans. Trans. Brit. Mycol. Soc. 17:195–202.
8. Yu, T. F. 1944. Fusarium diseases of broad bean. I. A. wilt of broad bean caused by Fusarium avenaceum var. fabae n. var. Phytopathology 34:385–93.
9. Yu, T. F. 1945. The red-spot disease of broad beans (Vicia faba L.) caused by Botrytis fabae Sardiña in China. Phytopathology 35:945–54.
10. Yu, T. F. 1946. Powdery mildew of broad bean caused by Erysiphe polygoni DC., in Yunnan, China. Phytopathology 36:370–78.
11. Yu, T. F. 1947. Ascochyta blight and leaf and pod spot of broad bean in China. Phytopathology 37:207–14.
12. Yu, T. F. 1947. Cercospora leaf spot of broad bean in China. Phytopathology 37:174–79.
13. Yu, T. F., and C. T. Fang. 1948. Fusarium diseases of broad bean. III. Root rot and wilt of broad beans caused by two new forms of Fusarium. Phytopathology 38:587–94.

Cabbage, *Brassica oleracea* Linnaeus

Soft rot, *Erwinia aroideae* (Townsend) Holland
Soft rot, *Erwinia carotovora* (L. R. Jones) Holland
Peppery leaf spot, *Pseudomonas maculicola* (McCulloch) F. L. Stevens
Black rot, *Xanthomonas campestris* (Pammel) Dowson
Clubroot, *Plasmodiophora brassicae* Woronin
Black root, *Aphanomyces raphani* Kendrick
Damping-off, *Pythium debaryanum* Hesse
Root rot, *Phytophthora megasperma* Drechsler
White rust, *Albugo candida* (Persoon ex Chevallier) Kuntze
Downy mildew, *Peronospora parasitica* (Persoon) Fries
Soft rot, *Rhizopus nigricans* Ehrenberg
Cottony rot, *Sclerotinia sclerotiorum* (Libert) De Bary
Ring spot, *Mycosphaerella brassicicola* (Fries ex Duby) Lindau
Powdery mildew, *Erysiphe polygoni* De Candolle
Smut, *Urocystis brassicae* Mundkur
Southern blight, *Botryobasidium rolfsii* (Saccardo) Venkatarayan
Bottom rot, *Thanatephorus cucumeris* (Frank) Donk
Black leg, *Phoma lingam* (Tode ex Fries) Desmazieres
Anthracnose, *Colletotrichum higginsianum* Saccardo
Anthracnose, *Gloeosporium concentricum* (Greville) Berkeley & Broome
Gray mold, *Botrytis cinerea* Persoon ex Fries
Leaf spot, *Cercosporella brassicae* (Fautrey & Roumeguère) Hoehnel
Leaf spot, *Cercospora brassicicola* P. Henning
Leaf spot, *Cercospora armoraciae* Saccardo
Leaf spot, *Ramularia armoraciae* Fuckel
Leaf spot, *Alternaria brassicicola* (Schweinitz) Wiltshire
Leaf spot, *Alternaria brassicae* (Berkeley) Saccardo
Leaf spot, *Alternaria raphani* Groves & Skolko
Yellows, *Fusarium oxysporum* f. sp. *conglutinans* (Wollenweber) Snyder & Hansen
Other fungi associated with cabbage

Soft rot, *Erwinia aroideae* (Towns.) Holland

Symptoms

This decay has been associated with a watery soft rot occurring mostly in transit and storage, not common in the field. Any part of the maturing plant may be infected, resulting in a wet, tissue-melting decay.

Etiology

The organism is a gram-negative rod, single or in pairs, forms no capsules, is motile by 2–8 peritrichous flagella, and measures $2–3 \times 0.5\ \mu$. Agar colonies are white to opalescent, glistening, and circular. Optimum temperature, 35C.

References

2, 4.

Soft rot, *Erwinia carotovora* (L. R. Jones) Holland

Symptoms

Soft rot, first described in 1901, causes a soft, wet, tissue-disintegrating, slimy decay. It is commonly encountered on this plant and many others and is initiated in wounds or in injured plant tissue during handling operations after removal from the field.

Etiology

The organism is a gram-negative rod, usually in chains, forms no capsules, is motile by 2–5 peritrichous flagella, and measures $1.5–2 \times 0.6–0.9\ \mu$. Agar colonies are whitish, glistening, entire, smooth, raised, and circular. Optimum temperature, 27C.

References

3, 4, 10.

Peppery leaf spot, *Pseudomonas maculicola* (McCull.) F. L. Stevens

Symptoms

Peppery spot, particularly associated with cauliflower, causes small brown to purple spots on the foliage. They range from small specks to spots several millimeters in diameter and cause malformation, if plentiful, and some defoliation. Stomatal infection occurs on either leaf surface. The parasite is largely seed disseminated.

Etiology

The organism is a gram-negative rod, produces no capsules, is motile by 1–5 polar flagella, and measures $1.5–3 \times 0.8–0.9\ \mu$. Agar colonies are white, glistening, smooth, round to irregular, and have wavy margins. A green fluorescent pigment is produced in culture. Optimum temperature, 24–25C.

References

4, 14.

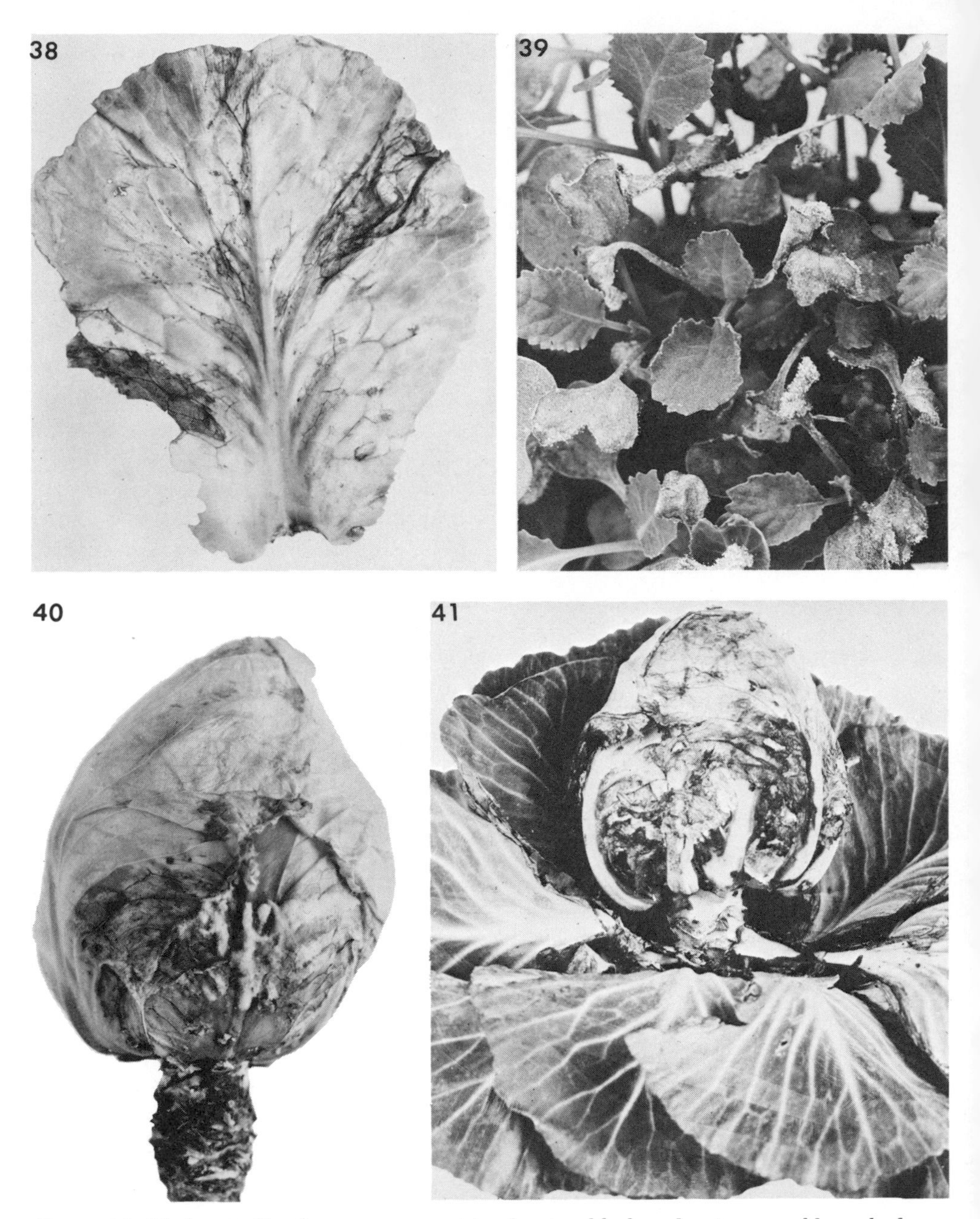

Figure 38. Black rot, *Xanthomonas campestris,* showing blackened veins on cabbage leaf.
Figure 39. Downy mildew, *Peronospora parasitica,* on cabbage seedlings. (Photograph by A. H. Eddins.)
Figure 40. Cottony rot, *Sclerotinia sclerotiorum,* on young cabbage head.
Figure 41. Bottom rot, *Thanatephorus cucumeris,* on cabbage foliage; leaves broken down for view.

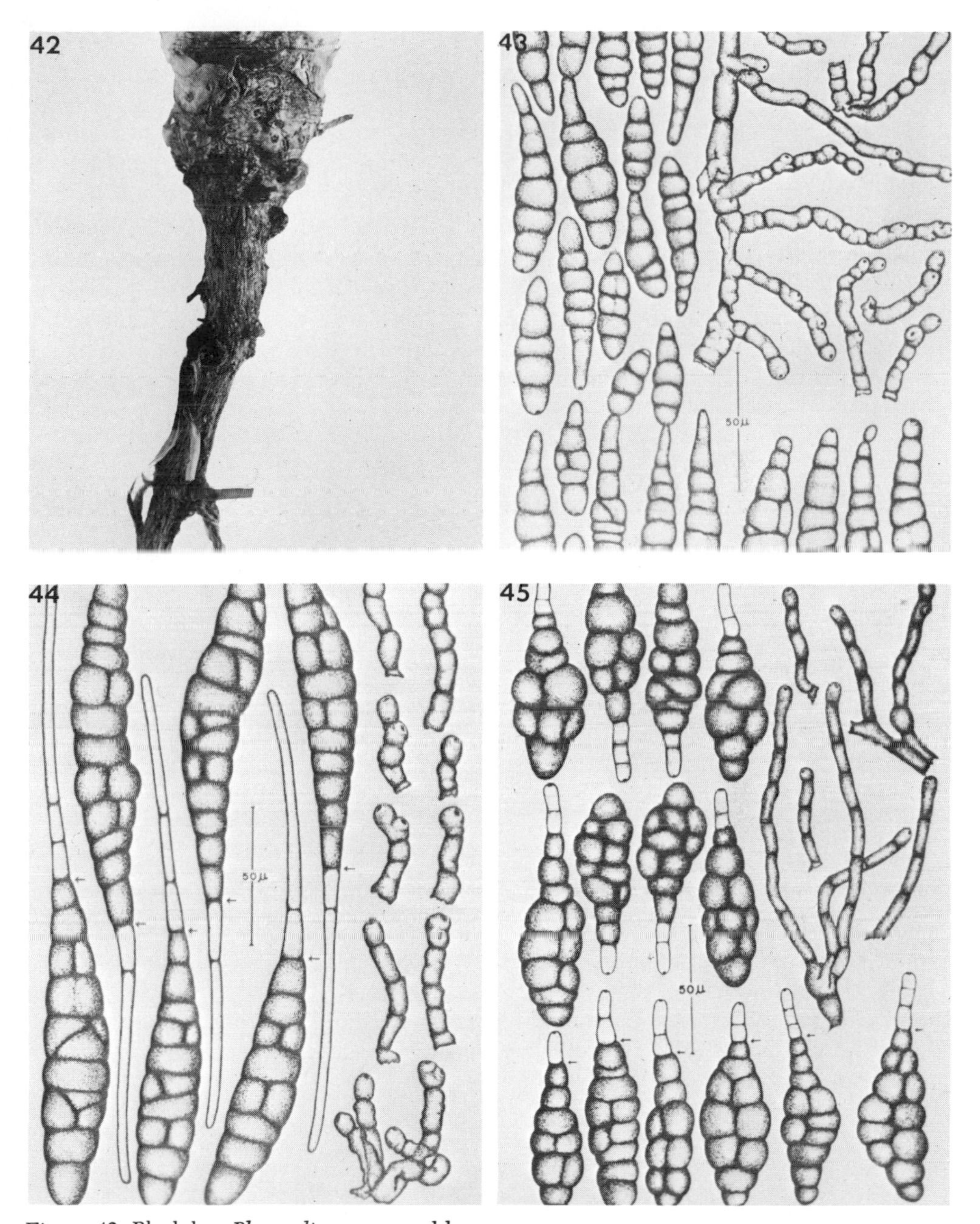

Figure 42. Black leg, *Phoma lingam,* on cabbage.
Figure 43. Conidia of *Alternaria brassicicola* (leaf spot). (Drawn by W. Changsri.)
Figure 44. Conidia of *Alternaria brassicae* (leaf spot). (Drawn by W. Changsri.)
Figure 45. Conidia of *Alternaria raphani* (leaf spot). (Drawn by W. Changsri.)

Black rot, *Xanthomonas campestris* (Pam.) Dows.

Symptoms

Black rot is a severe disease of many of the crucifers, causing extensive field losses from seedling damping-off, decay of mature heads, and infection of seed stalks. The disease is seed transmitted. Seedling infection develops on the cotyledons and may spread to the vascular tissues of the stem or, in instances of soil infestation, may result in marginal leaf invasion through the hydathodes to the vascular system. On growing plants this marginal avenue of ingress is most common and diagnostic, as the bacteria cause the veins to darken and become black, resulting in rounded to v-shaped, yellow and later brown, dead leaf areas. The stem invasion often results in the development of malformed heads.

Etiology

The organism is a gram-negative rod, often in chains, produces capsules, is motile by a single polar flagellum, and measures $0.3–0.5 \times 0.7–2\ \mu$. Agar colonies are pale yellow at first, darkening as they age, glistening, circular to somewhat irregular, wet, small, and not viscid. Optimum temperature, 30–32C.

References

10, 14.

Clubroot, *Plasmodiophora brassicae* Wor.

Symptoms

Clubroot is not considered a disease of plants in the tropics except possibly at high altitudes in certain countries. The disease is characterized by the formation of swollen, elongate areas on the main root and developing rootlets. These thickened areas are malformed, knobby to club-shaped. The aboveground symptoms are stunted growth and a droopy condition, followed by wilting and dying. The clubby roots consist of host cells containing spores of the fungus. Infection is through the root hairs.

Etiology

The fungus thallus is a plasmodium in the host tissue, developing within the host cells. The spores are hyaline, spherical, and about 4 μ in diameter. They germinate, forming an amoeboid zoospore body equipped with 2 flagella, and after an active period they come to rest, probably after fusion. This amoeboid stage enters cortical cells and the root hairs. The plasmodium increases in the host cells and finally develops into the clubs as it matures. Part of it migrates to adjoining host cells.

References

2, 14.

Black root, *Aphanomyces raphani* Kendr.

Symptoms

The disease is limited to radishes and is most severe on the long, white varieties. The fungus persists in the soil, and infection takes place where the secondary roots branch out from the main root. The parasite develops through the host cells, causing a dark discoloration and in early infections results in a constriction of the main root. The blackish tissue often encircles the root but usually does not extend lengthwise except through the vascular tissue, resulting sometimes in yellowing of lower leaves.

Etiology

The mycelium is hyaline, nonseptate, branched, and 4–8 μ in diameter. In water it forms zoosporangial structures which produce biciliate zoospores that germinate to form host-penetrating mycelium. The oogonium is 26–52 μ in diameter, and the oospore is 19–39 μ in diameter.

References

2, 14.

Damping-off, *Pythium debaryanum* Hesse

Symptoms

Damping-off of seedlings is a common occurrence. The fungus, a soil inhabitant, infects near the soil line of emerging seedlings, causing the main stem to be girdled and softened and resulting in a falling over of the plants. They become grown over by the mycelium of the fungus which grows out, usually contacting any adjacent plants with the same results.

Etiology

The hyphae are hyaline, branched, nonseptate, and about 5 μ in diameter. Sporangia are oval to spherical, terminal or intercalary, and not plentiful. Conidia are similar to sporangia and germinate directly. Oospores are smooth, spherical, and germinate directly.

References

2, 8, 14.

Root rot, *Phytophthora megasperma* Drechs.

Symptoms

This rot also affects cauliflower and is first noticed because of the reddish purple leaf margins progressively including the base of the leaves and resulting in leaf stunting, wilting, and defoliation along with root decay. The main root is often extensively decayed at its extremity and is brown; the root's vascular tissue and pith become involved. The cortex is soft, water-soaked, and often sloughs off. The upper margin of the decay is indicated by a darker zone next to the lighter-colored areas.

Etiology

The mycelium is hyaline, nonseptate, branched, and appears mealy in culture. The oogonia are spherical, short-stalked, and 32–43 μ in diameter. The oospores are yellowish, smooth, and 29–37 μ in diameter. Sporangia are obpyriform, nonpapillate, 41–56 $\times$ 28–40 μ, and produce up to 40 zoospores upon germination.

References

2, 12.

White rust, *Albugo candida* (Pers. ex Chev.) Kuntze

Symptoms

The rust is widespread, common, and is usually not of serious consequence. The infections cause local, raised, white, blisterlike sori on the stems and leaves, usually more abundant on the lower surfaces. The peridium covering the sporangia is colorless, thin, and shiny. The fungus also causes stunting, distortion, and malformation of stems, axillary growth, flower parts, and seedpods.

Etiology

The conidia are hyaline, spherical, and 14–16 $\times$ 16–20 μ. They germinate by germ tube or zoospore formation. The oogonia are formed in the host tissue and contain a single oospore which germinates by the formation of a vesicle containing biciliate zoospores.

References

2, 7, 14, 15.

Downy mildew, *Peronospora parasitica* (Pers.) Fr.

Symptoms

Mildew is worldwide on cabbage and most of the closely related plants. The disease is characterized by a white to lavender, loose, fluffy, downy mildew growth on the lower surfaces of leaves and on stems, petioles, peduncles, and seedpods. Frequently, dark slate-colored blotches develop on the fleshy roots of turnips and radishes. Cotyledons are often killed and shed, and seedling-killing is often important. On older plants the diseased leaves become blotched with pale green to yellow areas on the upper surface directly opposite the mildew on the lower surface. Maturing cabbage heads and cauliflower curds become infected and show irregular dark, sunken blotches, up to 1 in. in diameter. These lesions open the way for invasion by secondary soft rotting organisms.

Etiology

The mycelium-producing haustoria is hyaline, nonseptate, branched, and mostly intercellular. The conidiophores are hyaline, dichotomously branched, with acute tips upon which the conidia, which are hyaline, ellipsoidal, single-celled, and 20–22 × 16–20 μ, are produced. The conidia germinate directly. The oospores are mostly spherical, 26–43 μ in diameter, and develop imbedded in the host tissue.

References

2, 10, 14.

Soft rot, *Rhizopus nigricans* Ehr.

Symptoms

Soft rot is a storage and transit disease associated with wounded plant parts. It causes a brownish, soft, watery decay in warm, humid conditions.

Etiology

A black, fluffy mold may develop on any of the plant parts. The fungus produces rhizoids, stolons, and an abundance of mycelium. Erect sporangia develop, consisting of heads of black spores that measure 14 × 11 μ. Zygospores are black, warty, and up to 200 μ in diameter.

References

2, 14, 15.

Cottony rot, *Sclerotinia sclerotiorum* (Lib.) d By.

Symptoms

This rot rarely develops on cabbage plants until midseason, except for seedling infestation where circles of damping-off plants of varying diameter appear without warning in closely seeded beds. Lesions form in invaded stem tissue and white mycelium grows up into the leaf bases, destroying the blade tissue. Flabbiness of foliage, some general wilt, and defoliation also occur. Often the infection causes lopsided plants. The fungus usually overgrows the entire head, causing it to become yellow, and producing a slow decline.

Etiology

The large, shiny, black, irregular shaped sclerotia, diagnostic of the disease, may appear in the decayed stem tissues. These sclerotia survive in the soil and in plant debris between planting seasons. Upon germination they produce small, pale pink, saucer-shaped apothecia, up to 1 cm in diameter, from which hyaline, oval, single-celled ascospores, measuring 11–16 × 3–12 μ, are produced in abundance and which are the source of mycelium production as no true conidia are formed.

References

2, 10, 15.

Ring spot, *Mycosphaerella brassicicola* (Fr. ex Duby) Lindau

Symptoms

This disease has been found mostly on cabbage and cauliflower and appears as irregular, circular spots often bordered by a narrow yellow halo. The spots may also be recognized by an interrupted zonation, and the whole area becomes dotted with fine black specks. On the stems, branches, petioles, seedpods, and peduncles, elongate and light-colored, black-bordered lesions develop that are heavily marked with small, black spermagonia.

Etiology

The mycelium produces hyaline, 1-celled spermatia, measuring $2\text{–}5 \times 1\text{–}3\ \mu$. The perithecia are formed in the diseased host tissues in which the asci and ascospores of the fungus develop. The perithecia are black, partially imbedded, ostiolate, flask-shaped bodies, measuring $90\text{–}115\ \mu$ in diameter. The ascospores are hyaline, 2-celled, oblong to fusiform, often slightly curved, constricted, and $15\text{–}25 \times 3\text{–}6\ \mu$.

References

2, 14.

Powdery mildew, *Erysiphe polygoni* DC.

Symptoms

Powdery mildew does not cause serious damage to cabbage. It is very common on turnip, rutabaga, mustard, and other crucifers. The disease is recognized by the pale green to yellowish foliage and the chalky, white, dustlike conidia of the fungus which cover the leaf surface.

Etiology

The mycelium is hyaline, septate, branched, and mostly superficial, being supported by the production of haustoria in the epidermal cells. The conidiophores are hyaline, septate, and erect. The hyaline, oval to elongate, 1-celled conidia are produced consecutively, often in chains; they measure $26\text{–}52 \times 16\text{–}23\ \mu$. The perithecia are usually produced later in the season on deteriorating plants. They are black at maturity, spherical, with myceliumlike appendages, and measure up to $120\ \mu$ in diameter. The asci are numerous; each contains 2 ascospores that measure $24\text{–}28 \times 11\text{–}13\ \mu$.

References

2, 14, 15.

Smut, *Urocystis brassicae* Mund.

Symptoms

Diseased plants are recognized by their pale leaves, stunted appearance, and tendency to flower early. Galls of various sizes, up to 4 cm in diameter, develop on the roots. They are irregular in outline, wartlike, attached to the finer roots, light-colored when young, and lead gray when mature. The surface is smooth and shiny, and the contents of the galls is made up mostly of spore balls.

Etiology

Spores en masse are black, powdery, and compact; individual spore balls, composed of 1–5 fertile central spores surrounded by a layer of sterile cells, average about 38×32 μ. Individual spores are brown, 20×16 μ, and are surrounded by a layer of small, brown, sterile cells.

Reference

17.

Southern blight, *Botryobasidium rolfsii* (Sacc.) Venkat.

Symptoms

Southern blight appears frequently on the salad plants, such as mustard, Chinese cabbage, and turnip. The fungus infects the main root, and the mycelium grows up around the main stem and onto the leaf petioles and blades, causing them to become yellow and collapse.

Etiology

An abundance of white mycelium develops around the stem and into the soil immediately in contact with the stem. Usually numerous, light to dark brown, spherical sclerotia, less than 1 mm in diameter, form on the host tissue that has been overgrown by the mycelium. No conidia are produced and basidiospores are rare.

Reference

13.

Bottom rot, *Thanatephorus cucumeris* (Frank) Donk

Symptoms

This soil disease results from attacks on cabbage during several definite stages of its growth. The trouble is probably less severe on most of the other crucifers. Damping-off of seedlings, particularly in wet conditions, is frequent. The mycelium comes in contact with the seedlings and weakens the stem to such an extent that it falls over. Infection often takes place after the seedling stems become more woody, producing a condition known as wire stem. A root rot frequently occurs on plants up to heading time, particularly where the soil is waterlogged. The root system is killed and the aerial parts wilt. A typical bottom rot is produced, and under favorable conditions the entire head becomes overgrown. The outer head leaves are killed, turn brown, and remain in place, developing the head-rot stage of the disease.

Etiology

The fungus produces no conidia but produces an abundance of mycelium which is tan to brown, branched, septate, and often accumulates into brown, variously shaped, compact, superficial sclerotia. The basidiospore stage of the fungus sometimes is produced on the main stem or on the lower surfaces of foliage leaves as a frosty-colored, fine, thin, cobwebby network of hyphae, containing scattered basidia which measure 18–22 × 8–11 μ. The basidiospore is hyaline, 1-celled, oval to oblong, and 7–13 × 4–7 μ. The fungus is considered worldwide in distribution.

References

14, 15, 16.

Black leg, *Phoma lingam* (Tode ex Fr.) Desm.

Symptoms

The black discoloration is most conspicuous as a cankerous lesion on the main stem of cabbage. It is slightly sunken, tan to purple brown, linear, and may be so extensive as to completely girdle the stem. Black pycnidia develop in abundance, more or less scattered. On the foliage, especially the cotyledon, spots are light-colored and scattered. Wilting or leaf shedding does not occur often. Large cankers often weaken the stems so that the heads fall over. The fungus is common in seed-producing areas, and contaminated seed is a main source of dissemination in addition to local survival on plant debris.

Etiology

The mycelium is septate, branched, and hyaline, becoming smoky as it matures. The pycnidia are black, submerged or erumpent, and ostiolate; they vary in size from 200 to 450 μ and produce an abundance of pycnidiospores, in whitish masses or tendrils, that are hyaline, 1-celled, oval to slightly elongate, and 3–6 × 2 μ.

References

9, 14, 15.

Anthracnose, *Colletotrichum higginsianum* Sacc.

Symptoms

Anthracnose is associated with the thin-leafed, waxy-bloom deficient crucifers in contrast to cabbage, cauliflower, and others. The lesions appear on all aboveground parts as scattered, irregularly shaped, pale spots that are usually dry, papery, and shrunken with a darker narrow border. In wet weather and on older infections the centers often fall out. The infection is in the form of elongate, linear, dark lesions that develop on the midribs and petioles.

Etiology

The fungus mycelium is hyaline, septate, and branched. The small, erumpent acervuli are variously scattered over both surfaces of the lesions and consist of small stroma that contain conidiophores and several brown, slender, septate setae. The conidia are hyaline, elongate, 1-celled, and 11–24 × 2–6 μ.

References

14, 15.

Anthracnose, *Gloeosporium concentricum* (Grev.) Berk. & Br.

Symptoms

This disease is not common or important except on cauliflower, where it forms white spots measuring more than 1 in. in diameter.

Etiology

The acervuli lack setae and produce many hyaline, 1-celled conidia that measure $6–14 \times 2–3\ \mu$.

References

2, 14.

Gray mold, *Botrytis cinerea* Pers. ex Fr.

Symptoms

Mold is found mostly in transit and storage, following mechanical wounds. It is conspicuous because of the soft, watery decay and the abundance of gray, velvetlike conidiophores and conidia of the fungus.

Etiology

The dense gray tufts of conidia are terminal on long, branched conidiophores. The conidia are hyaline, 1-celled, oval, and $10–12\ \mu$ in diameter.

Reference

15.

Leaf spot, *Cercosporella brassicae* (Fautr. & Roum.) Hoehn.

Symptoms

The spotting develops mostly on the thin-leafed crucifers, although it does appear on cabbage and the waxy-leafed plants. The leaf spots on the thin-leafed plants are brown to almost white with dark margins, circular to irregular, and variously scattered. Infection on cotyledons is usually marginal. Seedlings are frequently killed. On the waxy-leafed plants the spots develop from small, dark flecks to lesions up to 4 mm in diameter. The spots become speckled with spore-producing structures of the fungus.

Etiology

The mycelium is hyaline, branched, septate, and up to $5\ \mu$ in diameter. It forms aggregates in substomatal chambers or under the epidermis, and produces conidiophores that protrude or rupture the epidermis. Conidia are hyaline, elongate, septate, and approximately $40–120 \times 1.5–3\ \mu$.

Reference

2.

Leaf spot, *Cercospora brassicicola* P. Henn.

Symptoms

This leaf spot is commonly found on the foliage of the nonwaxy crucifers. The spots are circular to angular, pale green to light brown with a dark border, and up to 4 mm in diameter. The disease is not important and seldom causes much leaf damage.

Etiology

The conidiophores, mostly on the lower surface, appear as up to a dozen olivaceous to brown fascicles, measuring $25–500 \times 4–7$ μ. The conidia are hyaline, acicular, curved, septate, and $25–200 \times 2–5$ μ.

Reference

2.

Leaf spot, *Cercospora armoraciae* Sacc.

Symptoms

The spot first appears as a pale circular area, enlarging rapidly to 1–2 cm in diameter; it may be zonate. The spots may coalesce and the dead central areas become gray and later dark. They are irregular in outline and frequently cause some withering of the leaf blade.

Etiology

The stromata are brown, and the conidiophores are olive brown, fasciculate in dense clusters, septate, geniculate, and $15–50 \times 4–7$ μ. The conidia are hyaline, obclavate, slightly curved, septate, and $25–150 \times 3–5$ μ.

Reference

7.

Leaf spot, *Ramularia armoraciae* Fckl.

Symptoms

The disease is restricted to horseradish, where pale green to yellow, circular to irregularly shaped spots up to 1 cm in diameter develop. The centers may drop out, causing the leaves to appear ragged, become yellow, and die.

Etiology

Conidiophores are colorless, erect, and fasciculate. The conidia are straight, with parallel sides, and rounded ends; they are hyaline, nonseptate, and $15–20 \times 4–5$ μ.

References

2, 7.

Leaf spot, *Alternaria brassicicola* (Schw.) Wiltsh.

Symptoms

This is one of the most frequently observed diseases of cabbage. The spots become very conspicuous as the plants mature. The fungus kills seedlings when contaminated seed are planted. On young plants the small spots are purplish green and may enlarge up to several centimeters, become circular, dry, and somewhat zonate.

Etiology

The olivaceous conidiophores are septate, branched, 35–45 × 5–8 μ, and support an abundance of dark-colored conidia that are linear to obclavate, muriform septate, and 50–85 × 11–20 μ. They are in chains of 8–12 spores, with 8–10 transverse cross walls and 1–5 longitudinal septations. The spore beaks are extremely short or nonexistent. The leaf spots are few, large, and black.

References

1, 14.

Leaf spot, *Alternaria brassicae* (Berk.) Sacc.

Symptoms

This spot is associated with the non-waxy-leafed members of the cabbage family. It is often of some importance as it forms seedling leaf spots that enlarge as the plants mature. The spots are tan to brown with yellow margins, smooth, circular, and may be up to 1 cm in diameter.

Etiology

The conidiophores are light brown, fasciculate, septate, constricted at the septations, and 14–74 × 4–8 μ. The conidia are olivaceous to tan, obclavate, smooth, constricted, muriform septate, with up to 15 cross walls and 1–8 longitudinal septations, and measure 100–300 × 16–30 μ. The beaks are hyaline, long, septate, and flexuous.

References

1, 2, 5, 15.

Leaf spot, *Alternaria raphani* Groves & Skolko

Symptoms

Leaf spot is associated with radish and is seldom found on other closely related plants. The spots are small and dark purplish on young plants; they become brown to dark-colored, with a distinct yellow halo, and reach a diameter of up to 1 mm.

Etiology

The conidiophores are brown to olivaceous, usually fasciculate, septate, and 29–160 × 4–8 μ. The conidia appear in short chains, obclavate to irregularly oval, tapering toward short beaks, smooth, constricted, muriform septate, with up to 10 cross walls and several longitudinal septations, and measuring 50–150 × 12–30 μ. Spore measurements are necessary for identification.

References

1, 14.

Yellows, *Fusarium oxysporum* f. sp. *conglutinans* (Wr.) Snyd. & Hans.

Symptoms

Yellows develops in the vascular tissue of many crucifers; cabbage is very susceptible. The fungus lives for indefinite periods in the soil, and infected plants show the first symptoms shortly after transplanting. The fungus causes the foliage to become pale yellow green. As the vascular system becomes involved, the older leaves begin to curl, warp, curve downward, turn partially brown, and shed. The main stem becomes barren of foliage and often curved to one side. The vascular tissue is dark-colored, and the root system becomes brown and reduced. Plants die before maturity.

Etiology

The mycelium is hyaline, septate, and branched. Chlamydospores are produced. Microconidia are hyaline, 1-celled, mostly oval to elongate, produced in abundance, and 6–15 × 2–4 μ. The macroconidia, not abundantly produced, are hyaline, 2–3-septate, elongate to fusiform, curved, and 25–33 × 3–6 μ. The disease is controlled by the development of resistant varieties.

References

2, 14.

Other fungi associated with cabbage

Ascochyta armoraciae Fuckel
Heterosporium variabile Cooke
Leptosphaeria olericola (Berkeley & Curtis) Saccardo
Olipidium brassicae (Woronin) Dangeard
Phyllosticta armoraciae (Cooke) Saccardo
Phyllosticta brassicola McAlpine
Phymatotrichum omnivorum (Shear) Duggar
Puccinia aristidae Tracy
Puccinia monoica (Peck) Arthur
Rhizopus stolonifer (Ehrenberg ex Fries) Lind
Sclerotinia minor Jagger
Streptomyces scabies (Thaxter) Waksman & Henrici
Verticillium albo-atrum Reinke & Berthold
Xanthomonas campestris (Pammel) Dowson var. *armoraciae* (McCulloch) Starr & Burkholder

References: cabbage

1. Changsri, W., and G. F. Weber. 1963. Three Alternaria species pathogenic on certain cultivated crucifers. Phytopathology 53:643–48.

2. Chupp, C., and A. F. Sherf. 1960. Vegetable diseases and their control. Ronald Press: N.Y., pp. 237–88.
3. Dowson, W. J. 1949. Manual of bacterial plant diseases. Macmillan: N.Y., pp. 68–183.
4. Elliott, C. 1951. Manual of bacterial plant pathogens. 2d ed. Chronica Botanica: Waltham, Mass.
5. Garrard, E. H. 1951. Bacterial diseases of plants. Bull. Agr. Coll., Guelph, Ontario 478:21.
6. Joly, P. 1964. Le genre Alternaria. Encyclopedie Mycologique 33:1–250.
7. Kadow, K. J., and H. W. Anderson. 1940. A study of horseradish diseases and their control. Illinois Agr. Expt. Sta. Bull. 469:531–78.
8. Matthews, V. D. 1931. Studies on the genus Pythium. University of North Carolina Press: Chapel Hill, pp. 82–87.
9. Pound, G. S. 1947. Variability in Phoma lingam. J. Agr. Res. 75:113–33.
10. Ramsey, G. B., and M. A. Smith. 1961. Market diseases of cabbage, etc. U.S. Dept. Agr., Agr. Handbook 184:1–21.
11. Ramsey, G. B., J. S. Wiant, and G. K. K. Link. 1938. Market diseases of fruits and vegetables: crucifers and cucurbits. U.S. Dept. Agr. Misc. Publ. 292:2–36.
12. Tompkins, C. M., C. M. Tucker, and M. W. Gardner. 1936. Phytophthora root rot of cauliflower. J. Agr. Res. 53:685–92.
13. Walker, J. C. 1948. Diseases of cabbage and related plants. U.S. Dept. Agr. Farmers' Bull. 1439:1–38.
14. Walker, J. C. 1952. Diseases of vegetable crops. McGraw-Hill: N.Y., pp. 123–72, 270.
15. Weber, G. F. 1932. Some diseases of cabbage and other crucifers in Florida. Florida Univ. Agr. Expt. Sta. Bull. 256: 1–62.
16. Wellman, F. L. 1932. Rhizoctonia bottom rot and head rot of cabbage. J. Agr. Res. 45:461–69.
17. Zundel, G. L. 1953. The Ustilaginales of the world. Penn. State Univ. Contribution 176:1–410.

Cacao, *Theobroma cacao* Linnaeus

Mealy pod, *Trachysphaera fructigena* Tabor & Bunting
Black pod, *Phytophthora palmivora* Butler
Witches'-broom, *Taphrina bussei* Tabor
Canker, *Ceratocystis fimbriata* Ellis & Halsted
Violet root rot, *Sphaerostilbe repens* Berkeley & Broome
Pod rot, *Nectria cacaoicola* Roger
Green point cushion-gall, *Calonectria rigidiuscula* (Berkeley & Broome) Saccardo
Collar rot, *Ustulina zonata* Léveillé
Root rot, *Rosellinia bunodes* (Berkeley & Broome) Saccardo
Root rot, *Rosellinia pepo* Patouillard
Felty fungus, *Septobasidium tanakae* (Miyabe) Boedijn & Steinmann
Thread blight, *Corticium invisum* Petch
Thread blight, *Ceratobasidium stevensii* (Burt) Venkatarayan
Pink disease, *Botryobasidium salmonicolor* (Berkeley & Broome) Venkatarayan
Brown root disease, *Fomes lamaensis* (Murrill) Saccardo & Trotter
White root, *Fomes lignosus* (Klotzsch) Bresadola
Brown crust, *Fomes noxius* (Berkeley) Corner
Collar crack, *Armillaria mellea* Vahl ex Fries
Brown thread, *Marasmius byssicola* Petch
South American witches'-broom, *Marasmius perniciosus* Stahel
White thread, *Marasmius scandens* Massee
Brown thread blight, *Marasmius trichorrius* Spegazzini
Pod rot, *Botryodiplodia theobromae* Patouillard
Gray pod rot, *Monilia roreri* Ciferri
Pod rot, *Thielaviopsis paradoxa* (De Seynes) Hoehnel
Sudden death, *Verticillium dahliae* Klebahn
Algal spot, *Cephaleuros virescens* Kunze
Other fungi associated with cacao

Mealy pod, *Trachysphaera fructigena* Tabor & Bunt.

Symptoms

A purplish brown discoloration may appear anywhere on the surface of the pod. It enlarges rapidly and under favorable conditions involves the entire pod. The skin darkens and hyphae collect beneath the epidermis. The fungus growth is at first whitish, then pink, and finally brown. Mealy masses are so dense that the pod becomes encrusted. This cumulative mass of fungus is the most characteristic symptom of the disease. Internally, the pericarp becomes discolored and decomposes. The vascular system turns reddish brown, and the inner white pulp is destroyed as well as the seed.

Etiology

The mycelium is hyaline, branched, nonseptate, and grows rapidly into all parts of the pod. Conidiophores are erect with a terminal vesicle supporting branches bearing conidia. The conidia are spherical, strongly echinulate, about 35 μ in diameter, and are borne on pedicels varying in length up to 30 μ. They are produced on the surface and in interior cavities. Chlamydospores are intercellular and echinulate; antheridia are present. Oogonia are pear-shaped, warty, and 40×24 μ.

References

1, 4, 23.

Black pod, *Phytophthora palmivora* Butl.

Symptoms

Cacao pods may be attacked in all stages of growth, and trunk lesions are frequently produced. A brown to black area may develop on the pod. It is circular in early infection, expands rapidly, and eventually involves the entire pod and its contents. Diseased pods become covered with a fine, mildew-mycelial growth, producing an abundance of sporangia and zoospores, especially during humid, rainy periods.

Etiology

The mycelium is normally nonseptate, hyaline, of variable diameter, branched, and develops abundantly below the epidermis. It eventually emerges by the rupture of the epidermis. A superficial, dense growth develops as the emerging sporangiophores become decumbent. The oval to lemon-shaped sporangia are produced in abundance. They are hyaline, papillate, 30–60 $\times$ 21–30 μ, and produce mycelium upon germination or give rise to 15–30 zoospores. The zoospores are 8×14 μ; they are active for a short period and germinate by producing mycelium. Chlamydospores are hyaline to yellowish, spherical, and 22–50 μ in diameter. Oospores are hyaline, spherical, smooth, and 18–41 μ in diameter.

References

4, 13, 18, 19, 24, 26.

Witches'-broom, *Taphrina bussei* Tabor

Symptoms

The fungus produces a witches'-broom that, once formed, spreads very little. The infection develops in the buds and causes abnormal branching, many adventitious buds, and hypertrophy. It is localized in the bark of the twigs and the leaf parenchyma. The filaments in the vessels expand and produce a faint yellow color. The fructifications begin on both surfaces, forming a gray, velvety area situated principally on the middle vein.

Etiology

The mycelium is hyaline, fine, intracellular, and scanty. The asci form in a continuous layer between the cuticle and epidermis. The asci are small, 15–17 $\times 5\ \mu$, and each contains 8 small ascospores measuring $2.5 \times 1.7\ \mu$.

References

20, 29.

Canker, *Ceratocystis fimbriata* Ell. & Halst.

Symptoms

The disease on aerial plant parts causes cankers on branches and trunk. The earliest indication is a sunken area in the bark; when cut open the wood is red to purple. The part of the tree above the canker dies, and the dead foliage remains attached. There is also a pod rot, usually introduced through wounds. The seeds often show a brown dry rot.

Etiology

A white mycelial growth and conidiophores may be observed on the diseased spots. The conidia are hyaline, elongate, cylindrical, continuous, and 10–50 μ long. Brown, oval chlamydospores, measuring $10\text{–}19 \times 7\text{–}11\ \mu$, are also present. The perithecia have black, long, hairlike beaks; they are usually topped by a gleaming bead of ascospores which are exuded under favorable conditions. The ascospores are hyaline, 1-celled, oval to allantoid, hat-shaped, and $4\text{–}8 \times 2\text{–}6\ \mu$.

References

15, 21, 22.

Violet root rot, *Sphaerostilbe repens* Berk. & Br.

Symptoms

There are very meager external symptoms. A root disease is evident as the entire plant begins to show a decline; the foliage becomes yellow, wilts, and finally is shed. The cortical tissues decay. The fungus forms plates or cords of rhizomorphlike material. The mycelium is white, becomes reddish brown and then black. It enters the roots, causing the tissue to become various light colors of yellow, dark blue, purple, and brown and is characterized by the distinctly acrid, foul odor.

Etiology

The conidia are produced on the crests of small, cylindrical columns or fascicles of conidiophores of 2–8 mm in height and 2–6 mm in diameter, in red, orange, or brown. The conidia are hyaline, oval to elongate, and $10–20 \times 6–9\ \mu$. The perithecia, appearing near the base of the stilbum, are red to brown, and measure 3–6 mm. The ascospores are brown, oval, 2-celled, constricted, and $16–21 \times 7–8\ \mu$. In most cases the fungus is considered a wound parasite.

Reference

20.

Pod rot, *Nectria cacaoicola* Roger

Symptoms

The lesions form on the pods and are often associated with stem cankers on the larger branches and on the trunk. The areas dry and may be variously colored in yellow, red, and brown. Sometimes the development of the disease becomes important following a wound.

Etiology

The mycelium is hyaline, usually light-colored en masse, plentiful, and septate. The microconidia are hyaline, oval, often 1-septate, and $4–7 \times 3–5\ \mu$. The macroconidia are hyaline, more or less cylindrical, straight or slightly crescent-shaped, 5–9-septate, and $55–80 \times 5–7\ \mu$. The perithecia are located in the stroma, reddish to yellow, often in clusters, spherical, and measure 250–350 μ in diameter. The cylindrical asci are pedicellate, straight, $50–60 \times 6–8\ \mu$, and contain 8 ascospores that are hyaline, 2-celled, oval, not restricted, and $10–14 \times 4–6\ \mu$. The conidial stage of the fungus is probably *Fusarium decemcellulare* or *F. theobromae*.

Reference

20.

Green point cushion-gall, *Calonectria rigidiuscula* (Berk. & Br.) Sacc.

Symptoms

The fungus is a wound parasite that is usually associated with cankers and dieback caused by other organisms. Large branches and often the crowns of trees are killed. The wood in the dieback areas is reddish-colored. Wounds offer an avenue of entry for this fungus, which penetrates deeper into the wood of the host causing acute or chronic dieback. The galls develop in axillary ends as a small group of multiple buds or a depressed branch system in the flower cushion, forming a rounded cauliflowerlike set of excrescences as a hypertrophied greenish florescence.

Figure 46. Canker, *Ceratocystis fimbriata,* killed cacao tree. (Photograph by E. Ampuero.)
Figure 47. Root rot, *Rosellinia pepo,* killed cacao tree.
Figure 48. Above: Gray pod rot, *Monilia roreri,* on cacao branch. Below: Healthy pod. (Photograph by E. Ampuero.)
Figure 49. Left: Healthy twig. Right: South American witches'-broom, *Marasmius perniciosus,* on cacao. (Photograph by E. Ampuero.)
Figure 50. South American witches'-broom, *Marasmius perniciosus,* on cacao: enlarged diseased branch. (Photograph by E. Ampuero.)
Figure 51. Gray pod rot, *Monilia roreri,* on cacao pods: external and sectional. (Photograph by E. Ampuero.)
Figure 52. Leaf spot, *Cercospora ceratoniae,* on carob foliage. (Photograph by H. Burnett.)

Etiology

The surface mycelium is usually scanty and produces a whitish bloom over the surface of the lesion. The microconidia are hyaline, mostly 1-celled, 8×4 μ, and coiled in heads at the tip of the conidiophore. The macroconidia are found on the older cankers or dead wood. The sporodochia are pale pink and emerge through cracks in the bark; the macroconidia are hyaline, sickle-shaped, up to 10-septate, and $53–97 \times 4–8$ μ. The perithecia are pinkish en masse and emerge through cracks in the bark. The ascospores are hyaline, slightly curved, 3-septate, with rounded ends, and measure 27×7 μ. The vegetative stages of the fungus are *Fusarium decemcellulare* or *F. avenaceum*.

References

3, 7, 10, 11.

Collar rot, *Ustulina zonata* Lév.

Symptoms

A diseased plant shows an unthrifty condition, off-color, poor growth, and frequently an acute collapse with no aerial indications of a pathogen. Root examination reveals no external mycelium. There are, however, white, fan-shaped areas of internal mycelium in the cambium and cortex. Above the soil surface, the fungus produces a sporophore of variable size and shape as a soft yellowish white cushion. It turns greenish gray and ultimately becomes black, hard, and brittle; a 1–3-mm-thick plate is formed.

Etiology

The mycelium is brownish, and the conidia are hyaline and $6–8 \times 2–3$ μ. The mature sporophore contains the perithecia that appear as small black points. The ascospores are dark and $30–38 \times 9–13$ μ.

References

4, 20.

Root rot, *Rosellinia bunodes* (Berk. & Br.) Sacc.

Symptoms

This disease is similar to any malady causing top deterioration because of root decay. The roots show an external mycelium that grows into the soil causing the soil to cling to the outer cortex in an uneven fashion. The cortex is discolored, and usually the cambium area is invaded by the internal mycelium that accumulates in parallel lines and in fan-shaped masses or clusters.

Etiology

The conidia, produced on a coremium on a velvety surface, are a mass of pink, 1-celled, oblong spores that measure $4–6 \times 2$ μ. Perithecia are black, small, globose, slightly roughened structures with smooth, apical, ostiolate protuberances. These perithecia produce brown ascospores, measuring $80–110 \times 7–12$ μ, that are narrow, cymbiform, with both extremities finely drawn-out to thread-like points up to 25 μ long.

References

1, 3, 18.

Root rot, *Rosellinia pepo* Pat.

Symptoms

The aboveground symptoms are more or less general; the plant shows a typical decline, sometimes slowly but often suddenly. It wilts, and the foliage becomes yellow, then brown, and often remains on the tree. Dark brown cankers may appear on the branches but are found more often on the lower trunk. Roots show fans or stars of white mycelium in the cambium area. The mycelium forms a gray coating over the roots that turns into a black, carbonous layer. The hyphae penetrate the cortical areas, cambium, and wood, causing a brown to reddish color in the greenish wood. The external hyphae encircle the stem of the plant, and may spread out to a slight extent, but more often completely girdle it.

Etiology

The fungus develops the graphium conidial stage. The black sporulating structure is 2–3 mm high, terminating in a swollen tip that is usually whitish because of exuded conidia, which are hyaline, oval, 1-celled, and 5×2 μ. The perithecia are not frequently found. They form in the carbonous stroma, imbedded at first, subglobose, and 2.5–3 mm in diameter. The asci are elongate, and paraphyses are present. Ascospores are brown, narrowly fusiform, pointed at each end, and 62–67×8–9 μ.

References

1, 18, 28.

Felty fungus, *Septobasidium tanakae* (Miy.) Boed. & Steinm.

Symptoms

The fungus produces a dull brown, stromalike coating over the surfaces of twigs. The margins are thin and gray while the older parts are thick and produce a sort of canopy over the outer cortex of the plant. The structure is soft, pliable, and quite permanent. The fungus is parasitic on certain sedentary insect mites that parasitize the cacao plant.

Etiology

On the exposed surface the fungus produces probasidia, which upon germination develop a cross septum and a 4-celled basidium. Lateral basidiospores are hyaline, slightly curved, and 27–40×4–8 μ.

References

6, 20.

Thread blight, *Corticium invisum* Petch

Symptoms

The hyphae of the fungus grow superficially over the cortical regions of branches and twigs. Some threads advance up the leaf petioles, fan out over the blade, usually the lower surface, and immediately cause the leaves to become yellow, then brown, and die. The fungus usually survives the dormant period on the wood in more compact masses that begin to produce mycelium soon after the first flush of new growth.

Etiology

The mycelium is light tan or hyaline and 4–6 μ in diameter. The basidia are subglobose and 8–9 μ in diameter. Basidiospores are hyaline, oval, and 5–6 × 3–4 μ. Some sclerotia may be formed.

Reference

20.

Thread blight, *Ceratobasidium stevensii* (Burt) Venkat.

Symptoms

The threadlike strands of light to dark brown mycelium grow externally over the cortex of branches and twigs. The strands grow along the lower side of stems and branch off on the leaf petioles. Upon reaching the blades, they fan out as a thin layer over the lower surface of the leaf. These leaves blacken and remain attached to the twigs, held by the threads.

Etiology

The basidia are formed among the brown, septate, branched, silky hyphae, singly, or in clusters resembling a hymenium. The basidiospores are hyaline, oval to elongate, and 8–12 × 4–5 μ. The fungus aggregates into thickened areas, forming sclerotialike masses.

References

2, 20.

Pink disease, *Botryobasidium salmonicolor* (Berk. & Br.) Venkat.

Symptoms

Fine, shining, light tan hyphae grow over the bark of twigs and branches. Hyphae collect over the lenticels, followed by the pink stage in which the hyphae cover the twig with a dense mycelial mass, producing bright orange red fertile areas. The pink area may fade to almost white and develop crevices or cracks. The fungus penetrates to the wood, killing the branch.

Etiology

The mycelium becomes pinkish as it accumulates to form the hymenium; the fruiting surface is thin, soft, and resupinate. The hymenium is composed of hyphae growing longitudinally and producing short, upright branches which form conidia that are catenulate and 14–18×7–8 μ. The basidiospores are hyaline, measuring 9–12×6–8 μ.

References

1, 5.

Brown root disease, *Fomes lamaensis* (Murr.) Sacc. & Trott.

Symptoms

Diseased plants appear to be unhealthy, suffering from the loss of root functions. Excavated roots show heavily encrusted areas of brown mycelium, producing a covering that gives an external blackish appearance. The fungus is closely confined to the plant roots and is mostly disseminated by root contact.

Etiology

The mycelium is brown, developing in adjacent soil and in the host cortex. The sporophores are up to several layers thick and usually appear on old stumps. They are purple brown to deep brown, sessile, bracketed, 3–4 in. wide, and relatively thin. The developing margin is creamy yellow. The pores vary from 80 to 110 μ in diameter, and the basidiospores measure 3–5×3–4 μ.

References

4, 20.

White root, *Fomes lignosus* (Klotzsch) Bres.

Symptoms

Diseased plants show off-green to yellow foliage and lack of growth associated with malnutrition. The roots are more or less covered with a dirty white to yellowish red external mycelium that aggregates into strands or rhizomorphic cords behind the advancing margin of fine hyphae. The fungus penetrates the cortex to the cambium, gradually developing a girdling effect.

Etiology

The fructifications of the fungus usually occur on dead stumps and logs cleared from the land. They are yellow, small, flat, cushionlike structures that develop into thin, sessile, bracket sporophores, measuring 4–8×2–4 in. Fructifications are orange on the upper surface with yellowish margins, and orange over the lower poroid surface. The mycelium penetrates the surrounding soil for several feet around a decaying stump. The hymenial pores are 40–85 μ in diameter. The basidiospores are hyaline, spherical, and 4–8 μ in diameter. Cystidia are present.

References

4, 20.

Brown crust, *Fomes noxius* (Berk.) Corner

Symptoms

In advanced cases the tree is killed because of trunk girdling. The normal nutrition is curtailed, and some stunting, yellowing, and even wilting occur. The penetration of the fungus mycelium into the phloem and xylem elements causes adjacent soil particles to be included in a thin, velvety crust. The upper edge of the crust develops a broad, creamy white margin which eventually becomes brown. The fungus is a wound parasite.

Etiology

The sporophores are usually yellowish brown, shelving or resupinate, somewhat bullate and irregular, and form on old crusts on trunks and stumps. The pores are dark brown and stratified. The basidiospores are hyaline, smooth, 1-celled, and 4×6 μ.

References

1, 25.

Collar crack, *Armillaria mellea* Vahl ex Fr.

Symptoms

The disease is root associated and causes a decline in vigor of the affected plant, a premature shedding of foliage, some dieback of twigs, cortex cracking near the soil line, and a general unhealthy condition. The fungus invades the tissue in one or more of the lateral roots. The cambium areas are killed and become water-soaked, and white mycelial mats or shields develop. Numerous black, shiny, terete, cordlike rhizomorphs occupy the cambium area and also appear in the dead roots and immediately surrounding soil.

Etiology

The sporophores, produced plentifully aboveground, occur around the dead plant on lateral roots and at the crown. They are honey-colored, up to 6 in. tall, with a 4–6-in. diameter cap. The gills are light-colored, and the stipe carries a typical annulus. The basidiospores are white, 1-celled, and $8\text{–}9 \times 5\text{–}6$ μ.

References

1, 4.

Brown thread, *Marasmius byssicola* Petch

Symptoms

The mycelium grows over the host tissue in a rather loose fashion, in the form of rhizomorphs attached by protuberances at various places on the host cortex.

Etiology

The mycelium is yellow brown, 2–4 μ in diameter, and filmy rather than compact. The sporophores are small, 1.5 mm in diameter, with fine, delicate lamellae producing the basidiospores.

Reference

20.

South American witches'-broom, *Marasmius perniciosus* Stahel

Symptoms

Hypertrophy of young shoots, hardening and malformation of pods, and the star blooms that develop on diseased cushions are outstanding features of this disease. The young shoots enlarge several times the diameter of healthy twigs, have a rough, more or less furrowed surface, and do not become woody but dry up and blacken. The hardened, often bumpy, malformed pods usually arise from infected cushions and never ripen. Pods that escape infection until mature are often affected with black spots. This disease is limited to the Western Hemisphere.

Etiology

The sporulating structures are of the mushroom type with a bell-shaped cap, which is at first somewhat spherical and later becomes parasol-shaped and even curved upward on the margins, with a diameter of 5–15 mm. The cap is relatively thin with 8–20 gills that are up to several millimeters wide and less than 1 mm thick. The upper surface is faintly reddish with radiating lines. The stipe is yellow, 5–10 mm long, with a swollen base, and develops from the mycelium. Basidiospores are white, 1-celled, and $10–11 \times 4–5\ \mu$.

References

1, 18, 20.

White thread, *Marasmius scandens* Mass.

Symptoms

The dirty white mycelial threads that grow along the lower side of twigs and branches kill the foliage, leaf after leaf. The thread fans out over the lower surface causing the leaf to become yellow, rapidly turn brown, and die; it usually continues to hang pendant from the shoot by the mycelial strands.

Etiology

The mycelium is white but darkens in many instances as it gets older. It produces many mushroomlike sporophores that are dirty white to brown, up to 8 mm in diameter, with short, dark, curved stipes. The basidiospores are white, 1-celled, oval, and $8–12 \times 4\ \mu$.

References

4, 20.

Brown thread blight, *Marasmius trichorrius* Speg.

Symptoms

The fungus is not an important parasite; in fact, it more frequently develops as a saprophyte, growing over the twigs, foliage, and fruit of plants. It is attached by protuberances to the cortex of the host at intermittent places. The mycelium then penetrates the tissue.

Etiology

The mycelium is light-colored and unites into rhizomorphs that soon become brown. The sporophores are small, pendant mushrooms appearing along the threads. They are light-colored or yellow to reddish brown, more or less membraneous, and 4–8 mm in diameter. The stipe is thin, 5–20 mm long, and fragile; the cap is hemispherical, and there are 5–8 lamellae. The basidia are $12\text{–}18 \times 4\text{–}6\ \mu$, and the 4 sterigmata are 2.5–3 μ long. The basidiospores are hyaline, oblong to oval, pointed at place of attachment, and $7\text{–}11 \times 4\text{–}6\ \mu$.

Reference

20.

Pod rot, *Botryodiplodia theobromae* Pat.

Symptoms

The disease begins as a small brown spot that increases in size and darkens to almost black, involving the entire pod. The surface of the pod becomes lumpy and roughened by black, small, hard, submerged, but later erumpent, fruiting bodies. An abundance of dark, sooty, dusty spores cover the entire pod, which dries up and remains a mummy. The fungus is generally considered a wound parasite.

Etiology

The mycelium is gray to slate in culture, grows rapidly, and is branched and septate. The pycnidia are black, spherical to subglobose, ostiolate, and become erumpent and protruding. The pycnidiospores are exuded from the pycnidia in tendrils. The 1-celled, oval spores are white at first, darken, and become septate and striate; they measure $24\text{–}30 \times 12\text{–}15\ \mu$. The ascospore stage of the fungus is *Physalospora rhodina.*

References

4, 27.

Gray pod rot, *Monilia roreri* Cif.

Symptoms

During humid weather, the fungus causes a pod rot resulting in a yellowish or gray spot that is irregular in outline, measures up to 70 mm, and is often poorly defined. The pods are eventually encased in a gray mold, turn black, become watery, and drop. Only the pods are infected. The mycelium grows into the older pods and around the seed.

Etiology

The hyphae are hyaline, branched, and guttulate. The conidia are hyaline, produced in groups or chains, simple or branched, oval to spherical, and 9–14 ×8–11 μ.

References

9, 12, 17, 20.

Pod rot, *Thielaviopsis paradoxa* (De Seyn.) Hoehn.

Symptoms

Pod infections develop into large brown lesions that are mostly soft and yield to slight pressure. The internal mycelium is grayish, and the decay produces a characteristic fresh odor. The fungus is a wound parasite.

Etiology

The mycelium is gray, septate, branched, and as it matures produces endoconidiophores that become septate from their extremity, each cell rounding out to become a 1-celled, vegetative conidium which is hyaline and 8–12×3–5 μ. A second kind of conidium is produced on conidiophores, acrogenously, in short chains; these conidia are smoky olive, ovoid, 1-celled, thick-walled, and 14–17×11 μ. The ascospore stage of the fungus is *Ceratocystis paradoxa.*

Reference

8.

Sudden death, *Verticillium dahliae* Kleb.

Symptoms

A tree in full vigor and apparent health may die suddenly with leaves hanging pendant. The dying process may be slower and may even apparently stop with only a portion of the aerial parts being killed. The wilting and browning start near the ends of branches and progress downward. Wood sections show black dot-speckles and irregular black streaks.

Etiology

The mycelium is dark, septate, branched, and produces minute sclerotoid aggregations in the wood. These aggregations are black, ovoid, and 80×40 μ. Conidia sometimes develop, measuring 3–12×3–5 μ. The chlamydospores are more or less irregular in shape in intermittent chains.

Reference

16.

Algal spot, *Cephaleuros virescens* Kunze

Symptoms

Algal spot commonly causes a dieback of twigs. It produces small, yellow red cushions of fine hairs up to 1 mm high on either leaf surface. On the twigs the cushions are dark, even blackish, but they become brown when the hairs develop and form small globular heads. The bark under the hairy areas cracks, becomes rough, and causes cankers to form prior to killing the twigs.

Etiology

There is no mycelium. The thallus is composed of algal cells that contain chlorophyll and extend in a growing fashion, producing new cells in a more or less flat plane. It grows mostly below the cuticle, forming spots up to 10 mm in diameter. The hairlike structures are the sporangiophores of the alga and produce 3–6 sporangia at the tips that later release small, oval, biciliate zoospores that swim about, finally germinate, and produce a new plant.

References

5, 18.

Other fungi associated with cacao

Calonectria flavida (Corda) Saccardo
Calospora theobromae Camara
Ceratocarpia theobromae Fabre
Colletotrichum theobromicolum Delacroix
Fusarium theobromae Appel & Strunk
Ganoderma pseudoferreum (Wakefield) van Overeem & Steinmann
Helminthosporium theobromae Turconi
Leptosphaeria theobromicola Ciferri & Fragoso
Meliola theobromae Fabre
Pestalotia theobromae Petch
Phomopsis theobromae (de Almeida & Carrara) Bondartzev
Septoria theobromicola Ciferri & Fragoso
Spicaria colorans De Jonge

References: cacao

1. Briton-Jones, H. R. 1934. The diseases and curing of cacao. Macmillan: London, pp. 1–114.
2. Briton-Jones, H. R., and R. E. D. Baker. 1934. Thread blights in Trinidad. Trop. Agr. (Trinidad) 11:55–67.
3. Brunt, A. A., and A. L. Wharton. 1962. Etiology of a gall disease of cocoa in Ghana caused by Calonectria rigidiuscula (Berk. & Br.) Sacc. Ann. Appl. Biol. 50:283–89.
4. Bunting, R. H., and W. W. Dade. 1924. Gold Coast plant diseases. Waterlow & Sons: London, pp. 25–36, 40–44, 51–53.
5. Butler, E. J. 1918. Fungi and diseases in plants. Thacker: Calcutta, India, pp. 413–42.
6. Couch, J. N. 1938. The genus Septobasidium. University of North Carolina Press: Chapel Hill.
7. Crowdy, S. H. 1947. Observations on the pathogenicity of Calonectria rigidiuscula (Berk. & Br.) Sacc. on Theobroma cacao L. Ann. Appl. Biol. 34:45–59.
8. Desrosiers, R. 1955. Thielaviopsis pod rot of cacao in Ecuador. FAO Plant Protect.

Bull. 3:154–55.
9. Desrosiers, R., et al. 1955. Effect of rainfall on the incidence of Monilia pod rot of cacao in Ecuador. FAO Plant Protect. Bull. 3:161–64.
10. Ford, E. J., J. A. Bourret, and W. C. Snyder. 1967. Biologic specialization in Calonectria (Fusarium) rigidiuscula in relation to green point gall of cocao. Phytopathology 57:710–12.
11. Hansen, A. J. 1963. The role of Fusarium decemcellulare and Fusarium roseum in the green point cushion gall complex of cacao. Inter-Am. Inst. Agr. Sci. Turrialba, Costa Rica 13:80-87.
12. Hardy, F. 1960. Cacao manual. Inter-American Institute of Agriculture: Turrialba, Costa Rica, pp. 240–92.
13. Hislop, E. C. 1963. Studies on the chemical control of Phytophthora palmivora (Butl.) Butl., on Theobroma cacao L. in Nigeria. Ann. Appl. Biol. 52:465–92.
14. Holliday, P. 1954. Control of witches' broom disease of cacao in Trinidad. Trop. Agr. (London) 31:312–17.
15. Hunt, J. 1956. Taxonomy of the genus Ceratocystis. Lloydia 19:1–58.
16. Leakey, C. L. A. 1965. Sudden death disease of cacao in Uganda associated with Verticillium dahliae Kleb. E. African Agr. J. 31:21–24.
17. Moreno, J. D. 1957. Observaciones sobre la incidencia de monilia en Ecuador. Inter-Am. Inst. Agr. Sci. 7:95–99.
18. Nowell, W. 1923. Diseases of crop-plants in the Lesser Antilles. West India Committee: London, pp. 126–46, 152–54, 162–72.
19. Rocha, H. M. 1965. Cacao varieties resistant to Phytophthora palmivora (Butl.) Butl. Cacao (Turrialba, Costa Rica) 10:1–9.
20. Roger, L. 1951. Phytopathologie des pays chauds. Encyclopedie Mycologique 17: 918–19, 948–55, 1049–50, 1053–56; 1073–78, 1087–89, 1091–94; 1953, 18: 1162–63, 1454–58, 1484–86, 1488–95, 1893–94.
21. Saunders, J. L. 1965. The Xyleborus ceratocystis complex of cacao. Cacao (Turrialba, Costa Rica) 10:7–13.
22. Schieber, E., and O. N. Sosa. 1960. Cacao canker in Guatemala incited by Ceratocystis fimbriata. Plant Disease Reptr. 44:672.
23. Tabor, R. J., and R. H. Bunting. 1923. On a disease of cacoa and coffee fruits caused by a fungus hitherto undescribed. Ann. Botany (London) 37: 153–57.
24. Thorold, C. A. 1955. Observations on black-pod disease (Phytophthora palmivora) of cacao in Nigeria. Trans Brit. Mycol. Soc. 38:435–52.
25. Thrower, L. B. 1965. Parasitism of cacao by Fomes noxius in Papua, New Guinea. Trop. Agr. (London) 52:63–67.
26. Tollenaar, D. 1958. Phytophthora palmivora of cocoa and its control. Neth. J. Agr. Sci. 6:24–38.
27. Voorhees, R. K. 1942. Life history and taxonomy of the fungus Physalospora rhodina. Florida Univ. Agr. Expt. Sta. Bull. 371:1–91.
28. Waterston, J. M. 1941. Observations on the parasitism of Rosellinia pepo Pat. Trop. Agr. (Trinidad) 18:174–84.
29. Watrous, R. C. 1950. Cacao: a bibliography on the plant and its culture and primary processing of the bean. U.S. Dept. Agr. Library List 53:23–29.

Carob, *Ceratonia siliqua* Linnaeus

Leaf spot, *Pestalotia zonata* Ellis & Everhart
Leaf spot, *Cercospora ceratoniae* Patouillard & Trabut
Other fungi associated with carob

Leaf spot, *Pestalotia zonata* Ell. & Ev.

Symptoms

Leaf spots are scattered, brown to almost black, and mostly inconspicuous except at maturity when the submerged pustules become erumpent and appear in concentrically zoned areas under the shredded epidermis. Spots measure up to 150 μ in diameter.

Etiology

The conidia, measuring $20–26 \times 6–8.5$ μ, are long, narrowly fusiform, and

5-celled. The 3 central cells are large and smoky to opaque while the 2 external cells are hyaline. The apical cell is conical, supporting 3 setulae, sometimes 2 or 4, each up to 25 μ long; the basal cell is attached to a pedicel 3–7 μ long.

Reference

3.

Leaf spot, *Cercospora ceratoniae* Pat. & Trabut

Symptoms

The leaf spots first appear as mere peppery specks which are usually widely scattered but frequently seem to be clumped in specific leaf areas. The spots enlarge, remain mostly dark brown to black, and may become more or less irregular in outline; they seldom exceed 1 cm in diameter.

Etiology

The stromata are brown, flattened, and support compact fascicles of conidiophores that are rarely septate, not branched, straight, light brown, and 10–40 ×2–3.5 μ. The conidia are pale olivaceous, cylindric to obclavate, mostly straight, up to 6-septate, and measure 30–95×4.5–5 μ.

References

1, 2.

Other fungi associated with carob

Diplodia ceratoniae Tassi
Oidium ceratoniae Comes
Pestalotia ceratoniae Maublanc
Phyllosticta ceratoniae Berkeley

References: carob

1. Chupp, C. 1953. A monograph of the fungus genus Cercospora. Published by the author: Ithaca, N.Y., pp. 291–92.
2. Condit, I. J. 1919. The carob in California. Calif. Univ. Agr. Expt. Sta. Bull. 309:431–40.
3. Guba, E. F. 1961. Monograph of Monochaetia and Pestalotia. Harvard University Press: Cambridge, Mass., p. 215.

Cashew, *Anacardium occidentale* Linnaeus

Decline, *Pythium spinosum* Saw.

Symptoms

A general decline of the entire plant is followed by defoliation and dieback and if not corrected usually results in the death of the plant. A careful examination of the root system reveals a decay of the small, fine, feeder roots. The age of the tree makes little difference in the symptoms and severity, although the disease is more severe during dry periods.

Etiology

The discolored roots yield a nonseptate mycelium which is hyaline, branched, and 3–5 μ in diameter. Sporangia are spherical to spindle-shaped, of various sizes, and produce mycelium upon germination. The oogonia are spherical, spiny, and about 18 μ in diameter. Oospores are spherical, smooth, and about 16 μ in diameter.

Reference

1.

Other fungi associated with cashew

Asterina carbonacea Cooke var. *anacardii* Ryan
Atichia millardeti Raciborski
Botryobasidium rolfsii (Saccardo) Venkatarayan
Botryobasidium salmonicolor (Berkeley & Broome) Venkatarayan
Botrytis anacardii Viégas
Cercospora anacardii Mueller & Chupp
Pestalotia heterocornis Guba
Pestalotia virgatula Klebahn
Uredo anacardii Mains

Reference: cashew

1. Ramakrishnan, T. S. 1955. Decline in cashew nut. Indian Phytopathol. 8:58–63.

Cassava, *Manihot esculenta* Crantz

Leaf spot, *Xanthomonas cassava* Wiehe & Dowson
Wilt, *Xanthomonas manihotis* (Arthaud-Berthet) Starr
Leaf spot, *Erwinia cassavae* (Hansford) Burkholder
Root rot, *Rosellinia bunodes* (Berkeley & Broome) Saccardo
White root rot, *Fomes lignosus* (Klotzsch) Corner
Leaf spot, *Phyllosticta manihotae* Viégas
Canker, *Diplodia theobromae* (Patouillard) Nowell
Withertip, *Gloeosporium manihotis* P. Henning
Dieback, *Colletotrichum gloeosporioides* Penzig f. *manihotis* Chevallier
Leaf spot, *Septogloeum manihotis* Zimmermann
Leaf spot, *Cercospora caribaea* Ciferri
Leaf spot, *Cercospora henningsii* Allescher
Leaf spot, *Helminthosporium heveae* Petch
Wilt, *Verticillium dahliae* Klebahn
Other fungi associated with cassava

Leaf spot, *Xanthomonas cassava* Wiehe & Dows.

Symptoms

Chlorotic yellow, more or less circular spots, up to 2 mm in diameter, appear on the leaves. The centers become brown, and as the spots enlarge they tend to become angular and surrounded by a yellow band. The veins in the areas are also brown beyond the edges of the spots. An exudate is found on the lower surface of the spots. The organism enters the host through wounds and stomata.

Etiology

The parasite is a short, gram-negative rod, motile with peritrichous flagella. Agar colonies are pale yellow and slimy.

Reference

9.

Wilt, *Xanthomonas manihotis* (Arth.-Ber.) Starr

Symptoms

A wilt of the aerial parts of the plant develops first on the outer older leaves and finally on the entire plant, which shows no additional external symptoms. Discoloration of the vascular tissue is dark in the root and main stem areas and less extensive in the upper branches. The bark is sufficiently water-soaked to become slightly transparent, showing the discolored vascular tissue beneath and frequently showing sticky bacterial exudate.

Etiology

The organism is a gram-negative rod, produces no capsules, is nonmotile, or motile by a single polar flagellum, and measures $1.4–2.8 \times 0.35–0.93\ \mu$. Agar colonies are white to hyaline and mucoid. Optimum temperature, 30C.

Reference

6.

Leaf spot, *Erwinia cassavae* (Hansf.) Burkh.

Symptoms

Small, angular, water-soaked, greenish spots rapidly enlarge, become confluent, and appear often more abundantly along the main veins. The disease may advance into the leaf petioles, killing the leaves, and into the stems that are killed and become brown. An extension to the roots is not frequent. The distribution of the disease on this host is limited to East Africa.

Etiology

The organism is a gram-negative rod, forming no spores or capsules, and motile by peritrichous flagella. Agar colonies are yellow, round, translucent, and smooth.

References

6, 7.

Root rot, *Rosellinia bunodes* (Berk. & Br.) Sacc.

Symptoms

This disease shows aboveground characteristics of root deficiencies. Later, fructifications appear on the collars of old stumps. Evidence of infection is the dark brown mycelium running through the cortex and wood.

Etiology

The fungus produces minute perithecia that are dark and spherical. The ascus is cylindrical, measuring $90 \times 14\ \mu$, and the 8 ascospores each measure $9 \times 6\ \mu$.

References

1, 3.

White root rot, *Fomes lignosus* (Klotzsch) Corner

Symptoms

This disease is characterized by the poor condition of lateral and taproots. They are covered with a growth of thick, dirty white or yellowish cords of mycelium, branching at the ends into fan-shaped networks of fine threads. Root tissues are destroyed, after which the entire tree falls over. Orange, semi-circular, leathery, thin-margined, and bracket-shaped fructifications, 6–12 in. in diameter, are usually seen on stumps of killed trees.

Etiology

Fungus grows from dead wood of stumps and logs of forest trees. It spreads a considerable distance by cords of mycelium that extend through the soil. The sporophores produce basidiospores in pores on their lower surfaces. Spores are disseminated by air currents.

Reference

1.

Leaf spot, *Phyllosticta manihotae* Viégas

Symptoms

The leaf spots often destroy an entire leaflet. When young, the spots average 3–10 mm and are surrounded by a yellow halo. They are zonate and appear mostly on the lower leaf surface, where they are yellow green to smoky green; on the upper surface they are yellow to chocolate. The leaves are often droopy, pendant, and reversely rolled.

Etiology

The pycnidia appear on both leaf surfaces. They are globose to subglobose, $88\text{–}144 \times 64\text{–}96\ \mu$, and are often in concentric rings around a typical spot. The conidiophores are hyaline, simple, short, and narrow. The pycnidiospores are hyaline, ovoid to elongate, 1-celled, usually biguttulate, sometimes irregularly shaped, and $7\text{–}12 \times 2\text{–}3\ \mu$.

Reference

3.

Canker, *Diplodia theobromae* (Pat.) Nowell

Symptoms

This disease is characterized by necrotic leaf spots containing pycnidia that are isolated or enclosed in minute stroma. It causes partial defoliation. All organs of the tree are attacked from leaf to root. On cuttings, infection causes the cork to become corrugated, dry, and detached by bits on which small swellings or depressions develop.

Etiology

The sporulating stage of the fungus usually shows 1–5 globular pycnidia imbedded in a single stroma in the outer cortex. Pycnidiospores are first unicellular and hyaline, then brown, and finally uniseptate with longitudinal striations. The ascospore stage is found on dead wood in the form of submerged, ostiolate, black perithecia which produce sacklike asci. The asci produce hyaline, oval, 1-celled ascospores as *Physalospora rhodina.*

References

3, 8.

Withertip, *Gloeosporium manihotis* P. Henn.

Symptoms

First symptoms appear always on young twigs a few centimeters from the tip. Affected twigs become devoid of chlorophyll, reduced in diameter, and develop lesions bordered by a very thin, brown line. Finally, drying of twigs, wilt, leaf drop, and death of terminal shoots occur. In severe cases, the plant soon dies, but it sometimes reacts to infection by producing axillary shoots.

Etiology

The fungus acervuli are light brown, erumpent, pulvinate, and $100 \times 120\ \mu$. The conidia are hyaline, oblong, ellipsoid, and $10–15 \times 4.5\ \mu$. The perfect stage is *Glomerella cingulata* (Ston.) Spauld. & Schrenk f. sp. *manihotis* P. Henn.

Reference

8.

Dieback, *Colletotrichum gloeosporioides* Penz. f. *manihotis* Chev.

Symptoms

The infection appears mostly on young branches. Lesions develop, up to several centimeters in length, restricting elongation. The curving of the twig into a crescent shape causes the extremity to die and thus is described as dieback.

Etiology

The acervuli are up to 150 μ in diameter and appear as black dots scattered over the cortex of the killed portion of the branch. They break out from beneath the cuticle and are supported by a stromata of fungus hyphae. The conidiophores are short and plump, developing between the black, long, slender, septate, curved setae. The conidia are hyaline, cylindric, 1-celled, rounded at the ends, and 12–16×4–6 μ. The ascospore stage is *Glomerella cingulata* f. sp. *manihotis.*

Reference

3.

Leaf spot, *Septogloeum manihotis* Zimm.

Symptoms

Grayish green, irregular leaf spots and patches with a thin margin of purplish discoloration advance for a short distance along adjacent veins. Brown spots on both surfaces of the leaf bear numerous clusters of brown conidiophores.

Etiology

The conidiophores are 20–50×4–6 μ and bear hyaline, cylindrical, straight or curved, 0–7-septate spores that measure 30–104×4–8 μ. The spores are obclavate, narrow at the top end, and taper abruptly to point of attachment.

References

1, 2.

Leaf spot, *Cercospora caribaea* Cif.

Symptoms

The spots are numerous on the leaves, are usually small, ranging up to 5 mm, circular to irregular, and surrounded by a thin, brown line. On the upper surface they are light to translucent; below, they are gray to green, with a yellow halo, and slightly raised.

Etiology

The conidiophores are in loose tufts, emerging from the stomata on the lower surface. They are olive brown, septate, irregularly extended, 1–15 geniculate, and 50–200×3–5 μ. The conidia are hyaline to faintly colored, 1–6-septate, straight, and 20–90×4–8 μ.

Reference

4.

Leaf spot, *Cercospora henningsii* Allesch.

Symptoms

Leaf spots are isolated or coalesce, almost circular, 1–8 mm in diameter, and irregular near large veins. In some areas fine veins become visible by their black color. Fructifications occur on the spots on the underside of the leaf. Spots are without halo but are surrounded by a very narrow brown line.

Etiology

The conidiophores, arising in clusters on a minute stroma, are light brown, simple, nonseptate at the apical region, and slightly tortuous. Conidia are borne singly. They are hyaline or very slightly colored, claviform or subcylindrical, 0–11-septate, and measure 30–50 $\times$ 4–7 μ. The ascospore stage is *Mycosphaerella manihotis.*

Reference

3.

Leaf spot, *Helminthosporium heveae* Petch

Symptoms

Spots on the leaves are 3–5 mm in diameter, circular to angular, and light-colored in the center with a reddish brown border. Spots are very important on the small branches which become dry, cracked, and wrinkled.

Etiology

The conidiophores appear in clusters in the spots, separated from the healthy tissue by a thin, light brown line. They are olivaceous, in fascicles or separate, simple, and 120–240 μ long. The conidia are light olive at first, becoming brown, with 8–11 septations, and measure 48–160 $\times$ 12–21 μ.

Reference

3.

Wilt, *Verticillium dahliae* Kleb.

Symptoms

A gradual wilting develops, and eventually the fungus involves the plant completely. It causes dark streaks in the vascular tissue in the wood, stems, and roots. Plants are usually infected during the first two months after planting. The wilt sometimes develops suddenly when the fungus rapidly girdles the stem.

Etiology

The mycelium is hyaline, fine, verticillate, septate, branched, and produces microsclerotia that are black, oval, and 80 $\times$ 40 μ. Conidia are produced in large numbers, clustered in a moist head; they measure 4–5 $\times$ 2–4 μ.

Reference

7.

Other fungi associated with cassava

Armillaria mellea Vahl ex Fries
Asterina manihotis Sydow
Botryobasidium rolfsii (Saccardo) Venkatarayan
Botryosphaeria ribis (Tode) Grossenbacher & Duggar
Cercosporella pseudoridum Spegazzini
Diaporthe manihoticola Viégas
Heterosporium luci Chevallier
Iremina entebbeenis Hansford & Stevens
Leptosphaeria petri Chevallier
Marasmius lyssicola Petch
Melanconium moreaui Chevallier
Mycosphaerella manihotis (Sydow) Saccardo
Oidium manihotis Henning
Phomopsis manihot (Spegazzini) Chevallier
Phyllosticta portoricensis Young
Sphaceloma manihoticola Bitancourt & Jenkins
Sphaerostilbe repens Berkeley & Broome
Stagonospora manihotis Chevallier
Uromyces janiphae (Winter) Arthur

References: cassava

1. Bunting, R. H., and H. A. Dade. 1924. Gold Coast plant diseases. Waterlow & Sons: London, pp. 40–41, 43, 70–71.
2. Butler, E. S. 1918. Fungi and diseases in plants. Thacker: Calcutta, India, p. 310.
3. Chevaugeon, J. 1956. Flore cryptogamique associee au manioc. Encyclopedie Mycologique 28:32–33, 36–38, 58–67, 70–78, 81–90, 92.
4. Chupp, C. 1953. A monograph of the fungus genus Cercospora. Published by the author: Ithaca, N.Y.
5. Chupp, C., and A. F. Sherf. 1960. Vegetable diseases and their control. Ronald Press: N.Y., pp. 597–99.
6. Elliott, C. 1951. Manual of bacterial plant pathogens. 2d ed. Chronica Botanica: Waltham, Mass.
7. Hansford, C. G. 1945. Uganda plant diseases. E. African Agr. J. 10:147–51.
8. Nowell, W. 1923. Diseases of crop-plants in the Lesser Antilles. West India Committee: London, pp. 235–335.
9. Wiehe, P. O., and W. J. Dowson. 1953. A bacterial disease of cassava (Manihot ultissima) in Nyasaland. Empire J. Exptl. Agr. 21:141–43.

Castorbean, *Ricinus communis* Linnaeus

Leaf spot, *Xanthomonas ricinicola* (Elliott) Dowson
Bacterial wilt, *Pseudomonas solanacearum* E. F. Smith
Seedling blight, *Phytophthora parasitica* Dastur
Damping-off, *Pythium ultimum* Trow
Graymold, *Sclerotinia ricini* Godfrey
Scab, *Sphaceloma ricini* Jenkins & Cheo
Leaf spot, *Alternaria ricini* (Yoshii) Hansford
Seedling blight, *Alternaria compacta* (Cooke) McClellan
Leaf spot, *Cercospora ricinella* Saccardo & Berlese
Charcoal rot, *Sclerotium bataticola* Taubenhaus
Other fungi associated with castorbean

Leaf spot, *Xanthomonas ricinicola* (Elliott) Dows.

Symptoms

Small, irregular, water-soaked lesions appear on the foliage and generally enlarge to 5 mm in diameter. The brown spots cause some yellowing. When numerous, they coalesce, resulting in extensive premature shedding. Affected parts are mostly in the parenchyma tissues, sometimes limited by the veins. Vascular invasion and wilting are not observed.

Etiology

The organism is a gram-negative rod, produced in short chains, with capsules, motile by polar flagella, and measuring $1.3–2.6 \times 0.4–0.9$ μ. Agar colonies are lemon yellow, darkening to brown, and circular. Optimum temperature, 29–30C.

Reference

6.

Bacterial wilt, *Pseudomonas solanacearum* E. F. Sm.

Symptoms

Plants of various sizes and ages show a wilting of certain leaves, petioles, branches, and inflorescences while in the green condition with no previous yellowing. The vascular tissue is a distinctive brown color. Where infection is less severe, dwarfing may result, or sometimes wilt appears on only one side of a plant. An exudate may be observed when the plants are still wet with dew. The disease is considered vascular.

Etiology

The organism is a gram-negative rod, produces no capsules, is motile by a single polar flagellum, and measures 1.5×0.5 μ. Agar colonies are opalescent becoming brown, usually small, wet, shiny, smooth, and irregular. Optimum temperature, 35–37C.

References

6, 14.

Seedling blight, *Phytophthora parasitica* Dast.

Symptoms

On seedlings 6–8 in. high, a round, dull green patch or spot appears on either surface of the cotyledons, causing them to droop. After this the petiole becomes involved, and the terminal portion of the seedling is killed. On older plants the disease is mostly localized on the leaf blades where the spots become yellowish brown, concentrically zoned, often confluent, and cause leaf shedding and stunted plants.

Etiology

The nonseptate mycelium appears on the lower surfaces of the leaf spots where conidiophores, measuring 100–300 μ, emerge through the stomata. The sporangia are pear-shaped, measuring 25–50 × 20–40 μ; upon germination 15–30 zoospores emerge, each measuring 8–12 × 5–8 μ. They are motile for short periods and then germinate by producing mycelium. Oospores are hyaline, spherical, and 15–20 μ.

References

1, 5.

Damping-off, *Pythium ultimum* Trow

Symptoms

There are no aboveground symptoms of this disease because the seeds are invaded by the fungus. The fungus retards the emerged seedlings. The development of the disease depends largely on the proper maturity of the seed relative to the health of the parent plant and to the temperature prevalent at the time of planting. Low temperatures at planting and germination time produce the best stands.

Etiology

Hyphae are branched, 3–8 μ in diameter, and nonseptate. The conidia are spherical, terminal, and about 20 μ in diameter; germination is direct. Oogonia are spherical, smooth, and average 20.6 μ in diameter; oospores are spherical, 14–19 μ in diameter, and germinate directly.

References

10, 16, 17.

Graymold, *Sclerotinia ricini* Godfrey

Symptoms

Infection appears as irregular spots on leaves, stems, branches, and panicles. Spots are pale blue, variable in size, and often blotchlike. The inflorescence becomes covered with a drab, gray mold, soft and velvety in texture. The spots on other parts generate a loose superficial mycelium that spreads rapidly. The flowers are highly susceptible and become water-soaked, wilted, and densely covered with mold. Older plant parts are more resistant to infection. The flower stalk may be destroyed with little or no additional infection of the plant. As the season progresses, the plants remain barren although extensively branched by a new growth. The killed floral structures and pods are sources of the fungus, which penetrates the wood and survives there. Sclerotia are produced that give rise to the apothecia, which contain asci and ascospores.

Etiology

The mycelium, produced abundantly and rapidly, is olive drab to hyaline, branched, and septate; the conidia are borne on stigmata compactly grouped. Microspores are hyaline, globose, and 2–4 μ in diameter. Conidiophores develop from hyphal strands and produce conidia that are hyaline, globose, smooth, and 6–12 μ in diameter. The sclerotia are black, elongate to irregular, and up to 25 mm long, mostly 3–9 mm. The apothecia are brown, paraphysate, discoid, stalked, and 6–15 mm high. Asci are cylindrical, $50–110 \times 6–10\ \mu$, with ascospores that are hyaline, elliptical, continuous, and $9–12 \times 4–5\ \mu$. The conidial stage is *Botrytis cinerea.*

References

7, 8, 11.

Scab, *Sphaceloma ricini* Jenkins & Cheo

Symptoms

Spots are circular, usually scattered over the leaf blade, and more conspicuous above than below. They may include the leaf margins and the veins of the blade as reddish brown, water-soaked areas, finally becoming buff to brown, and up to 3 mm in diameter. On the lower leaf surface, lesions on veins are 1×10 mm, yellow to brown with slightly depressed lighter centers.

Etiology

The conidiophores that appear on the scabby cankers are yellow to amber or dark gray, awl-shaped to cylindrical, unbranched, and smooth. Conidia are hyaline, oblong to ovoid, elliptical to fusiform, and $10–15 \times 2–5\ \mu$.

Reference

2.

Leaf spot, *Alternaria ricini* (Yoshii) Hansf.

Symptoms

Cotyledons of young seedlings are spotted, stunted, and malformed, resulting in a blight. When severe, the plant dies. If the plant survives with a light infection, spores are often found on the diseased parts. The fungus appears on the seedpods when they are about half grown. The florescence wilts and becomes brown, the peduncles collapse, and the seed are partially filled. If the infection occurs when capsules are more mature, the invasion is less severe. The seed develop well, but the outside walls of the pods become blackened from the dark mycelium and conidia of the fungus. Leaf spotting is common but not serious. Some defoliation results.

Etiology

The mycelium is dark. There are 1 or several dark conidiophores to a fascicle; they are generally single but may be branched, septate, and measure $70–128 \times 4–6\ \mu$. Conidia are dark olive en masse, obclavate, and $47–96 \times 15–29\ \mu$, averaging $70 \times 22\ \mu$.

References

12, 13, 15.

Seedling blight, *Alternaria compacta* (Cke.) McClellan

Symptoms

The disease produced on the foliage of seedlings and older plants develops as a severe necrosis in which large leaf areas become flaccid and harbor many dark brown, zonate spots of small diameter. Some leaf shedding occurs. Seed infection develops in maturing capsules in which the fungus invades the caruncles more or less completely, depending on stages of their maturity. Such diseased seed results in seedling blight when planted. Some seed fail to germinate; others develop weak plants that die by so-called damping-off; others suffer only cotyledon injury.

Etiology

The mycelium is dark, septate, branched, and produces brown, erect, short, conidiophores, bearing muriform conidia that measure 14–39$\times$7–20 μ, and have beaks that are shorter than the spore body.

Reference

9.

Leaf spot, *Cercospora ricinella* Sacc. & Berl.

Symptoms

The leaf-spotting disease is probably found on this host wherever it is grown. The spots are usually small but may be up to 1 cm in diameter. They are circular with light-colored centers and dark brown borders. Although the spores of the fungus are produced on both leaf surfaces, they occur mostly on the undersurface.

Etiology

Conidiophores are brown, usually in clusters. Conidia are hyaline, cylindrical to obclavate, mostly straight, septate, and 15–120$\times$2–5 μ.

Reference

3.

Charcoal rot, *Sclerotium bataticola* Taub.

Symptoms

This disease under average conditions may appear in the early seedling stage, when plants are girdled and killed. At this time there may be some reduction in growth; some lower leaves become yellow and may be shed. A slight discoloration of the stem occurs near the soil line and may extend up the stalk, depending on favorable weather conditions.

Etiology

Many small, black sclerotia are imbedded in the epidermis, cortex, and stem. They are also found densely scattered in the wood and pith. The abundance of sclerotia causes the cankered areas to become slate-colored or blackish. These sclerotia are angular, seldom spherical, and 32–150×22–25 μ. The conidial stage of the fungus is *Macrophomina phaseolina.*

Reference

4.

Other fungi associated with castorbean

Botryodiplodia curta Saccardo
Botryosphaeria gregaria Saccardo
Cercospora canescens Ellis & Martin
Cercosporina ricinella (Saccardo & Berlese) Spegazzini
Clitocybe tabescens (Scopoli ex Fries) Bresadola
Colletotrichum ricini Petch
Diaporthe ricini Spegazzini
Diplodia ricinicola Saccardo
Erysiphe ricini Speschnew
Fusarium sambricinum Fuckel
Gloeosporium ricini Maublanc
Macrophoma ricini (Saccardo) Berlese & Voglino
Melampsorella ricini (de Bivona-Bernardi) de Toni
Phomopsis ricini Grove
Phyllosticta ricini Rostrup
Tuberculina ricini (Cocconi) Saccardo & Sydow

References: castorbean

1. Butler, E. J. 1918. Fungi and disease in plants. Thacker: Calcutta, India, pp. 326–30.
2. Cheo, C. C., and A. E. Jenkins. 1945. Elsinoë and Sphaceloma diseases in Yunnan, China, particularly hyacinth bean scab and scab of castor bean. Phytopathology 35:339–52.
3. Chupp, C. 1953. A monograph of the fungus genus Cercospora. Published by the author: Ithaca, N.Y., p. 229.
4. Cook, A. A. 1955. Charcoal rot of castor bean in the United States. Plant Disease Reptr. 39:233–35.
5. Dastur, J. F. 1913. On Phytophthora parasitica nov. spec., a new disease of the castor oil plant. Mem. Dept. Agr. India, Botan. Ser. 5:177–226.
6. Elliott, C. 1951. Manual of bacterial plant pathogens. 2d ed. Chronica Botanica: Waltham, Mass., pp. 136–37, 139–42.
7. Godfrey, G. H. 1923. Gray mold of castor bean. J. Agr. Res. 23:679–715.
8. Groves, J. W., and C. A. Loveland. 1953. The connection between Botryotinia fuckeliana and Botrytis cinerea. Mycologia 45:415–25.
9. McClellan, W. D. 1944. A seedling blight of castor bean, Ricinus communis. Phytopathology 34:223–29.
10. Matthews, V. D. 1931. Studies on the genus Pythium. University of North Carolina Press: Chapel Hill, pp. 79–82.
11. Orellana, R. G. 1959. Botrytis leaf blight of Ricinus communis L. Plant Disease Reptr. 43:363–64.
12. Pawar, V. H., and M. K. Patel. 1957. Alternaria leaf spot of Ricinus communis L. Indian Phytopathol. 10:110–14.
13. Singh, R. S. 1955. Alternaria blight of castor plants. J. Indian Botan. Soc. 34: 130–39.
14. Smith, E. F., and G. H. Godfrey. 1921. Bacterial wilt of castor bean (Ricinus communis L.). J. Agr. Res. 21:255–62.
15. Stevenson, E. C. 1945. Alternaria ricini (Yoshii) Hansford, the cause of a serious disease of the castor-bean plant (Ricinus communis L.) in the United States. Phytopathology 35:249–56.
16. Thomas, C. A. 1960. Relations of variety,

temperature, and seed immaturity to pre-emergence damping-off of castor-bean. Phytopathology 50:473–74.
17. Trow, H. 1901. Observations on the biology and cytology of Pythium ultimum n. sp. Ann. Botany (London) 15: 269–312.

Chayote, *Sechium edule* (Jacquin) Swingle

Downy mildew, *Pseudoperonospora cubensis* (Berkeley & Curtis) Rostowzew
Fruit rot, *Glomerella cingulata* (Stoneman) Spaulding & Schrenk
Black rot, *Mycosphaerella citrullina* (C. O. Smith) Grossenbacher
Southern blight, *Botryobasidium rolfsii* (Saccardo) Venkatarayan
Leaf spot, *Phyllosticta sechii* E. Young
Anthracnose, *Colletotrichum lagenarium* (Passerini) Ellis & Halsted
Leaf spot, *Cercospora cucurbitae* Ellis & Everhart
Leaf spot, *Cercospora sechii* Stevenson
Leaf spot, *Helminthosporium sechicola* Stevenson

References: chayote

1. Chupp, C., and A. F. Sherf. 1960. Vegetable diseases and their control. Ronald Press: N.Y., p. 599.
2. Cook, O. F. 1901. The chayote: a tropical vegetable. U.S. Dept. Agr. Div. Botany Bull. 28:1–30.
3. Hoover, L. G. 1923. The chayote: its culture and uses. U.S. Dept. Agr. Circ. 286:1–11.

Chickpea (gram), *Cicer arietinum* Linnaeus

Rust, *Uromyces ciceris-arietini* (Grognot) Jacquin
Charcoal rot, *Rhizoctonia bataticola* (Taubenhaus) Butler
Leaf spot, *Phyllosticta rabiei* (Passerini) Trotter
Blight, *Ascochyta rabiei* (Passerini) Labrousse
Leaf spot, *Stemphylium sarcinaeforme* (Cavara) Wiltshire
Wilt, *Fusarium lateritium* (Nees) Snyder & Hansen f. *ciceri* (Padwick) Erwin
Wilt, *Verticillium albo-atrum* Reinke & Berthold

Rust, *Uromyces ciceris-arietini* (Grog.) Jacq.

Symptoms

The leaves first show small, cinnamon brown, circular to oval pustules resembling small raised blisters. The lesions may coalesce, but usually cause little damage, unless plentiful. The pustules occur on both leaf surfaces, but sparingly on the upper.

Etiology

The uredospores are yellowish brown, irregularly spherical, 20–28 μ in diameter, with very fine echinulations, and several germ pores. The telia are dark brown, circular to oval, often angular, with a thickened apex, and an echinulate wall; they measure $18–30 \times 18–24$ μ and are produced on short stipes. Other spore stages or hosts are not known.

References

2, 9.

Charcoal rot, *Rhizoctonia bataticola* (Taub.) Butl.

Symptoms

Leaf bronzing of part or most of the foliage is an early symptom. The leaves finally become yellow and brown and are shed, leaving the upright leafstalk. No lesions appear above the soil line. The roots blacken and shrivel.

Etiology

The mycelium is hyaline, branched, septate, and invades the main root and stem. The fungus can usually be identified by the black, small, shiny, variably irregular, hard sclerotia, measuring less than 1/2 mm in diameter. Sclerotia are imbedded in the cortex and wood of the host. The spore-producing stage of this fungus is *Macrophomina phaseolina* (Tassi) Goid.

Reference

4.

Leaf spot, *Phyllosticta rabiei* (Pass.) Trott.

Symptoms

Circular yellow spots with dark borders are most characteristic of the disease, but oblong necrotic lesions occur on the stems and may cause some dieback on young shoots. Small, black specks on these spots are the pycnidia of the fungus.

Etiology

The pycnidia are scattered, submerged in host tissue, with ostioles protruding. The pycnidiospores are hyaline, 1-celled, oval, and sometimes exuded in long, silvery tendrils. The fungus is characterized by nonseptate pycnidiospores that measure $10\text{–}14 \times 3\text{–}5\ \mu$.

References

8, 11.

Blight, *Ascochyta rabiei* (Pass.) Labr.

Symptoms

All parts of the plant are susceptible to this disease in which the leaves show circular, yellow spots with a dark border. Stems and young shoots show somewhat elongate spots, while the younger parts may be girdled and the top killed. The pods may exhibit several infections. The spots on all parts eventually show numerous black specks within the killed area in which the spores of the fungus are produced. On the pods the diseased tissue may extend to the seed, causing seedling infection and damping-off.

Etiology

The mycelium is hyaline, septate, and branched; pycnidia are black, spherical to obovate, immersed, ostiolate, and up to 250 μ in diameter. The pycnidiospores are hyaline, oblong, 1-septate, sometimes constricted, with rounded ends, and measure 9–20×3–6 μ. The fungus is characterized by the 2-celled pycnidiospores. The ascospore stage of the fungus is *Mycosphaerella rabiei.*

References

2, 6, 10.

Leaf spot, *Stemphylium sarcinaeforme* (Cav.) Wiltsh.

Symptoms

Oval leaflet lesions, 6×3 mm in diameter, include necrotic tissue. The spots are brown in the center surrounded by a wide gray border. Spots on stems are usually smaller, brown, and elongate.

Etiology

Conidiophores are dark brown, scattered, numerous, 2–5-septate, and 37–70 ×5–7 μ. The conidia are brown, borne singly, subglobose to ovoid, smooth, restricted at the middle septum, contain other crosswise or oblique septations, and measure 20–30×17–27 μ.

Reference

3.

Wilt, *Fusarium lateritium* (Nees) Snyd. & Hans. f. *ciceri* (Padw.) Erwin

Symptoms

Leaves turn gray green followed by dull yellow, become weak, wilt, and usually die. During stages in the development of the disease, plants may show wilt on one side. Sections of the root tissue are most often dark brown to black. This discoloration is variable and may extend up to the stem in the xylem tissue; the phloem is seldom discolored.

Etiology

Mycelium is branched, septate, and normally hyaline or white. In artificial media it is variously colored rose, purple, or yellow. Microconidia are many, hyaline, 1-celled, and 5–12×2–3.5 μ. Macroconidia are few, hyaline, 3-septate, slightly curved, and 25–46×3–4 μ. There are no sporodochia, pionnotes, or sclerotia. Chlamydospores are globose to ovoid, 1-celled, and 7–10 μ in diameter.

References

1, 5, 7.

Wilt, *Verticillium albo-atrum* Reinke & Berth.

Symptoms

The leaves become dull green, soon changing to yellow, followed by wilting, stunting, browning, and death. Sometimes the disease is more or less on one side of the plant, resulting in distortion and the production of darkened vascular streaks along the stem.

Etiology

Mycelium is hyaline, septate, branched, and becomes darkened in older growth or by sclerotial production. The conidiophores are hyaline, elongate, narrow, and produce branches in whorls that abscise spores one at a time from each terminal. Conidia are hyaline, oval, and 1-celled. The conidia may appear in long, winding chains or in small heads held in droplets of moisture.

Reference

5.

References: chickpea

1. Afzal, M. 1965. Studies into the causes of gram wilt. W. Pakistan J. Agr. Res. 3:96–102.
2. Butler, E. J. 1918. Fungi and disease in plants. Thacker: Calcutta, India, pp. 268–71.
3. Das, G. N., and P. K. Sen Gupta. 1961. A stemphylium leaf spot disease of gram. Plant Disease Reptr. 45:979.
4. Dastur, J. F. 1935. Gram-wilts in the central provinces. Agr. Livestock India 5:615–27.
5. Erwin, D. C. 1958. Fusarium lateritium f. ciceri, incitant of fusarium wilt of Cicer arietinum. Phytopathology 48:498–501.
6. Luthra, J. C., A. Sattar, and K. S. Bedi. 1935. Life-history of gram blight Ascochyta rabiei (Pass.) Trot. on gram (Cicer arietinum L.) and its control in the Punjab. Agr. Livestock India 5: 489–98.
7. Manucheri, A., and G. Mesre. 1966. Fusarium wilt of chickpea in Iran. Iranian J. Plant Pathol. 3:11–19.
8. Roger, L. 1953. Phytopathologie des pays chauds. Encyclopedie Mycologique 18: 1344–46.
9. Saksena, H. K., and R. Prasada. 1955. Studies in gram rust, Uromyces cicerisarietini (Grogn.) Jacq. Indian Phytopathol. 8:94–98.
10. Sattar, A. 1933. A comparative study of the fungi associated with blight diseases of certain cultivated leguminous plants. Trans. Brit. Mycol. Soc. 18:276–301.
11. Sprague, R. 1930. Notes on Phyllosticta rabiei on chick-pea. Phytopathology 20:591–93.

Cinchona (quinine), *Cinchona officinalis* Linnaeus and *C. calisaya* Weddell

Stripe canker, *Phytophthora cinnamomi* Rands
Girdle canker, *Phytophthora parasitica* Dastur
Collar rot, *Phytophthora quininea* Crandall
Scab, *Elsinoë cinchonae* Jenkins
Leaf spot, *Guignardia yersini* Vincens
Sooty mold, *Capnodium brasiliense* Puttemans
Root rot, *Rosellinia arcuata* Petch
Black root rot, *Rosellinia bunodes* (Berkeley & Broome) Saccardo
Felty fungus, *Septobasidium bogoriense* Patouillard
Damping-off, *Thanatephorus cucumeris* (Frank) Donk
Pink disease, *Botryobasidium salmonicolor* (Berkeley & Broome) Venkatarayan
Brown root rot, *Fomes lamaensis* (Murrill) Saccardo & Trotter
Red rot, *Ganoderma pseudoferreum* Wakefield
Mushroom root rot, *Clitocybe tabescens* (Scopoli ex Fries) Bresadola
Shoestring root rot, *Armillaria mellea* Vahl ex Fries
Leaf spot, *Phyllosticta cinchonae* Patouillard
Dieback, *Diplodia cinchonae* Koorders
Leaf spot, *Pestalotia cinchonae* Zimmermann

Anthracnose, *Colletotrichum cinchonae* Koorders
Leaf spot, *Cercospora cinchonae* Ellis & Everhart
Algal spot, *Cephaleuros virescens* Kunze
Other fungi associated with cinchona (quinine)

Stripe canker, *Phytophthora cinnamomi* Rands

Symptoms

Vertical, slightly sunken stripes of dead bark, varying from thread lines to 1 in. in width, develop from a few inches to almost the full length of the tree. The disease originates in the roots and spreads upward. In certain instances one to several such stripes develop around the tree. There is a dark exudate on the diseased area. The foliage shows some deterioration as the disease develops, and severely affected trees die.

Etiology

The mycelium is hyaline, branched, nonseptate, and 4–6.5 μ in diameter. Chlamydospores are hyaline, terminal on short lateral branches, spherical to globose, thin-walled, and 30–50 μ in diameter. Swollen vesicles are often produced in culture, and conidiophores are myceliumlike. Sporangia are rarely produced in nature or in culture, except when mycelium is transferred to water. They are hyaline, borne terminally, ovoid to elongate, thin-walled, papillate, and 38–84 $\times$ 27–39 μ. Sporangia germinate in water by producing 8–40 kidney-shaped zoospores, with 2 flagella attached to the concave side; the zoospores measure 11–18 μ and germinate by producing a germ tube and mycelium.

References

9, 21.

Girdle canker, *Phytophthora parasitica* Dast.

Symptoms

Brown spots appear on sprouts growing at the base of the trunk and on foliage, petioles, and stems. Girdling often takes place, resulting in the wilting of growth beyond that point. Cankers may appear where top blight advances to more mature wood. Often a sap exudate appears on the outer bark. Under the bark the wood is various shades of brown. Cankers spread and eventually girdle the trunk. Killed trees usually produce root or trunk sprouts below the girdle. Most trunk infections occur within a 2-foot area above the soil. Entrance is gained through wounds in the bark.

Etiology

The hyaline, nonseptate, branched mycelium measures 3–6 μ in diameter and is almost entirely submerged. The conidia or sporangia show prominent papillae, are lemon-shaped, develop on elongate conidiophores, and measure 14–50 $\times$ 25–35 μ. They are yellowish and germinate by producing 15–20 zoospores. Chlamydospores are yellow, mostly spherical, thick-walled, and 19–26 μ.

References

2, 8, 19.

Collar rot, *Phytophthora quininea* Crandall

Symptoms

The loss of the normal green color followed by a pink or rose tint on the leaves is the first indication of the disease. Wilting and dropping of leaves also occur. The terminal leaves remain longest. The trees usually become affected when about 3 years old. Plants in a number of plantations have been almost completely killed in 3 or 4 years following infection. No symptoms other than the decline are evident on the aboveground parts with the possible exception of a ground canker or a collar canker that appears several meters up the stem. The root system may also become infected. The cankered tissue is dark brown, and the roots become soft and die.

Etiology

The mycelium is hyaline, nonseptate, branched, and 3.7–8 μ in diameter. Sporangia are papillate, ovate to pear-shaped, and 45–67 × 25–38 μ. They germinate in water by producing 14–20 zoospores, measuring 8–10 μ. Chlamydospores are dark brown, terminal or not, spherical, and 45–90 μ. Oospores are hyaline to light tan, spherical, and 48–64 μ.

References

5, 6.

Scab, *Elsinoë cinchonae* Jenkins

Symptoms

Spots are usually circular but may be irregular along margins or angular when the veins become involved. The lesions may be slightly raised and are often crowded on the main veins and midrib. They are light or dark brown and mostly about 1–2 mm in diameter. A single spot may be detected from both leaf surfaces. Infections occur on branches, rachises, florescences, and capsules, forming elongate to irregular lesions up to 5 mm in diameter. Capsules are often distorted when severely spotted.

Etiology

The ascomata are scattered over the lesions as black, small, raised areas, round to elongate, 300 × 75 μ, and crowded with asci that are spherical to ellipsoid and 18–28 μ. Ascospores are hyaline, 1–3-septate, and 15 × 15 μ. The conidial stage is found on the lesions as a sporodochium with dark conidiophores, which are pointed, often 1-septate, and 3–5 × 8–15 μ. The conidial stage is *Sphaceloma* sp.

Reference

13.

Leaf spot, *Guignardia yersini* Vincens

Symptoms

Circular brown spots appear on the foliage, and deformed, cankerlike, suberose formations evolve on the bark.

Etiology

The perithecia are partially submerged and 200–300 μ in diameter; the asci measure 70–90 × 15–17 μ, and the ascospores are hyaline, elliptical, and 18–22 × 8–10 μ.

Reference

18.

Sooty mold, *Capnodium brasiliense* Putt.

Symptoms

A black, sooty film develops over the foliage and stems, mostly on the upper surfaces, composed of dark mycelial strands and threads supported by the honeydew secretion of insects.

Etiology

The fungus produces dark-colored, elongate, cylindrical, 5–6-septate conidia that measure 40–60 × 3–4 μ. The black perithecia are 36–60 μ in diameter. The ascospores are hyaline to olivaceous, elongate, 3-septate, and 13–16 × 4–7 μ.

Reference

18.

Root rot, *Rosellinia arcuata* Petch

Symptoms

The diseased host in a growing condition shows indications of malnutrition. The foliage first appears off-color, and there may follow a lack of growth or stunting, some shedding of leaves, and a gradual deterioration ending in the death of the tree. Environmental conditions often hasten the development of the disease, and a plant may succumb in a few weeks from the time of the first visible symptom. The roots become covered with loose mycelial strands that produce a wooly appearance and often are purplish to black. Young plants are frequently killed before they are 3 years old. More mature plants are less frequently damaged. Under favorable conditions the fungus grows several inches up the stems of the attacked plants.

Etiology

The vegetative spore-producing stage is a *Graphium* sp. in which erect, hairlike, pycnidial beaks exude droplets of hyaline, oval, 1-celled spores that measure 4–6 × 2 μ. The perithecia are black, embedded becoming erumpent, gregarious, globose, and up to 3 mm in diameter. The asci are cylindric, paraphysate, ostiolate, and $300 \times 8\ \mu$; the ascospores are black, fusiform, pointed, and $40–47 \times 5–7\ \mu$.

Reference

18.

Black root rot, *Rosellinia bunodes* (Berk. & Br.) Sacc.

Symptoms

Black root rot has been reported to cause a disease similar to the root rot caused by *R. arcuata.* One may not be able to distinguish the two diseases in the conidial stage. Both produce mycelium over the roots; it may be light-colored when young but eventually forms black strands.

Etiology

The conidial stage is a *Graphium* sp., producing conidia measuring $4–6 \times 2\ \mu$. The ascospores are distinct brown to black, very narrow, cymbiform, and $80–110 \times 7–12\ \mu$. At each end there is a fine drawn-out thread up to 25 μ long. These two fungi can be distinguished by the characteristic morphologically different ascospores.

References

7, 10, 12.

Felty fungus, *Septobasidium bogoriense* Pat.

Symptoms

A gray to brown, feltlike, soft, resilient, superficial growth may appear on twigs and branches, extending for several inches in a resupinate position. The growth is circular to elongate, dense in the middle area and thinning at the margins, and protects cavernous areas above the outer cortex. The fungus is a parasite of insects that have become stationary parasites of the host. The fungus is more or less permanent, depending on the insect longevity. Damage by it to the plant is entirely secondary.

Etiology

The fungus is a basidiomycete and develops haustoria within the body of the parasitized insect, which, in turn, is more or less protected by the felty growth. The insect produces migratory young. The fungus develops basidiospores that can parasitize them.

References

4, 14.

Damping-off, *Thanatephorus cucumeris* (Frank) Donk

Symptoms

The soil-inhabiting fungus attacks seedlings of many plants, but may kill them at germination or before emergence. The young plants are invaded by the mycelium of the fungus, which weakens the tissue and rapidly overgrows the diseased and dying plant. Reduction of emergence is the first indication of disease. Young infected plants often show a drooping of the cotyledons.

Etiology

The mycelium of the fungus is brown, septate, branched, and grows rapidly over killed seedlings. It produces brown, flat, irregularly shaped, thin sclerotia on dead plant parts, and is seldom found otherwise. The basidia are produced on the mycelium, often grow on older plants in a superficial network, and produce basidiospores that are hyaline, 1-celled, and 7–13×4–7 μ.

References

14, 15.

Pink disease, *Botryobasidium salmonicolor* (Berk. & Br.) Venkat.

Symptoms

This disease is readily recognized by the appearance of the characteristic pink to pale rose or white, resupinate fungus growth, which is soft, resilient when fresh, and crusty and cracked when mature and dry. The submerged mycelium penetrates the cortex, invades the cambium, and girdles the branch. Young plants are frequently killed.

Etiology

The fungus first appears as superficial mycelial strands followed by the formation of white nodules composed of coils of hyphae, usually over the lenticels. The pink fruiting stage develops after cortical penetration. The *Necator decretur* stage is composed of a layer, up to 20 μ thick, of loosely interwoven, suberect hyphae, producing catenulate spores measuring 14–18×7–8 μ. The basidiospores are hyaline and 9–12×6–8 μ.

References

12, 14, 16.

Brown root rot, *Fomes lamaensis* (Murr.) Sacc. & Trott.

Symptoms

A type of slow decline of trees indicates some nutritional inefficiencies. Root infections are associated with such plant symptoms. The larger roots become more or less encrusted with soil held together and adhering to the roots, in layers up to 4 mm thick, by white to brown mycelium, which appears brown to black due to contact with the soil. The organism penetrates the cortex and cambium.

Etiology

The mycelium is brown, septate, and branched. The sporophores are perennial, layered, zonate, sessile, hornlike in texture, up to 25 cm wide and half as deep, and generally somewhat hoof-shaped. The spores are hyaline to tan, small, and 3–5×2–4 μ.

References

14, 18.

Red rot, *Ganoderma pseudoferreum* Wakef.

Symptoms

Diseased trees exhibit off-color foliage followed by yellowing, stunting, lack of new growth, shedding of leaves, some dieback, and death. The fungus invades the cortex and penetrates to the cambium, resulting in a browning and a killing of roots. The fungus produces rhizomorphs that are wine red to a brilliant and velvety red.

Etiology

The sporophores are brown above, sessile, shelving, zonate, and up to 30 cm in diameter. The hymenium is a dull white to gray. The pores are 90–160 μ in diameter. Spores are hyaline at first, later becoming tan to brown, sometimes roughened, and 6–9.5×2–4 μ.

References

14, 18.

Mushroom root rot, *Clitocybe tabescens* (Scop. ex Fr.) Bres.

Symptoms

The disease has been reported as a seedling damping-off disease in which the soil-inhabiting fungus contacts tender seedlings, girdling them. The fungus is readily identified by its white color, abundant mycelium, and septate condition. It also functions as a root rot, usually gaining entrance to the plant crown directly or through wounds. It proceeds into the cambium where cankers develop. Diseased trees show symptoms of malnutrition, in various stages, and of decline, and they eventually die. The fungus develops white, platelike shields of matted mycelium in the cortex and cambium. Their presence is usually diagnostic.

Etiology

The fungus develops sporophores that are up to 6 or 8 in. tall; gills are strongly decurrent, and the honey-colored stipitate pileus is from 2 to 4 in. in diameter. The stipe is central, of medium diameter, smooth, and contains no annulus. The basidiospores are hyaline, 1-celled, and 8–10×4–6 μ.

Reference

17.

Shoestring root rot, *Armillaria mellea* Vahl ex Fr.

Symptoms

The disease is confined to plants past the seedling stage and symptoms are similar in many ways to those of other root diseases. As the disease develops, it causes root decay, crown lesions, and trunk cankers, killing the tree.

Etiology

The sporophores are honey-colored, 6–8 in. tall, stipitate, and the stipe is annulated. The fungus also produces the black, shiny rhizomorphs that develop on killed trees primarily between the wood and the loosening cortex. The sporophores are usually more or less scattered around diseased trees rather than in dense clumps. The fungus favors cool temperatures, appearing in temperate zones or at high elevations in contrast to more humid tropical environments.

References

7, 14, 17, 20.

Leaf spot, *Phyllosticta cinchonae* Pat.

Symptoms

Leaf spots are yellow to tan, almost borderless, usually circular, up to 1/2 cm in diameter, and show on each leaf surface.

Etiology

The pycnidia are scattered, globose to ovoid, and about 120 μ in diameter. Conidia are hyaline, 1-celled, cylindrical, oblong, usually straight, and $8\text{–}10 \times 3\text{–}4$ μ.

References

14, 18.

Dieback, *Diplodia cinchonae* Koord.

Symptoms

The disease usually is observed in the form of dying back of twigs, small killed areas on the larger branches, or large cankerous conditions anywhere on the aboveground parts. The initial infection is usually through poorly developed twigs or mechanical injury.

Etiology

The infected parts often show the black, ostiolate pycnidia that are the fruiting structures of the fungus. The pycnidial spores are brownish black, oval to elongate, 2-celled, and $18\text{–}28 \times 10\text{–}14$ μ. The fungus is probably synonymous with *Diplodia natalensis* Evans, and the ascospore stage is *Physalospora rhodina* (Berk. & Curt.) Cke., of which *P. cinchonae* Vinc. is a synonym.

References

14, 18.

Leaf spot, *Pestalotia cinchonae* Zimm.

Symptoms

Brown, small, circular to irregular leaf spots develop from infections, and the fruiting bodies of the fungus are found on dead twigs and on the dead tissue in the leaves.

Etiology

The pycnidia are black, small, and almost superficial; they produce 5-celled pycnidiospores that measure 20–24 × 6 μ. The 3 center cells are colored, and the end cells are hyaline. The basal cell has a 4-μ-long pedicel, and the terminal cell has 3 spreading setulae, 20–30 μ long.

Reference

11.

Anthracnose, *Colletotrichum cinchonae* Koord.

Symptoms

The fungus produces light brown spots on the leaves and branches.

Etiology

The acervuli are scattered over the spots, and each spot measures up to 100 μ in diameter. The setae are yellowish brown, multiseptate, straight or curved, and 60–75 × 4–7 μ. The conidia are hyaline, 1-celled, cylindric with round ends, and measure 10–12 × 3–4 μ.

Reference

18.

Leaf spot, *Cercospora cinchonae* Ell. & Ev.

Symptoms

The leaf spots are reddish brown, circular, up to 4 mm in diameter, and often have slightly raised margins.

Etiology

The fungus sporulates on the upper surface where brown, small, globular stromata develop, producing dense clusters of conidiospores. Conidia have a faint olive tint, are linear, cylindrical, undulating, septate, and 25–80 × 2–3 μ.

Reference

3.

Algal spot, *Cephaleuros virescens* Kunze

Symptoms

The leaf spots are light reddish brown scattered with green, slightly raised, circular, appear mostly on the upper surfaces, and measure from minute points to spots up to 2 mm in diameter. They usually appear on opposite sides of the leaf when over 1 cm in diameter.

Etiology

The algae form a round disk, extend in all directions, and become more than 1 layer thick with radiating, dichotomous branches and many sterile hairs. Fertile filaments are branched with capitate swellings, including sporangia measuring $25–30 \times 38–40\ \mu$. Zoospores are biciliate, ovoid, and $5–8\ \mu$.

References

1, 14.

Other fungi associated with cinchona

Botryobasidium rolfsii (Saccardo) Venkatarayan
Ceratobasidium stevensii (Burt) Venkatarayan
Cladosporium herbarum (Persoon) Link
Cladosporium oxysporum Berkeley & Curtis
Dasyscypha warburgiana Henning
Dendrophoma cinchonae Vincens
Fusarium vasinfectum Atkinson
Macrosporium phaseoli (Maublanc) Ashby
Myriangium cinchonae Rehm
Phoma cinchonae Vincens
Sporodesmium cinchonae Koorders
Tulasnella cinchonae Raciborski
Uredo cinchonae P. Henning

References: cinchona

1. Butler, E. J. 1918. Fungi and disease in plants. Thacker: Calcutta, India, pp. 413–22.
2. Celino, M. S. 1934. Blight of cinchona seedlings. Philippine Agriculturist 23: 111–27.
3. Chupp, C. 1953. A monograph of the fungus genus Cercospora. Published by the author: Ithaca, N.Y., p. 493.
4. Couch, J. N. 1938. The genus Septobasidium. University of North Carolina Press: Chapel Hill.
5. Crandall, B. S. 1947. A new Phytophthora causing root and collar rot of cinchona in Peru. Mycologia 39:218–23.
6. Crandall, B. S. 1950. Cinchona root and collar rot in Peru and Bolivia. U.S. Dept. Agr. Circ. 855:1–16.
7. Crandall, B. S., and W. C. Davis. 1944. Occurrence of cinchona root rots in the Americas. Plant Disease Reptr. 28:926–27.
8. Crandall, B. S., and W. C. Davis. 1945. Phytophthora wilt and stem canker of cinchona. Phytopathology 35:138–40.
9. Darley, E. F., and M. A. Flores. 1951. Two cankers of cinchona in Guatemala caused by Phytophthora cinnamomi and P. parasitica. Phytopathology 41:641–47.
10. Davis, W. C., and B. S. Crandall. 1944. Some cinchona diseases in the Western Hemisphere. Plant Disease Reptr. 28: 996–97.

11. Guba, E. F. 1961. Monograph of Monochaetia and Pestalotia. Harvard University Press: Cambridge, Mass., p. 216.
12. Hubert, F. P. 1957. Diseases of some export crops in Indonesia. Plant Disease Reptr. 41:55–64.
13. Jenkins, A. E. 1945. Scab of cinchona in South America caused by Elsinoë. J. Wash. Acad. Sci. 35:344–52.
14. Lombard, F. F. 1947. Review of literature on cinchona diseases, injuries and fungi. U.S. Dept. Agr. Bibliog. Bull. 9: 4, 5–7, 28, 30–32, 37.
15. Madarang, S. A. 1941. Rhizoctonia damping-off of cinchona seedlings. Philippine J. Forestry 4:105–21.
16. Rant, A. 1912. Uber die djanoer-alpas krankheit und uber das Corticium javanicum Zimm. Buitenzorg Jard Botan. Bull. (Ser. 2) 4:1–50.
17. Rhoads, A. S. 1945. A comparative study of two closely related root-rot fungi, Clitocybe tabescens and Armillaria mellea. Mycologia 37:741–66.
18. Roger, L. 1951. Phytopathologie des pays chauds. Encyclopedie Mycologique 17:1042–44, 1049–52, 1293–94, 1351–54, 1376–94, 1625–29, 1644–65, 1856.
19. Szkolnik, M. 1951. Phytophthora parasitica diseases of cinchona in Central America field plantings. Plant Disease Reptr. 35:16–24.
20. Thomas, H. E. 1934. Studies on Armillaria mellea (Vahl) Quel., infection, parasitism, and host resistance. J. Agr. Res. 48:187–218.
21. Tucker, C. M. 1931. Taxonomy of the genus Phytophthora de Bary. Missouri Univ. Agr. Expt. Sta. Res. Bull. 153:5–196.

Cinnamon, *Cinnamomum zeylanicum* Nees

Stripe canker, *Phytophthora cinnamomi* Rands
Sooty mold, *Armatella cinnamomi* Hansford & Thirumalachar
Leaf spot, *Leptosphaeria cinnamomi* Shirai & Hara
Root rot, *Ganoderma pseudoferreum* Wakefield
Other fungi associated with cinnamon

Stripe canker, *Phytophthora cinnamomi* Rands

Symptoms

Long-killed areas of the bark up to 5 cm wide along the main trunk are the first indications of the disease. These black zones are parallel, usually narrow and somewhat sunken, and sometimes exceed 10 m in length. The interior discolorations are characteristically brown and gummy, sometimes reddish, and exude an amber liquid. The fungus is aided in infection principally through wounds.

Etiology

The mycelium is hyaline, branched, nonseptate, and occupies the host cells. The terminal chlamydospores are hyaline, globose to pyriform, 28–60 μ in diameter, and germinate by germ tubes. Conidiophores are myceliumlike. The sporangia have not been observed in nature or in ordinary artificial cultures, but they are produced abundantly on mycelial transfers to water. They are hyaline, borne terminally, ovoid to elongate, with inconspicuous papillae, and average 57×33 μ. Germination is by production of 8–40 zoospores per sporangium, each with 2 flagella, and measuring 10–11 μ at time of germination. Oogonia are hyaline, spherical, and average 32 μ in diameter; oospores are 25–27 μ in diameter.

Reference

3.

Sooty mold, *Armatella cinnamomi* Hansf. & Thirum.

Symptoms

Black, irregular, confluent, effused, thin fungus mycelium covers the lower leaf in areas from small patches to most of the surface.

Etiology

The mycelium is deep brown to black, irregular to undulating, 4–5 μ in diameter, branched, septate, with cells 20–30 μ long. Hyphopodia, 1 to each cell, are mostly alternate, about 15–20 μ long, and have no setae. Perithecia are black, scattered, globose, and up to 140 μ in diameter. The ascospores are pale brown, 1-septate, constricted, smooth, and 20–22×9–11 μ; the upper cell is larger.

Reference

1.

Leaf spot, *Leptosphaeria cinnamomi* Shirai & Hara

Symptoms

Round to elongate, irregular spots that are a dull brown with a narrow dark border are produced on leaves. These spots are often invaded by any of several other fungi.

Etiology

The pycnidia are light-colored, subglobose to spherical, ostiolate, and produce hyaline, 4-septate conidia about 20 μ long. The perithecia are black, spherical, subepidermal, and contain asci measuring 70 μ in length. Each ascus contains 8 ascospores that are brown, 3–5-septate, and constricted.

Reference

2.

Root rot, *Ganoderma pseudoferreum* Wakef.

Symptoms

Roots are readily infected under conditions of excessive moisture and rapidly decay. Slower growth usually results in poor top growth and contributes to the apparent slow decline. The mycelium developing from young infections is at first white and compact, later becoming reddish brown.

Etiology

The sporophores are not abundant and develop on the older, invaded dead parts at the base of the trunk. They are reddish brown above, grayish white below, shelving or sessile, and sometimes zonate. The context is pale yellow, and the pores are an average of about 130 μ in diameter, circular, and contain a light-colored hymenium. The basidiospores are brown, ovoid, 1-celled, and 6–8×4–5 μ.

Reference
2.

Other fungi associated with cinnamon

Cytosporella cinnamomi Turconi
Leptosphaeria almeidae d'Camara
Meliola zigzag Berkeley & Curtis
Mycosphaerella cinnamonicola Ciferri & Fragoso
Pestalotia cinnamomi d'Haan
Pestalotia funnerea Desmazieres
Phyllosticta cinnamomi (Saccardo) Lindau
Physalospora cinnamomi d'Camara

References: cinnamon

1. Hansford, C. G., and M. J. Thirumalachar. 1948. Fungi of south India. Farlowia 3:285–314.
2. Roger, L. 1951. Phytopathologie des pays chauds. Encyclopedie Mycologique 17: 1042–45; 1953, 18:1258–59.
3. Tucker, C. M. 1931. Taxonomy of the genus Phytophthora de Bary. Missouri Univ. Agr. Expt. Sta. Res. Bull. 153:5–197.

Citrus spp.

Canker, *Xanthomonas citri* (Hasse) Dowson
Blast, *Pseudomonas syringae* van Hall
Bacterial spot, *Erwinia citrimaculans* (Doidge) Magrou
Brown rot, *Phytophthora citrophthora* (R. E. Smith & E. H. Smith) Leonian
Foot rot, *Phytophthora parasitica* Dastur
Yeast spot, *Nematospora coryli* Peglion
Scab, *Elsinoë fawcetti* Bitancourt & Jenkins
Cottony rot, *Sclerotinia sclerotiorum* (Libert) De Bary
Melanose, *Diaporthe citri* Wolf
Dothiorella rot, *Botryosphaeria ribis* Grossenbacher & Duggar
Diplodia rot, *Physalospora rhodina* (Berkeley & Curtis) Cooke
Knot, *Sphaeropsis tumefaciens* Hedges
Root rot, *Ustulina zonata* Léveillé
Black spot, *Guignardia citricarpa* (McAlpine) Kiehly
Areolate leaf spot, *Leptosphaeria bondari* Bitancourt & Jenkins
Sooty mold, *Capnodium citri* Berkeley & Desmazieres
Fly speck, *Leptothyrium pomi* (Montagne & Fries) Saccardo
Sooty blotch, *Gloeodes pomigena* (Schweinitz) Colby
Root rot, *Rosellinia bunodes* (Berkeley & Broome) Saccardo
Felty fungus, *Septobasidium pseudopedicellatum* Burt
Southern blight, *Botryobasidium rolfsii* (Saccardo) Venkatarayan
Pink disease, *Botryobasidium salmonicolor* (Berkeley & Broome) Venkatarayan
Thread blight, *Ceratobasidium stevensii* (Burt) Venkatarayan
Damping-off, *Thanatephorus cucumeris* (Frank) Donk
Leaf spot, *Corticium areolatum* Stahel
Shoestring root rot, *Armillaria mellea* Vahl ex Fries
Mushroom root rot, *Clitocybe tabescens* (Scopoli ex Fries) Bresadola
Ganoderma root rot, *Ganoderma lucidum* (Leysser ex Fries) Karsten
Wood rots on *Citrus* spp.
Mal secco, *Deuterophoma tracheiphila* Petri
Fruit pit, *Septoria citri* Passerini
Black spot, *Phoma citricarpa* McAlpine

Bark blotch, *Ascochyta corticola* McAlpine
Ashy stem, *Macrophomina phaseolina* (Tassi) Goidanich
Bleach spot, *Pestalotia citri* Mundkur & Kheswalla
Anthracnose, *Colletotrichum gloeosporioides* Penzig
Lime anthracnose, *Gloeosporium limetticola* Clausen
Black rot, *Alternaria citri* Ellis & Pierce
Sour rot, *Geotrichum candidum* Link ex Persoon
Greasy spot, *Cercospora citri-grisea* Fisher
Tar spot, *Cercospora gigantea* Fisher
Black root rot, *Thielaviopsis basicola* (Berkeley & Broome) Ferraris
Fruit rot, *Candelospora citri* Fawcett & Klotz
Black mold, *Aspergillus niger* van Tieghem
Green mold, *Penicillium digitatum* Saccardo
Blue mold, *Penicillium italicum* Wehmer
Chocolate rot, *Trichoderma lignorum* Tode
Algal spot, *Cephaleuros virescens* Kunze
Other fungi associated with citrus

Canker, *Xanthomonas citri* (Hasse) Dows.

Symptoms

Citrus canker, a disease eradicated from Florida at a cost of several million dollars and the loss of several million trees, was introduced on imported budwood. The disease may be found on any part of the tree aboveground including leaves, fruit, twigs, and branches, and is most characteristic on the foliage and fruit as small, watery, translucent, dark green to brown spots, each showing a raised, rough surface on opposite sides of the leaf. These symptoms are accentuated as the lesion becomes older. The spots are concentrically zoned, develop a narrow, yellowish halo, and average 1/4 in. in diameter. The bacterium does not survive in the soil.

Etiology

The organism is a gram-negative rod, found in chains; it produces capsules, is motile by a single polar flagellum, and measures $1.5–2.0 \times 0.5–0.75\ \mu$. Hard agar poured plates produce straw-colored to light yellow colonies that are glistening, opaque, slightly raised, entire, and viscid. Optimum temperature, 30–34C.

References

10, 11, 12, 33, 37.

Blast, *Pseudomonas syringae* van Hall

Symptoms

This organism causes black pit of the fruit and is most frequently found associated with some injury such as a torn leaf or thorn puncture. A dark brown color develops at the injury and extends rapidly. When the base of the petiole is involved, the brown discoloration extends into the twigs; the leaves often remain attached. The black pit of fruit found mostly on lemon develops from thorn pricks; the area becomes sunken, brown to black, and usually averages less than 1/2 in. in diameter.

Etiology

The parasite is a gram-negative rod, developing singly, in pairs, or in chains; it forms capsules, is motile by 1 or several polar flagella, and measures 0.75–1.5 × 1.5–3 μ. Agar colonies are white, transparent, spreading, thin, convex, circular, and mostly smooth. A green fluorescent pigment is produced in culture. Optimum temperature, 28–30C.

References

11, 12, 21.

Bacterial spot, *Erwinia citrimaculans* (Doidge) Magrou

Symptoms

Sunken spot develops in the form of pits on the fruit. The dark-colored spots are circular, tough, and leathery at first, later becoming hard, and measure 3–10 mm in diameter. Infections probably originate from an injury. Stems are attacked around the leaf bases, and the leaf usually falls. Dark brown leaf spots often fall out.

Etiology

The organism is a gram-positive rod, produces capsules, is motile by 5–10 peritrichous flagella, and measures 1–4 × 0.4–0.7 μ. Agar colonies are yellow with dense centers, spreading, and mostly circular. Optimum temperature, 35C.

References

11, 14.

Brown rot, *Phytophthora citrophthora* (R. E. Sm. & E. H. Sm.) Leon.

Symptoms

Brown rot is a fruit decay. Fruit that has dropped or that hang within a foot or two of the soil may be infected. The diseased fruit show a slight brown discoloration and may appear somewhat water-soaked. Sporangia are readily produced in large numbers. On the larger branches, crown, and roots, a gummosis and foot rot develop. Diseased trees often show a gradual deterioration and decline, dieback, and death. Additional *Phytophthora* spp. have been associated with the disease.

Etiology

The mycelium is hyaline, nonseptate, branched, and produces hyaline, ovate to lemon-shaped sporangia, sometimes elongate or even spherical, with apical papillae. They are terminal on the conidiophore and measure 60–90 × 20–30 μ. They germinate by producing about 30 zoospores, each measuring 10–16 μ in diameter, which are motile by 2 lateral cilia. The active zoospore period is short and is followed by germination, giving rise to mycelium.

References

12, 21, 24, 41.

Foot rot, *Phytophthora parasitica* Dast.

Symptoms

Foot rot occurs on the crown and larger roots of trees in various stages of maturity. Dark spots may appear on the bark, become water-soaked, and color the outer bark surface. As the disease develops, the bark splits or shreds, becomes dry, and often curls or becomes loosened from the wood. During the early development, gumming may appear. The lesions continue to expand, and a type of dieback may appear. Usually the trunk becomes girdled and the tree finally dies.

Etiology

The mycelium is hyaline, branched, nonseptate, and up to 9 μ in diameter. The sporangia are hyaline, ovoid to slightly elongate, 25–50 $\times$ 20–40 μ, and are borne on long, slender sporangiophores. They germinate by producing about 30 zoospores, measuring 8–12 $\times$ 5–8 μ, which in turn produce mycelium. Chlamydospores are pale yellow and 20–60 μ in diameter. The oospores are hyaline, spherical, thick-walled, and 15–30 μ in diameter.

References

12, 37.

Yeast spot, *Nematospora coryli* Pegl.

Symptoms

With some exceptions, yeast spot is seldom of serious importance. There are no distinct external diagnostic symptoms. A slight depression of the rind may be evident, and some discoloration may appear. Upon removing the rind and outer albedo, reddish brown areas may be observed directly below the external depression. The tissue is often dry and fibrous and may extend into the juice sacks in the individual sections. The yeast in culture produces a heaped-up, glistening colony, entire, opaque, and granular, reminding one of wet, undissolved fine sugar.

Etiology

This yeast develops free-floating cells in nature. Very pauperate mycelium is developed sometimes. The vegetative cells are hyaline, spherical, slightly granular, and often vacuolate. They increase in number by budding and may develop short chains of cells. The ascus develops from vegetative cells and is free-floating. It is rather thick and elongate, with rounded ends. Each ascus measures 65–100 $\times$ 6–20 μ, contains, at opposite ends, two groups of 4 ascospores that are hyaline, 2-celled, and taper from the larger cell to an elongate, whiplike second cell. Ascospores measure 42–65 $\times$ 2–4 μ.

References

12, 33.

Scab, *Elsinoë fawcetti* Bitanc. & Jenkins

Symptoms

Scab is characterized by the production of wartlike lesions, on fruit twigs and stems, that are irregular in size and shape, somewhat raised above the general contour, and hard and horny as they grow older. They eventually appear on all plant parts aboveground. The leaf lesions begin as small, water-soaked, circular, somewhat translucent areas that soon develop raised excrescenses on one side of the leaf.

Etiology

The mycelium is hyaline, very scanty, septate, and short-branched. In culture, the growth is very slow and produces a piled-up rose- to purple-colored colony, well raised above the agar surface, and covered with short, erect, dense hyphae. Conidiophores are hyaline, short, compact, up to 3-celled, and $12–22 \times 3–4\ \mu$. The conidia are hyaline to dusky, reniform to oval, often develop successively, and measure $5–10 \times 2–5\ \mu$. The ascomata are scattered, circular to elliptical, and $38–106 \times 36–80\ \mu$; the epithelium is dark and thick. Asci, imbedded in stromata, are hyaline, scattered, globose to ovoid, and $12–16 \times 5\ \mu$. The ascospores are hyaline, oblong, 2–4-celled, constricted, and $10–12 \times 5\ \mu$. The imperfect stage is *Sphaceloma fawcetti.*

References

3, 12, 18, 29.

Cottony rot, *Sclerotinia sclerotiorum* (Lib.) d By.

Symptoms

This disease is particularly destructive to lemon. It forms a white, fluffy mycelial growth over the fruit under moist conditions. In drier conditions the fruit becomes tan to brown, with less mycelial growth. The black, irregular sclerotia develop on the surface of the mycelium and are up to 1 cm long and half as wide. Infection may be through wounds due to close contact in storage or shipment. The fungus also attacks foliage, twigs, and bark under favorable conditions, but it is not normally considered to be of great importance.

Etiology

The mycelium is often superficial and abundant. It is hyaline, branched, septate, and clumps together in small tufts, where it forms black, firm sclerotia which may be up to 2 cm long. These resting bodies germinate seasonally, producing 1 to several stipitate apothecia that measure 4–8 mm in diameter. The asci are cylindrical, $130 \times 10\ \mu$, and are compact in the upper area of the apothecia. The ascospores, arranged uniseriately, are hyaline, oval, 1-celled, and $9–13 \times 4–7\ \mu$; paraphyses are present.

References

12, 21.

Melanose, *Diaporthe citri* Wolf

Symptoms

Melanose is first observed as small, black, pointlike specks, circular, slightly sunken, water-soaked, and gradually becoming colored, with a yellow halo. As the spots enlarge they become dark brown, measuring up to 1/32 in. in diameter, and raised, feeling like coarse sandpaper. They appear on either leaf surface and when numerous may cause distortion and even shedding. The twigs, branches, and fruit are similarly attacked. The fruit displays, in addition, tear staining and mud-cake crusts. A stem-end soft decay is often very severe following removal of the fruit from the tree.

Etiology

The mycelium is hyaline, white in culture, branched, septate, and compact. Pycnidia are produced on dead twigs, old peduncles, and decayed fruit. They are dark, submerged becoming erumpent, ovoid, thick-walled, and 200–400 μ in diameter. The pycnidiospores are hyaline, ovate, guttulate, and 5–9 × 2–4 μ. The stylospores are produced in the same or similar pycnidia; they are hyaline, hooked, and 20–30 × 1–2 μ. The perithecia are immersed in black stromata, 125–160 μ in diameter. The long, tapering beaks are black and up to 1 mm high. The asci are clavate, sessile, and 50–55 × 9–10 μ. The ascospores are hyaline, 2-celled, constricted, guttulate, and 11–15 × 3–6 μ.

References

12, 29, 37.

Dothiorella rot, *Botryosphaeria ribis* Gross. & Dug.

Symptoms

This rot is most frequently found causing a rather dry, stem-end decay of lemon. On lemon a brown area appears around the stem end and enlarges. The fungus extends beneath the rind and through the center, causing an amber to mummy brown color. The outer texture is tough, pliable, dry, and eventually becomes olivaceous to black. On oranges, the color is cinnamon brown.

Etiology

The mycelium is hyaline to smoky, septate, branched, and mostly submerged but produces stromata from which pycnidia become erumpent. Pycnidia are black, ostiolate, flattened, globular, and produce an abundance of hyaline, fusoid, 1-celled pycnidiospores, measuring 16–25 × 4–8 μ. The perithecia develop in the same stromata and appear botryose. The asci are clavate, measuring 80–120 × 17–20 μ. The paraphyses are elongate-slender. The ascospores are hyaline, 1-celled, oval to elliptical, and 16–23 × 5–7 μ. The imperfect stage is *Dothiorella gregaria.*

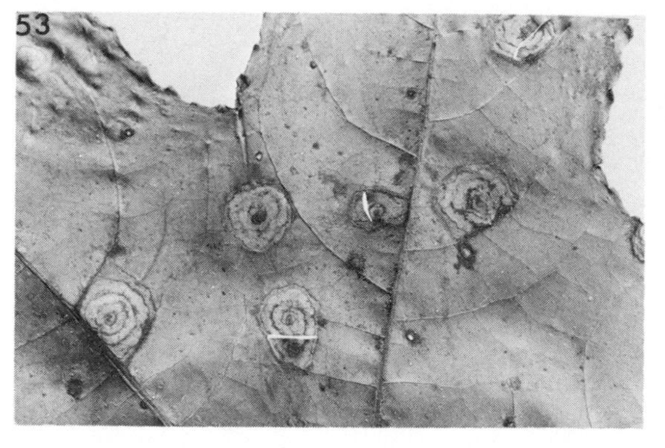

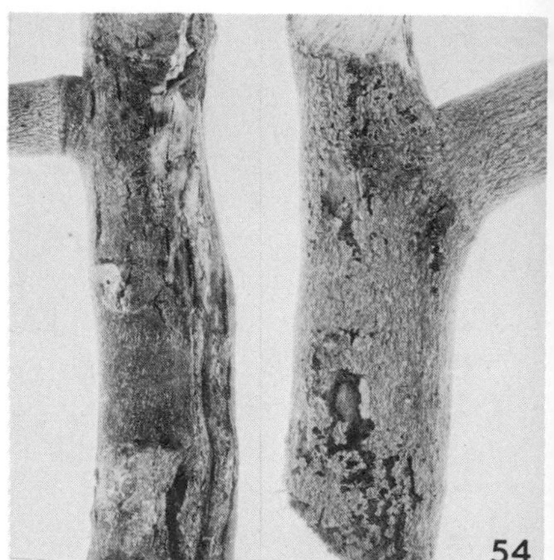

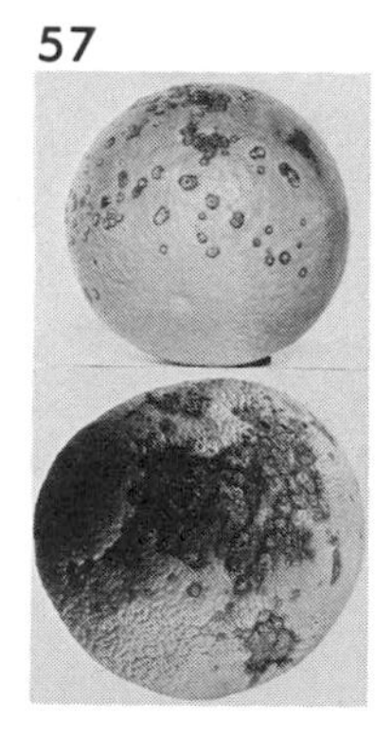

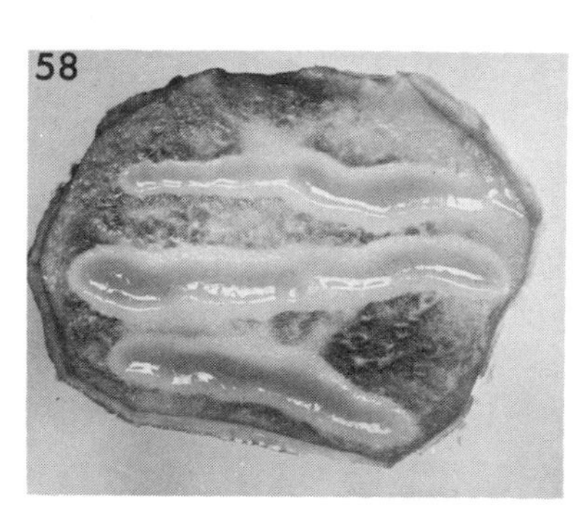

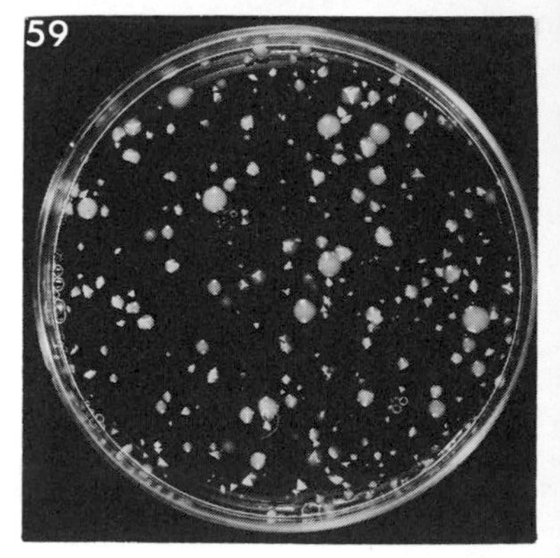

Figure 53. Leaf spot, *Alternaria ricini,* on castorbean.
Figure 54. Dothiorella rot, *Botryosphaeria ribis.* Left: On lime. Right: On orange branches.
Figure 55. Canker, *Xanthomonas citri,* on orange foliage and twigs.
Figure 56. Scab, *Elsinoë fawcetti,* on lemon. (Photograph by A. S. Rhoads.)
Figure 57. Canker, *Xanthomonas citri,* on orange and grapefruit.
Figure 58. Canker, *Xanthomonas citri,* pure culture on sterile, raw potato section.
Figure 59. Yeast spot, *Nematospora coryli,* on PDA, one week old, showing surface and submerged colonies.

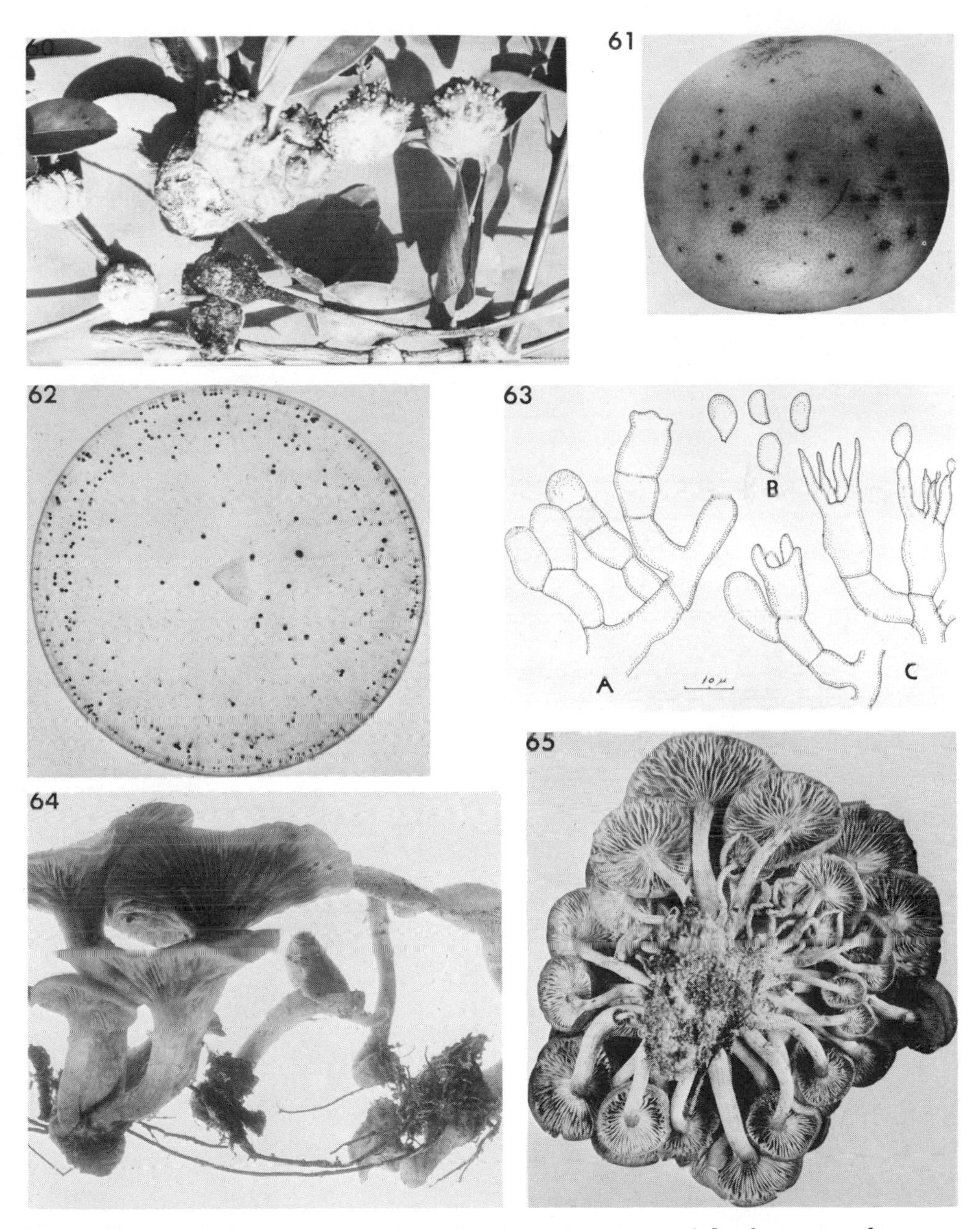

Figure 60. Knot, *Sphaeropsis tumefaciens*, showing various stages of development on lime. (Photograph by C. H. Blazquez.)

Figure 61. Black spot, *Guignardia citricarpa*, showing phoma stage on grapefruit.

Figure 62. *Botryobasidium rolfsii* (southern blight) culture showing sclerotia formation and scanty mycelium.

Figure 63. Damping-off, *Thanatephorus cucumeris*. A. Mycelium. B. Spores. C. Basidia. (Photograph by C. C. Tu.)

Figure 64. Rhizomorphs and sporophores of *Armillaria mellea* (shoestring root rot).

Figure 65. Bottom view of typical clustered nonannulated sporophores of *Clitocybe tabescens* (mushroom root rot), causing crown rot of guava. (Photograph by A. S. Rhoads.)

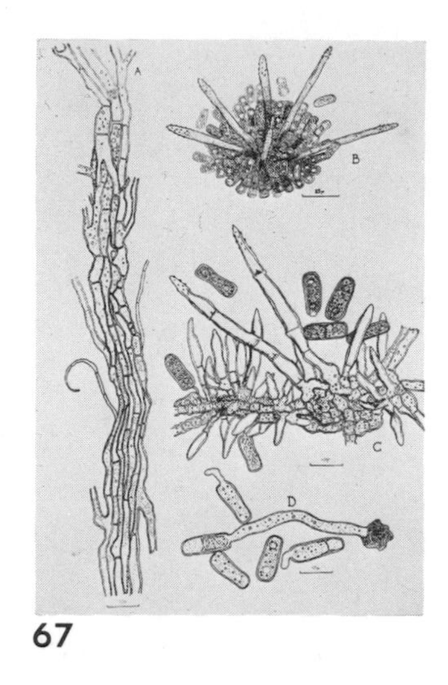

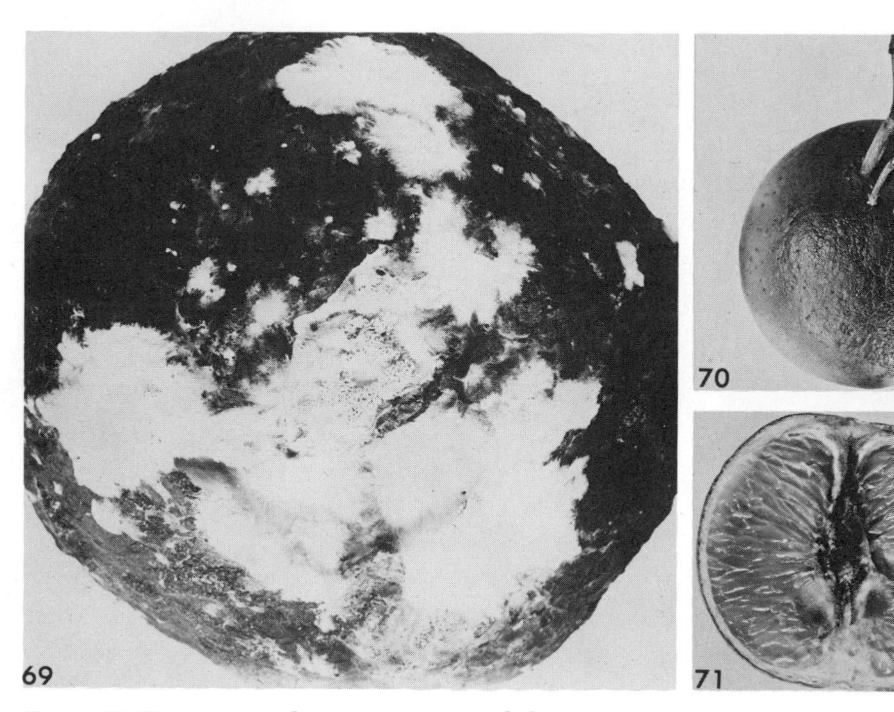

Figure 66. *Fomes* sp. on decaying orange tree bole.

Figure 67. Anthracnose, *Colletotrichum gloeosporioides*. A. Mycelium. B. Acervulus and setae. C. Setae and conidia. D. Conidia germinating. (Photograph by S. T. Doeung.)

Figure 68. Greasy spot, *Cercospora citri-grisea,* on orange foliage. (Photograph by F. Fisher.)

Figure 69. Tuckahoe from orange tree root showing sporulating *Poria cocos.*

Figure 70. Anthracnose, *Glomerella cingulata,* showing infection pattern on orange.

Figure 71. Black rot, *Alternaria citri,* on orange, showing internal decay.

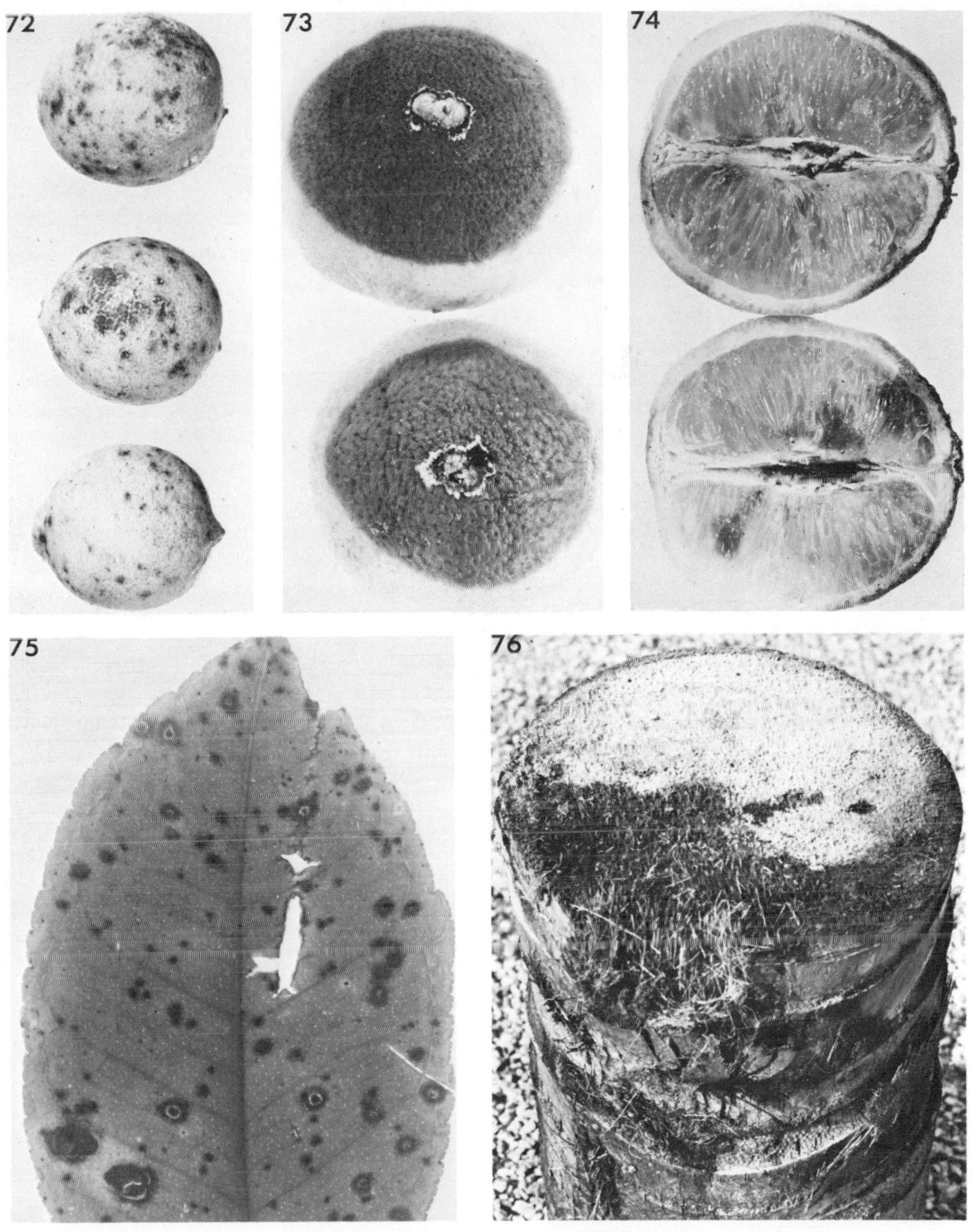

Figure 72. Tar spot, *Cercospora gigantea,* on lemon. (Photograph by F. Fisher.)
Figure 73. Chocolate rot, *Trichoderma lignorum,* on orange.
Figure 74. Chocolate rot, *Trichoderma lignorum,* on orange, showing internal decay.
Figure 75. Tar spot, *Cercospora gigantea,* on orange leaf. (Photograph by F. Fisher.)
Figure 76. Bleeding and decay of bole of coconut, resulting from mechanical injury and infection by *Endoconidiophora paradoxa.*

References

12, 21.

Diplodia rot, *Physalospora rhodina* (Berk. & Curt.) Cke.

Symptoms

Diplodia invades tissue through injuries, wounds, or weakened plants and does not infect foliage. The fungus exhibits weak pathogenic tendencies. A root rot on plants in an unfavorable environment causes dieback of the tops. The affected parts are black on the outside, and the wood is brown. Aboveground the disease is frequently associated with trunk cankers appearing at crotches and at the tips of twigs that have been injured. On the fruit, a stem-end rot develops by advancing from a twig injury or through the clipped peduncle. The fruit decay after removal from the tree is often destructive.

Etiology

The mycelium is usually plentiful and in culture is hyaline at first and gradually darkens to a grayish black. It is septate and branched. It invades host tissues and produces scattered pycnidia under the epidermis. They erupt exposing a flattened, globose, ostiolate, fruiting structure, measuring 150×180 μ. The pycnidiospores are dark-colored, oval to elliptical, 1-septate, and 24×15 μ. The spores may be hyaline and 1-celled in juvenile condition. The perithecia are black, usually in clusters, and become erumpent at maturity. The asci are hyaline, clavate, and about 100 μ long. The ascospores are hyaline, 1-celled, oval to elliptical, and $24–42 \times 7–17$ μ. The imperfect stage is *Diplodia natalensis*

References

12, 37, 40.

Knot, *Sphaeropsis tumefaciens* Hedges

Symptoms

Citrus knot may appear on trees of any age, producing irregularly spherical protuberances on the branches at various intervals, often correlated with leaf traces. Internally, the knot tissue is soft and crumbly compared to healthy tissue. The protuberances are woody in texture, bark covered, smooth or fissured, sometimes darkened, and up to 2 in. in diameter. They are mostly spherical but may be elongate, especially on lime, and are frequently associated with witches'-brooms.

Etiology

The mycelium is mostly contained in the knots and adjoining host tissue. It is hyaline to brown, multibranched, septate, and up to 4 μ in diameter. Pycnidia are brown to black, papillate, crowded, subglobose, imbedded becoming erumpent, and 200×150 μ. Usually scarce, the pycnidiospores are pale yellow, ovoid to elongate, with obtuse and acute ends, and measure $16–32 \times 6–12$ μ.

References

12, 37.

Root rot, *Ustulina zonata* Lév.

Symptoms

This disease has been reported to cause some root damage. It gains entrance to the crown and produces a black, carbonous stromata containing many perithecia and ascospores. The tree may decline over a long time and eventually die.

Etiology

The mycelium is brown, septate, branched, and produces fructifications of various sizes and shapes on the host above the soil level. They are soft, yellowish plates with concentric furrows on their exposed surfaces. They become gray with the formation of conidia and finally black, hard, and brittle; they are about 3 mm thick and several inches in length and width. The conidia are produced superficially on the surface of the stromata. They are hyaline, oval, 1-celled, and 6–8 × 2–3 μ. The perithecia are submerged with ostioles protruding, black, ovoid, and 1–5 mm in diameter. The asci are long and slender, and the ascospores are brown, oval to elongate, and 30–40 × 8–13 μ.

Reference

5.

Black spot, *Guignardia citricarpa* (McAlp.) Kiehly

Symptoms

This disease occurs in three stages. (a) Shothole spot—lesions several millimeters in diameter develop on mature fruit. As the lesions age there is little expansion, but they become deeper, craterlike, with a light-colored center upon which pycnidia develop. The rim of the crater is black surrounded by green on an orange background. (b) Freckle spot—fifty to several hundred orange to brick red specks, each about 1 mm in diameter, appear on the exposed surface of the fruit in 2–4 days. They enlarge to 3 mm, turn brown, and appear to cause no further change. (c) Virulent spot—this stage develops on mature fruit during warm weather. The lesions expand rapidly; one or several may involve most of the fruit. Spots become irregular in outline and appear to become active after a certain stage of fruit maturity is reached. Leaf lesions are very rare (they may be found on lemon) and seldom bear pycnidia.

Etiology

The inoculum originates in the fallen leaves on the soil under bearing trees and not from diseased fruit to young fruit on the same tree. Perithecia are solitary or grouped, carbonous, globose, ostiolate, and submerged to erumpent; each measures 125–135 μ in diameter. Asci are clavate, cylindrical, 8-spored, uniseriate, and 50–85 × 12–15 μ. Ascospores are hyaline, elliptical to ovate, nonseptate, 8–17 × 3–8 μ, with gelatinous polar caps. There are no paraphyses. The vegetative stage is *Phoma citricarpa.*

References

7, 19, 20, 25, 26.

Areolate leaf spot, *Leptosphaeria bondari* Bitanc. & Jenkins

Symptoms

This leaf spot appears on the foliage in the form of concentric circles or parts of circles. They extend from a central point for distances up to 10 mm with the succeeding circles 1/2 mm apart or less. The spots are almost white above and yellowish below, surrounded by a wide halo or band. The disease causes some defoliation of nursery stock plants.

Etiology

The pycnidia are lighter-colored than surrounding walls, subglobose to spherical, and ostiolate. They produce hyaline, 4-septate conidia, measuring up to 20 μ in length. The perithecia are black, spherical, subepidermal, and contain asci up to 70 μ long, bearing 3- to 5-celled ascospores which are brown and constricted at the septa.

References

2, 12.

Sooty mold, *Capnodium citri* Berk. & Desm.

Symptoms

This superficial black mold develops on the upper surface of foliage, fruit, and twigs on honeydew excreted by insects such as aphids. The fungus is not a parasite and is readily removed.

Etiology

The mycelium is brown to black, branched, septate, superficial, and densely interwoven. The individual cells of the hyphae vary from 8 to 10 μ in length to 6–10 μ in width. A papery covering of the leaf surface is produced. The conidia are cylindrical cells abscised from the hyphae. Stylospores are produced in elongated, flask-shaped structures. The pycnidia are black, about 40 μ in diameter, and form many hyaline, 1-celled, oval spores. The perithecia are black, mostly spherical, and about 80 μ in diameter. The asci each contain 8 yellowish ascospores; each ascospore is 4–6-celled with both transverse and longitudinal septations.

References

12, 37.

Fly speck, *Leptothyrium pomi* (Mont. & Fr.) Sacc.

Symptoms

Small spots have been frequently observed on the rind of fruit, but the disease causes no damage in the form of decay.

Etiology

This unimportant fungus produces black, small, circular, raised, superficial specks, averaging about 200 μ in diameter and often barely visible without magnification. The pycnidia develop in the stromata, and are 25–100 μ in diameter. The pycnidiospores are elliptic, measuring 12–14 $\times$ 2–3 μ.

Reference

24.

Sooty blotch, *Gloeodes pomigena* (Schw.) Colby

Symptoms

This is a smoky black, superficial fungus mold developing in irregular areas over the fruit. Sometimes it is dense, but it is usually thin and filmy and composed of many dark hyphal threads. It can be readily removed in the packing process. It is not important.

Etiology

The thallus is composed of black, profusely branched, septate, hyphal threads with stubby lateral branches. Pycnidia are formed in the superficial thallus. They are flattened, with oval tops, and are 76–100 μ in diameter and about half as tall. The pycnidiospores are hyaline, 2-celled, oblong to cylindrical, with pointed ends, and measure 10–12 $\times$ 2–3 μ.

References

24, 29.

Root rot, *Rosellinia bunodes* (Berk. & Br.) Sacc.

Symptoms

This uncommon root disease causes trees, especially lime, to appear in decline resulting from root depletion. White mycelium invades and kills the cortical tissue. Blackened roots become covered with ropy strands of the mycelium. The imperfect stage is *Graphium* spp.

Etiology

The mycelium is black, branched, septate, and forms strands that are closely applied to the root and eventually form a thick mat. Conidia are formed on erect, dark hyphae and are hyaline, 1-celled, ovoid, and about 5 μ long. Perithecia are black and clustered on stromata. Asci are elongate, narrow, and contain 8 fusiform ascospores, each measuring 60–70 $\times$ 8–9 μ.

Reference

12.

Felty fungus, *Septobasidium pseudopedicellatum* Burt

Symptoms

This is a frequently encountered fungus on many hosts in the tropics and is only of passing interest. A brown to gray, soft, resilient, hyphal, feltlike covering appears on twigs, branches, foliage, and fruit in rather small superficial patches or extending for 8–10 in. It usually surrounds the branch or expands over the leaf blades and fruit. The fungus is a parasite of certain sedentary, plant-parasitizing insects and could be considered somewhat symbiotic. The fungus growth is usually raised, producing cavernous spaces below, and may vary from tan to dark brown externally.

Etiology

The external hyphae are tan to brown. The mycelium is hyaline, resupinate, septate, branched, and forms matlike structures protecting, so to speak, the sedentary insects parasitizing the host. These insects are in turn the hosts of the fungus. The basidia form submerged in the hymenium, and basidiospores develop on the external surface of the felty covering. The basidiospores are hyaline, elongate, $16–23 \times 3–5\ \mu$, and upon germination infect the young, crawling insects.

References

9, 12.

Southern blight, *Botryobasidium rolfsii* (Sacc.) Venkat.

Symptoms

This disease causes damping-off and girdling of seedlings up to a year old. It attacks the young plants, girdling them at the soil line at the time of emergence and causing them to collapse. Plants that have escaped early infection may be girdled when several inches tall, resulting in killing and shredding of the cortex and the cambium. These plants usually wither, turn gray to pale brown, die, and remain standing. The white mycelium surrounds the stem and grows into the soil. Brown, spherical sclerotia are produced at the soil line.

Etiology

The mycelium is hyaline, forming white clumps on the host and in adjacent soil. It is septate and branched. No conidia are formed. The mycelium forms small, whitish projections that soon turn into small, stipitate sclerotia. The sclerotia become brown, almost spherical, and remain attached to the hyphae by slender threads. They are smooth, hard, and up to 1 mm in diameter. The basidial stage of the fungus is rarely found. The basidia are hyaline, short, barrel-shaped, and produce hyaline, 1-celled basidiospores that measure $5–10 \times 3–6\ \mu$. The imperfect stage of the fungus is *Sclerotium rolfsii.*

References

12, 29.

Pink disease, *Botryobasidium salmonicolor* (Berk. & Br.) Venkat.

Symptoms

The disease appears on the twigs and branches first as a brownish green drying of the cortex. The fungus surviving on the aerial parts of the host invades the cortex and cambium, killing the tissue. It forms a pinkish, resupinate, soft mycelial mat over the diseased area that advances in both directions along the twig or branch. The submerged mycelium invades the cambium and girdles the branch. The pinkish area often dries, develops cracks, and bleaches to a dull white color.

Etiology

The mycelium is hyaline to pink, branched, septate, and forms sporodochia along the host twig in the bark. They are orange red, scattered, usually few, and sessile. The conidia are hyaline, irregular in shape, catenulate, thin-walled, and 14–18 $\times$ 7–8 μ. The basidial stage is very conspicuous because of the extensive pink to rose, resupinate, membranous hymenophore on one side of or encircling the branch. The basidia and basidiospores are not frequently observed. The basidiospores are hyaline, 1-celled, oval to elongate, and 9–12 $\times$ 6–8 μ. The imperfect stage is *Necator decretus*.

Reference

12.

Thread blight, *Ceratobasidium stevensii* (Burt) Venkat.

Symptoms

The fungus survives in the form of brown, small, loaflike, firm sclerotia adhering to the outer cortex of the host. It is usually associated with lenticels, leaf traces, or buds. The sclerotia produce mycelium, under favorable conditions, that extends along the branches and new flush of growth, up the petioles and peduncles to the leaf blade or fruit. There it fans out invading the tissue, killing it, or marking the fruit. Ropelike strands of brown hyphae extend along the branch from sclerotia to leaf blades. When killed, this thread often prevents the leaf from falling. The disease is diagnosed by the superficial threads and the characteristic sclerotia.

Etiology

The mycelium is hyaline to light brown, septate, branched, and forms fine, silky threads over leaf blades. Sclerotia form on the leaf traces, over lenticels, and around terminal ends. The basidia are not often observed. Basidiospores are hyaline, oval, and 10–12 $\times$ 4 μ. The imperfect stage is *Hypochnus stevensii.*

References

12, 29.

Damping-off, *Thanatephorus cucumeris* (Frank) Donk

Symptoms

Citrus seedlings are killed by this disease from the time they emerge until they are almost a year old. The soil-inhabiting fungus girdles the plant at the soil line, causing it to fall over in the very young stage. Older plants usually remain standing after being girdled and killed. The lesion at the soil line exposes bare wood; the cortex and cambium are brown and shredded. The fungus produces a scanty amount of brown mycelium. Sclerotia are very rarely found on these plants but may develop on adjacent debris in the form of brown, irregularly shaped, thin structures.

Etiology

The mycelium is pale brown, septate, branched, and survives in the soil as mycelium or as sclerotia when no plants are present. The sclerotia germinate by producing infective mycelium. There are no vegetative spores; the basidia are frequently produced on stems and foliage of seedlings as a thin network of hyphae. The hyphae appear white to silvery, short, thick, and aggregated. Sterigmata develop at their tips. Basidiospores often appear in pairs; they are hyaline, 1-celled, oval to elongate, and $7\text{–}13 \times 4\text{–}7\ \mu$.

References

24, 29.

Leaf spot, *Corticium areolatum* Stahel

Symptoms

During rainy weather, the first symptoms appear as light green spots, 0.5–1 mm in size, on young, expanding leaves. Should the rainy period continue too long, the infected leaves are shed. Spots begin to enlarge during wet weather and add a new ring each day. During dry weather the expansion is topped by a gum barrier developed by the host. Each spot is usually composed of from 10 to 20 brown rings with dark brown ridges. Frequently, a whitish mildew can be seen on the lower surfaces of diseased leaves similar to a fungus growth found on the earth side of fallen leaves.

Etiology

In culture a rapidly growing light brown to gray fungus develops. Sclerotia that form in culture are brown and very loose in structure. They are not seen in nature. The mycelium produces basidia containing 4 sterigmata, each supporting a hyaline, smooth, papillate basidiospore, measuring $5 \times 8\text{–}9\ \mu$.

Reference

35.

Shoestring root rot, *Armillaria mellea* Vahl ex Fr.

Symptoms

This root rot is much more severe on a large number of plants than on citrus which is somewhat resistant. The aboveground symptom is usually a decline and failure to show healthy growth. By the time aerial symptoms are recognized, the root system may be extensively invaded and partially killed. The fungus penetrates the cortex to the wood, girdling the root and rendering it useless. As it advances, a complete girdle at the crown kills the plant.

Etiology

The fungus is recognized by the production of brown to black shoestringlike rhizomorphs that penetrate the surrounding soil and can be found under the cortex next to the wood on killed trees. It produces honey-colored mushrooms several inches high and up to 5 in. across on terete stems around each of which is a collarlike ring or annulus. The basidiospores are hyaline, 1-celled, and in dense layers appear white. No additional spore forms are produced.

References

12, 21.

Mushroom root rot, *Clitocybe tabescens* (Scop. ex Fr.) Bres.

Symptoms

This disease manifests itself as a typical root rot, causing the aboveground parts to become unhealthy, decline, and appear off-color. The fungus develops on the roots, killing them by invading the cambium and cortical zones and advancing toward the crown. Removal of the root cortex reveals a white mycelial shield or mat of the fungus. This fungus does not produce rhizomorphs other than modified white layers of mycelium in strands or plates.

Etiology

The sporophores are honey-colored, usually in clusters, each 4–8 in. tall and half as wide. There is no annulus on the terete stem. The basidiospores are hyaline, white en masse, 1-celled, oval, and $8–10 \times 4–6\ \mu$. No other spores are produced.

References

12, 29.

Ganoderma root rot, *Ganoderma lucidum* (Leyss. ex Fr.) Karst.

Symptoms

The root rot disease causes trees to decline by means of yellow, thinning foliage, some dieback, and undersized, early colored fruit. The brown and decaying roots become covered with a furry mycelial growth that clings to the adjacent soil and forms an earthy covering. The bark and wood are brown, dry, and crumbly.

Etiology

The fungus produces characteristic stipitate or shelving, diagnostic sporophores that are tan to reddish brown with a light-colored margin, a creamy white, pored hymenium, and a shiny, varnishlike covering of the pileus and stipe. The sporophores range in diameter from 2 to 5 in. and may be sessile or several inches high on sturdy stipes. Basidiospores are hyaline, 1-celled, oval to elongate, and $5–8 \times 4–5\ \mu$.

References

4, 12.

Wood rots on *Citrus* spp.

Symptoms

Various well-known hymenomycetes are usually considered cellulose and lignin destroyers but are not frequently proven to be pathogenic on healthy, growing trees. All of them could be classed as wound invaders that have gained entrance to heartwood through dead branch stobs, insect wounds, machinery in grove work, pruners, pickers, or because of detrimental temperatures, water supply, fertilizer, and fungicides.

Etiology

The following fungi have been reported to cause wood decay of citrus: *Daldinia concentrica* (Bolton) Cesati & de Notaris, *Fomes applanatus* (Fries) Gillet, *Fomes igniarius* (Linnaeus ex Fries) Kicks, *Ganoderma lucidum* (Leysser ex Fries) Karsten, *Phellinus lamaensis* (Murrill) Heim, *Polyporus fulvus* Fries, *Polyporus squamosus* (Hudson) Fries, *Polystictus versicolor* (Linnaeus) Saccardo, *Poria vaporaria* (Persoon ex Fries) Cooke, *Schizophyllum commune* Fries, *Stereum purpureum* Persoon, and *Trametes obstinatus* Cooke.

References

12, 24.

Mal secco, *Deuterophoma tracheiphila* Petri

Symptoms

This disease is frequently associated with lemon in the Mediterranean region and causes drying of foliage, wilting, and dieback. The affected wood often shows a pale red to pink discoloration. Infection may develop from twig tips or any place along the larger branches, trunk, or roots. Local line of discoloration can sometimes be traced up the tree from diseased roots.

Etiology

The fungus mycelium is hyaline, branched, septate, and very fine in texture. The pycnidia are black, subepidermal, numerous, small, spherical, and less than 100 μ in diameter. The pycnidiospores are hyaline, 1-celled, and $2.5–4 \times 1–1.5\ \mu$.

References

12, 21, 23.

Fruit pit, *Septoria citri* Pass.

Symptoms

Septoria has not been reported as common or important on the fruit and is rarely associated with foliage. Small, circular depressions, 1–2 mm in diameter, appear on the green fruit on the tree. As the color changes in ripening, those spots remain green but later become reddish brown. At first they are shallow, but some larger spots may include some of the albedo tissue.

Etiology

Pycnidia, formed in the larger pits, are black, globose, 50–80 μ, and produce conidia that are 12–20 × 2–3 μ.

References

12, 21.

Black spot, *Phoma citricarpa* McAlp.

Symptoms

This spotting, found mainly on the fruit, has been observed infrequently on foliage and twigs. The spots on the fruit are at first reddish brown, later turn black, and enlarge from 1 to 2 mm in diameter up to 10 mm. The margins are somewhat raised, but the centers gradually become sunken, often forming pits up to 2 mm deep.

Etiology

The pycnidia develop as black, minute, ostiolate pustules in the center of the pits; they contain hyaline, oval, 1-celled pycnidiospores, measuring 8–11 × 4–6 μ. The ascospore stage is *Guignardia citricarpa*.

References

12, 19.

Bark blotch, *Ascochyta corticola* McAlp.

Symptoms

The disease consists of dark brown areas with slightly raised margins appearing on the bark of branches and stems. The interior bark dies, and the margin extends until the branch is girdled. There is some gum exudate.

Etiology

Later the killed area becomes dotted with black, imbedded, papillate pycnidia. They contain hyaline, 2-celled, smooth, elliptical spores, measuring 7–9 × 2–3 μ.

Reference

12.

Ashy stem, *Macrophomina phaseolina* (Tassi) Goid.

Symptoms

This seedling disease attacks the stems of young plants and has been observed to be severe in poorly drained nurseries. The fungus invades the stems at the soil line, produces brown stem lesions, causes the cortex to shred, girdles the stem, and kills the plant.

Etiology

The mycelium is pale to brown, very fine, branched, and septate. The fungus forms black, minute, hard, spherical to irregular sclerotia that measure 32–150 $\times$22–25 μ and are imbedded in the cortex and wood. Pycnidia are seldom found. The sterile stage of this fungus is *Sclerotium bataticola.*

Reference

12.

Bleach spot, *Pestalotia citri* Mund. & Kheswalla

Symptoms

This spotting develops on the foliage, forming light-colored, irregular areas with raised margins in which the tissue is killed.

Etiology

Black, minute, dotlike acervuli appear in the center of diseased areas. The conidia are 5-celled; the end cells are hyaline, taper toward the tips, and measure 10–20$\times$5–6 μ. The central 3 cells are slate- to brown-colored, not constricted; the 3 or 4 setulae are divergent and 10–20 μ long. The pedicel is short, blunt, and about 4 μ long.

Reference

17.

Anthracnose, *Colletotrichum gloeosporioides* Penz.

Symptoms

This disease is found on all parts of the plant aboveground, causing fruit spots, tear staining, russeting, twig dieback, and leaf spots, depending largely on wounds. The dieback of twigs follows poor nutrition and cold and lightning injury; certain fruit lesions develop on sun- or spray-burned tissue. The tear staining forms from germinating conidia washed down over the fruit by rain or dew from dieback twigs or infected peduncles. The fungus remains inactive on leaves until some injury permits a place of entry followed by leaf-spot formation.

Etiology

The mycelium is hyaline, septate, branched, and often becomes dark-colored. The acervuli develop readily, producing hyaline, 1-celled, elongate conidia, measuring 10–16 × 5–7 μ, that are scattered among numerous black, stiff, septate setae that measure 60–160 μ. The ascospore stage is *Glomerella cingulata.*

References

12, 37.

Lime anthracnose, *Gloeosporium limetticola* Clausen

Symptoms

Although limited in range, this is probably the most important disease of lime. Leaves, shoots, twigs, and fruit are subject to attack when in the succulent stage and become immune when expanded and hardened. The shoots, twigs, and leaves wilt, shrivel, and die. The fruit become spotted by brown, dry, hard, raised, scattered, corky regions, rather small in area, and usually less than 1 cm in diameter.

Etiology

The acervuli are flesh-colored to salmon pink or dark, measure 50–100 μ, and appear in the sunken center of the spots. The conidiophores are 13–30 × 3–5 μ. The conidia are hyaline, 1-celled, elongate, several times as long as wide, and 12–20 × 3–6 μ. There are no setae. The sexual stage is not described.

References

12, 37.

Black rot, *Alternaria citri* Ell. & Pierce

Symptoms

Fruit affected by this fungus show premature orange coloring and usually can be readily spotted on the tree a couple of weeks ahead of normal coloring. Diseased fruit often show a tan to brown, smooth spot of variable size, up to 1 cm in diameter, at the blossom end. The inner contents reveal a distinct blackening of the core and radiations toward the perimeter. The blackened area is most intense and inclusive at the blossom end. If the fungus extends to the stem end, the fruit usually drop.

Etiology

The mycelium is dark to black, septate, branched, usually plentiful, with erect, slightly swollen tips. The conidia are formed at the hyphal tips in short chains. The conidia are obovate to oblong-elliptical, and 16–22 × 8–15 μ. They are multiseptate with 1 or more longitudinal septations. The conidia continue to increase somewhat in size and septations with maturity.

References

12, 37.

Sour rot, *Geotrichum candidum* Lk. ex Pers.

Symptoms

The decay is confined to injured fruit; it develops rapidly, producing a sour vinegar, pungent odor, and causes the tissues to soften and become watery to the extent of leaking. The color of the fruit slowly fades to a pale, water-soaked condition. The disease may develop on other fruit in close contact. The fruit finally melt down to a soft mushy condition and become attractive to small vinegar flies. The odor is so characteristic that one can usually recognize it diagnostically.

Etiology

The fungus produces aerial mycelium upon which chains of hyaline, oval to oblong, 1-celled, rectangular spores develop, measuring $6–12 \times 3–6\ \mu$.

References

12, 21.

Greasy spot, *Cercospora citri-grisea* Fisher

Symptoms

On artificially inoculated plants, water-soaked lesions develop in 2 weeks. Black areas appear in these lesions, the leaves begin to droop, and after 2 months typical greasy spots are evident, similar to those found in nature. Greasy spot usually appears on both leaf surfaces as a dark, slightly raised, semitransparent, shiny area, more or less circular, and often irregular. It is prominent on the lower side. The areas may be yellow, through shades of brown to black. Some defoliation may result.

Etiology

Stroma are dark brown to black; conidiophores are fasciculate, unbranched, olivaceous, amphigenous, and $60–80 \times 6–4\ \mu$. Conidia are olivaceous, slender, indistinctly septate, usually curved with a truncated base and subacute tips, and measure $25–200 \times 1.5–3\ \mu$.

References

15, 16.

Tar spot, *Cercospora gigantea* Fisher

Symptoms

The disease spots are dark, raised, and appear shiny or greasy. They can be distinguished from greasy spot by a narrow, mahogany red circle inside the margin of the spot. The diseased leaves do not become yellow. Tar spot is found on fruit as well as leaves and stems. It produces black, circular, raised spots on the fruit in an irregular pattern.

Etiology

Stromata are dark brown to black and $60–90\ \mu$ in diameter; the conidiophores are fasciculate, 3–6-septate, not geniculate, unbranched, amphigenous, and 90–160 μ. The conidia are olive brown, 3–12-septate, straight, and 80–180 × 6–8 μ.

Reference

15.

Black root rot, *Thielaviopsis basicola* (Berk. & Br.) Ferr.

Symptoms

Aboveground symptoms depend upon the severity of the root disease. Very light infections produce no marked effects except some stunting of growth. However, heavily infected root systems resulted in marked stunting of tops and roots and yellowing of the foliage. Chlorosis of the leaf veins, indicating malnutrition, is characteristic. Infection is mainly in fibrous roots. Heavily infected roots are sparingly branched and covered with dark lesions of various sizes and shapes. The root tips are usually free from lesions.

Etiology

The mycelium is hyaline, soon becoming gray then black, septate, and branched. On the older hyphae, brown to black, large, oval conidia form in chains. They vary somewhat in size, measuring 10–17 × 10–11 μ, and are produced acrogenously. The hyaline conidia are produced on endoconidiophores exuded from thin tips, successively. They are hyaline, rectangular becoming oval, 1-celled, and 8–12 × 3–5 μ.

References

30, 40.

Fruit rot, *Candelospora citri* Fawc. & Klotz

Symptoms

A brown, sunken, spreading, rather firm decay of the fruit is probably initiated through wounds or subnormal rind tissue. The brown area, if initiated at the blossom or stem ends, appears to spread inside the rind into the albedo, down the central core, and partially between the division of the segments.

Etiology

The fungus produces light tan to brown mycelium characterized by elongate, erect, swollen-tipped, aerial hyphae, measuring 200–300 μ in height, upon which conidiophores develop. The conidiophores are 14–23 μ long and divide at the tips into metulae. The metulae produce the rami from which sterigmata arise terminated by the production of 3-septate, cylindrical conidia with obtuse ends and measuring 43–48 × 4–5 μ.

Reference

13.

Black mold, *Aspergillus niger* v. Tiegh.

Symptoms

This mold, known as a secondary invader and wound parasite, has been found to flourish under certain highly humid conditions. The disease develops rapidly on the fruit as a soft, water-soaked decay.

Etiology

The fungus produces mycelium, conidiophores, and conidia in abundance. The spores are black, spherical to oval, 1-celled, powdery, $2–5 \times 4$ μ, and produced in long chains on the surface of swollen-tipped conidiophores.

References

12, 21.

Green mold, *Penicillium digitatum* Sacc.

Symptoms

Green mold is the most frequently observed disease of citrus fruit. The decay originates through mechanical wounds located anyplace on the fruit. The fungus develops rapidly from a mere puncture and in a few days causes a soft, colorless decay.

Etiology

The mycelium of the fungus grows over the fruit, radiating from the point of infection. After a day or two, the green spores appear on erect conidiophores rising from the mycelium. The spores are produced in chains or clusters. They are pale to green, spherical to elongate, 1-celled, and $4–7 \times 6–8$ μ.

References

12, 21, 29.

Blue mold, *Penicillium italicum* Wehmer

Symptoms

This mold is a wound parasite and may cause decay on adjacent fruit in close contact. The decay is a soft, water-soaked area originating from an infected wound. The fungus produces a white mycelium over the diseased area, which in a couple of days becomes bluish green because of the production of chains of spores.

Etiology

The conidiophores are short, arise from the mycelium, and branch several times, finally developing conidia in elongate chains. The conidia are spherical to oval, slightly greenish individually, bluish en masse, and $2–3 \times 3–5$ μ.

References

12, 29.

Chocolate rot, *Trichoderma lignorum* Tode

Symptoms

This disease is considered to be a wound parasite, which often begins as a dark brown decay where it enters fruit through the clipped stem. The decay is rather dry and leathery. A white mycelium may appear at the margins of the diseased area. When the entire fruit is brown, supporting, sporulating hyphae and many small heads of spores give the surface a dark green color.

Etiology

The mycelium is hyaline becoming dull green, septate, branched, supporting multibranched fertile tips with clusters of conidia. The spores are hyaline to green, slightly oval, and 2×3 μ. They are produced in chains or in moist heads.

References

12, 21.

Algal spot, *Cephaleuros virescens* Kunze

Symptoms

This alga forms a circular leaf spot, varying in size up to 1 cm in diameter, on the upper surface of old leaves that are shaded. The spots are at first green, become brown as they enlarge, and take on a reddish tinge when they produce the fruiting structures. The organism is more destructive on twigs and small branches where the bark becomes overgrown with a reddish brown felt, dries out, cracks, and dies. On the fruit, the spots are small, scattered, numerous, and seem to be slightly sunken.

Etiology

The algal thallus, originating from a germinating zoospore, expands centrifugally, up to 1 cm in diameter, by repeated dichotomous divisions. It becomes covered with a dense mass of orange filaments, which are the stalks of the sporangia. The sporangiophores are thick, rigid, septate, hairlike structures about 1 mm high and 20 μ wide, with swollen tips upon which rest 3–6 oval sporangia, averaging about 35 μ in length. Each sporangium produces a number of orange, round, biciliated zoospores.

References

12, 24, 29, 37.

Other fungi associated with citrus

Amazonia butleri (Sydow) Stevens
Anthina citri Sawada
Botrytis cinerea Persoon ex Fries
Candida krusei Castellani & Berkhout
Cephalosporium deformans Crandall
Cephalosporium omnivorum Crandall & Guiscafre

Chaetothyrium hawaiiensis Mendoza
Didymella citri Noack
Elsinoë australis Bitancourt & Jenkins
Exosporina fawcetti E. E. Wilson
Fomes lamaensis (Murrill) Saccardo & Trotter
Fusarium solani Martins
Hendersonula toruloidea Nattrass
Limacinia citri (Briosi & Passerini) Saccardo
Meliola citri Saccardo
Meliola citricola Sydow
Mycosphaerella aurantiorum Ruggieri
Omphalia flavida (Cooke) Maublanc & Rangel
Phyllosticta citricola Hori
Physoderma citri Childs
Poria cocos (Schweinitz) Wolf
Pythium ultimum Trow
Ramularia citrifolia Sawada
Rosellinia pepo Patouillard
Sphaceloma fawcetti scabriosa Jenkins
Sphaceloma fawcetti viscosa Jenkins
Sphaerostilbe repens Berkeley & Broome
Torula dimidiata Penzig
Verticillium albo-atrum Reinke & Berthold

References: citrus

1. Baker, R. E. D. 1935. Citrus fruit-rots in Trinidad. Trop. Agr. (Trinidad) 12: 145–52.
2. Bitancourt, A. A., and A. E. Jenkins. 1935. Areolate spot of citrus caused by Leptosphaeria bondari. Phytopathology 25: 884–86.
3. Bitancourt, A. A., and A. E. Jenkins. 1937. Sweet orange fruit scab caused by Elsinoë australis. J. Agr. Res. 54:1–18.
4. Blackford, F. W. 1944. A ganoderma root rot of citrus. Queensland J. Agr. Sci. 1: 77–81.
5. Bunting, R. H., and W. W. Dade. 1924. Gold Coast plant diseases. Waterlow: London.
6. Butler, E. J. 1918. Fungi and disease in plants. Thacker: Calcutta, India.
7. Cobb, N. A. 1897. Letters on the diseases of plants: black-spot of the orange. Agr. Gaz. N. S. Wales 8:229–31.
8. Cook, M. T. 1913. The diseases of tropical plants. Macmillan: London.
9. Couch, J. N. 1938. The genus Septobasidium. University of North Carolina Press: Chapel Hill.
10. Dopson, R. N., Jr. 1964. The eradication of citrus canker. Plant Disease Reptr. 48:30–31.
11. Elliott, C. 1951. Manual of bacterial plant pathogens. 2d ed. Chronica Botanica: Waltham, Mass.
12. Fawcett, H. S. 1936. Citrus diseases and their control. 2d ed. McGraw-Hill: N.Y.
13. Fawcett, H. S., and L. J. Klotz. 1937. A new species of Candelospora causing decay of citrus fruits. Mycologia 29: 207–15.
14. Fawcett, H. S., and H. A. Lee. 1926. Citrus diseases and their control. McGraw-Hill: N.Y.
15. Fisher, F. E. 1961. Greasy spot and tar spot of citrus in Florida. Phytopathology 51:297–303, 1070–74.
16. Fisher, F. E. 1966. Tar spot of citrus and its chemical control in Florida. Plant Disease Reptr. 50:357–59.
17. Guba, E. F. 1961. Monograph of Monochaetia and Pestalotia. Harvard University Press: Cambridge, Mass., p. 106.
18. Jenkins, A. E. 1931. Development of the citrus-scab organism, Sphacelonia fawcettii. J. Agr. Res. 42:545–58.
19. Kiehly, T. B. 1948. Preliminary studies on Guignardia citricarpa n. sp., the

acigerous stage of Phoma citricarpa McAlp., and its relation to black spot of citrus. Proc. Linnaean Soc. N. S. Wales 73:249–92.
20. Kiehly, T. B. 1950. Control and epiphytology of black spot of citrus. N. S. Wales Dept. Agr. Sci. Bull. 71:1–66.
21. Klotz, L. J. 1961. Color handbook of citrus diseases. 3d ed. California Agricultural Experiment Station Publications: Riverside.
22. Klotz, L. J., and H. S. Fawcett. 1948. Color handbook of citrus diseases. 2d ed. University of California Press: Berkeley.
23. Knorr, L. C. 1965. Serious diseases of citrus foreign to Florida. Florida Dept. Agr. Div. Plant Ind. Bull. 5:1–60.
24. Knorr, L. C., R. F. Suit, and E. P. DuCharme. 1957. Handbook of citrus diseases in Florida. Florida Univ. Agr. Expt. Sta. Bull. 587.
25. McAlpine, D. 1899. Fungus diseases of citrus trees in Australia and their treatment. Brain: Melbourne, Australia.
26. McOnie, K. C. 1964. The latent occurrence in citrus and other hosts of a Guignardia easily confused with G. citricarpa, the citrus black spot pathogen. Phytopathology 54:40–43.
27. Nowell, W. 1923. Diseases of crop-plants in the Lesser Antilles. West India Committee: London.
28. Reinking, O. A. 1921. Citrus diseases of the Philippines, southern China, Indo-China, and Siam. Philippine Agriculturist 9:121–79.
29. Rhoads, A. S., and E. F. DeBusk. 1931. Diseases of citrus in Florida. Florida Univ. Agr. Expt. Sta. Bull. 229.
30. Roger, L. 1953. Phytopathologie des pays chauds. Encyclopedie Mycologique 18: 1300–1308.
31. Rose, D. H., et al. 1943. Market diseases of fruits and vegetables: citrus and other subtropical fruits. U.S. Dept. Agr. Misc. Publ. 498:1–35.
32. Schüepp, H. 1960. Untersuchungen uber Guignardia citricarpa Kiehly, den erreger der schwarzfleckenkrankheit auf citrus. Phytopathol. Z. 40:258–71.
33. Sinclair, J. B. 1968. Eradication of citrus canker from Louisiana. Plant Disease Reptr. 52:667–70.
34. Stahel, G. 1940. Corticium areolatum, the cause of the areolate leaf spot of citrus. Phytopathology 30:119–30.
35. Stevenson, J. A. 1918. Citrus diseases of Porto Rico. Porto Rico Dept. Agr. 2: 43–123.
36. Suit, R. F. 1949. Parasitic diseases of citrus in Florida. Florida Univ. Agr. Expt. Sta. Bull. 463:1–112.
37. Tsao, P. H., and S. D. Van Gundy. 1962. Thielaviopsis basicola as a citrus root pathogen. Phytopathology 52:781–86.
38. Voorhees, R. K. 1942. Life history and taxonomy of the fungus Physalospora rhodina. Florida Univ. Agr. Expt. Sta. Bull. 371:1–91.
39. Wardlaw, C. W. 1961. Diseases of the banana and of the Manila hemp plant. Macmillan: London, pp. 289–92.
40. Weber, G. F. 1933. Occurrence and pathogenicity of Nematospora spp. in Florida. Phytopathology 23:384–88.
41. Whiteside, J. O. 1970. Factors contributing to the restricted occurrence of citrus brown rot in Florida. Plant Disease Reptr. 54:608–12.

Clove, *Syzygium aromaticum* Merrill & Perry

Dieback, *Cryptosporella eugeniae* Nutman & Roberts
Leaf spot, *Mycosphaerella caryophyllata* Bouriquet & Heim
Sudden death, *Valsa eugeniae* Nutman & Roberts
Other fungi associated with clove

Dieback, *Cryptosporella eugeniae* Nutm. & Roberts

Symptoms

The fungus is a widely distributed wound parasite that causes dieback. A reddish brown discoloration appears in bark and wood. The development of pycnidia soon after infection aids in diagnosis of the disease. Their ostioles are often exerted, producing a rough surface; sometimes long, yellowish pycnidiospore tendrils are observed.

Etiology

Pycnidia are scattered or somewhat linear, usually unilocular, with necks up to 1 mm long. Conidiophores are hyaline and unbranched. Pycnidiospores are hyaline, yellowish en masse, oval, 1-celled, and 3–5 × 2 μ. Perithecial stroma contain 2–10 perithecia that appear in linear fissures in the bark. Asci are hyaline, sessile, 8-spored, and 24–30 × 4–6 μ. Ascospores are hyaline, oval to spherical, 1-celled, and 7–8 × 3–4 μ; paraphyses are absent.

References

2, 4.

Leaf spot, *Mycosphaerella caryophyllata* Bour. & Heim

Symptoms

This important leaf spot appears on the foliage as very small areas, gradually enlarging from 4 to 10 mm, sometimes up to 16 mm in diameter; it is oval and yellow to rose with a reddish border. The upper surface shows a scattering of black, immersed, punctate, pyriform perithecia.

Etiology

The perithecia are ostiolate, lack paraphyses, and measure 58–68 × 13–17 μ. They are narrow at the base and somewhat bulbous near the middle, extending to a narrow, round-tipped terminal. The ascospores, measuring 21–27 × 3–7 μ, are 2-celled, the upper cell wider and shorter than the elongated, narrow bottom cell.

Reference

1.

Sudden death, *Valsa eugeniae* Nutm. & Roberts

Symptoms

The fungus has been constantly found associated with killed clove trees and is an apparent wound parasite. A brilliant saffron yellow stain is often manifested in the affected wood which is separated from healthy wood by a narrow, bluish gray zone. A black zone line is often present.

Etiology

The pycnidia are black, unilocular, about 250 μ in diameter and up to twice as high. The conidiophores are hyaline, branched, and 19 × 1–2 μ. The pycnidiospores are hyaline, 1-celled, bean-shaped, and 2–4 × 1 μ. The perithecial stroma is ash gray, imbedded in host tissue. The perithecia are black, grouped 4–12, globular or compressed in host tissue, with necks up to 5 mm. Asci are sessile, subclavate, indistinct, 8-spored, 12–26 × 4–6 μ, and have paraphyses. Ascospores are hyaline, 1-celled, allantoid, and 4.5 × 1–2 μ. Fruiting structures appear in the cortex and on cut surfaces soon after trees die.

References

2, 3.

Other fungi associated with clove

Capnodium brasiliensis Puttemans
Cylindrocladium quinqueseptatum Boedijn & Reitsma
Diplodia eugenioides Melsford
Gloeosporium piperatum Ellis & Everhart
Mycosphaerella vexans Massee
Peziza caryophyllata Welsford

References: clove

1. Bouriquet, G. 1946. Maladies des plantes cultivees a Madagascar. Encyclopedie Mycologique 12:237–52.
2. Nutman, F. J., and F. M. Roberts. 1953. Two new species of fungi on clove trees in the Zanzibar protectorate. Trans. Brit. Mycol. Soc. 36:229–34.
3. Nutman, F. J., and F. M. Roberts. 1954. Valsa eugeniae in relation to the sudden-death disease of the clove tree. (Eugenia aromatica). Ann. Appl. Biol. 41: 23–44.
4. Tidbury, G. E. 1949. The clove tree. C. Lockwood: London, pp. 146–48.

Coconut, *Cocos nucifera* Linnaeus

Bacterial bud rot, *Xanthomonas vasculorum* (Cobb) Dowson
Bud rot, *Phytophthora palmivora* Butler
Petiole rot, *Phytophthora parasitica* Dastur
Bleeding, *Ceratocystis paradoxa* (Dade) Moreau
Tar spot, *Catacauma torrendiella* Chaves Batista
Thread blight, *Corticium penicillatum* Petch
Stem rot, *Ganoderma lucidum* (Leysser) Karsten
Dieback, *Diplodia natalensis* Evans
Leaf spot, *Pestalotia palmarum* Cooke
Leaf rot, *Gliocladium roseum* (Link) Thom
False smut, *Graphiola cocoina* Patouillard
Leaf spot, *Exosporium durum* Saccardo
Other fungi associated with coconut

Bacterial bud rot, *Xanthomonas vasculorum* (Cobb) Dows.

Symptoms

The disease is manifested by wilting, yellowing, and drooping of the leaves which gradually turn brown. Closer examination often reveals a limited amount of gum exudate in association with cut vascular bundles. A dissection of the bud area shows a complete softening of the young tissue at the growing point, extending downward often more than a foot where contact is made with the frond bases. The youngest unfurled fronds lean over or fall out of the top of the plant. The decay usually harbors many secondary organisms, causing foul odor.

Etiology

The organism is a gram-negative rod, single, in pairs, or in chains, producing no capsules, motile by a single polar flagellum, and measuring 0.4–0.5 $\times$ 1–1.5 μ. Agar colonies are yellow and circular to spreading. Optimum temperature, 28C.

References

6, 13.

Bud rot, *Phytophthora palmivora* Butl.

Symptoms

This disease is diagnosed by the withering and dying of the heart leaf, which decays at its base, falls over, or usually can be pulled out. Successively the next younger leaf becomes weakened and dies, leaving a ring of outer, older leaves prior to the complete defoliation and death of the plant. Serious outbreaks result in many leafless, standing, dead trunks.

Etiology

The fungus produces stout, nonseptate, intercellular mycelium with fingerlike haustoria penetrating the host cells. Chlamydospores are hyaline, spherical, and 22–50 μ in diameter. Sometimes mycelium shows as a white, mildewlike web, mat, or felt on the exposed areas of the host. Oval to lemon-shaped sporangia develop under moist, humid conditions. They are hyaline, papillate, 32 $\times$ 58 μ, and germinate directly or by zoospore formation. The 15–30 zoospores each measure 8–15 μ, are biciliate, active for short periods, and germinate by producing mycelium. Oospores are hyaline, spherical, smooth, 18–41 μ in diameter, and not frequently seen.

References

1, 2, 15, 18, 19.

Petiole rot, *Phytophthora parasitica* Dast.

Symptoms

Earliest symptoms appear as yellowing and withering of the tips of the leaves on the outside of the crown. The disease gradually spreads to the adjoining leaves, and finally most of the outer leaves break down. It progresses slowly, sparing only the bud. Eventually bunches of nuts fall. Stalks of attacked leaves show dark brown, somewhat sunken spots on upper- and undersurfaces. The disease occurs in trees from 10 to 15 years of age, but the greatest number of affected trees are those 5 years of age which are beginning to bear.

Etiology

The fungus produces stout, sparingly septate, intercellular mycelium with few branches. Sporangia develop abundantly on material transferred to a moist chamber, and chlamydospores and oospores occur in culture. Sporangia are decidedly more globose than in *P. palmivora*, with narrower papilla, and show less tendency to break away with an attached stalk.

References

2, 12.

Bleeding, *Ceratocystis paradoxa* (Dade) Moreau

Symptoms

Infection is characterized by a liquid exudate that dries to a reddish brown stain on the fruit. Removal of rind reveals a patch of sodden tissue, yellow or reddish discoloration, or browning as decay progresses. A small amount of gumming is often associated with typical infection, thus the name "bleeding disease." Decay is slow, and there is no apparent effect on the health and bearing of the tree, but severe infection is followed by death.

Etiology

Hyphae are creeping and subhyaline; conidiophores are simple and septate. There are two kinds of conidia: namely, macroconidia that are catenulate, ovate, and fuscous; and microconidia that are hyaline, cylindric, and catenulate within the conidiophore. Macrospores measure $16–19 \times 10–12\ \mu$ and microspores $10–15 \times 3.5–5\ \mu$. Perithecia are brown to black, immersed, and ostiolate; hyphae are hyaline to brown, and erect. Ascospores are hyaline, ellipsoidal, curved, in a gelatinous matrix, and $7–10 \times 2–4\ \mu$.

References

2, 4, 5, 8, 14.

Tar spot, *Catacauma torrendiella* C. Batis.

Symptoms

Yellow spots that eventually become black appear on the foliage, leaflets, rachis, fruit, and peduncles and develop a stroma of variable dimensions up to 1 mm in diameter and half as thick. The spots are rough or verrucose. The leaflets lose their healthy green color, fade, and remain attached to the rachis. On the peduncles and nuts, the spots are the centers of a gummy exudate, and they become detached.

Etiology

The stromata are superficial, sometimes divided into separate parts, and may be in concentric circles. The perithecia are formed in groups, submerged. The asci are hyaline, straight or curved, and $48–62 \times 15–20\ \mu$. The ascospores are hyaline, oval, 1-celled, and $12–20 \times 6–10\ \mu$.

Reference

17.

Thread blight, *Corticium penicillatum* Petch

Symptoms

The mycelium of the fungus in the form of flattened strands grows along the lower surface of the leaf. It ascends the petiole and continues along the rachis in a superficial manner. The main strands branch at the base of each leaflet or pinnae; certain strands continue up the rachis, while others grow onto the lower surface of the leaflet where they fan out over the entire lower surface, killing tissue. Leaves are killed and become brown. Successive leaves become involved. The fungus may grow from the leaf to other parts by contact.

Etiology

The mycelium is brownish, septate, slightly branched on the petiole but profusely branched on the pinnae. The margin of the mycelium advances rapidly. Basidia develop over the lower surface of the leaflets, scattered or in groups. They are elongate to oval, $8–12 \times 4–5\ \mu$, and produce 4 basidiospores. The spores are hyaline, oval, and $4–5 \times 2–4\ \mu$.

References

2, 3.

Stem rot, *Ganoderma lucidum* (Leyss.) Karst.

Symptoms

There is drooping, yellowing, and drying of lower leaves, sometimes followed by reduction of crown size. Trees continue to bear nuts. A brownish, gummy exudate appears at the base of the trunk. Fruiting bodies often develop after the palm is cut down. In the interior of the trunk, from the base for several feet upward, there is a dark brown coloration producing a musty odor. Roots are discolored, dry, and brittle.

Etiology

The fungus produces a reddish brown, varnished or lacquerlike surface. The sporophore is usually large and thick, measuring several inches in length and width. These conks are produced annually. The hymenium is light gray, and light-colored basidiospores are produced in abundance within the poroid surface.

References

2, 9, 20.

Dieback, *Diplodia natalensis* Evans

Symptoms

The fungus is generally considered a wound parasite, developing on tissues of the coconut that have been injured or are deteriorating as a result of any environmental condition. It affects the foliage, florescence, young and maturing nuts, as well as the trunk and roots. The disease is often associated with bleeding, the dying back of the foliage, and a dry decay of the husks surrounding the nuts. The fungus can usually be observed in the pycnidial stage in the form of imbedded black points in dead tissues.

Etiology

The pycnidia contain pycnidiospores in abundance; when mature the spores are dark, 2-celled, oval to slightly elongate, and 24–30×12–15 μ. The presence of pycnidiospores is definitely diagnostic of the fungus. The sexual stage of the fungus is *Physalospora rhodina*. There are several *Diplodia* spp. reported on coconuts.

References

11, 21.

Leaf spot, *Pestalotia palmarum* Cke.

Symptoms

The fungus is usually considered a secondary invader. A general yellow color may include the leaves, while the individual leaflets have yellow, brown, or gray patches scattered over them with dead areas at their tips and along their margins. Spots are usually speckled with black pycnidia of the fungus scattered on both surfaces.

Etiology

Yellow or brown oval spots are 1 cm or more in diameter. Conidia are formed in subepidermal pustules on both sides of the leaf. Each conidium has 5 cells of which the 3 median ones are brown with thicker walls than the other 2. Terminal cells are hyaline, the superior one bearing 3 threadlike appendages up to 16 μ long and the inferior one prolonged to about 6 μ. Conidia are 16–22×5–7 μ.

References

7, 12.

Leaf rot, *Gliocladium roseum* (Lk.) Thom

Symptoms

A blackening and shriveling of the distal ends of leaflets occur in some of the inner whorls of leaves. The ends are dry and are broken off successively, bearing an abbreviated fanlike leaf. Reddish brown spots and patches appear on the tender leaves, and as they enlarge they penetrate for some distance into the interior of the shoot. A soft rot of the central shoot develops, eventually blackens, and falls away. If unchecked, each leaf becomes diseased and growth is halted. A palm is seldom killed but vitality is reduced, mostly on young trees.

Etiology

Mycelium is colorless to pink, floccose, simple or ropelike; conidiophores are perpendicular to hyphae, and 45–125 μ high. Penicilli are up to 140 μ and verticillately branched. Sterigmata and verticils bear conidia in balls. Conidia are pink, elliptical, smooth, and 5–7×3–4 μ.

Reference

11.

False smut, *Graphiola cocoina* Pat.

Symptoms

The disease is readily detected on the foliage by blackish, small, hard pustules that are slightly raised, making the leaflet surface feel like sandpaper.

Etiology

The fruiting structures do not occupy extensive surface area unless they are exceedingly numerous. At certain times each pustule produces a quantity of yellow, dusty spore masses. The spores are spherical to elongate, measuring 3–6 μ in diameter.

References

2, 16.

Leaf spot, *Exosporium durum* Sacc.

Symptoms

The spots on the foliage are more or less water-soaked at first, gradually enlarging up to a diameter of 1 cm, usually elongate–lens-shaped, light-colored in the center, surrounded by a dark band that borders on the green tissue. The disease does not usually cause extensive damage.

Etiology

The conidia are produced in small, compact sporodochia that appear in the central portion of the leaflet as dark, raised structures. The conidia are dark, septate, somewhat tapered toward the ends, thicker in the middle, and 80–100 $\times$ 6–12 μ.

References

2, 17.

Other fungi associated with coconut

Aspergillus niger Tieghem
Capnodium footii Berkeley & Desmazieres
Cercospora palmicola Spegazzini
Cytospora palmicola Berkeley & Curtis
Epicoccum coccos Stevens
Fomes noxius Corner
Gloeosporium cocophilum Wakefield
Helminthosporium incurvatum Bern
Marasmius palmivorus Sharples
Oidium coccocarpum Stevens

Phomopsis cocoes Petch
Phyllosticta coccophila Passerini
Poria ravenalae Berkeley & Broome
Ramularia eriodendri Raciborski
Rosellinia cocoes Henning
Sphaeropsis palmarum Cooke
Sphaerulina cocophila (Cooke) Arnaud

References: coconut

1. Ashby, S. F. 1929. Strains and taxonomy of Phytophthora palmivora Butler (P. faberi Maubl.). Trans. Brit. Mycol. Soc. 14:18–38.
2. Briton-Jones, H. R. 1940. The diseases of the coconut palm. Bail, Tin and Cox: London.
3. Bryce, G. 1924. Coconut thread blight, Corticium penicillatum. Dept. Agr. New Guinea. Leaflet 5, pp. 1–2. (R.A.M. 4:278–79.)
4. Copeland, E. B. 1921. The coconut. 2d ed. Macmillan: London.
5. Davidson, R. W. 1935. Fungi causing stain in logs and lumber in the southern states, including five new species. J. Agr. Res. 50:789–807.
6. Elliott, C. 1951. Manual of bacterial plant pathogens. 2d ed. Chronica Botanica: Waltham, Mass., pp. 146–49.
7. Guba, E. F. 1961. Monograph of Monochaetia and Pestalotia. Harvard University Press: Cambridge, Mass., pp. 107–71.
8. Hunt, J. 1956. Taxonomy of the genus Ceratocystis. Lloydia 19:1–58.
9. McFadden, L. A. 1959. Palm diseases. Principes 3:69–75.
10. Maramorosch, K. 1964. A survey of coconut diseases of unknown etiology. FAO: Rome, pp. 1–39.
11. Menon, K. P., and K. M. Pandalai. 1960. The coconut palm, a monograph. Indian Central Coconut Comm.: Ernakulam, South India.
12. Nowell, W. 1923. Diseases of crop-plants in the Lesser Antilles. West India Committee: London.
13. Orian, G. 1947. Bud rot of the royal palm in Mauritius. Rev. Agr. Ile Maurice 26:223–58.
14. Protacio, D. B. 1960. The Certostomella (Thielaviopsis) paradoxa leaf spot of coconut new in the Philippines. Philippine J. Agr. 25:67–77.
15. Reinking, O. A. 1923. Comparative study of Phytophthora faberi on coconut and cacao in the Philippine Islands. J. Agr. Res. 25:267–84.
16. Roger, L. 1951. Phytopathologie des pays chauds. Encyclopedie Mycologique 17: 793–96.
17. Roger, L. 1953. Phytopathologie des pays chauds. Encyclopedie Mycologique 18: 1449–50, 2201–3.
18. Seal, J. L. 1928. Coconut bud rot in Florida. Florida Univ. Agr. Expt. Sta. Bull. 199:5–87.
19. Tucker, C. M. 1926. Phytophthora bud rot of coconut palms in Porto Rico. J. Agr. Res. 32:471–98.
20. Venkatarayan, S. F. 1936. The biology of Ganoderma lucidum on areca and cocoanut palms. Phytopathology 26: 153–75.
21. Voorhees, R. K. 1942. Life history and taxonomy of the fungus Physalospora rhodina. Florida Univ. Agr. Sta. Bull. 371:1–91.

Coffee, *Coffea arabica* Linnaeus

Fruit spot, *Trachysphaera fructigena* Tabor & Bunting
Yeast spot, *Nematospora coryli* Peglion
Brown eyespot, *Mycosphaerella coffeicola* Cooke
Canker, *Ceratocystis fimbriata* Ellis & Halsted
Anthracnose, *Glomerella cingulata* (Stoneman) Spaulding & Schrenk
Root rot, *Rosellinia bunodes* (Berkeley & Broome) Saccardo
Sooty mold, *Capnodium brasiliense* Puttemans
Sooty mold, *Capnodium coffeae* Patouillard
Rust, *Hemileia vastatrix* Berkeley & Broome

Gray rust, *Hemileia coffeicola* Maublanc & Roger
Pink disease, *Botryobasidium salmonicolor* (Berkeley & Broome) Venkatarayan
Thread blight, *Ceratobasidium stevensii* (Burt) Venkatarayan
Damping-off, *Botryobasidium rolfsii* (Saccardo) Venkatarayan
Collar canker, *Helicobasidium compactum* Boedijn
Damping-off, *Thanatephorus cucumeris* (Frank) Donk
Web blight, *Rhizoctonia ramicola* Roberts & Weber
Cobweb disease, *Rhizoctonia* sp.
Shoestring root rot, *Armillaria mellea* Vahl ex Fries
Mushroom root rot, *Clitocybe tabescens* (Scopoli ex Fries) Bresadola
American leaf spot, *Mycena citricolor* (Berkeley & Curtis) Saccardo
Derrite, *Phyllosticta coffeicola* Delacroix
Brown spot, *Phoma coffeicola* Tassi
Leaf blight, *Ascochyta tarda* Stewart
Leaf spot, *Pestalotia coffeae* Zimmermann
Warty disease, *Botrytis cinerea* Persoon ex Fries var. *coffeae* Hendricks
Mold, *Sarcopodium coffearum* Nag Raj & George
Leaf spot, *Myrothecium adriena* Saccardo
Brown eyespot, *Cercospora coffeicola* Berkeley & Cooke
Zonate leaf spot, *Cephalosporium sacchari* Butler
Bark disease, *Fusarium stilboides* Wollenweber
Algal spot, *Cephaleuros virescens* Kunze
Other fungi associated with coffee

Fruit spot, *Trachysphaera fructigena* Tabor & Bunt.

Symptoms

The fungus causes spots on cherries at all ages but more frequently as they mature. The fruit become brown, shrunken, dry, and often remain attached.

Etiology

The fungus produces chlamydospores, oospores, and spherical, echinulate sporangia measuring about 35 μ in diameter. Germination is by formation of mycelium.

References

4, 37.

Yeast spot, *Nematospora coryli* Pegl.

Symptoms

The only external indication of infection is a small sunken spot caused by the feeding puncture of an insect. Upon examination of the internal parts, considerable discoloration is usually evident, accompanied by some shrinkage and poor filling if infection is early.

Etiology

The yeast is readily isolated from the discolored areas and forms a white, circular, heaped up, granular colony of hyaline, variously sized cells, spherical, or attached in small groups, and about 20 μ in diameter. The asci are hyaline, cylindrical with round ends, and $60\text{–}85 \times 10\text{–}12$ μ. They are free-floating and not too persistent. The 8 ascospores are produced in groups of 4, one group in each end of the ascus with the spore appendages overlapping in the center.

The ascospores are hyaline, 1-septate, 40–46 × 2–3 μ, with a rounded apical cell; the other cell is attenuated in a long, nonmotile appendage.

References

22, 29, 42, 45.

Brown eyespot, *Mycosphaerella coffeicola* Cke.

Symptoms

Spots are common on upper surfaces of leaves and rare on flowers or berries. On leaf blades, they are circular, isolated, nonlimited by veins, and 4–7 mm in diameter. They are brown at first, later fade, and ultimately have a white center and a halo. On fruit there appear brown to black circular spots, which become light and have no halo. Fructifications usually appear on upper surfaces of limbs and in the center of discolored areas, which contain concentric, minute, dark specks.

Etiology

The conidial stage, *Cercospora coffeicola* Berk. & Cke., is most destructive. Conidia are hyaline, very long, tapered at the end, almost filiform, multiseptate, and 75 × 3 μ. Perithecia are scant, partly imbedded in leaf tissue, and form minute black specks. Asci contain 8 ascospores which are hyaline, fusiform, bicellular, slightly constricted at the septation, and biguttulate.

References

10, 13, 22.

Canker, *Ceratocystis fimbriata* Ell. & Halst.

Symptoms

The earliest indication of disease is chlorosis of the leaves on the stems near the canker. Chlorotic leaves drop and the stem dries. Branches both above and below the original canker may be killed, and the affected tree may die within a few months or 1 year. Removal of the bark reveals a red brown to black discoloration in the wood. Cankers are of variable sizes and shapes, appearing on any woody part of the tree, but commonly in proximity to pruning wounds. On trees 5 years old, affected stems or branches are often completely girdled.

Etiology

Conidia are hyaline, oval, cylindrical, continuous, and 9–50 μ long. Chlamydospores measure 10–19 × 7–11 μ. Asci are oblong to ovate; they disintegrate early, freeing the 8 ascospores which are hyaline, 1-celled, oval to allantoid, and 4–8 × 2–6 μ. The spores are exuded in viscid droplets or in tendrils from the fimbriate ostioles.

References

11, 24, 36.

Anthracnose, *Glomerella cingulata* (Ston.) Spauld. & Schrenk

Symptoms

On leaves the spots are roundish or irregular, those at the margin often elongated. The largest spots are 1 in. long, at first brown, then gray. The acervuli are visible as small, black dots, mostly on the upper surface of the leaf. On berries the spots are more limited, and setae are well developed, causing depression or deformation of berries. Dark-colored, elongate, lesions develop on the internodal areas between fruit clusters.

Etiology

Stromata are formed by a collection of hyphae under the cuticle. At the surface, a layer of erect, almost colorless conidiophores appear. They are cylindrical, 18–20 $\times$ 4 μ, and each bears a single conidium. Characteristic stiff bristles or setae are present that are nonseptate or have 1 or 2 cross walls and are pointed at tips. Conidia are hyaline singly, but pale pink en masse, irregularly cylindrical, rounded at the ends, often somewhat curved or indented, and measure 12–18 $\times$ 4–5 μ. As they develop, they rupture the cuticle of the leaf and appear on the surface. The conidial stage is *Colletotrichum coffeanum* var. *virulans.*

References

2, 23, 26, 33.

Root rot, *Rosellinia bunodes* (Berk. & Br.) Sacc.

Symptoms

Plants appear to be dying out in patches as from drought or lack of nourishment. Leaves wither, turn brown, and drop. Some plants die so rapidly the leaves are not shed. As plants showing the first symptom die, those in close proximity begin to show the early indication of the disease. Examination of the plant crown and principal roots reveals a rough, brownish black surface to which many clusters of soil adhere in an irregular fashion. Upon loosening the soil, the dark-colored, mostly dead cortex is exposed. Imbedded in the layers of cells and in the cambium are numerous flat, dendritic mycelial mats, white or variously colored, that are easily peeled off.

Etiology

The conidia-producing areas are somewhat velvety crusts with black, short conidiophores, producing hyaline, small, 1-celled spores about 5 μ long. The perithecia are black, round, carbonaceous, ostiolate, clustered, imbedded in the velvety covering, up to 2 mm in diameter, and paraphysate. The ascospores are brown, elongate, fusiform, and 40–45 $\times$ 12 μ.

References

21, 22, 44.

Sooty mold, *Capnodium brasiliense* Putt.

Symptoms

This mold is characterized by a black, sooty, more or less dense covering on leaves and twigs of insect-infested plants and has at first the appearance of a thin coating. It appears chiefly on the upper surface of leaves, increases in density, and becomes granular and woolly. It dries up and peels off in large flakes like pieces of paper.

Etiology

The hyphae are greenish black, composed of unequal cells, constricted at the septa, and 6–10 μ in diameter or more regularly cylindrical. The pycnidial form is found on the affected leaves and twigs and consists of elongated, spindle-shaped or almost cylindrical bodies. Each ascus has 4–8 ascospores, is first hyaline then dark smoky green, and 3-septate with the second cell from the top broader than the rest. The imperfect stage is *Torula* spp.

References

5, 14.

Sooty mold, *Capnodium coffeae* Pat.

Symptoms

The mycelium is brown, branched, multicellular, superficial, and usually abundant.

Etiology

Spores are brown and several-celled. Perithecia are black and 200 μ high; asci are irregularly oval, and ascospores are 4-celled.

Reference

29.

Rust, *Hemileia vastatrix* Berk. & Br.

Symptoms

As a rule, the disease is confined to leaves and only rarely is it found on fruit and near the tips of very young branches. The first symptom is the appearance of small, yellowish spots, 1–2 mm in diameter, on the underside of leaves. Spots later become covered with an orange powder composed of spores of the fungus. There is no discoloration on the upper surface at first, but livid or greasy brownish patches eventually appear, corresponding to spots below. In later stages, the lower sides of the leaf spots turn gray or brown with a yellow margin. With severe attack, the affected leaf turns brown. Mature leaves are less readily infected than young ones.

Etiology

The fungus is found in the uredinia stage only. Mycelium develop in discolored spots, ramifying to form moderately thick, usually colorless hyphae between cells of leaf parenchyma and haustoria. Each stalk produces several echinulate uredospores, attached by short sterigmata. Stalks are less distinct with age, forming compound columns. Spores are hyaline, roundish at first, and smooth-walled; later they elongate, with the outer rough side convex and the inner smooth surface concave. Spores measure $30–40 \times 27–30\ \mu$.

References

2, 3, 7, 9, 43, 49.

Gray rust, *Hemileia coffeicola* Maubl. & Roger

Symptoms

Leaf symptoms reveal a more or less general spreading of the fungus in the leaf tissue with no definite delimited areas, but a rather general spreading from leaf tip or edges on the underside, gradually involving the entire leaf.

Etiology

Spores are produced on the lower surfaces as gray, thick dust. Severe attacks may result in complete defoliation. Spores are smaller than those of *H. vastatrix.* The disease is of minor importance.

References

17, 48.

Pink disease, *Botryobasidium salmonicolor* (Berk. & Br.) Venkat.

Symptoms

This disease is found chiefly on the twigs or smaller branches, but also on the main stem. The first noticeable symptom is the loss of green color in leaves, which wither and turn brown. The fungus appears in different forms, but the most easily recognized and destructive is the vegetative stage, which occurs as a pale rose or whitish crust, often much cracked somewhat like badly dried whitewash. At later stages, the bark splits and peels away from the wood. Fungus spreads rapidly over the surface after the bark is killed. Prevalence not only of pink disease, but also canker, is very largely influenced by the extent of the shade, high temperature, and adequate moisture.

Etiology

Fungous mycelium with a loose, basal layer of interwoven hyphae develops into a much firmer basidial layer, composed of a dense mass of parallel basidia at the surface. Individual basidia are bulb-shaped, with 4 narrow sterigmata that are $4–6\ \mu$ long. Spores are hyaline, pear-shaped, with a little apiculus at the narrow end, and measure $9–12 \times 6–7\ \mu$. The basidial stage is usually preceded by the white, nodular stage, but seems to have no relationship in time with the necator stage and often occurs without the latter.

Figure 77. Root rot, *Rosellinia bunodes,* on coffee, showing girdled root. (Photograph by L. Abrego.)

Figure 78. Damping-off, *Botryobasidium rolfsii,* of coffee seedlings. (Photograph by L. Abrego.)

Figure 79. Left: Soil fumigation produced healthy coffee seedlings. Right: Damping-off, *Thanatephorus cucumeris,* of coffee seedlings. (Photograph by L. Abrego.)

Figure 80. Damping-off, *Thanatephorus cucumeris,* at soil line of coffee seedlings. (Photograph by L. Abrego.)

Figure 81. American leaf spot, *Mycena citricolor,* on coffee. (Photograph by L. Abrego.)

Figure 82. Brown eyespot, *Cercospora coffeicola,* on coffee. (Photograph by L. Abrego.)

References

3, 5.

Thread blight, *Ceratobasidium stevensii* (Burt) Venkat.

Symptoms

The diagnostic characteristic of the disease is the brown, undulating, coarse, ropy, superficial thread associated with dying and dead foliage, usually present and destructive on year-old wood and the current flush. The fungus originates from surviving fungus sclerotia on the year-old twigs. It grows along the twigs and up leaf petioles to the leaf blade where delicate mycelial threads kill the tissue, resulting in a dead leaf that is shed but continues to remain suspended by the threads of the fungus. As the fungus growth continues along the new host, flush-leaves, flowers, and young fruit are successively involved. The berries are overgrown, turn black, die, and shell off the tree.

Etiology

The fungus produces fertile cells among the fine, silky threads that spread over the leaf blades. The basidia are mostly scattered, not formed in a definite layer, and produce small, turgid sterigmata and 4 basidiospores that are hyaline, oval to elongate, thin-walled, and $10\text{–}12 \times 4\ \mu$. Sclerotia usually associated with fungi of this genus are not formed by this fungus on coffee.

References

5, 13, 40.

Damping-off, *Botryobasidium rolfsii* (Sacc.) Venkat.

Symptoms

Coffee seedlings in seedbed formation show a typical type of damping-off at the soil line. The stems at that point show some browning but are in a state of collapse and shrivel over an area of several millimeters to 2 cm. Wilting of cotyledons occurs in late stages of the disease. Few seedlings topple over, but instead die standing erect.

Etiology

The soil-inhabiting fungus produces hyaline, branched, septate mycelium that accumulates into white matting around plant crowns and in seedbeds on seedlings. No conidia are produced. The sclerotia are usually plentiful at the soil line. They are brown, spherical, hard, and less than 1 mm in diameter. The basidiospores are hyaline, ellipsoidal to oblong, and $7\text{–}13 \times 4\text{–}7\ \mu$.

Reference

46.

Collar canker, *Helicobasidium compactum* Boed.

Symptoms

In early stages plants fail to show good growth, tend to be chlorotic, appear somewhat stunted, and produce smaller foliage. Later, the plants tend to show some dieback. The decline is mostly a slow process, except on seedlings.

Etiology

The fungus mycelium is hyaline, usually becomes colored as it matures, septate, branched, and 5–7 μ in diameter. It becomes somewhat compact in brown cushions on the main stem near the base or sometimes higher up the trunk. The cushions appear mostly as a superficial growth. The loosely semicompact stromata are more or less resupinate, usually elongate, parallel with the tree trunk, and measure up to several inches in length and much less in width. The basidia develop at the hyphal tips near the surface of the cushions or stromata and are hyaline, curved, up to 3-septate, and 28–40 × 5–8 μ. They produce basidiospores that are hyaline, 1-celled, oval to elongate, and 12–20 × 4–5 μ.

References

29, 30.

Damping-off, *Thanatephorus cucumeris* (Frank) Donk

Symptoms

Dark, chocolate brown lesions appear on the stems at the soil line. Seedlings in their youngest stages usually wilt; the upper portions bend over, and the lower stem remains erect. Cotyledon infection is manifested by dark, chocolate brown infection blotches which expand rapidly until the cotyledons are completely involved. They dry rapidly and remain attached to the stem. At some times, the basidial stage of the fungus develops on the lower surface of partially killed cotyledons, producing typical basidiospores in abundance.

Etiology

The vegetative stage of the fungus produces brown, septate, branched mycelium and brown, irregular, superficial sclerotia which survive on or away from the plant in the soil. Under favorable conditions, it produces new hyphae that are highly infectious. The basidiospores are produced on basidia formed on a loose network of hyphae, usually on plant stems or on any prominence in the soil.

References

12, 39, 46.

Web blight, *Rhizoctonia ramicola* Rob. & Weber

Symptoms

A silky thread blight, this disease causes considerable defoliation and shedding of fruit of coffee plants. Early symptoms include a yellowish green color of foliage which later loses its chlorophyll, becomes yellow spotted with brown necrotic spots, and finally becomes completely brown and is shed. The young and half-mature berries also become brown, remain stunted, and finally drop.

Etiology

The fungus is composed of very delicate, fine, silky hyphae. It grows from young stems to petioles, peduncles, and onto the leaf blades and fruit. Accumulations of hyphae often occur at the point of union of stem and petiole or peduncle, also where infected leaves touch, at which point the hyphae grow from a diseased leaf to a nondiseased one, sort of tying them together. The sexual stage is *Ceratobasidium ramicola.*

References

38, 46.

Cobweb disease, *Rhizoctonia* sp.

Symptoms

This fungus is found on twigs and leaves, forming white, thick, branching strands that spread out on the underside of the leaf into a fine cobweblike layer that can be peeled off when moist. Affected leaves lose color, turn flaccid, and fall. The disease extends to other leaves and twigs by contact, and a large part of the bush may be destroyed, the twigs dying as a result of loss of leaves but not seeming to be directly injured by the fungus.

Etiology

The superficial mycelium extends as a mat of long, straight, parallel hyphae, from which a finer network spreads over the leaf surface. Hyphae of a second kind forming on the leaf are larger, straight, and little branched. No reproductive stage is known, but specialized bodies, composed of peculiarly branched, thick-walled cells or anchor cells, form as an elongated, somewhat spindle-shaped cell on the ends of the hyphae.

References

5, 48.

Shoestring root rot, *Armillaria mellea* Vahl ex Fr.

Symptoms

There is often a sudden wilting and yellowing of foliage followed by death of the tree, or there may be a slower decline in which the tree appears unthrifty, produces little growth, and continues to survive. Cracks or cankers appear on the lower trunk or major roots, and some sap exudate may appear.

Etiology

White mycelium and often black rhizomorphs are present. Honey-colored sporophores up to 6 in. tall with a cap up to 5 in. in diameter often appear attached to the base of diseased trees. These mushrooms produce white spores on the gills, and an annulus surrounds the stipe.

Reference

17.

Mushroom root rot, *Clitocybe tabescens* (Scop. ex Fr.) Bres.

Symptoms

Plants of different ages frequently appear to be stunted, fail to put out good growth, show some yellowing of the foliage and leaf shedding, or, in severe cases, exhibit a cankerous girdling of the main trunk at or near the soil line or crown of the plant. A diseased root system accounts for the slow decline. A trunk girdle usually results in a rapid collapse.

Etiology

The fungus mycelium is white, septate, branched, and grows through the soil. The mycelium involves the cortex and with gained potential invades and kills the cambium. Mycelial mats or plates develop between the wood and cortex. Their presence is usually diagnostic. The spores are white, 1-celled, and oval; they are produced in mushroomlike sporophores that are honey-colored, stipitate, 1–8 in. high, and 2–5 in. wide. The stipe lacks an annulus.

References

6, 27.

American leaf spot, *Mycena citricolor* (Berk. & Curt.) Sacc.

Symptoms

Dark-colored, water-soaked, circular spots appear on the foliage, ranging from small, recognizable areas up to 12 mm. They may be scattered or very numerous and become lighter-colored with age on older leaves. Similar spots appear on the berries. On the twigs the lesions are elongate, and the cortex becomes rough. Usually only a few leaf spots cause the leaf to abscise. Defoliation is the most important manifestation of the disease.

Etiology

The vegetative stage appears as one to several, yellow, almost transparent, hairlike structures from 1 to 4 mm tall with capitate, oval to flattened heads, or gemmae, of the same color. The entire structure resembles short pins. The heads are loosely attached and thus readily disseminated. Under moist conditions the detached heads generate hyphae which in turn initiate infection and a new lesion within a few days. The sexual stage of the fungus, a yellow, stipitate mushroom with frail cap and gills, about 1/2 in. high, is produced infrequently on diseased, cast leaves. Basidiospores are unimportant in the spread of the disease compared to the gemmae.

References

13, 22, 48.

Derrite, *Phyllosticta coffeicola* Dela.

Symptoms

This disease is characterized by the blight of the new shoots and leaves. Leaf lesions are at the tips or along the margins where somewhat circular spots show concentric ridges. The lesions are black at first and frequently harbor a small, light-colored, central area. The affected parts change to a brownish slate color and shrink, causing malformed leaves. Fruiting structures of the fungus can be found on these spots. The apical bud of twigs is often involved and killed. The diseased area includes the tip of the stem and the region past the next two leaves at the node.

Etiology

The mycelium is hyaline, septate, and branched; it produces pycnidia that are black, subcuticular becoming superficial, spherical to oval, and 65–70 μ in diameter. The pycnidiospores are hyaline, ovoid, 1-celled, 2-guttulate, and measure 2–3×1–2 μ.

Reference

28.

Brown spot, *Phoma coffeicola* Tassi

Symptoms

This disease is mostly observed on branches and small twigs as an intermittent browning and drying out.

Etiology

There are numerous pycnidia, appearing under the cuticle of the spots and measuring 60–110 μ in diameter. The pycnidiospores are oval, 2-guttulate, and 4–5×1.5–2 μ.

Reference

29.

Leaf blight, *Ascochyta tarda* Stewart

Symptoms

The leaf lesions, more prominent on the upper surface, are tan to brown, circular to mostly irregular, frequently originate at the apex, and develop concentric rings. The margin is water-soaked, somewhat indefinite, and becomes reddish brown. A blackening, withering, and drying—a typical dieback—occurs on the new growth. The disease is more severe on young plants up to 4 years old and on those growing in deep shade.

Etiology

The pycnidia are brown, mostly spherical, appear on the upper leaf surface, and measure 70–110 μ in diameter. The pycnidiospores are hyaline, cylindrical, straight, 2-celled, and 9–14×2–3 μ.

Reference

34.

Leaf spot, *Pestalotia coffeae* Zimm.

Symptoms

This disease appears on leaves as reddish brown, round spots, 1 cm or more in diameter, concentrated on the edge of a leaf. Lesions have slightly raised borders, are concentrically zoned, and the spots are often surrounded by a limited halo. Lesions are usually on the upper surface. The later appearance of black dots arranged concentrically are fungus fructifications arising from beneath the epidermis.

Etiology

Pustules are black, subcuticular, erumpent, and bear multicellular conidia. Conidia are oblong-fusiform, 5-celled, and 25×5–$6\ \mu$. The 3 central cells are brown and slightly constricted; the terminal cells are hyaline. The apical cell has 3 hyaline setae, 15–20 μ long, and a basal cell with a short pedicel.

References

3, 16.

Warty disease, *Botrytis cinerea* Pers. ex Fr. var. *coffeae* Hendr.

Symptoms

This is principally a disease of the fruit, which become brown, rough, and verrucose and are usually covered with a gray mycelium. The seeds do not form in the berries where there is only a black, deformed mass. The peduncles are not affected, and as a result the mummy berries remain attached to the tree. The mycelium enters the young fruit through the blossom end.

Etiology

The rapidly growing mycelium is hyaline, branched, septate, 18–20 μ in diameter, and produces dense gray tufts that turn brown. The conidiophores are erect and branched, and the conidia are hyaline, 1-celled, subglobose to oval, and 10–12 μ in diameter.

References

6, 48.

Mold, *Sarcopodium coffearum* Nag Raj & George

Symptoms

Leaf spots are brown, oval to round, often zonate, delimited by a white fringe of mycelium, and up to 3 mm in diameter.

Etiology

Spore production occurs on each surface of the spots, interspersed with brown setae. Phialides are hyaline and continuous. The conidia, produced on irregularly branched conidiophores, are hyaline, 1-celled, cylindrical, guttulate, and 5–7 × 1–2 μ.

Reference

19.

Leaf spot, *Myrothecium adriena* Sacc.

Symptoms

The circular leaf spots appear soaked in texture and turn brown within a few days under favorable conditions. They are small at first, about 1 mm in diameter, but may reach a diameter of 1 in. A concentric zonation in a target-board pattern may develop on the upper surface of the spots. Black fruiting bodies appear in profusion along the rings on the lower leaf surface. These spots often coalesce, forming large, irregular blotches. Some defoliation may result. Twigs and fruit are not often attacked.

Etiology

The mycelium is hyaline, branched, and found in the host tissue. The sporodochia, arising from a stroma, are black with ciliate margins, hypophyllous, superficial, gregarious, and 160–350 μ. The conidiophores are hyaline, septate, branched, with terminal phialides which are light-colored at first but soon darken. The olivaceous black masses of conidia are produced acrogenously, and the conidia are cylindrical, smooth, 1-celled, usually 2-guttulate, and 5.6–7 × 1.4 μ.

References

18, 31.

Brown eyespot, *Cercospora coffeicola* Berk. & Cke.

Symptoms

Brown eyespot is most abundant and devastating in seedbeds and less so on bearing plants. Spots on leaves are scanty, round, 7–10 mm in diameter, and visible on both sides. Spots are brown at first, later becoming almost white in the center but retaining a reddish brown ring with a sharply defined outer margin. Some defoliation may result. On berries, black blotches cover as much as half of the surface and are most fully developed on the upper side, where the sun causes earlier ripening.

Etiology

The fungus is visible on the surface of spots as small, greenish olive tufts made up of the simple conidiophores on which are formed terminal conidia. Conidia are colorless, long, narrow, obclavate, multiseptate, and 75 × 3 μ. The sexual stage is *Mycosphaerella coffeicola.*

References

13, 22.

Zonate leaf spot, *Cephalosporium sacchari* Butl.

Symptoms

The spot is characterized by its tendency to develop concentric rings. Such rings are sometimes incomplete at the outside of the spot, and often begin as entirely separate spots which increase until united with the central mass. On the underside of older spots, a thin, white mold appears which bears the spores of the fungus.

Etiology

The spores are produced within tissues of the host plant. The conidia are borne on short hyphae within large vessels. Spores are small, elliptical or curved, formed apically, usually unicellular, and $4\text{–}9 \times 2\text{–}4\ \mu$. They are generally held together in a drop of liquid at the tip of the hyphae to form a mass.

References

5, 13.

Bark disease, *Fusarium stilboides* Wr.

Symptoms

Bark disease is most frequently observed on suckers where slightly depressed, dark brown lesions, bordered by a yellow halo, cause a definite constriction. The leaves wilt, become dark, brown, and brittle, but remain attached. Wilt may occur suddenly and death of the branch results. If the sucker extends into the main branch, a collar rot develops. The first symptom associated with the disease is a general lack of nutrition, producing a plant that appears weak. An overall yellowing is followed by wilting and finally death. Examination shows a constriction of the stem at the soil line and often a definite lesion. Usually collar rot develops very slowly over several months. Scaly bark first shows a depressed lesion on the main stem, extending longitudinally up and down the stem from the base of pruned branches. If the lesion extends down the stem to the soil line, it may result in a girdle and be recognized as collar rot. Occasionally foliage and berries become infected.

Etiology

The parasite formerly known as *Fusarium lateritium* Nees var. *longum* Wr., and also designated as *F. stilboides* Wr., has, by cross inoculation experiments, finally been proven to be the latter. The fungus sporulates on the bark and sometimes on foliage and berries in the form of pinkish compact sporodochia. Fungi isolated from each of the so-called bark diseases in different localities have produced cultures composed of mycelium and conidia that are comparable, showing no significant differences in spore size or septation. The ascospore stage of the fungus is *Gibberella stilboides.*

References

25, 32, 35.

Algal spot, *Cephaleuros virescens* Kunze

Symptoms

The parasite produces slightly raised spots or patches on the upper surface of foliage and twigs. These patches are greenish gray, velvety, usually darker than adjoining tissue, and appear in an irregular pattern. The entire surface becomes covered with reddish brown, hairlike structures, which are the sporulating sporangia of the algae and when mature discharge numerous motile spores.

Etiology

The thallus develops from the spores, which divide dichotomously and extend in various directions above or below the cuticle. Penetration may be through the cuticle, epidermis, palisade cells, and even completely through the leaf. The algal cells are orange yellow in color and derive nourishment from the host by osmosis. The 3–6 sporangia are produced on vertical stalks up to 1 mm high and 50 μ wide, with swollen tips. The sporangia are 40–50 μ in diameter.

References

5, 29.

Other fungi associated with coffee

Aschersonia coffeae Henning
Ashbya gossypii Guilliermond
Bacillus coffeicola Steyaert
Cylindrocarpon radicicola Wollenweber
Fomes lamaensis (Murrill) Saccardo & Trotter
Fomes lignosis (Klotzsch) Bresadola
Guignardia coffeicola Spegazzini
Irenina coffeae Roger
Macrophomina phaseolina (Tassi) Goidanich
Marasmius scandens Massee
Meliola coffea Hansford
Nectria coffeigena Averna-Saccá
Ophiobolus coffeae Patouillard
Peniphora coffeae Zimmermann
Phoma coffeicida Spegazzini
Sclerotium coffeicolum Stahel
Septobasidium coffeicola Henning
Stereum coffeanum Berkeley & Curtis
Ustulina vulgaris Tulasne

References: coffee

1. Bock, K. R. 1962. Dispersal of uredospores of Hemileia vastatrix under field conditions. Trans. Brit. Mycol. Soc. 45: 63–74.
2. Bock, K. R. 1963. The control of coffee berry disease in Kenya. Empire J. Exptl. Agr. 31:97–107.
3. Bouriquet, G. 1946. Les maladies des

plantes cultivees a Madagascar. Encyclopedie Mycologique 12:137–68, 175–77, 180–82.
4. Bunting, R. H., and H. A. Dade. 1924. Gold Coast plant diseases. Waterlow & Sons: London, pp. 55–57.
5. Butler, E. J. 1918. Fungi and diseases of plants. Thacker: Calcutta, India, pp. 247, 477–89, 497, 499–506.
6. Coste, R. 1955. Cafetos y cafes en el mundo. Vol. 1. Maisonneuve & Larose: Paris, pp. 251–81.
7. Cramer, P. J. S. 1957. In F. L. Thellman, ed., A review of literature in coffee research in Indonesia. Inter-American Institute of Agricultural Science: Turrialba, Costa Rica, pp. 41–49.
8. Cummins, G. B. 1959. Illustrated genera of rust fungi. Burgess Co.: Minneapolis, Minn., p. 68.
9. De Oliviera, B. 1959. Selection of coffee types resistant to the Hemileia rust. Coffee Tea Ind., pp. 78–83.
10. Echandi, E. 1959. La chasparria de los cafetos causada for el hongo Cercospora coffeicola B. & C. Turrialba 9:54–67.
11. Echandi, E., and R. H. Segall. 1956. Trunk, branch, and stem canker of coffee trees. Plant Disease Reptr. 40:916–18.
12. Exner, B. 1953. Comparative studies of four rhizoctonias occurring in Louisiana. Mycologia 45:698–719.
13. Fawcett, G. L. 1915. Fungus diseases of coffee in Porto Rico. Porto Rico Agr. Expt. Sta. Bull. 17:7–27.
14. Goncalves, G. J., and D. B. Pickel. 1940. Catalogo das bacterias e dos fungos do cafeeiro. Instituto Biológico: São Paulo, Brazil.
15. Gopalkrishnan, K. S. 1951. Notes on the morphology of the genus Hemileia. Mycologia 43:271–83.
16. Guba, E. F. 1961. Monograph of Monochaetia and Pestalotia. Harvard University Press: Cambridge, Mass., pp. 177–78.
17. Haarer, A. E. 1956. Modern coffee production. L. Hill Ltd.: London, pp. 284, 293–94.
18. Nag Raj, T. R., and K. V. George. 1958. "Target leaf spot" disease of coffee. Indian Phytopathol. 11:153–58.
19. Nag Raj, T. R., and K. V. George. 1960. A new species of Sarcopodium on coffee from India. Current Sci. (India) 29: 192–93.
20. Newhall, A. G., and F. T. Orillo. 1971. Coffee rust control experiments in the Philippines. Plant Disease Reptr. 55: 216–19.
21. Nowell, W. 1917. Rosellinia root diseases in the Lesser Antilles. West Indian Bull. 16:31–71.
22. Nowell, W. 1923. Diseases of crop-plants in the Lesser Antilles. West India Committee: London, pp. 29–32, 126–46, 226–29.
23. Nutman, F. J., and F. M. Roberts. 1960. Investigations on a disease of Coffea arabica caused by a form of Colletotrichum coffeanum Noack. Trans. Brit. Mycol. Soc. 43:489–505, 643–59; 44: 511–21.
24. Pontis, R. E. 1951. A canker disease of the coffee tree in Colombia and Venezuela. Phytopathology 41:178–84.
25. Rahman, M. U., and S. Subramaniam. 1967. A new fusarium wilt of coffee (Coffea arabica) in South India. Plant Disease Reptr. 51:758–59.
26. Raymer, R. W. 1955. Coffee berry disease. Kenya Dept. Agr. Ann. Rept., 1954, pp. 105–7.
27. Rhoads, A. S. 1945. A comparative study of two closely related root-rot fungi, Clitocybe tabescens and Armillaria mellea. Mycologia 37:741–66.
28. Rodríguez, R. A., et al. 1957. Studies on the control of "derrite" disease of coffee caused by Phyllosticta coffeicola Del. Plant Disease Reptr. 41:560–63.
29. Roger, L. 1951. Phytopathologie des pays chauds. Encyclopedie Mycologique 17:908–11; 1953, 18:1150–59, 1621–25, 1674–75, 1910–12; 1954, 19:2260–65.
30. Schieber, E., and G. A. Zentmyer. 1967. Collar canker of coffee trees in Guatemala. Plant Disease Reptr. 51:267–69.
31. Schieber, E., and G. A. Zentmyer. 1968. Myrothecium stem necrosis and leaf spot: an important coffee disease in Guatemala. Plant Disease Reptr. 52: 115–17.
32. Siddiqi, M. A., and D. C. M. Corbett. 1963. Coffee bark diseases in Nyasaland. Trans. Brit. Mycol. Soc. 46:91–101.
33. Small, W. 1926. On the occurrences of a species of Colletotrichum. Trans. Brit. Mycol. Soc. 11:112–37.
34. Stewart, R. B. 1957. Leaf blight and stem dieback of coffee caused by an undescribed species of Ascochyta. Mycologia 49:430–33.
35. Storey, H. H. 1932. A bark disease of coffee in East Africa. Ann. Appl. Biol. 19:173–84.
36. Szkolnik, M. 1951. Coffee trunk and stem canker in Guatemala. Plant Disease Reptr. 35:500–501.
37. Tabor, R. J., and R. H. Bunting. 1923. On a disease of cocoa and coffee fruits

caused by a fungus hitherto undescribed. Ann. Botany (London) 37: 153–57.
38. Tu, C. C., D. A. Roberts, and J. W. Kimbrough. 1969. Hyphal fusion, nuclear condition, and perfect stages of three species of Rhizoctonia. Mycologia 61:775–83.
39. Valdez, R. B., and J. R. Acedo. 1963. An evaluation of fungicides for the control of damping-off of coffee seedlings. Plant Disease Reptr. 47:176–79.
40. Venkatarayan, S. V. 1949. The validity of the name Pellicularia koleroga Cooke. Indian Phytopathol. 2:186–89.
41. Villares, J. D. 1927. O cafe sua producao e exportacao. Instituto de Cafe do Estado de São Paulo: Brazil.
42. Wallace, G. B. 1931. A coffee bean disease. Trop. Agr. (Trinidad) 8:14–17.
43. Wallis, J. A. N., and I. D. Firman. 1962. Spraying arabica coffee for the control of leaf rust. E. African Agr. Forestry J. 28:89–104.
44. Waterston, J. M. 1941. Observations on the parasitism of Rosellinia pepo Pat. Trop. Agr. (Trinidad) 18:174–84.
45. Weber, G. F. 1933. Occurrence and pathogenicity of Nematospora spp. in Florida. Phytopathology 23:384–88.
46. Weber, G. F., and L. Abrego. 1958. Damping-off and thread blights of coffee in El Salvador. Plant Disease Reptr. 42:1378–81.
47. Wellman, F. L. 1950. Dissemination of omphalia leafspot of coffee. Vol. 1. Inter-Am. Inst. Agr. Sci.: Turrialba, Costa Rica, pp. 12–27.
48. Wellman, F. L. 1961. Coffee: botany, cultivation, and utilization. Interscience: N.Y., p. 264.
49. Wellman, F. L. 1970. The rust Hemileia vastatrix now firmly established on coffee in Brazil. Plant Disease Reptr. 54:539–41. (Abstr.)
50. Wheeler, B. E. J. 1969. An introduction to plant diseases. John Wiley & Sons: London, pp. 122–23.

Cola, *Cola acuminata* Schott & Endlicher

Sooty mold, *Meliola pterospermi* F. L. Stevens
White thread, *Marasmius scandens* Massee
Brown thread blight, *Marasmius trichorrhizus* Spegazzini
Other fungi associated with cola

Sooty mold, *Meliola pterospermi* F. L. Stevens

Symptoms

This fungus produces a black mycelial growth on the upper surfaces of foliage, petioles, and stems, where there is an accumulation of honeydew produced by certain insects.

Etiology

The mycelium is black, superficial, septate, branched, and develops in circular mats up to 10 mm in diameter. The perithecia are flat to globular, and the ascospores are brown, 4-septate, restricted, and $30–41 \times 14–18\ \mu$.

Reference

1.

White thread, *Marasmius scandens* Mass.

Symptoms

The disease is indicated by portions of leaves or entire leaves killed successively along the branches of the tree. On the lower surface of the twigs there is a dirty white, irregularly undulating, superficial thread composed of many hyphal strands, growing from the larger branches onto the smaller ones where the foliage is attacked; the hyphae fan out over the leaf blade from its attachment to the petiole or from contact with an adjacent diseased leaf or stem.

Etiology

The vegetative thallus is up to 1 mm in diameter and consists of whitish threads when young and somewhat darker threads when older. Along these threads on the foliage there are produced many creamy white, small, delicate, mushroomlike sporophores about 3–4 mm in diameter, without a stipe attached directly to the mycelium. There are 3–5 lamellae, producing basidia and hyaline, ellipsoidal basidiospores that measure $6–8 \times 4$ μ.

Reference

1.

Brown thread blight, *Marasmius trichorrhizus* Speg.

Symptoms

The mycelium forms into rhizomorphic filaments that are dark brown to blackish and grow along the exterior of twigs. They are attached randomly to the different organs of the host by light brown to chestnut clamp protuberances. The mycelium very slightly penetrates the leaves, fruit, and young stems.

Etiology

The sporophores are small caplike mushrooms, 4–8 mm in diameter, light-colored at first, becoming brown. There are 5–8 creamy white lamellae, widely spaced, and thin. The basidiospores are hyaline, oval to pointed at one end, and $7–11 \times 4–6$ μ.

Reference

1.

Other fungi associated with cola

Armillaria mellea Vahl ex Fries
Botryobasidium salmonicolor (Berkeley & Broome) Venkatarayan
Botryodiplodia theobromae Patouillard
Marasmius byssicola Petch
Phyllosticta colae Werwoerd & Duplessis

Reference: cola

1. Roger, L. 1951. Phytopathologie des pays chauds. Encyclopedie Mycologique. 17: 1091–93; 1953, 18:1087–89, 1646–50.

Corn, *Zea mays* Linnaeus

Bacterial wilt, *Xanthomonas stewartii* (E. F. Smith) Dowson
Stalk rot, *Erwinia dissolvens* (Rosen) Burkholder
Stalk rot, *Pseudomonas lapsa* (Ark) Burkholder
Leaf blight, *Pseudomonas alboprecipitans* Rosen
Bacterial stripe, *Pseudomonas andropogoni* (E. F. Smith) Stapp
Brown spot, *Physoderma maydis* Miyabe
Pythium stalk rot, *Pythium aphanidermatum* (Edson) Fitzpatrick
Pythium root rot, *Pythium debaryanum* Hesse
Crazy top, *Sclerophthora macrospora* (Saccardo) Thirumalachar, Shaw, & Narasimhan
Downy mildew, *Sclerophthora rayssiae* var. *zeae* Payak & Renfro
Downy mildew, *Sclerospora graminicola* (Saccardo) Schroeter
Downy mildew, *Sclerospora philippinensis* Weston
Seedling blight, *Rhizopus nigricans* Ehrenberg
Ergot, *Claviceps gigantea* Fuentes, Isla, Ullstrup, & Rodriguez
Gray ear rot, *Physalospora zeae* Stout
Stalk rot, *Physalospora zeicola* Ellis & Everhart
Ear rot, *Gibberella fujikuroi* (Sawada) Wollenweber
Root rot, *Gibberella zeae* (Schweinitz) Petch
Stalk rot, *Rhopographus zeae* Patouillard
Tar spot, *Phyllachora maydis* Maublanc
False smut, *Ustilaginoidea virens* (Cooke) Takahashi
Head smut, *Sphacelotheca reiliana* (Kuehn) Clinton
Smut, *Ustilago maydis* (De Candolle) Corda
Rust, *Puccinia sorghi* Schweinitz
Southern rust, *Puccinia polysora* Underwood
Tropical rust, *Physopella zeae* (Mains) Cummins & Ramachar
Southern blight, *Botryobasidium rolfsii* (Saccardo) Venkatarayan
Damping-off, *Thanatephorus cucumeris* (Frank) Donk
Banded disease, *Thanatephorus sasakii* (Shirai) Donk
Tuckahoe, *Poria cocos* (Schweinitz) Wolf
Dry rot, *Diplodia zeae* (Schweinitz) Léveillé
Dry rot, *Diplodia macrospora* Earle
Anthracnose, *Colletotrichum graminicola* (Cesati) G. W. Wilson
Charcoal rot, *Macrophomina phaseolina* (Tassi) Goidanich
Ear mold, *Aspergillus flavus* Link ex Fries
Kernel rot, *Penicillium oxalicum* Currie & Thom
Cob rot, *Nigrospora oryzae* (Berkeley & Broome) Petch
Black bundle, *Cephalosporium acremonium* Corda
Zonate leaf spot, *Gloeocercospora sorghi* D. Bain & Edgerton
Leaf spot, *Helminthosporium carbonum* Ullstrup
Southern leaf spot, *Helminthosporium maydis* Nisikato & Miyake
Root rot, *Helminthosporium pedicellatum* Henry
Northern leaf spot, *Helminthosporium turcicum* Passerini
Gray ear rot, *Macrophoma zeae* Tehon & Daniels
Leaf spot, *Cercospora zea-maydis* Tehon & Daniels
Sclerotial rot, *Rhizoctonia zeae* Voorhees
Other fungi associated with corn

Bacterial wilt, *Xanthomonas stewartii* (E. F. Sm.) Dows.

Symptoms

This is a common disease of corn east of the hundredth meridian from the Gulf States to the Great Lakes in the United States. It survives from season to season in hibernating flea beetles. Its appearance is seasonal. Diseased plants frequently wilt and die. Surviving plants are usually stunted. Leaf symptoms often consist of long, irregular, pale green to yellow streaks. Sometimes entire leaves die and dry up. The vascular bundles are filled with yellow bacteria.

Etiology

The organism is a gram-negative, nonmotile rod, forming capsules and short chains; it measures 1–2×0.5–0.8 μ. Agar colonies are yellow, circular, flat, small, and slow growing. Optimum temperature, 30C.

References

9, 13, 22.

Stalk rot, *Erwinia dissolvens* (Rosen) Burkh.

Symptoms

This rot is not frequently found and is favored by an oversupply of moisture. The stalk and sheaths become water-soaked and collapse at the point of infection, usually at the soil line and up to 10–12 in. above. The stalk tissues decay in a wet rot, showing only slight brown discoloration.

Etiology

The organism is a gram-negative rod, in pairs or sometimes chains, forms capsules, is nonmotile or motile by a single flagellum, and measures 0.7–1.2× 0.5–0.9 μ. Agar colonies are white, glistening, opaque, circular, entire, and have a putrid odor. Optimum temperature, 30–32C.

References

8, 23.

Stalk rot, *Pseudomonas lapsa* (Ark) Burkh.

Symptoms

This rot develops as a soft decay of the stalk near the soil line; it shows little or no color other than a water-soaked green. Severely diseased stalks fall over. It is not frequently found.

Etiology

The organism is a gram-negative rod, motile by 1–4 polar flagella, and measures 1.55×0.56 μ. It produces a green fluorescent pigment in culture.

Reference

9.

Leaf blight, *Pseudomonas alboprecipitans* Rosen

Symptoms

Blight appears in the form of dark green to water-soaked, small, elliptical spots that may develop into long, narrow stripes on the leaves. These may coalesce, involving most of the leaf and resulting in some shredding. Badly diseased leaves become yellow, tan, and dry, dying in place. Stalk lesions occur on the parts above the node where the ear is attached, causing a soft, dark, decay which kills the upper portion.

Etiology

The parasite is a gram-negative rod, motile, single or in pairs, producing capsules, and 1.8×0.6 μ. Colonies are white, entire, smooth, raised, and sticky. Optimum temperature, 30–35C.

References

9, 27.

Bacterial stripe, *Pseudomonas andropogoni* (E. F. Sm.) Stapp

Symptoms

Stripe is recognized by the olive brown to red color of the long, narrow, water-soaked lesions between the veins on the leaves and originates from leaf spots resulting from stomatal infections. The disease is favored by warm, wet weather and first appears on the lower leaves, spreading upward.

Etiology

The organism is a gram-negative rod, forming capsules, motile by 1 or more bipolar flagella, and measuring 1.3–2.5×0.4–0.8 μ.

References

9, 13.

Brown spot, *Physoderma maydis* Miy.

Symptoms

This disease causes considerable losses in the southern United States and in warm, moist, tropical regions. On the leaf blades it takes the form of yellowish, small, slightly water-soaked spots about 1 mm in diameter, turning brown, and often exhibiting successive bands across the blade. Dark, chocolate brown areas develop on the midrib, the sheath, and often on the stalk. When dry, the epidermis of the sheath ruptures, freeing the abundant brown, powdery resting spores.

Etiology

The fungus produces no mycelium. However, there are infective hyphae emerging from the germinating zoospores and the rudiments of hyphae in infected host cells. This is a sort of connective hyphae within a host cell. The sporangia, produced abundantly, are brown, smooth, with a flattened side, and 20–30×18–24 μ. The sporangium lid opens upon germination, freeing hyaline, uniciliate zoospores that measure 5–7×3–4 μ. The fusion of gametes initiates the resting spores which are usually plentiful in old, mature lesions. These resting spores are brown, thick-walled, spherical, and survive in the soil between corn plantings. They initiate infection the following year.

References

5, 7, 15.

Pythium stalk rot, *Pythium aphanidermatum* (Edson) Fitzp.

Symptoms

This rot usually produces stalk lesions slightly above the soil line in warm, humid areas, causing the plant to fall over. When severe, it results in extensive lodging of plants. The lesion is usually limited to an internodal area that becomes brownish, soft, and water-soaked.

Etiology

The mycelium is hyaline, branched, and nonseptate. It forms lobulate, inflated, branched sporangia that germinate by zoospore production. The oogonia are spherical and usually terminal, measuring 27×22 μ. The oospores are intercalary and 12–20 μ in diameter. The fungus is strictly soil-inhabiting.

References

5, 8.

Pythium root rot, *Pythium debaryanum* Hesse

Symptoms

This pythium is associated with a typical root decay and damping-off. The small feeder roots become brown and functionless, and as the decay continues they are destroyed, causing lodging of the plant. Aboveground infections rarely occur. The fungus is favored by cool, wet soil.

Etiology

The mycelium is hyaline, branched, nonseptate, and about 5 μ in diameter. Sporangia are oval. Conidia are numerous and similar to the sporangia. Oogonia are smooth, spherical, and 15–25 μ in diameter. Oospores are smooth, 10–18 μ in diameter, and germinate directly. There are 1–6 antheridia to an oogonium.

References

5, 19.

Crazy top, *Sclerophthora macrospora* (Sacc.) Thirum., Shaw, & Naras.

Symptoms

Crazy top deals mostly with the malformation of bud proliferation of the aboveground parts centered in the growing tip and florescence. The plants are stunted, show a shortening of the internodes, and produce excessive tillers. The leaves are thicker than normal and appear straplike. The tassel is often composed of layer after layer of suppressed leaves, forming a dusterlike brush composed entirely of vegetative tissue.

Etiology

The mycelium is hyaline, nonseptate, branched, and almost entirely submerged. The sporangiophores emerge through the stomata and bear hyaline to granular, oval, papillate sporangia that measure 60–100×43–64 μ. They germinate by freeing a dozen or more biflagellate zoospores. The oospores germinate in place in the host tissue by giving rise to sporangia similar to the asexual sporangia. Oospores are 60–65 μ in diameter.

References

5, 11, 26, 27, 28.

Downy mildew, *Sclerophthora rayssiae* var. *zeae* Payak & Renfro

Symptoms

This form of mildew, found in India, is characterized by a severe leaf striping. The stripes are yellowish to variable in color, narrow, 3–7 mm wide, elongate, and parallel with each other and with the leaf veins by which they are delineated. The maturing stripes become reddish to purple and extend laterally. Reddish to rusty-colored blotches involve large leaf areas that die prematurely, frequently resulting in early death of the plant. No shedding takes place, only the leaves show the disease, and no malformation occurs.

Etiology

Sporangiophores and sporangia are produced on both surfaces of the leaves. The sporangia are hyaline, 1-celled, ovate to obclavate or cylindrical, smooth, with persistent peduncles, and measure 29–67×18–26 μ. They germinate by the production of 4–8 zoospores that are hyaline, spherical, and 7–11 μ in diameter.

Reference

21.

Downy mildew, *Sclerospora graminicola* (Sacc.) Schroet.

Symptoms

The grass mildew has a wide host range on millets and on many grass weeds, but is rarely found on corn. The fungus causes a local and systemic disease. The local spots on blooming plants elongate, are delimited in width by the veins, and appear scattered over the blade. In systemic infections, the tips or a large portion of the leaf blades lose their chlorophyll and become yellowish white streaked with green. The color is often associated with seedling infection, half grown stunted plants, and the mature plants that also often display a frizzled terminal. The tissue disintegration frees the resting spores of the fungus.

Etiology

The mycelium is submerged in the host tissue; it is hyaline, branched, and nonseptate. The conidiophores protrude through the stomatal openings. They are short and stubby with thick stalks and many terminal branches, each supporting a conidium or sporangium. The sporangia are hyaline to granular, oval to elliptical, papillate, smooth, and 14–23 × 11–17 μ. Zoospores in small numbers are produced upon germination of the sporangia. They are biciliate, bean-shaped, and 10 × 5 μ. Oospores are produced in the parenchyma tissue between the vascular bundles. They are brown, thick-walled, spherical, smooth, and 30–60 μ in diameter. They remain viable for more than a year.

References

5, 26, 27.

Downy mildew, *Sclerospora philippinensis* Weston

Symptoms

This mildew forms whitish yellow, linear to irregular spots that may involve a portion or most of an individual plant. There are several additional fungi of this genus found on gramineous hosts in various parts of the tropics. They produce very similar symptoms, but are basically distinguished by geographical distribution, morphology relating to conidiophores, conidia, and oospores, measurements, germination, and host relations.

Etiology

The conidiophores emerge from basal cells through the stomata. They are supported by hyaline, branched, nonseptate, submerged mycelium and are erect, 50–200 μ tall, and often dichotomously branched at the tips. The conidia are hyaline, oval to elliptical, borne on long sterigmata, and measure 17–57 × 11–27 μ. They germinate directly. The oospore stage is unknown.

Reference

5.

Seedling blight, *Rhizopus nigricans* Ehr.

Symptoms

Blight develops under highly favorable conditions, for the fungus, that prevent germination and rapid growth of the host. Seed often decay in the soil before emerging, and the emerged plants are often seriously weakened and develop poorly should they survive the detrimental effect of the fungus.

Etiology

The mycelium is hyaline, branched, and usually luxuriant. It supports, through rhizoids, sporangiophores that are hyaline, erect, sometimes branched, and surmounted by a sporangium producing many dark-colored conidia that are striate and measure 14 × 11 μ. Zygospores are 150–200 μ in diameter.

Reference

16.

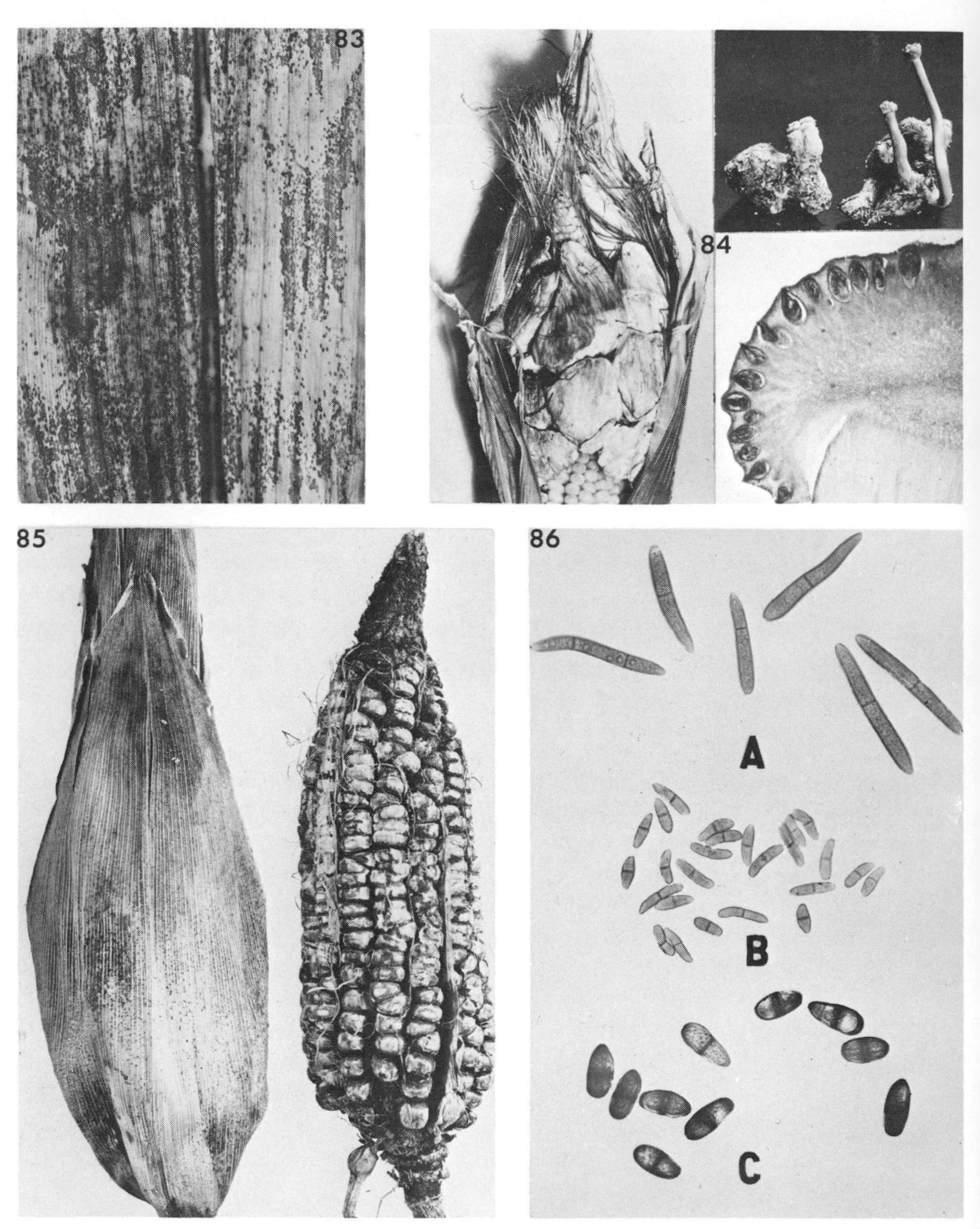

Figure 83. Brown spot, *Physoderma maydis*, on corn leaf. (Photograph by A. H. Eddins.)
Figure 84. Ergot, *Claviceps gigantea*, on corn, showing sclerotia on ear, germinating sclerotia, and perithecial stroma. (Photograph by A. S. Ullstrup.)
Figure 85. Dry rot of corn, caused by the *Diplodia zeae* stage of *Physalospora zeae*. (Photograph by A. H. Eddins.)
Figure 86. Pycnidiospores of *Diplodia* on corn. A. *D. macrospora*. B. *D. zeae*. C. *D. frumenti*. (Photograph by A. H. Eddins.)

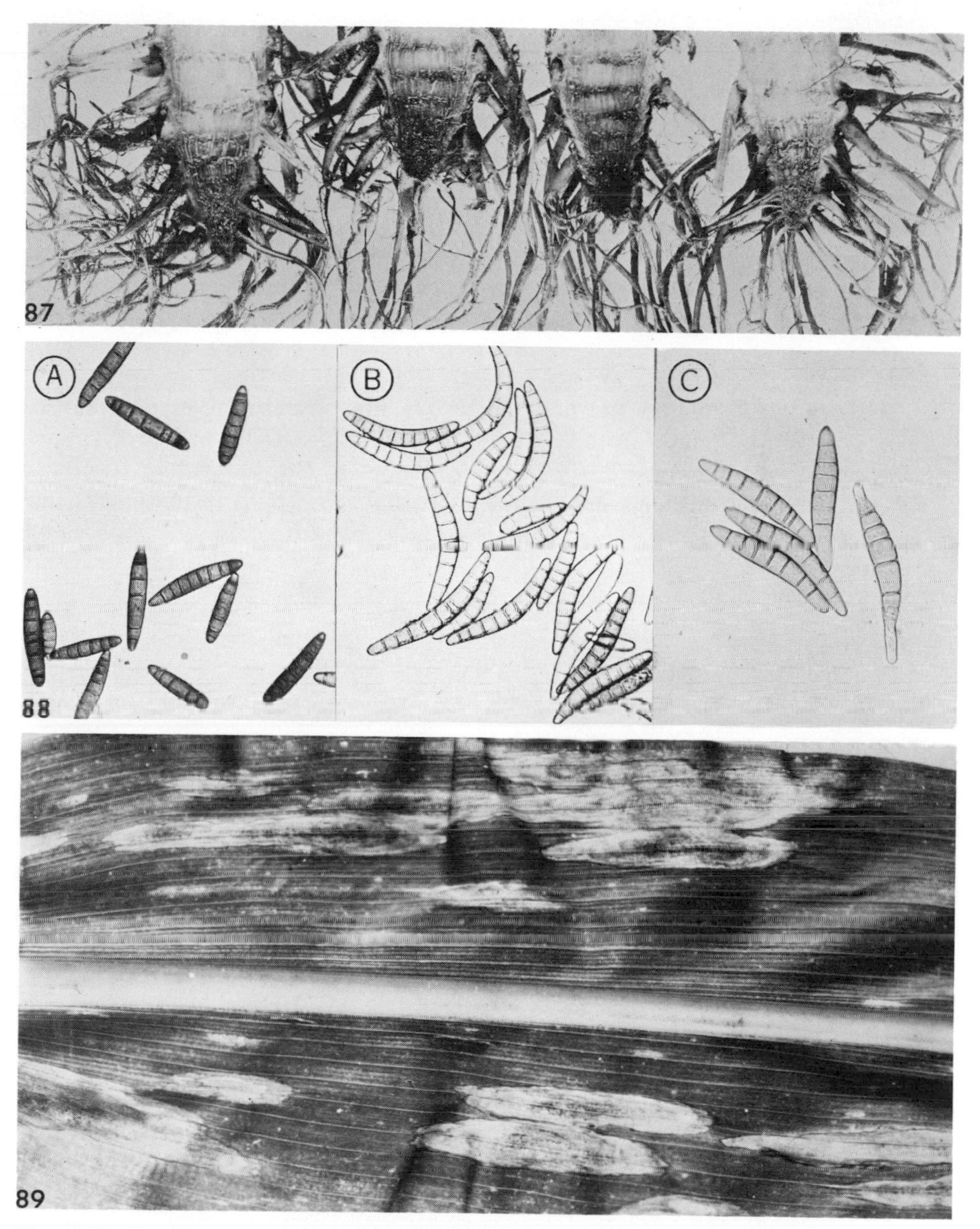

Figure 87. *Fusarium* root rot of corn, caused by *Gibberella zeae.* (Photograph by A. H. Eddins.)

Figure 88. Conidia of *Helminthosporium* of corn. A. *H. carbonum.* B. *H. maydis.* C. *H. turcicum.* (Photograph by A. S. Ullstrup.)

Figure 89. Northern leaf spot, *Trichometasphaeria turcica,* on corn, caused by *Helminthosporium turcicum.* (Photograph by R. K. Voorhees.)

Ergot, *Claviceps gigantea*, Fuentes, Isla, Ullstrup, & Rodrig.

Symptoms

Sclerotia are produced on the pistillate florescence of the host, replacing the kernels on the cob. They are of a slightly yellowish color, soft, and mostly hollow, but as they mature the walls thicken and the outside becomes hard, covering a pale pink to lavender interior. They vary in size from a few millimeters in length to 8×5 cm. In the younger stages an abundance of so-called honeydew is produced by the young sclerotia, which contain many hyaline, small, ovoid spores. This exudate is the sphacelial stage.

Etiology

Sclerotia collected in the field germinate after manipulation of temperatures, moisture, light, and time. They produce long-stiped perithecial stroma. Stipes are 2–6 cm long and 0.2–0.4 cm in diameter, with protruding ostioles of the perithecia. Perithecia are flask-shaped, imbedded in the periphery, and measure $338–444 \times 152–164$ μ. Asci are cylindrical and hyaline. Ascospores are hyaline, filiform, nonseptate, and $176–186 \times 1.5$ μ. Macroconidia measure $8–27 \times 2–6$ μ, and the microconidia measure $4–7 \times 2–4$ μ.

Reference

12.

Gray ear rot, *Physalospora zeae* Stout

Symptoms

Gray ear rot is so named because of the mold that develops between the kernels, next to the cob. The mold frequently includes the inner husks that are sealed by the fungus hyphae. Such ears are poorly filled and light in weight. The cob often reveals black, small sclerotia when broken across.

Etiology

The mycelium is hyaline to smoky or even black. The pycnidia are black, globose, submerged in the leaf tissue, become erumpent, and produce hyaline, 1-celled, oval pycnidiospores. The black perithecia develop in the leaves and contain the asci, which are long and slender, measuring $85–175 \times 17–22$ μ; they contain hyaline to amber, 1-celled ascospores, measuring $19–25 \times 6–8$ μ.

Reference

5.

Stalk rot, *Physalospora zeicola* Ell. & Ev.

Symptoms

This rot develops from infected seed that reduce the stand in the field in the seedling stage. Surviving plants produce an irregular growth and may wilt and die during the season because of root rot. A stalk rot and ear decay are also attributed to this disease. A dark brown to black mold envelops the ear, beginning at the shank. It often causes blackened kernels.

Etiology

The mycelium is amber to black, septate, branched, and produces submerged, obovate pycnidia, wherein the 2-celled, 25–30×6 μ, oval, dark, striate pycnidiospores develop in abundance. The perithecia are black, clustered, and subepidermal. The asci are cylindrical, sessile, and 90–140×10–13 μ. The ascospores are hyaline, oval, 1-celled, and 20–30×8–9 μ.

Reference

5.

Ear rot, *Gibberella fujikuroi* (Saw.) Wr.

Symptoms

This disease produces a kernel and stalk rot wherever corn is grown, particularly in the tropics. The fungus causes a decay of individual kernels or an immediate group of kernels. They become pink to reddish brown with conspicuous mycelium of the fungus, or they may show no external symptom. Root lesions and slender, weakened plants are often found among seedlings.

Etiology

The mycelium is hyaline, septate, and branched. Sometimes the color of the seed coat is the only indication of the disease. The 1-celled microconidia are borne in chains or in wet, false heads on hyphal branches. The macroconidia are hyaline, elongate, curved, 3–5-septate, and sparingly produced. The perithecia are blue black, more or less superficial, smooth, and globose. The ascospores are straight or tapering to the tips, 1–3-septate, and slender. The imperfect stage of the fungus is *Fusarium moniliforme*.

References

5, 16.

Root rot, *Gibberella zeae* (Schw.) Petch

Symptoms

This fungus causes a pink ear rot, stalk rot, root rot, and occasionally seedling blight. The ear rot progresses from the tip downward toward the shank. The stalk rot results in the stalk breaking at the internodes at reddish lesions. The root rot predisposes the plant to premature ripening, with rapid advance of stalk and ear rot on ripe, dead stalks before harvest. The husks often adhere tightly because of pink mold. Bluish black perithecia may be found at the nodes on dead stalks.

Etiology

The mycelium is hyaline or white and pink. It is septate, branched, and produces conidia that are hyaline, sickle-shaped, tapering, not constricted, mostly

5-septate, and 41–60 × 4–6 μ. They develop in sporodochia or in pionnotes. There are no chlamydospores. The perithecia are bluish to purple black, superficial, scattered, frequently imbedded in mycelium, and ovoid to conical. The ascospores are fusiform, slightly curved, mostly 3-celled, and 20–30 × 3–5 μ. The imperfect stage is *Fusarium graminium.*

References

5, 16.

Stalk rot, *Rhopographus zeae* Pat.

Symptoms

This is not an important disease. It has been found almost entirely on mature plants. The disease is inconspicuous, characterized by black, hysteriform, slit-like stromata that are imbedded in the rind.

Etiology

The perithecia are dark, obovate, paraphysate, and ostiolate. They are distributed in more or less parallel lines, 1–3 mm long. The asci are cylindrical, clavate, and 100–150 × 20 μ. The ascospores are hyaline, 3–5-septate, constricted, and 30–52 × 6–10 μ. The vegetative stage is *Clasterosporium longisporum* Voorhees.

Reference

29.

Tar spot, *Phyllachora maydis* Maubl.

Symptoms

Tar spot is distinguished from the many diseases of grasses by the shiny black stromata. It is almost entirely limited to tropical areas.

Etiology

The black, irregularly circular stromatic spots are up to 2 mm in diameter and often confluent. The perithecia are immersed in the black stromata. They are paraphysate, spherical, with ostioles opening on both leaf surfaces. The asci are cylindrical, and the ascospores are hyaline, ovoid, 1-celled, and 9–12 × 5–7 μ.

Reference

5.

False smut, *Ustilaginoidea virens* (Cke.) Tak.

Symptoms

This disease is found only on the staminate florescence, the tassel, where it forms olive green to black sclerotialike structures in the staminodia of the individual flowers and not in the host tissue. The structures are velvety individually and rough en masse.

Etiology

The conidia are produced over the surface of the fertile area. An abundance of dark-colored, smooth to warty, spherical spores are formed, measuring 4–6 μ in diameter. The sclerotia protrude and measure approximately 4×2 mm. They are greenish, oval, and overgrow the glumes.

Reference

5.

Head smut, *Sphacelotheca reiliana* (Kuehn) Clint.

Symptoms

Head smut appears first on the tassels and ears, which may be partially or almost completely utilized and converted to the production of dark blackish spores. They form within the peridium, which is broken open to free the spores. The vascular bundles of the host persist in the gall. Some leafy proliferations have been observed on diseased tassels.

Etiology

The chlamydospores are reddish brown, spherical, echinulate, and 9–12 μ in diameter.

Reference

5.

Smut, *Ustilago maydis* (DC.) Cda.

Symptoms

Common smut causes losses of from 2 to 6 per cent annually; all aboveground parts of the plant are susceptible to infection and the production of the black, dusty galls of various dimensions. The galls, when immature, appear shiny under a whitish membrane which later dries, breaks open, and liberates the chlamydospores. The galls on the ears are often large, pushing open the husks as they develop, and may measure up to 8 or 10 in. in diameter. The galls on the stalk sheaths and leaf margins are small. The vascular tissue of the host does not appear in the galls or sori.

Etiology

The chlamydospores are black, spherical, echinulate, and 8–11 μ in diameter. They germinate by forming a basidium-bearing sporidium. These in turn increase by budding.

References

5, 26, 27.

Rust, *Puccinia sorghi* Schw.

Symptoms

Rust has been found almost everywhere the host is grown, and only rarely has it been an important economic factor. Infection on seedlings has been observed to be the cause of defoliation and stunting. Usually rust appears about tasseling time, causing cinnamon brown to black pustules and linear streaks on both leaf surfaces. The sori form under the epidermis and break through to disperse the echinulate uredospores. The rust cycle is completed on *Oxalis* spp. as the alternate host, developing pycniospores and aeciospores.

Etiology

The fungus is an obligate parasite. Spores from the aecia are air disseminated and cause infection on corn. The uredia develop on this host, producing dark brown, 1-celled, almost spherical, echinulate uredospores. The telia are scattered and usually appear in the vicinity of the uredinia. They are black, often subepidermal to erumpent, and contain the teliospores that are dark brown, 2-celled, stipitate, oblong, constricted, and $29\text{–}45 \times 16\text{–}23\ \mu$. The aeciospores are pale yellow, globoid, and $18\text{–}26 \times 13\text{–}19\ \mu$. The uredospores are brown, ellipsoidal, and $26\text{–}32 \times 23\text{–}29\ \mu$. Basidiospores are the forms resulting from germinating teliospores and infect *Oxalis* spp.

References

1, 5.

Southern rust, *Puccinia polysora* Underw.

Symptoms

This rust is distinguished primarily by the longer retention of the epidermis over the sori and also by the fact that the sori are small and circular. They are scattered over both surfaces of the leaf and are light-colored. The rust is favored by a higher temperature and has been most frequently observed in tropical areas. The alternate host is not known.

Etiology

The uredia sori develop on either leaf surface. The uredospores are yellow to pale brown, echinulate, ovoid, 1-celled, and $27\text{–}34 \times 24\text{–}29\ \mu$. The telia are usually closely associated with the uredia and are almost black. The teliospores are dark brown to black, ellipsoidal, 2-celled, borne on short pedicels, and measure $29\text{–}40 \times 18\text{–}26\ \mu$. No pycnia or aecia sori are known.

References

1, 26, 27.

Tropical rust, *Physopella zeae* (Mains) Cumm. & Ramachar

Symptoms

This rust, as the common name indicates, has been found only in warm, tropical areas and not in the temperate zones. The uredia sori occur on the upper leaf surface below the epidermis; they are pale yellow, turn pale brown, and measure 16–30 μ in diameter.

Etiology

The telia are dark brown, usually develop around the uredia, and remain subepidermal until the epidermis disintegrates. The teliospores are brown, with no stipe, 1-celled, oblong to angular, and borne 2 in a chain. The alternate host is not known. The rust has been known as *Angiospora zeae*.

References

4, 5.

Southern blight, *Botryobasidium rolfsii* (Sacc.) Venkat.

Symptoms

This blight can be found in areas of high temperatures, adequate moisture, and poor ventilation. The fungus is soil-inhabiting and attacks the brace roots and crown of plants at the soil line where they may be girdled, killed, and fall or are easily blown over.

Etiology

The mycelium is white, usually plentiful, dense, and supports external sclerotia that are light-colored at first and dark brown finally, spherical, and up to 1 mm in diameter. The rarely observed basidia are hyaline, stubby, and produce 4 basidiospores that are 1-celled and hyaline.

References

5, 6.

Damping-off, *Thanatephorus cucumeris* (Frank) Donk

Symptoms

Rhizoctoniose is encountered mostly in areas of excessive moisture and causes damping-off of seedlings. It occasionally causes a root rot and stalk decay, depending on environmental conditions. The fungus is soil-inhabiting and under favorable conditions causes variable types of lesions on plants at the soil line.

Etiology

The mycelium is brown, branched, septate, not abundant, and seldom produces sclerotia on the living host but rather on the dead host or adjacent material. The vegetative stage of the fungus is *Rhizoctonia solani* Kuehn, which is its parasitic stage. There are no vegetative spores produced. The basidia frequently found associated with the disease each produce 4 hyaline, 1-celled basidiospores.

Reference

5.

Banded disease, *Thanatephorus sasakii* (Shirai) Donk

Symptoms

This mildew develops under rainy and humid conditions in hot, tropical areas in dense growth that lacks sufficient ventilation. The mycelium of the fungus grows up from the soil surface onto the plant, mostly superficially, until it contacts the leaf blades. There it develops a water-soaked, scalded appearance and is covered by fine, faintly colored, branched, septate, rapidly growing hyphae. The growth expands at night and is almost stationary during the day, resulting in a banding condition that is characteristic of the disease.

Etiology

The mycelium is pale lavender to brown, septate, branched, and 4–12 μ wide. It is abundant under favorable conditions, external at first, but rapidly penetrating the host. The hymenium is white and downy. The basidia are hyaline, clavate, have 2–4 sterigmata, and measure 10–15×7–9 μ. The basidiospores are hyaline, oval to elliptical, smooth, 1-celled, and 8–11×5–7 μ.

Reference

5.

Tuckahoe, *Poria cocos* (Schw.) Wolf

Symptoms

Poria parasitizes corn plants and forms sclerotia of various sizes and shapes within the stalk. The sclerotia are brown externally and white-starchy within.

Etiology

The sporophore develops as a brown, resupinate, pored hymenium on the outer, corky, sclerotial wall. The tubes are 2–3 mm long, angular to labyrinthiform, and form no cystidia. Basidia are clavate, measuring 20–25×6–8 μ, and basidiospores are hyaline, measuring 7–8×3–4 μ.

Reference

31.

Dry rot, *Diplodia zeae* (Schw.) Lév.

Symptoms

This disease, found as a seedling blight, is frequently characterized by a brown dry rot initiated below the soil surface from diseased seed. The stalk rot develops mostly after midseason from reddish-purple-brown blotches on the leaf sheaths, extending into the nodes. It is also found on the brace roots and lower stalk where lesions hasten ripening and cause some stalk breakage and moldy ears.

Etiology

The pycnidia are black, globose, and submerged with ostioles protruded. The pycnidiospores are olivaceous, ovate, straight, 2-celled, and 25–30 × 6 μ.

References

5, 6.

Dry rot, *Diplodia macrospora* Earle

Symptoms

Dry rot causes a seedling blight and stalk rot, the former resulting from the planting of diseased seed and the latter from wounds and accumulations in leaf sheaths. The fungus invades the pith and extends through the shank, causing ear mold which also develops from ear injury primarily by insects.

Etiology

The pycnidiospores are dark-colored, straight, septate, and 70–80 × 6–8 μ.

References

5, 6.

Anthracnose, *Colletotrichum graminicola* (Ces.) G. W. Wils.

Symptoms

This disease causes considerable damage to many cereals but is not of considerable importance on corn. It is characterized by tan to brown, oval to elliptical, elongate leaf and sheath spots, which are usually darkened by the production of acervuli and the accompanying black clusters of setae.

Etiology

The acervuli are superficial with dark stromata and septate, tapering setae. The conidia are hyaline, spindle-shaped, slightly curved, 1-celled, and 18–26 × 3–4 μ.

References

5, 27.

Charcoal rot, *Macrophomina phaseolina* (Tassi) Goid.

Symptoms

This ashy gray rot often develops on seedlings and growing plants as well as on mature ones. The dark, discolored diseased areas contain the numerous, black, small, hard sclerotia of the fungus. Often the stalk is involved from the base upward, and the pith is destroyed, leaving the rind and vascular tissue. The disease is found on many other hosts.

Etiology

The pycnidia are black, obovate, ostiolate, and 100–200 μ. Conidia are hyaline or light-colored, ovate to elongate, 1-celled, and $12–34 \times 6–12\ \mu$. The vegetative, sterile stage of the fungus is *Sclerotium bataticola.*

References

5, 26.

Ear mold, *Aspergillus flavus* Lk. ex Fr.

Symptoms

This mold, associated with discolored kernels, frequently damages seedlings germinated from diseased seed. Contamination may take place from exposure after maturity. The seedling disease is similar to that observed under wet, cold, growing periods conducive to detrimental effects caused by many weakly parasitic fungi.

Etiology

Conidial heads vary greatly in dimensions, the conidiophores averaging 500–800 μ in height and up to 15 μ in diameter. The conidia are yellow green, subglobose, and $3–4 \times 4–5\ \mu$.

References

5, 16.

Kernel rot, *Penicillium oxalicum* Currie & Thom

Symptoms

This soil-inhabiting fungus is widely distributed, causing a decay of the cob and infection of the seed detected mostly on the germinator. Seedlings growing from diseased seed often show a yellow green color toward the leaf base, followed by a drying of the leaf margins, the entire leaf, and other leaves until the plant dies. Infection develops around the embryo and mesocotyl.

Etiology

The fungus is blue green to olive and produces no exudate or odor. Conidiophores are $100–200 \times 3–5\ \mu$ and produce conidia in long chains from tapered tips. Conidia are elliptical, smooth, and $5–6 \times 3–4\ \mu$.

References

5, 16.

Cob rot, *Nigrospora oryzae* (Berk. & Br.) Petch

Symptoms

This cob rot is well distributed in corn-growing areas and has been observed on plants that show a stalk rot and shank breakage. Ears are usually lightweight and appear partly bleached. The cob is often shredded and broken, and the chaff is brown and dark-colored instead of a bright color.

Etiology

The fungus produces abundant, black, spherical spores, measuring 13–15 μ in diameter, in the rotted cob and at the base of mature kernels. No sexual stage is known.

Reference

5.

Black bundle, *Cephalosporium acremonium* Cda.

Symptoms

There are no definite symptoms during the first half of the season; gradually, stripelike lesions may appear on the leaves, followed by off-color, barren stalks, nubbin ears, suckering, and internally blackened fibrovascular bundles. In extreme cases, all the leaves become reddish purple. Seed taken from these plants produce diseased seedlings.

Etiology

The mycelium is hyaline, septate, and branched. The conidiophores are 10–50 μ high and form spores by abstriction of their tips in mucus-forming heads 10–35 μ in diameter. The spores are 4–15 × 2–6 μ and are usually 1-celled but may become septate.

Reference

16.

Zonate leaf spot, *Gloeocercospora sorghi* D. Bain & Edg.

Symptoms

Characteristic targetlike spots form on the leaves. The young spots are reddish brown, small, water-soaked lesions at first, enlarging, rapidly becoming darker-colored, and forming wide, circular to semicircular concentric bands. Lesions often extend across the leaf blade and are up to 2 in. in diameter.

Etiology

The conidia are produced in sporodochia in a mucus matrix. The spores are hyaline, filiform, septate, and 20–195 × 1.4–3.5 μ. Only the vegetative stage is known. It is seed-borne.

Reference

5.

Leaf spot, *Helminthosporium carbonum* Ullstrup

Symptoms

The disease develops sparingly on the foliage, forming tan, oval to circular lesions up to 1 in. in diameter. A target design frequently is observed. Under favorable conditions the fungus sporulates abundantly on the foliage and sheaths and may cause a dark mold on the kernels.

Etiology

The conidia are usually slightly curved, tapering from the middle to rounded ends, often multiseptate, up to 12, and 25–100 × 7–18 μ. A sexual stage has been described as *Cochliobolus carbonus.*

References

5, 26, 27.

Southern leaf spot, *Helminthosporium maydis* Nisik. & Miyake

Symptoms

This widely disseminated disease is most commonly found in the warmer corn-growing areas. Typical lesions on the leaves are tan to gray, generally narrow, limited somewhat in width by the leaf veins, and range from minute spots to those more than 1 in. in length.

Etiology

The conidia are produced on the lesions on olivaceous, erect conidiophores that arise from the stomata and are up to 170 μ high. The spores are brown, elongate, slightly curved, tapering toward the rounded ends, up to 12-septate, and 30–115 × 10–17 μ. The sexual stage of the fungus is *Cochliobolus heterostrophus.*

References

5, 26, 27.

Root rot, *Helminthosporium pedicellatum* Henry

Symptoms

The primary and secondary roots are at first partially discolored by light brown lesions that are elongate, sunken, become dark-colored as they enlarge, encircle the root, and destroy the cortical tissue and eventually the vascular tissue. The aboveground parts of diseased plants show no symptom with the possible exception of rapidly produced secondary roots.

Etiology

The mycelium is septate, branched, and plentiful. Conidia are greenish brown, abundant, up to 9-septate, usually 5–7, and 46–90 × 18–30 μ. The basal portion of the spore is abruptly narrowed to an elongate, blunt terminal in the attachment to the conidiophore where clusters of them appear as black masses. The sexual stage has been described as *Trichometasphaeria pedicellata.*

References

20, 31.

Northern leaf spot, *Helminthosporium turcicum* Pass.

Symptoms

Leaf spot is a common disease of the host wherever grown, becoming very severe under favorable conditions of temperature and moisture. It is recognized by the development on the leaves of long, narrow, stripelike lesions that are often 6 in. long and 1 in. wide at their greatest diameter. The diseased tissue is light tan to white and, when abundant, changes the general color of a field. It is most severe at maturity but may seriously affect seedlings.

Etiology

Spores, produced abundantly on the lower surfaces on long conidiophores, are olivaceous, slightly curved, widest at the middle, tapering toward the rounded ends, 3–8-septate, and $45\text{–}132 \times 15\text{–}25\ \mu$. The sexual stage has been described as *Trichometaspheria turcica.*

References

5, 18, 20, 27.

Gray ear rot, *Macrophoma zeae* Tehon & Daniels

Symptoms

Ear rot shows a premature bleaching and growing together of the outer husks followed by an ear mold and gray discoloration of the cob. Black sclerotia develop on the kernels. The disease is generally distinguished by the gray cob color, some shank shredding, and the presence of the dark sclerotia.

Etiology

The pycnidia are mostly oval to spherical, ostiolate, and $65\text{–}120\ \mu$ in diameter. Spores are hyaline, 1-celled, fusiform, and $17\text{–}31 \times 6\text{–}10\ \mu$. The sexual stage has been described as *Physalospora zeae.*

Reference

5.

Leaf spot, *Cercospora zea-maydis* Tehon & Daniels

Symptoms

Destructive foliage spotting in certain instances is not generally considered a serious disease. Leaf spots are pale brown or tan, elongate streaks parallel with the veins, and produce spores abundantly on the lower surface.

Etiology

Conidiophores are pale brown and appear in fascicles of 3–12. Conidia are hyaline, obclavate, slightly curved, 3–10-septate, and $30\text{–}95 \times 5\text{–}9\ \mu$. The sexual stage is not known.

Reference

2.

Sclerotial rot, *Rhizoctonia zeae* Voorhees

Symptoms

This rot usually involves the entire ear and husks in light, pinkish, mycelial mold that becomes gray in later stages. The exposed areas become covered with small, immature, white to pink sclerotia which turn brown as they mature. The mycelium engulfs the kernels and causes a softening of the cob.

Etiology

The mycelium is hyaline to pink and plentiful, producing sclerotia over the exposed surfaces. The sclerotia are usually scattered, hard, homogeneous in structure, and 0.5–1 mm in diameter. No spores have been associated with this fungus. In culture, clamp connections between hyphal cells are frequent.

Reference

30.

Other fungi associated with corn

Alternaria tenuis Nees ex Corda
Ascochyta maydis Stout
Botrytis cinerea Persoon ex Fries
Cercospora sorghi Ellis & Everhart
Chaetomium globosum Kunze ex Fries
Chaetophoma maydis Spegazzini
Cladosporium herbarum Persoon ex Link
Curvularia lunata (Wakker) Boedijn
Dendrophoma zeae Tehon
Diaporthe incongrua Ellis & Everhart
Epicoccum nigrum Link ex Wallroth
Leptosphaeria maydis Stout
Leptothyrium zeae Stout
Mycosphaerella zeicola Stout
Phoma zeicola Ellis & Everhart
Phyllosticta zeae Stout
Pyrenochaeta terrestris (Hansen) Gorenz, J. C. Walker, & Larson
Sclerotium maydis Preuss
Septoria zeae Stout
Trichoderma viride Persoon ex Fries

References: corn

1. Arthur, J. C., and G. B. Cummins. 1962. Manual of the rusts in United States and Canada. Hafner Co.: N.Y.
2. Chupp, C. 1953. A monograph of the fungus genus Cercospora. Published by the author: Ithaca, N.Y.
3. Cummins, G. B. 1941. Identity and distribution of three rusts of corn. Phytopathology 31:856–57.
4. Cummins, G. B., and P. Ramachar. 1958. The genus Physopella (uredinales) replaces Angiospora. Mycologia 50:741–44.
5. Dickson, J. G. 1956. Diseases of field

crops. 2d ed. McGraw-Hill: N.Y., pp. 74–114.
6. Eddins, A. H. 1930. Corn diseases in Florida. Florida Univ. Agr. Expt. Sta. Bull. 210:5–34.
7. Eddins, A. H. 1933. Infection of corn plants by Physoderma zeae-maydis Shaw. J. Agr. Res. 46:241–53.
8. Elliott, C. 1943. A pythium stalk rot of corn. J. Agr. Res. 66:21–39.
9. Elliott, C. 1951. Manual of bacterial plant pathogens. 2d ed. Chronica Botanica: Waltham, Mass.
10. Fischer, G. W. 1953. Manual of the North American smut fungi. Ronald Press: N.Y.
11. Frederiksen, R. A., and A. J. Bockholt. 1969. Sclerospora sorghi, a pathogen of corn in Texas. Plant Disease Reptr. 53: 566–69.
12. Fuentes, S. F., et al. 1964. Claviceps gigantea, a new pathogen of maize in Mexico. Phytopathology 54:379–81.
13. Garrard, E. H. 1951. Bacterial diseases of plants. Bull. Agr. Coll., Guelph, Ontario 478:1–69.
14. Holbert, J. R., et al. 1924. Corn root, stalk, and ear rot diseases and their control thru seed selection and breeding. Illinois Univ. Agr. Expt. Sta. Bull. 255: 239–472.
15. Karling, J. S. 1950. The genus Physoderma (Chytridiales). Lloydia 13:29–72.
16. Koehler, B., and J. R. Holbert. 1930. Corn diseases in Illinois. Illinois Univ. Agr. Expt. Sta. Bull. 354.
17. Leukel, R. W., and V. F. Tapke. 1954. Cereal smuts and their control. U.S. Dept. Agr. Farmers' Bull. 2069:1–28.
18. Luttrell, E. S. 1958. The perfect stage of Helminthosporium turcicum. Phytopathology 48:281–87.
19. Matthews, V. D. 1931. Studies on the genus Pythium. University of North Carolina Press: Chapel Hill, pp. 82–87.
20. Nelson, R. R. 1965. The perfect stage of Helminthosporium pedicellatum. Mycologia 57:665–68.
21. Payak, M. M., and B. L. Renfro. 1967. A new downy mildew disease of maize. Phytopathology 57:394–97.
22. Robert, A. L. 1955. Bacterial wilt and Stewarts leaf blight of corn. U.S. Dept. Agr. Farmers' Bull. 2092:1–13.
23. Rosen, H. R. 1926. Bacterial stalk rot of corn. Arkansas Agr. Expt. Sta. Bull. 209: 1–28.
24. Shepherd, R. J., E. E. Butler, and D. H. Hall. 1967. Occurrence of a root rot disease of corn caused by Helminthosporium pedicellatum. Phytopathology 57: 52–56.
25. Sprague, R. 1950. Diseases of cereals and grasses in North America. Ronald Press: N.Y.
26. Ullstrup, A. J. 1943. Diseases of dent corn in the United States. U.S. Dept. Agr. Circ. 674:2–34.
27. Ullstrup, A. J. 1961. Corn diseases in the United States and their control. U.S. Dept. Agr., Agr. Handbook 199:1–29.
28. Ullstrup, A. J., and M. H. Sun. 1960. The prevalence of crazy top of corn in 1968. Plant Disease Reptr. 53:246–50.
29. Voorhees, R. K. 1934. Rhopographus zeae on corn. Mycologia 26:115–17.
30. Voorhees, R. K. 1934. Sclerotial rot of corn caused by Rhizoctonia zeae n. sp. Phytopathology 24:1290–1303.
31. Wolf, F. A. 1922. The fruiting stage of the tuckahoe, Pachyma cocos. J. Elisha Mitchell Sci. Soc. 38:127–37.

Cotton, *Gossypium hirsutum* Linnaeus

Angular leaf spot, *Xanthomonas malvacearum* (E. F. Smith) Dowson
Boll rot, *Rhizopus nigricans* Ehrenberg
Mold, *Choanephora cucurbitarum* (Berkeley & Ravenel) Thaxter
Damping-off, *Pythium debaryanum* Hesse
Stain, *Nematospora coryli* Peglion
Fiber stain, *Ashbya gossypii* (Ashby & Nowell) Guilliermond
Leaf spot, *Mycosphaerella gossypina* (Atkinson) Earle
Rust, *Cerotelium desmium* (Berkeley & Broome) Arthur
Rust, *Puccinia schedonnardi* Kellerman & Swingle
Rust, *Puccinia stakmanii* Presley
Sore shin, *Thanatephorus cucumeris* (Frank) Donk
Southern blight, *Botryobasidium rolfsii* (Saccardo) Venkatarayan
Root rot, *Phymatotrichum omnivorum* (Shear) Duggar
Blight, *Ascochyta gossypii* Woronin
Boll rot, *Diplodia gossypina* Cooke

Anthracnose, *Colletotrichum gossypii* Southworth
Black boll rot, *Aspergillus niger* Tieghem
Leaf spot, *Corynespora cassiicola* (Berkeley & Curtis) Wei
Areolate mildew, *Ramularia areola* Atkinson
Leaf spot, *Helminthosporium gossypii* Tucker
Wilt, *Verticillium albo-atrum* Reinke & Berthold
Wilt, *Fusarium oxysporum* f. sp. *vasinfectum* (Atkinson) Snyder & Hansen
Leaf spot, *Alternaria gossypina* (Thuemen) Hopkins
Stem blight, *Alternaria macrospora* Zimmermann
Black root rot, *Thielaviopsis basicola* (Berkeley & Broome) Ferraris
Ashy stem, *Macrophomina phaseolina* (Tassi) Goidanich
Other fungi associated with cotton

Angular leaf spot, *Xanthomonas malvacearum* (E. F. Sm.) Dows.

Symptoms

First evidence of the disease is seen on the undersurface of the cotyledons, young leaves, and stems as green, round, water-soaked, translucent spots of varying size, appearing within a week or 10 days after planting of seeds. Later infections penetrate to the upper side of the leaf, forming elongated, angular, irregular areas between the veins or along the main leaf vein. Infected areas become dry, sunken, and reddish brown with a rust red to purplish margin. On Sea Island and Egyptian varieties, leaves turn yellow, curl, and fall.

Etiology

The organism is a gram-negative rod, forming no capsules, motile by 1 polar flagellum, and $1.3–2.7 \times 0.3–0.6\ \mu$. Colonies in agar cultures are pale to deep yellow, circular, thin, raised, entire, smooth, wet, and shining. The pathogen lives over on seed and diseased plant parts. It does not survive in the soil. Optimum temperature, 25–30C.

References

10, 22, 28.

Boll rot, *Rhizopus nigricans* Ehr.

Symptoms

Affected portions of the boll are olive green and become darker only when the decayed parts dry up. Spore masses are not dense; they form a dark gray or blue gray mold over the boll.

Etiology

Aerial stolons develop from mycelium anchored by rhizoids. Sporangiophores arise from stolons with terminal, multispored sporangia. The columellae are hemispherical; the spores are oval or angular, smooth, and $14 \times 11\ \mu$. Zygospores are $150–200\ \mu$ in diameter.

Reference

27.

Mold, *Choanephora cucurbitarum* (Berk. & Rav.) Thaxt.

Symptoms

The disease starts as blossom blight, with fading of the corolla after opening. The corolla becomes covered with a white, dense fungus growth, consisting chiefly of immature conidiophores. Conidial heads develop promptly, becoming purple black at maturity.

Etiology

Conidiophores are unbranched, with spherical heads bearing sterigmata on which lemon-shaped, continuous, striate conidia form, measuring $15–25 \times 7–11$ μ. Sporangiophores are unbranched, recurved at the tips, and bear sporangia on the columella. Sporangiospores are ovoid to fusiform, continuous, with a cluster of very fine radiating appendages at each end, and measure $18–30 \times 10–15$ μ. The mycelium contains intercalary chlamydospores with more or less thickened walls. Zygospores are 50–90 μ in diameter.

Reference

35.

Damping-off, *Pythium debaryanum* Hesse

Symptoms

Seedling hypocotyls assume a pale color due to the destruction of chlorophyll. Invaded tissue is dirty white, shrivels, and becomes constricted just above ground line, causing young plants to fall over.

Etiology

Pythium, with colorless, nonseptate, much-branched mycelium, penetrates the epidermis of the seedling hypocotyl. Asexual reproduction occurs by formation of conidia and zoospores, whereas sexual reproduction is by oospores. Incidence of the disease on seedlings is favored by abundance of moisture, compact and poorly aerated soil, and thick stands.

Reference

14.

Stain, *Nematospora coryli* Pegl.

Symptoms

No visible, external symptoms appear on infected bolls early in their growth except possibly miniature epidermal wounds or scars. The disease is directly associated with bug infestation. Later, however, killing of seed, immature and brownish lint, discolored sheaths, and shedding of affected bolls occur.

Etiology

Yeastlike cells, showing a wide variation in form, are present in the host tissue. Colonies of large, spherical cells, 15–20 μ in diameter, single or attached in small groups as in typical budding, appear in the lint. Asci are hyaline, cylindrical, elongate, free living, with rounded ends, thick walls, and measure 60–85 × 10–12 μ. The 8 ascospores are spindle-shaped with an acute apex, septate, with a long, threadlike appendage, and measure 40–56 × 2–3 μ. They are formed in 2 groups of 4 in each ascus.

References

13, 22.

Fiber stain, *Ashbya gossypii* (Ashby & Nowell) Guillierm.

Symptoms

There is little or no external indication of the infection of the bolls. The fibers and seed of the cotton bolls show different degrees of discoloration resulting from the presence of the fungus. The disease causes a drying out of the tan to brown tissues. The parasite is associated with wounds caused by the feeding of the cotton stainer bug.

Etiology

The scanty mycelium is hyaline; it is usually nonseptate except during spore formation and general budding. Certain structures continue to bud off new cells. The asci are formed in the mycelial strands in great numbers. They are elongate, smooth, often intercalary, and contain 8 ascospores, 4 in a sheath in each end of the ascus. The ascospores are spindle-shaped, 1-septate, blunt at one end, attenuated at the other end into a whiplike appendage, and 27–35 × 4.2 μ, excluding appendage.

References

22, 26.

Leaf spot, *Mycosphaerella gossypina* (Atk.) Earle

Symptoms

The disease is characterized by the appearance of spreading, roundish or irregular spots on both surfaces of the leaf. Spots are first yellowish brown, then whitish in the center, and up to 10 mm in diameter. The margin is formed by a distinct, dark brown or blackish rim. Adjacent spots may unite, causing large patches of the leaf to wither. In old spots, centers may crack and break away, leaving a perforation.

Etiology

Intercellular mycelium is irregular, branched, septate, and produces a tuberculate stroma. Conidiophores emerge in clusters through the stomata and are dark brown, flexuous, septate, and irregularly bent near the tip. Conidia are nearly colorless, long, slender, curved, narrow above and rounded below, 5–7-septate, and 70–180 × 3 μ. Perithecia are black, ovate, partly immersed, with mouth projecting to either surface. Asci are 40–45 × 8–10 μ; ascospores are greenish hyaline, elliptical or broad fusoid, 1-septate, narrow at the septum, with unequal cells, and measure 15–18×3–4 μ. The imperfect stage is *Cercospora gossypina* Cke.

References

6, 26.

Rust, *Cerotelium desmium* (Berk. & Br.) Arth.

Symptoms

The disease is confined to green parts, especially the leaves. The uredo stage is yellowish brown, developing as small pustules at first, then becoming more powdery. On the upper surface of the leaf, the pustules are deeply immersed in host tissue. The upper surface is often marked by small, purplish brown spots, which may coalesce into large patches, causing defoliation.

Etiology

Uredosori develop under the epidermis or are more deeply immersed in the mesophyll and surrounded by a ring of colorless, clavate paraphyses. Uredospores arise directly from the surface cells of the stromatic layer at the base of the sorus with no definite stalk. They are light yellow, oval and broad, pear-shaped, with short but distinct spines on the wall, and measure 19–27×16–19 μ. Other spore stages are not known.

References

2, 6.

Rust, *Puccinia schedonnardi* Kell. & Swing.

Symptoms

This fungus causes orange, large, raised spots on leaves, bolls, and involucral bracts.

Etiology

Pycnia and aecia are associated; aeciospores are globoid, 15–21×16–26 μ, finely verrucose, and appear on malvaceous hosts. Uredospores are cinnamon brown, globoid, finely echinulate, and 18–26×19–30 μ. Teliospores are broadly oblong, with rounded or narrow ends, and measure 27–45×15–24 μ. The chestnut brown pedicel is longer than the spore.

Reference

2.

Rust, *Puccinia stakmanii* Presley

Symptoms

Orange yellow spots develop on the lower leaf surfaces, beginning as small areas which are lighter green than the healthy leaf. They are raised and conspicuous and enlarge rapidly up to 5 mm in diameter, gradually becoming yellowish as the spores are shed.

Etiology

The pycnia and aecia appear on the cotton plant. The former are small and inconspicuous compared to the aecia, which are colorful and easily observed. The aeciospores are yellow, oblong to subglobose, verrucose, and 15–21 × 18–26 μ. The uredia develop on grass hosts, appearing as brown spots containing the cinnamon brown, echinulate, globoid spores that measure 18–26 × 19–30 μ. The telia are produced on the same host, intermixed with the uredia or separate. Teliospores are chocolate brown, oblong, 2-celled, and 15–24 × 27–45 μ, with pedicels up to 90 μ long.

Reference

7.

Sore shin, *Thanatephorus cucumeris* (Frank) Donk

Symptoms

There may develop a pre-emergence decay or a postemergence damping-off of seedlings, first indicated by a drooping of cotyledons on older plants and the appearance of dark to reddish brown sunken cankers on the stem near or below the soil line. In severe cases, cankers completely encircle the stems and cause plants to fall over and die.

Etiology

The fungus produces an abundance of brown, septate, branched mycelium and, under certain conditions, many dark brown, irregularly shaped, flat or oval sclerotia. The basidia, produced on a loose hyphal network, measure 18–23 × 8–11 μ. Hyaline, oblong, 1-celled basidiospores measure 7–13 × 4–7 μ. The parasitic form of the fungus, *Rhizoctonia solani*, is sterile.

References

1, 5, 8.

Southern blight, *Botryobasidium rolfsii* (Sacc.) Venkat.

Symptoms

Plants from the seedling stage to maturity are subject to attack characterized by wilting and a flaccid condition of the foliage. Gradually, wilting becomes more severe resulting in loss of leaves, blossoms, or bolls and death of the plant. The roots are usually free of the disease, but a girdling canker develops on the main stem or crown of the plant near the soil line or below, depending on soil moisture. White, cottony mycelium may be present, and many sclerotia adhere to the plant.

Etiology

The mycelium is white, septate, branched, and supports many sclerotia that are small, white at first and turning brown, spherical, loosely attached to the mycelium, usually smooth, and variable in size from 0.6 to 1.6 mm in diameter The basidia are hyaline, short, clavate, and produce 4 sterigmata, bearing hyaline, 1-celled, ovate to elongate basidiospores that measure 5–10×3.5–6 μ. The imperfect stage is *Sclerotium rolfsii.*

References

3, 32, 34.

Root rot, *Phymatotrichum omnivorum* (Shear) Dug.

Symptoms

This disease appears during the summer as spore mats on the surface of alkaline soils. First symptoms are slight yellowing or bronzing of foliage, followed by sudden wilting and death of the plant. Root-rot-affected areas assume a reddish brown to black color due to dead plants and present a striking contrast to adjoining, green, healthy plants. Whitish growth of the fungus forms on plant sufaces or on root wartlike wefts appearing at the lenticels.

Etiology

The fungus is found surviving on roots of infected plants as slender, fuzzy strands branching at right angles to the main axis, as a sclerotial form. They are of considerable size, usually round, ovoid, flattened, and become yellow to reddish brown with age. Sclerotia in a chainlike arrangement are found in soil 2–30 in. from colonies. Mats, white to buff, are up to 16 in. in diameter and 1/2 in. thick. The spores are smooth, globose, and 6–8×4–6 μ. The vegetative stage of the fungus is *Ozonium omnivorum.*

References

5, 8, 19, 30.

Blight, *Ascochyta gossypii* Woron.

Symptoms

Large, rounded to irregular spots 3–10 mm in diameter form on the leaves. Spots enlarge rapidly in humid weather and are light brown or tan with reddish brown borders. On bolls and stems, similar spots are darker. Stem infection is the most conspicuous and often occurs at the base of leaf petioles. Spots at first are dark brown, enlarge rapidly, and become sunken and light brown in the center. Early decay and falling out of centers of diseased stem spots are characteristic.

Etiology

Pycnidia are amphigenous, ostiolate, lens-shaped to globose, and 80–100 μ in diameter. Spores are hyaline, oblong or sharply cylindrical, rounded at both ends, 1-septate at or about the middle, not at all or barely constricted, and measure 8–10×2–4 μ.

Reference

11.

Boll rot, *Diplodia gossypina* Cke.

Symptoms

Disease attacks the bolls from the stem end and enters the tissues through weevil punctures or weevil injuries. Tiny blisters appear on the surface, and, from these, black sooty masses of spores are released that cover the boll surfaces. Lint of infected bolls is usually matted, smutty, and worthless. Frequency of attack is high on lower bolls in contact with soil. Bolls become dry, black, and fail to open.

Etiology

Pycnidia are carbonous, black, immersed becoming erumpent, and usually ostiolate papillate. Conidia are dark, ellipsoid or ovate, 2-celled, striate, and $17\text{–}35 \times 9\text{–}23\ \mu$. The ascospore stage of the fungus is *Physalospora rhodina.*

References

9, 22, 26, 33.

Anthracnose, *Colletotrichum gossypii* Southworth

Symptoms

Anthracnose may appear on all aboveground parts of the plant. Spots on radicle and cotyledons at the margins or in the blade are light brown to reddish. Damping-off on seedling stems at or just below soil level shows a reddish brown discoloration. Seeds are also infected. On mature plants, the attack is through the stem and is not common on the leaf. On the bolls small reddish or red brown spots, slightly depressed in the center, appear. With enlargement, the spot center turns black, with a reddish margin. In damp weather, the center of the spot becomes covered by salmon-colored masses of moist spores, drying into a crust. Deformation of the boll occurs if it is attacked at an early stage of development.

Etiology

The mycelium is hyaline, branched, septate, and later darkens. Sporophores are equal to or twice the length of the spores, broader below than above, and usually nonseptate. Setae are scarce, colored, and septate. Spores are hyaline, unicellular, straight, smooth, entire, rounded at the ends, and $11\text{–}20 \times 4\text{–}9\ \mu$. The perfect stage is *Glomerella gossypii.*

References

4, 6, 9, 24.

Black boll rot, *Aspergillus niger* v. Tiegh.

Symptoms

Rot begins as a soft, pinkish spot either on the side of the boll or near its base. The older decayed area turns from pink to brown. Disease may involve and destroy all parts of the boll including seed and lint. Abundant production of black spores in the affected area gives a smutty appearance. The fungus is usually considered a wound parasite.

Etiology

This fungus is characterized by spherical heads arising from substratum hyphae. Chains of conidia radiate from the swollen knob of the conidiophore. There are primary and secondary sterigmata. Conidia are black, globose, spinulose at maturity, and usually less than 5 μ in diameter.

Reference

27.

Leaf spot, *Corynespora cassiicola* (Berk. & Curt.) Wei

Symptoms

Leaf spots are circular to irregular with light brown centers, a narrow, dark brown margin, and vary from 2 to 6 mm in diameter, larger when they coalesce. Mature spots show a very distinctive zonation consisting of brown rings alternating with light brown areas. The spots show on both surfaces. The centers often fall out. In severe cases, some leaf yellowing occurs.

Etiology

Usually, dark brown conidiophores can be seen on both leaf surfaces in groups of up to 6. They appear on the darker areas of the zonate spots, are up to 20 septate, mostly 3–5, and measure 125–200 × 8 μ. The conidia are brown, elongate, tapering toward the tip, mostly 10-septate, and 120 × 17 μ. They are straight or slightly curved with a basal hilum and germ pore.

Reference

16.

Areolate mildew, *Ramularia areola* Atk.

Symptoms

Blight is also called frosty mildew. The leaf spots are small, angular, limited by the veins, and typically white on the undersides of leaves, owing to the presence of fruiting bodies of the fungus.

Etiology

The mycelium is intercellular and hyaline. The hyaline, branched conidiophores penetrate the epidermis, emerging to the surface through stomata. The conidia are separate or in chains, oblong, and measure 14–30 × 4–5 μ. The ascospore stage is considered to be *Mycosphaerella areola*.

Reference

26.

Leaf spot, *Helminthosporium gossypii* Tucker

Symptoms

Leaves, flower bracts, and bolls are attacked. Spots on leaves and bracts are numerous and 1–8 mm in diameter. They are light red, gradually darken to deep purple, and have brown centers that fall out. Bolls are infected when young but cause no damage to fiber. Only small, dotlike, purplish lesions appear on hard bolls. Damage is confined to leaf destruction and defoliation.

Etiology

The conidiophores are brown, amphigenous, and septate on ashen gray tissue in the leaf spots, arising from stomata. The conidia, measuring 87×15.3 μ, are 1–8-septate, smooth, rounded at the ends, elliptical, usually curved but occasionally straight, light to dark fuliginous, densely granulated, and thick-walled.

References

25, 31.

Wilt, *Verticillium albo-atrum* Reinke & Berth.

Symptoms

This disease appears about blossoming time, following rainy weather. It first develops on lower leaves and spreads to middle and upper leaves later in the season. Leaves become mottled with pale yellowish, irregular areas at the margins and between principal veins. These areas eventually become brown. A longitudinal cut into the wood at the base of the main stalk reveals a slight browning of the vascular system. Plants are not dwarfed or killed during the season, but often almost complete shedding of leaves occurs.

Etiology

This fungus is a soil inhabitant, entering plants through roots. Conidia are ellipsoidal, unicellular, $4\text{–}11 \times 1.7\text{–}4.2$ μ, and may or may not be collected in heads on sterigmatic tips. Mycelium is hyaline, brown with age, septate, and sometimes swollen into chlamydosporelike chains of closely septate, knotted masses forming sclerotia.

References

7, 23, 26.

Wilt, *Fusarium oxysporum* f. sp. *vasinfectum* (Atk.) Snyd. & Hans.

Symptoms

The disease usually appears about blossoming time, although it also causes a damping-off in the seedling stage. Infected plants appear dwarfed, and leaves are yellow at the margins and between veins. A premature wilting and an internal darkening of the vascular areas develop in affected stalks. Vascular tissues connecting leaves and seeds darken and normal functions are damaged, resulting in shorter taproots, fewer lateral roots, excessive shedding of leaves, and bare stalks.

Etiology

The fungus produces sporodochia or pionnotes from the mycelium. Microconidia are inconspicuous, continuous on pionnotes, and 2–4×5–12 μ. Macroconidia are hyaline, somewhat narrow, mostly 3-septate, sickle-shaped, and 27–28×3–3.75 μ. Chlamydospores and sclerotia are formed. The sexual stage is not known.

Reference

20.

Leaf spot, *Alternaria gossypina* (Thuem.) Hopkins

Symptoms

This disease is observed on cotyledons, seedlings, leaf petioles, and stems as well as on mature leaves. Spots are brown, usually irregularly circular, of papery consistency, usually marked by a series of concentric rings, and 1/2 in. in diameter. Spores are borne on these spots, giving a black color to each one. Severely infected leaves shrivel and fall off. Defoliation is severe in moist conditions. The fungus is extremely abundant late in the season on mature leaves and causes them to shed early.

Etiology

The mycelium is dark, septate, branched, and mostly internal. The conidiophores are black, usually short, septate, and produce conidia at their tips, usually single but sometimes catenulate, 2–3 in chains. The conidia are dark-colored, 3–7-septate, with 1 to several longitudinal septations, 32–46×12–16 μ, and with a variable beak length, up to 60 μ.

Reference

15.

Stem blight, *Alternaria macrospora* Zimm.

Symptoms

On the stem, twig, and leaf petiole of the mature cotton plant, the disease first appears as dark brown, somewhat circular spots, enlarging rapidly under favorable conditions, with centers deeply sunken. Later, elliptical to oval cankers develop with axes parallel to the stem. Infected tissue usually splits longitudinally or cracks into small pieces. Finally, the diseased stem or twig breaks off at the canker.

Etiology

The mycelium is dark, septate, and branched; the septate conidiophores are brown. The conidia are brown, obclavate, with 3–13 transverse and 3–5 longitudinal septa, constricted at each septation, and provided with hyaline, long, filiform, and septate beaks. Conidia are borne singly on conidiophores and measure $57 \times 18\ \mu$ with beaks about $75\ \mu$ long.

References

18, 24.

Black root rot, *Thielaviopsis basicola* (Berk. & Br.) Ferr.

Symptoms

Seedlings are stunted, and leaves are pale green, small, cupped, develop purplish borders with marginal browning, and are frequently shed. In mature plants, there is a swelling downward of the main root beginning at the collar. Cortical tissue appears healthy. Vascular tissue becomes black or purple. The lines of demarcation between healthy and diseased tissues are distinct.

Etiology

The mycelium is $3–7\ \mu$ in diameter, hyaline to slightly colored, and produces endoconidia from hyaline mycelium which are cylindrical and variable in size, measuring $15–25 \times 4–6\ \mu$. The exogenous conidia are light brown, change to dark brown, and become opaque. They are thick-walled and appear scattered or clustered in chains that finally separate into a single cell or groups of many cells. The chlamydospores are cylindrical to ovoid or round, measuring $25–65 \times 10–12\ \mu$, or in units $40–50\ \mu$ long. They produce a black crust on the roots. No sexual stage develops.

Reference

17.

Ashy stem, *Macrophomina phaseolina* (Tassi) Goid.

Symptoms

The root, crown, and stem rot of plants from the seedling stage to the blossoming stage is frequently severe. The plants show drooping and flabby foliage. As the disease becomes more advanced, growth is retarded and stunting is evident. Yellowed leaves are shed and plants die, but usually remain standing. Roots are decayed and stem lesions develop near the soil line, extending several inches up the stem. These lesions are grayish white and speckled with black, minute sclerotia imbedded in the cortex and wood.

Etiology

The sclerotia are black, small, hard, smooth to angular, well distributed in the tissue, and $50–150\ \mu$ in diameter. Their presence is diagnostic of the disease. The pycnidia are not frequently present; they are black and submerged becoming erumpent. The conidia are hyaline, 1-celled, filiform, straight or curved with acute and obtuse ends, and measure $15–30 \times 5–8\ \mu$. The imperfect stage is *Sclerotium bataticola*.

References

7, 12, 29.

Other fungi associated with cotton

Botryosphaeria ribis Grossenbacher & Duggar
Calonectria rigidiuscula (Berkeley & Broome) Saccardo
Cercospora gossypina Cooke
Cladosporium gossypii Jacquin
Doassansia gossypii Lagerheim
Erysiphe malachrae Seaver
Gibberella fujikuroi (Sawada) Wollenweber
Gloeosporium gossypii Averna-Saccá
Nigrospora gossypii Jacquin
Ovulariopsis gossypii Wakefield
Papulospora polyspora Hotson
Peronospora gossypina Averna-Saccá
Pestalotia gossypii Hori
Phoma roumii Fron
Phyllosticta gossypina Ellis & Martin
Phytophthora parasitica Dastur
Sclerotinia sclerotiorum (Libert) Massee
Septoria gossypina Cooke
Spermothora gossypii Ashby & Nowell

References: cotton

1. Al-Beldawi, A. S., and J. A. Pinckard. 1970. Control of Rhizoctonia solani on cotton seedlings by means of benomyl. Plant Disease Reptr. 54:76–80.
2. Arthur, J. C., and G. B. Cummins. 1962. Manual of the rusts in United States and Canada. Hafner Co.: N.Y., pp. 62, 143–44.
3. Aycock, R., et al. 1961. Symposium on Sclerotium rolfsii. Phytopathology 51: 107–28.
4. Bagga, H. S. 1970. Pathogenicity studies of organisms involved in the cotton boll-rot complex. Phytopathology 60:158–60.
5. Brown, H. B., and J. O. Ware. 1958. Cotton. 3rd ed. McGraw-Hill: N.Y., pp. 181–87.
6. Butler, E. J. 1918. Fungi and diseases in plants. Thacker: Calcutta, India, pp. 363–70.
7. Dickson, J. G. 1956. Diseases of field crops. 2d ed. McGraw-Hill: N.Y., pp. 418–24.
8. Duggar, B. M. 1909. Fungous diseases of plants. Ginn Co.: N.Y., pp. 444–52.
9. Edgerton, C. W. 1912. The rots of the cotton boll. Louisiana Agr. Expt. Sta. Tech. Bull. 137:20–69.
10. Elliott, C. 1951. Manual of bacterial plant pathogens. 2d ed. Chronica Botanica: Waltham, Mass., pp. 122–25.
11. Elliott, J. A. 1922. A new ascochyta disease of cotton. Arkansas Univ. Agr. Expt. Sta. Bull. 178:1–18.
12. Ghaffar, A., and G. A. Zentmyer. 1968. Macrophomina phaseoli on some new weed hosts in California. Plant Disease Reptr. 52:223.
13. Guilliermond, A., and F. W. Tanner. 1920. The yeasts. John Wiley & Sons: N.Y., pp. 290–91.
14. Heald, F. D. 1943. Introduction to plant pathology. 2d ed. McGraw-Hill: N.Y., pp. 91–94.
15. Hopkins, J. C. F. 1931. Alternaria gossypina (Thuem.) Comb. nov. causing a leaf spot and boll rot of cotton. Trans. Brit. Mycol. Soc. 16:136–44.
16. Jones, J. P. 1961. A leaf spot of cotton caused by Corynespora cassiicola. Phytopathology 51:305–8.
17. King, C. J., and J. T. Presley. 1942. A root rot of cotton caused by Thielaviopsis basicola. Phytopathology 32:752–61.
18. Ling, I., and J. Y. Yang. 1941. Stem blight of cotton caused by Alternaria macrospora. Phytopathology 31:664–71.
19. McNamara, H. C., R. E. Wester, and K. C. Gunn. 1934. Persistent strands of the cotton root-rot fungus in Texas. J. Agr. Res. 49:531–38.
20. Neal, D. C. 1928. Cotton diseases in

Mississippi and their control. Mississippi State Coll. Agr. Expt. Sta. Bull. 248:1–30.
21. Neal, D. C., and W. W. Gilbert. 1935. Cotton diseases and methods of control. U.S. Dept. Agr. Farmers' Bull. 1745:1–34.
22. Nowell, W. 1923. Diseases of crop-plants in the Lesser Antilles. West India Committee: London, pp. 29–31, 263–76.
23. Presley, J. T. 1950. Verticillium wilt of cotton with particular emphasis on variation of the causal organism. Phytopathology 40:497–511.
24. Rane, M. S., and M. K. Patel. 1956. Diseases of cotton in Bombay. I. Alternaria leaf spot. Indian Phytopathol. 9:106-13.
25. Rane, M. S., and M. K. Patel. 1956. Diseases of cotton in Bombay. II. Helminthosporium leafspot. Indian Phytopathol. 9:169–73.
26. Roger, L. 1953. Phytopathologie des pays chauds. Encyclopedie Mycologique 18:133, 1150–58, 1377–84, 1398–1403, 1924–32, 1944–45.
27. Shapovalov, M. 1927. The two most common decays of cotton bolls in the southwestern states. J. Agr. Res. 35: 307–12.
28. Smith, E. F. 1920. An introduction to bacterial diseases of plants. W. B. Saunders: Philadelphia, pp. 314–39.
29. Sprague, R. 1950. Diseases of cereals and grasses in North America. Ronald Press: N.Y.
30. Streets, R. B. 1937. Phymatotrichum (cotton or Texas) root rot in Arizona. Arizona Univ. Agr. Expt. Sta. Tech. Bull. 71:299–410.
31. Tucker, C. M. 1926. A leaf, bract, and boll spot of Sea-Island cotton caused by Helminthosporium gossypii n. sp. J. Agr. Res. 32:391–95.
32. Venkatarayan, S. V. 1950. Notes on some species of Corticium and Pellicularia. Indian Phytopathol. 3:81–86.
33. Voorhees, R. K. 1942. Life history and taxonomy of the fungus Physalospora rhodina. Florida Agr. Expt. Sta. Bull. 371:4–91.
34. Weber, G. F. 1931. Blight of carrots caused by Sclerotium rolfsii with geographical distribution and host range of the fungus. Phytopathology 21:1129–40.
35. Wolf, F. A. 1917. A squash disease caused by Choanephora cucurbitarum. J. Agr. Res. 8:319–28.

Cucumber, *Cucumis sativus* Linnaeus, and other cucurbits

Soft rot, *Erwinia carotovora* (L. R. Jones) Holland
Wilt, *Erwinia tracheiphila* (E. F. Smith) Holland
Angular leaf spot, *Pseudomonas lachrymans* (E. F. Smith & Bryan) Carsner
Bacterial spot, *Xanthomonas cucurbitae* (Bryan) Dowson
Damping-off, *Pythium debaryanum* Hesse
Fruit rot, *Pythium aphanidermatum* (Edson) Fitzpatrick
Blossom end rot, *Pythium ultimum* Trow
Blight rot, *Phytophthora capsici* Leonian
Downy mildew, *Pseudoperonospora cubensis* (Berkeley & Curtis) Rostowzew
Fruit rot, *Choanephora cucurbitarum* (Berkeley & Ravenel) Thaxter
Metallic mold, *Blakeslea trispora* Thaxter
Soft rot, *Rhizopus nigricans* Ehrenberg
Stem rot, *Sclerotinia sclerotiorum* (Libert) De Bary
Powdery mildew, *Erysiphe cichoracearum* De Candolle
Gummy stem blight, *Mycosphaerella melonis* (Passerini) Chiu & Walker
Stem-end rot, *Physalospora rhodina* (Berkeley & Curtis) Cooke
Soil rot, *Thanatephorus cucumeris* (Frank) Donk
Southern blight, *Botryobasidium rolfsii* (Saccardo) Venkatarayan
Anthracnose, *Colletotrichum lagenarium* (Passerini) Ellis & Halsted
Leaf spot, *Septoria cucurbitacearum* Saccardo
Wilt, *Verticillium albo-atrum* Reinke & Berthhold
Scab, *Cladosporium cucumerinum* Ellis & Arthur
Blight, *Alternaria cucumerina* (Ellis & Everhart) J. A. Elliott
Blight, *Stemphylium cucurbitacearum* Osner
Leaf spot, *Cercospora citrullina* Cooke
Leaf spot, *Cercospora cucurbiticola* P. Henning
Blight, *Corynespora melonis* (Cooke) Lindau

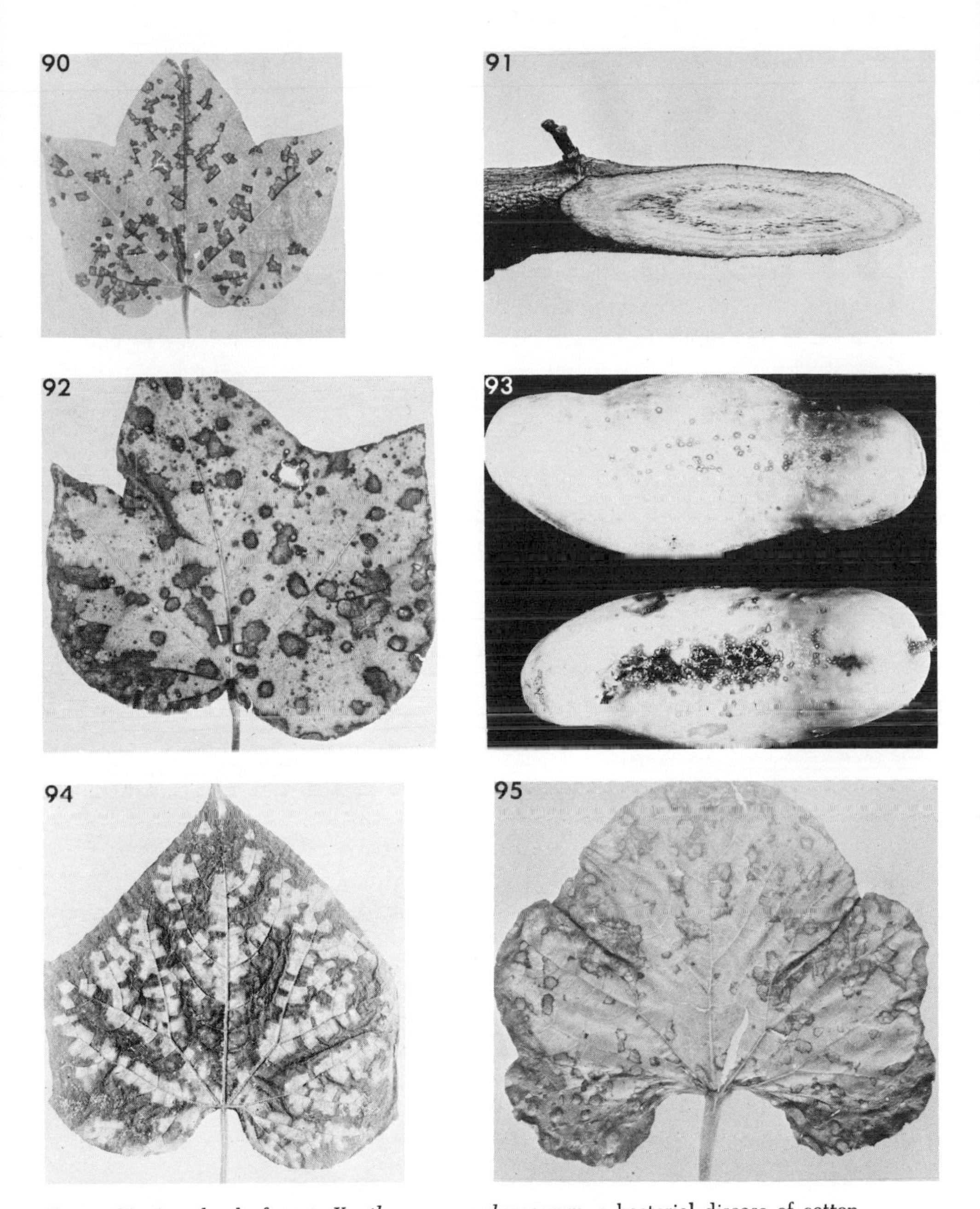

Figure 90. Angular leaf spot, *Xanthomonas malvacearum,* a bacterial disease of cotton.
Figure 91. Wilt, *Fusarium oxysporum* f. sp. *vasinfectum,* on cotton, showing vascular darkening.
Figure 92. Leaf spot, *Alternaria gossypina,* on cotton.
Figure 93. Angular leaf spot, *Pseudomonas lachrymans,* causing cucumber fruit decay.
Figure 94. Downy mildew, *Pseudoperonospora cubensis,* on cucumber.
Figure 95. Downy mildew, *Pseudoperonospora cubensis,* on cantaloupe.

Wilt, *Fusarium oxysporum* f. sp. *niveum* (E. F. Smith) Snyder & Hansen
Wilt, *Fusarium oxysporum* f. sp. *melonis* (Leach & Currey) Snyder & Hansen
Wilt, *Fusarium oxysporum* f. sp. *cucumerinum* Owen
Foot rot, *Fusarium solani* f. sp. *cucurbitae* Snyder & Hansen
Other fungi associated with cucurbitae

Soft rot, *Erwinia carotovora* (L. R. Jones) Holland

Symptoms

Soft rot is frequently found where soil and surface refuse is abundant. Infection follows field, packing house, and transit operations. Mechanical injury is mostly responsible for the soft, watery, rapidly developing decay of cucumber, cantaloupe, melon, and others. The bacteria, entering through wounds, spread rapidly, resulting in a soft, wet, often leaking fruit.

Etiology

The organism is a gram-negative rod, in chains, produces no capsules, is motile by 2–8 peritrichous flagella, and measures $1.5–2 \times 0.6–0.9\ \mu$. Agar colonies are dirty white, glistening, raised, round, and smooth. Optimum temperature, 27C.

References

4, 5, 10.

Wilt, *Erwinia tracheiphila* (E. F. Sm.) Holland

Symptoms

Wilt is scattered throughout the world, affecting watermelon, cantaloupe, and cucumber. It causes a characteristic drooping and flaccid condition of individual leaves and parts of the vine. The plant eventually becomes completely wilted and dies. It is a vascular wound parasite transmitted by insects. The sources of inoculum for primary infection are the striped and spotted cucumber beetles that harbor the bacteria in their bodies during hibernation. Upon sectioning the stems of wilted plants, a sticky exudate oozes from the vascular tissue.

Etiology

The organism is a gram-negative rod, sometimes in pairs, forms capsules, is motile by 4–8 peritrichous flagella, and measures $1.2–2 \times 0.5–0.7\ \mu$. Agar colonies are white, glistening, small, circular, smooth, and viscid. Optimum temperature, 25–30C.

References

2, 4, 5, 10.

Angular leaf spot, *Pseudomonas lachrymans* (E. F. Sm. & Bryan) Carsner

Symptoms

This disease is disseminated almost entirely by contaminated seed from which cotyledon infection takes place. It is mostly associated with cucumber, cantaloupe, and squash. Infected cotyledons show circular, sunken lesions often involving most of the tissue. On the leaves, the spots are at first small, circular, and water-soaked. They rapidly enlarge and, when scattered, become angular, being slightly yellow and delimited by the veins. When numerous they coalesce, forming large, dead blotches that dry and may fall away. This is characteristic of old, scattered leaf spots. On the cucumber fruit, the points of infection are less than 1 mm in diameter, slightly sunken, and circular. Fruit infection is often overlooked because of the small size of the spots and lack of differential color characters of the disease. The penetration is into the loose, fleshy tissue of the placentae where extension is rapid, resulting in the decay of the entire fruit. In older seed fruit, seed become inoculated in this manner. The organism enters the host through the stomata in both foliage and fruit.

Etiology

The organism is a gram-negative rod, in chains, forms capsules, is motile by 1–5 polar flagella, and measures 1.2×0.8 μ. Agar colonies are white, glistening, smooth, round, and entire. A green fluorescent pigment is produced in culture. Optimum temperature, 25–27C.

References

4, 5, 10, 12.

Bacterial spot, *Xanthomonas cucurbitae* (Bryan) Dows.

Symptoms

This disease has been mostly associated with squash, where small, water-soaked, angular spots appear on the lower surface of the foliage. Fruit are not affected. On the upper leaf surface the areas are less definite, except for a yellow halo, although they may enlarge up to almost 1 cm in diameter and have thin, brown, translucent centers. These centers remain intact. Splitting of the stems and petioles sometimes occurs.

Etiology

The organism is a gram-negative rod, in pairs or short chains, forms capsules, is motile by a single polar flagellum, and measures $0.5–1.3 \times 0.4–0.6$ μ. Agar colonies are yellow, opalescent, round, and slightly striate. Optimum temperature, 25–30C.

References

2, 4, 10.

Damping-off, *Pythium debaryanum* Hesse

Symptoms

Damping-off may be found on seedlings in the cotyledon stage on most of the cucurbits, destroying them during post- or pre-emergence periods. The cotyledon tips first display a droopy condition and cease to grow, showing a dull, darker green color than healthy seedlings. The wilting condition continues, and the seedling falls over, becomes dry, and withers. The fungus is soil-inhabiting, attacking the root system near the soil surface and penetrating the young succulent stem, softening it sufficiently to cause it to collapse.

Etiology

The fungus mycelium is hyaline, branched, nonseptate, and plentiful. Mycelium is visible on diseased parts in early morning and measures about 5 μ in diameter. Sporangia are spherical, 15–26 μ in diameter, and germinate by germ tube or zoospores. Zoospores are reniform, biciliate, and 8–12 μ. Oospores are smooth, 12–20 μ in diameter, and germinate by forming germ tubes. Usually there are several antheridia per oogonium.

References

7, 10.

Fruit rot, *Pythium aphanidermatum* (Edson) Fitzp.

Symptoms

This decay is destructive particularly on cucumber and squash, where it may be found producing an abundance of white, cottony mycelium on fruit of all ages. The fungus penetrates fruit in contact with the soil and causes watery decay. The young fruit are sometimes brownish-colored. Roots are also frequently rotted.

Etiology

The fungus mycelium is hyaline, branched, nonseptate, and 4–6 μ in diameter. Sporangia are filamentous and composed of a lobulate, inflated mass of branches. Zoospores, 15–40 in a vesicle, are reniform, biciliate, and 6×12 μ. Oospores are spherical, smooth, 12–28 μ in diameter, thick-walled, and germinate by germ tubes. There is one or sometimes two antheridia per oogonium.

References

2, 7, 10.

Blossom end rot, *Pythium ultimum* Trow

Symptoms

Blossom end rot most frequently occurs on watermelon fruit, beginning shortly after the blossoming period. Browning of the developing fruit continues to maturity, is initiated possibly in the fading flowers, and gradually advances into the melons, resulting in a blackening of the host tissue. The young, severely affected fruit are usually shed. The disease may be found on more mature melons in various stages from small, depressed areas to a decay that involves some or most of the melon.

Etiology

The mycelium is hyaline, branched, and nonseptate. The sporangia are hyaline, spherical, usually terminal, germinate directly, and measure 16–22 μ in diameter. Oospores are spherical, thick-walled, germinate directly, and are up to 20 μ in diameter. This fungus is distinguished by the presence of 1 antheridium per oogonium.

References

7, 10, 11.

Blight rot, *Phytophthora capsici* Leonian

Symptoms

This blight causes extensive damage on squash, pumpkin, and certain other related plants in the field and on melon and cucumber fruit in the field and in transit. A soft, watery breakdown of petioles, stems, and young fruit develops in the field, and a root and crown infection usually kills the plants. In transit, the disease develops on watermelons as surface, water-soaked spots that enlarge and result in extensive brown areas involving surface areas of the rind. In advanced stages, the fungus mycelium overgrows the central portions of the external brown patches. The rind tissue becomes soft but shrinks very little, retains its normal shape, and produces no marked discoloration or odor.

Etiology

The mycelium is white or hyaline, branched, and nonseptate. The sporangia are faintly tinged with yellow, widely variable, elongate to almost spherical, papillate, and average 36×28 μ. The oogonia are spherical with slender stalks, amphigenous antheridia, and measure about 30 μ in diameter. Oospores are spherical, thick-walled, and 22–32 μ in diameter.

References

2, 9, 10.

Downy mildew, *Pseudoperonospora cubensis* (Berk. & Curt.) Rostow.

Symptoms

Mildew reported on cucumber, cantaloupe, squash, and watermelon, in order of importance, is often severe and destructive. The foliage of cucumber shows pale yellow, angular areas in contrast to the otherwise green blade. Eventually the entire leaf becomes yellow. On the lower surface of the yellow areas, the fungus produces a fine-threaded mildew that is hyaline until the conidia, which are lavender to pale purple, develop in abundance. The disease spreads during the growing season by air dissemination of spores, and finally the plants become unproductive. On watermelon, the leaf spots are dark slate to black, less prominent, and only infrequently destructive.

Etiology

The fungus mycelium is hyaline, branched, nonseptate, intercellular, and produces haustoria in the cells. Sporangiophores emerge through the stomata with stem and acute tipped branches. The sporangia are 1-celled, oval, thin-walled, with papillae, and measure 21–39 × 14–23 μ. Zoospores are biciliate and 10–13 μ in diameter. Zoospores germinate by producing mycelium. Oospore formation is rare; the fungus survives on wild hosts between seasons.

References

2, 10, 12.

Fruit rot, *Choanephora cucurbitarum* (Berk. & Rav.) Thaxt.

Symptoms

Blossom infection can be found on day-old or fading blossoms, mostly on squash and pumpkin. The fungus appears as a mold beginning on the margins of the corollas and gradually overgrowing them and continuing onto the young squash. It enters the blossom end, causing a water-soaked area that becomes pale yellow to brown and soft and watery.

Etiology

The entire fruit decays and becomes overgrown by the white metallic-luster mycelium that produces clusters in heads of black spores. The spores are 1-celled, striate, lemon-shaped, with terminal tufts of 12–20 hairlike structures, and measure 18–30 × 10–15 μ. The zygospores are 50–90 μ in diameter.

References

2, 10, 14.

Metallic mold, *Blakeslea trispora* Thaxt.

Symptoms

This mold is a rarely observed growth associated with fading squash blossoms. It begins in the fading corollas and continues to the fruit, causing the young fruit to become discolored, leathery, and mummified. The mold is found on other weak and dying parts of the plant.

Etiology

The fungus produces purplish brown, striate spores on sporangioliferous heads, 3 per sporangiolum, with fine terminal hairs at the poles. The spores measure 8–18 × 5–8 μ. Sporangia, chlamydospores, and zygospores are produced, the last measuring 38–81 × 45–63 μ.

Reference

13.

Soft rot, *Rhizopus nigricans* Ehr.

Symptoms

Soft rot has become almost entirely a disease of the fruit of the Cucurbitaceae in transit, storage, and the market. The affected parts usually become soft and water-soaked. The fungus enters through mechanical injuries. The decay rapidly extends internally and soon involves the entire fruit. The mycelium and spore development is most abundant on exposed, fleshy parts in contrast to the unbroken epidermis.

Etiology

The hyaline, branched mycelium forms tufts of fungus mold supporting the terminal sporangiophores. The spores produced in the sporangia cause it to appear as a black, terminal head with an enclosing wall that eventually breaks, freeing them. The spores are dark, globose to oval, 1-celled, and measure 6×17 μ. The zygospores are dark brown, mostly subspherical, with rough, warty walls, and measure 160–220 μ in diameter.

References

2, 9, 10.

Stem rot, *Sclerotinia solorotiorum* (Lib.) d By.

Symptoms

Stem canker at the soil line in the field, principally on cucumber and cantaloupe, results in a stem girdling. Some decay takes place where the fruit are in contact with the soil, and here there usually develops an abundance of white mycelium. In greenhouses, frequent ascospore dissemination causes infection and killing on trellised vines.

Etiology

Black, large, irregular, hard, shiny sclerotia, measuring up to 2 cm, are produced by the mycelium in or on the tissue. On germination, apothecia are produced and ascospores are formed. The ascospores are hyaline, oval, 1-celled, continuous, and $11–15 \times 5–8$ μ. No oonidia are produced. The microconidia are formed in chains.

References

2, 10, 12.

Powdery mildew, *Erysiphe cichoracearum* DC.

Symptoms

This mildew normally develops on the upper surface of leaves of cucumber, cantaloupe, and squash. A white, superficial, weblike mycelium produces circular blotches that may coalesce and in a short time produce white spores that accumulate on the leaf surface in the form of dust, dense enough to obscure the green color. Diseased leaves curl upward from the edges cuplike, change from green to yellow to tan brown, die, and remain adhering to the upright petioles. Severely diseased plants deteriorate rapidly, and production of marketable fruit ceases.

Etiology

The fungus mycelium is mostly superficial, hyaline, branched, septate, and produces haustoria in epidermal host cells. Erect conidiophores support hyaline, oval, 1-celled conidia in short chains, each spore measuring 28–60 × 11–28 μ. The ascospore stage is formed in cleistothecia which are superficial, appendaged, scattered over the leaf surface, measure 80–140 μ in diameter, and contain 10–30 asci; each ascus contains 2–3 ascospores which are hyaline, 1-celled, elongate oval, and 20–28 × 12–20 μ.

References

2, 10.

Gummy stem blight, *Mycosphaerella melonis* (Pass.) Chiu & Walker

Symptoms

This blight is found frequently on watermelon, but also develops on cucumber and cantaloupe to a much lesser degree. The fungus is seed-borne and is often first observed as brown, circular lesions on the cotyledons. Infections follow on the main stalk or stem just above the soil line. As the vines develop, infections continue to appear along the runners, producing a yellow green color and extensive, longitudinal cracks that produce a yellow, sappy exudate that dries into a brown gum. The foliage lesions are dark brown to slate-colored, small and circular at first, becoming blotchy as they expand. Diseased areas on the leaves often fall out, leaving irregular, torn and tattered leaf blades. No exudate appears on the blades. All diseased parts become speckled with black, small, spherical, ostiolate, partially imbedded pycnidia, measuring up to 300 μ in diameter. This is followed later by the production of perithecia that appear on the stems and crown of the plant.

Etiology

Pycnidiospores are hyaline, mostly 1-septate, elongate, and of two sizes, the smaller ones measuring 4–10 × 2–5 μ and the larger, 5–14 × 2–7 μ. Perithecia are black, rough, scattered, imbedded, erumpent, and 100–165 μ in diameter. The ascospores are hyaline, 2-celled, oblong, fusoid, restricted, 8–16 × 4–9 μ, and an important source of inoculum. The asexual stage is *Diplodina citrullina.*

References

10, 11, 12.

Stem-end rot, *Physalospora rhodina* (Berk. & Curt.) Cke.

Symptoms

This decay is a wound disease occurring primarily on watermelons at harvest time and in storage, transit, and the market. In the field, some loss may result from primary root infection at the crown of the plant where it has been injured or through lesions caused by another fungus. The melons may become infected in the same manner. The excessive losses from stem-end injury are the important part of the disease. Infection takes place through the cut stems from inoculum originating in the field. The fungus invades the melon from the cut stem end and rapidly causes softening, shrinking, water-soaking, and dark discoloration.

Etiology

The mycelium is dark, septate, branched, and mostly internal. Some pycnidia may form over the outer rind. They are subglobose, submerged becoming erumpent, ostiolate, and 120–180 μ. Pycnidiospores are elliptical to oval with longitudinal striations, 1-septate, slightly restricted, and 15×24 μ. The perithecial stage is not common. The asci are stubby, saclike, and contain 8 hyaline, clustered, 1-celled, irregularly elliptical ascospores that measure 24–42×7–17 μ. The imperfect stage is *Diplodia natalensis.*

References

2, 10, 11.

Soil rot, *Thanatephorus cucumeris* (Frank) Donk

Symptoms

The fungus causes damping-off of young seedlings soon after emergence, resulting in poor stands. The plants are killed following early drooping of cotyledons and softening of the stems. The vines frequently show internal lesions on the soil side, and the fruit are frequently pitted with yellow, calloused, rough areas on the bottom. Infection seldom penetrates the depth of the thick rind.

Etiology

The fungus hyphae are yellow brown, branched, septate, and may form brown, irregular sclerotia. There are no vegetative spores. The sexual basidiospores are not frequently found on the plants. The vegetative stage is *Rhizoctonia solani.*

References

2, 11.

Southern blight, *Botryobasidium rolfsii* (Sacc.) Venkat.

Symptoms

This disease is not usually serious but causes damage to watermelon. The fungus girdles the main stem at the soil line thus killing the plant at any stage of development. It also attacks the melon on the soil side, softening the rind and permitting a gradual collapse and melting down of the fruit.

Etiology

An abundance of white mycelium grows over the outside of the melon from the soil area around it. Many brown, spherical, hard sclerotia, measuring 0.5–

1 mm in diameter, are found on the mycelium. There are no vegetative spores produced. The sexual stage is not common. The fungus is soil-inhabiting and is disseminated by the mycelium and sclerotia. The vegetative stage is *Sclerotium rolfsii.*

References

10, 11.

Anthracnose, *Colletotrichum lagenarium* (Pass.) Ell. & Halst.

Symptoms

Anthracnose is most destructive on watermelon, cantaloupe, and cucumber, showing similar but distinctive foliage characteristics on each. On cucumber the leaf spots begin as small, water-soaked areas, enlarging to brown, circular spots up to 1–2 cm. Similar spots develop on cantaloupe but are usually less than 1 cm in diameter. On watermelon the spots are up to several millimeters in diameter on the foliage and are slate to black in contrast to the brown color on other cucurbits. They are sunken and sometimes appear angular in contrast to the circular spots on the other hosts. The fruit are very susceptible to infection at about the time of ripening. Lesions first appear as circular, watery, and sunken; they expand rapidly in the field and after removal from the field. The sunken spots, often up to 1 cm in diameter, have dark centers and pink spore masses in the oldest parts. Lesions continue to enlarge in storage.

Etiology

The conidia develop in acervuli that break through the cuticle, become surrounded by black, hairlike setae, and appear as pink masses among them. Spore production is very plentiful, and the spores are readily disseminated. The sexual stage of the fungus, *Glomerella cingulata,* is not frequently found.

References

9, 10, 11, 12.

Leaf spot, *Septoria cucurbitacearum* Sacc.

Symptoms

This disease is rarely destructive even though it is widely distributed. The spots on most hosts except watermelon are scattered, less than several millimeters in diameter, and usually located in the center of small, bleached areas. Foliage is seldom killed and plants are not materially weakened.

Etiology

The pycnidia are black, small, globose structures forming in the stomatal areas and measuring 80–120 μ in diameter. The pycnidiospores are hyaline, septate, needle-shaped, often expelled in twisting, undulating tendrils, and measure $55–75 \times 1.5–3$ μ.

References

2, 10.

Wilt, *Verticillium albo-atrum* Reinke & Berth.

Symptoms

This wilt is usually of minor importance. It causes a yellowing of diseased plants. The soil-inhabiting fungus enters the root system and slowly progresses through the crown into the runners. The older leaves droop, wilt, turn brown, and are shed. Plants are not usually killed but are stunted and unproductive.

Etiology

The mycelium is branched, septate, hyaline at first and darkens later. The conidia are oval, 1-celled, and 5–7 × 3 μ. They are produced in chains at the tips of the nodal whorls of branches of the conidiophores; often the conidia ball up in water droplets.

References

2, 10.

Scab, *Cladosporium cucumerinum* Ell. & Arth.

Symptoms

Scab is of importance on most cucurbits except watermelon. It produces lesions on all aboveground parts. Leaf spots are pale green, water-soaked areas between the veins and somewhat elongate areas on petioles; they later turn brown. The dark color is due to the formation of an olive brown mold which is the spore-producing structure of the fungus. The fruit are most severely infected when young. The disease continues to maturity. The infections are small, up to 3 mm in diameter, and slightly sunken. Often an exudate occurs that dries to an amber color and fills the cavity. The depressed tissue darkens, and the spores of the fungus are produced in abundance, resembling a soft brown to black mold. In severe infections these lesions are deep, and some drying and shrinkage of the fruit take place.

Etiology

The mycelium is septate, branched, hyaline at first becoming dark greenish black. The conidia develop successively on the tips of branched conidiophores; they are brown, oval to oblong, 1-celled, and measure 18–23 × 4–6 μ.

References

2, 10.

Blight, *Alternaria cucumerina* (Ell. & Ev.) J. A. Elliott

Symptoms

Leaf spot, which is distributed throughout the world, is most serious on cantaloupe and watermelon. The earliest spots form on the upper surface and are small, yellow to brown with light green halos. As they enlarge they become brown. Distinct concentric zones appear, but with age leaves become torn and ragged. On watermelon, the lesions are dark brown to black, smaller, usually less than 2 mm in diameter, with dark halos and marked zonation. Melons are sometimes infected, causing black, moldy patches to develop on exposed upper areas.

Etiology

The mycelium is hyaline at first becoming dark, septate, and branched. The conidia are dark, obclavate, muriform, with 5–13 cross septations and 1–10 longitudinal septations, and measure $30–75 \times 15–25\ \mu$. The conidial beaks are variable, measuring $106–121 \times 2–3\ \mu$.

References

2, 6, 10.

Blight, *Stemphylium cucurbitacearum* Osner

Symptoms

On cucumber, blight forms leaf spots which at first are about 1 mm in diameter and yellow brown with a light yellow halo. They are somewhat angular and may expand up to 1 cm in diameter, forming a sort of blotch.

Etiology

The spores are produced at the apex on brown conidiophores. They are dark-colored, almost globose, muriform, very loosely grouped, and composed of from 5 to 20 cells each. Spores form mostly on the lower leaf surfaces and are 10–18 μ in diameter.

Reference

10.

Leaf spot, *Cercospora citrullina* Cke.

Symptoms

This leaf spot is pale brown with a dark brown to black margin and whitish center, small, circular, and seldom exceeds 1 cm in diameter. Numerous spots cause the leaves to become yellow. It is found mostly on cucumber. The foliage is the principal part of the host infected, although some petiole and stem lesions occur.

Etiology

The fungus produces small stromata. The conidiophores are brown, curved or bent, septate, and usually in fascicles of 2–5 in a cluster. The conidia are hyaline, acicular, cylindrical to obclavate, septate, and $22–50 \times 2–4\ \mu$.

References

2, 11.

Leaf spot, *Cercospora cucurbiticola* P. Henn.

Symptoms

These spots are usually very indefinite and blotchy and seldom cause more than a faintly detectable yellowing of the foliage. The disease is not severe except under highly favorable circumstances.

Etiology

The fungus produces minute specks on the lower leaf surface in the form of stromata 0.5–3 mm in diameter. Conidiophores are pale fuliginous, short, septate, and 10–40 × 4–6 μ. The conidia are pale, smoky, cylindric, curved, septate, and 20–75 × 3–6 μ.

Reference

1.

Blight, *Corynespora melonis* (Cke.) Lindau

Symptoms

This blight is rather rare as a recognized disease of cucumber. It produces brown, circular to angular leaf spots surrounded by a yellow halo. Seriously diseased leaves become yellow and die. Some stem and petiole lesions develop. A certain amount of blossom blight may develop, and, as a result, small fruit may be infected and shed.

Etiology

The fungus produces dark brown conidiophores, which appear on both leaf surfaces in small fascicles. They are multiseptate and 125–200 × 6–10 μ. The conidia are brown, elongate, tapering, obclavate, up to 10-septate, usually curved, and measure 120 × 17 μ.

Reference

2.

Wilt, *Fusarium oxysporum* f. sp. *niveum* (E. F. Sm.) Snyd. & Hans.

Symptoms

This wilt attacks watermelon and is found on few if any additional hosts. The fungus is soil-inhabiting and invades the plant through the root system. Watermelons are susceptible at all stages of development. Some seedlings damp-off, wilt, and die and others are stunted and show a slow wilt, which results in dying leaves and long, leafless stems. The plants are barren and generally die; the interior of stems is often discolored.

Etiology

The fungus develops hyaline, septate, branched mycelium containing terminal or intercalary chlamydospores and hyaline, curved, 3–5-septate conidia that measure 28–64 × 3–5 μ.

References

2, 8, 10, 11.

Wilt, *Fusarium oxysporum* f. sp. *melonis* (Leach & Curr.) Snyd. & Hans.

Symptoms

This wilt, which develops only on cantaloupe, causes a typical damping-off and a pre-emergence seed decay. The usual wilt symptoms are characteristic of diseases caused by *Fusarium* spp. The fungus enters the plant from the soil, progresses through the crown of the plant into the branches, and slowly invades the leaves successively along the vines. The vascular areas may show some discoloration, and finally the plant dies.

Etiology

The mycelium is hyaline, septate, branched, and contains spherical to oval chlamydospores that are terminal or intercalary and measure 5–15 μ in diameter. The conidia are elongate, curved, 2–5-septate, mostly 4-septate, and 30–60 × 3–5 μ.

References

2, 8, 10.

Wilt, *Fusarium oxysporum* f. sp. *cucumerinum* Owen

Symptoms

This wilt is specific to cucumber with little or no infection taking place on cantaloupe or watermelon. The disease is a typical wilt in which the soil-inhabiting fungus gains entrance through the root system, invades the crown of the plant, and advances slowly into the runners, causing the oldest leaves to become yellow, droop, wilt, die, and shed. Plants may be stunted and some damping-off may occur in the seedling stage. There is usually a slight discoloration of the vascular tissue, especially in the taproot.

Etiology

The fungus mycelium is hyaline, septate, branched, and contains globose to ovoid, terminal or intercalary chlamydospores that are smooth and 6–13 × 5–10 μ in diameter. Microconidia are hyaline, numerous, mostly oval, and nonseptate. The macroconidia are elongate, curved, 1–4-septate, and 20–48 × 3.5 μ.

Reference

8.

Foot rot, *Fusarium solani* f. sp. *cucurbitae* Snyd. & Hans.

Symptoms

Foot rot, found principally on squash, may affect pumpkin but is rarely found on other cucurbits. The fungus is seed transmitted and causes pre-emergence decay and damping-off of seedlings. Older plants suddenly wilt and die. Close examination shows a tan to brown discoloration of the older taproot and a deep lesion at the soil line of the main stem. This area is softened, and the stem is girdled at this point. The vascular tissue is invaded and causes the sudden collapse of the entire plant. The white mycelium may appear around the diseased stem. The fruit are invaded and softened in a wet decay, and seed contamination takes place.

Etiology

The fungus is highly variable in pathogenicity. Mycelium is abundant and varies in color from white to purple. The macroconidia are hyaline, forming pink masses, abundant, slightly curved, vary in septations up to 5, mostly 4, and measure 20–68 $\times$ 3–6 μ. The sexual stage of the fungus is *Hypomyces solani* f. *cucurbita.*

References

2, 10.

Other fungi associated with cucumber

Cercospora sechii Stevenson
Coniosporium fairmani Saccardo
Curvularia lunata (Wakker) Boedijn
Helminthosporium sechicola Stevenson
Macrophomina phaseolina (Tassi) Goidanich
Marssonina melonis Dolan
Pestalotia torulosa Berkeley & Curtis
Phoma lagenariae (Thuemen) Saccardo
Phoma lagenicola Saccardo
Phoma subvelata Saccardo
Phyllosticta sechii E. Young
Phymatotrichum omnivorum (Shear) Duggar
Rhizopus stolonifer (Ehrenberg ex Fries) Lind
Septoria vestida Berkeley & Curtis
Thielaviopsis basicola (Berkeley & Broome) Ferraris
Trichothecium roseum Link ex Fries

References: cucumber

1. Chupp, C. 1953. A monograph of the fungus genus Cercospora. Published by the author: Ithaca, N.Y., p. 186.
2. Chupp, C., and A. F. Sherf. 1960. Vegetable diseases and their control. Ronald Press: N.Y., pp. 289–340.
3. Drechsler, C. 1943. Two species of Pythium occurring in southern states. Phyto-

pathology 33:261–99.
4. Elliott, C. 1951. Manual of bacterial plant pathogens. 2d ed. Chronica Botanica: Waltham, Mass.
5. Garrard, E. H. 1951. Bacterial diseases of plants. Bull. Agr. Coll., Guelph, Ontario 478:1–69.
6. Jackson, C. R. 1959. Symptoms and host-parasite relations of the Alternaria leaf-spot disease of cucurbits. Phytopathology 49:731–33.
7. Matthews, V. D. 1931. Studies on the genus Pythium. University of North Carolina Press: Chapel Hill.
8. Owen, J. H. 1956. Cucumber wilt, caused by Fusarium oxysporum f. cucumerinum n. f. Phytopathology 46:153–57.
9. Ramsey, G. B., and M. A. Smith, 1961. Market diseases of cabbage, cauliflower, turnips, cucumbers, melons and related crops. U.S. Dept. Agr., Agr. Handbook 184:21–49.
10. Walker, J. C. 1952. Diseases of vegetable crops. McGraw-Hill: N.Y., pp. 173–208.
11. Walker, M. N., G. F. Weber, and G. K. Parris. 1949. Diseases of watermelon in Florida. Florida Univ. Agr. Expt. Sta. Bull. 459:5–46.
12. Weber, G. F. 1929. Cucumber diseases in Florida. Florida Univ. Agr. Expt. Sta. Bull. 208:5–48.
13. Weber, G. F., and F. A. Wolf. 1927. Heterothallism in Blakeslea trispora. Mycologia 19:302–7.
14. Wolf, F. A. 1917. A squash disease caused by Choanephora cucurbitarum. J. Agr. Res. 8:319–38.

Cumin, *Cuminum cyminum* Linnaeus

Wilt, *Fusarium oxysporum* f. sp. *cumini* Patel & Prasad

Symptoms

In the seedling stage, plants wilt suddenly and die. In older plants, where infection is delayed, the first indication of disease is a drooping of the tips of leaves. Wilt becomes progressively worse until the plants collapse and quickly die. The root system is destroyed, and diseased plants can be pulled up easily. Infection of mature plants results in a yellowing of the foliage progressively from the bottom upward and general wilting. Sometimes only one side of the plant shows infection.

Etiology

The mycelium is white, abundant, aerial, septate, branched, and becomes pinkish. The microconidia are hyaline, 1-celled, and ovoid to ellipsoid. There are no sporodochia or pionnotes. The macroconidia are few in number, 2–3-septate, slightly curved, with a beak, and measure $34–44 \times 3–4$ μ. Terminal or intercalary chlamydospores are present, averaging 8–9 μ in diameter.

References

1, 2, 3.

References: cumin

1. Joshi, N. C., and J. P. Agnihortri. 1958. Studies on wilt disease of cumin (Cuminum cyminum L.) in Ajmer State, India. Lloydia 21:29–33.
2. Mathur, B. L., and N. Prasad. 1964. Studies on wilt disease of cumin caused by Fusarium oxysporum f. sp. cumini. Indian J. Agr. Sci. 34:131–37.
3. Patel, P. N., et al. 1957. Fusarium wilt of cumin. Current Sci. (India) 26:181–82.

Dasheen, *Colocasia esculenta* (Linnaeus) Schott

Soft rot, *Erwinia carotovora* (L. R. Jones) Holland
Blight, *Phytophthora colocasiae* Raciborski
Root rot, *Pythium debaryanum* Hesse
Southern blight, *Botryobasidium rolfsii* (Saccardo) Venkatarayan
Java black rot, *Diplodia natalensis* Evans
Leaf spot, *Phyllosticta colocasiicola* von Hoehnel
Leaf spot, *Cercospora colocasiae* (von Hoehnel) Chupp
Powdery dry rot, *Fusarium solani* (Martius) Appel & Wollenweber
Other fungi associated with dasheen

Soft rot, *Erwinia carotovora* (L. R. Jones) Holland

Symptoms

A watery, slimy, odoriferous decay characterizes this disease in which tubers and corms are invaded, resulting in little or no change in color. Ingress is usually associated with wounds or proximity to decaying vegetables under favorable moisture and temperature conditions.

Etiology

The organism is a gram-negative rod, single or in chains, and forms no capsules; it is motile by 2–5 peritrichous flagella, and measures $1.5–2 \times 0.6–0.9\ \mu$. Agar colonies are grayish white, glistening, round, entire, smooth, and raised. Optimum temperature, 27C.

References

3, 5.

Blight, *Phytophthora colocasiae* Rac.

Symptoms

This disease is primarily associated with the leaves, where purplish brown spots appear. The spots are circular to irregular, water-soaked at first, later becoming brown, and varying in size up to 2 cm, except where several may coalesce. Under humid conditions the spotting of the blade extends to the petiole and then to the corm, where a soft rot develops.

Etiology

The mycelium of the fungus can be observed in early morning along the margin of the leaf spot on the lower leaf surface. The mycelium is hyaline, nonseptate, branched, produces haustoria, and is plentiful both aerially and submerged. The conidiophores are produced on the aerial hyphae. They are hyaline, short, stalky, and produce the conidium at their apex. The conidia are oval to lemon-shaped, with blunt papilla, and measure $38–67 \times 28\ \mu$. Spherical chlamydospores develop in the host tissue and measure $17 \times 29\ \mu$. Oospores are spherical. Zoospores are bean-shaped to irregularly oval and $6–10\ \mu$.

References

1, 4, 9.

Root rot, *Pythium debaryanum* Hesse

Symptoms

The decay of the roots of maturing plants is frequently so severe that none remains in a healthy condition. The decay of the corm however is more spectacular. The root system becomes darkened and flaccid and may be so depleted that the plants fall over and may float loose from the soil. Such plants are usually stunted with shortened, curled or crinkled petioles and yellowish spotted leaves. The diseased corms show a decay starting at the base, gradually involving the entire fleshy portion, and producing various colors from yellow to purple. Partially decayed parts may show distinct demarcation along the line of healthy tissue. Planting stock should not be taken from plants in diseased areas. The fungus is soil-inhabiting.

Etiology

The mycelium is hyaline, branched, nonseptate, and about 5 μ in diameter. The sporangia are hyaline, spherical to oval, terminal or not, usually germinating by germ tube formation, and 15–26 μ in diameter. The oospores are hyaline, spherical, smooth, do not fill the oogonium, germinate directly after rest period, and are up to 21 μ in diameter. Antheridia, 1–6 to an oogonium, arise from the same filament.

References

6, 7.

Southern blight, *Botryobasidium rolfsii* (Sacc.) Venkat.

Symptoms

There are no indications of the disease in the field. The fungus is soil-inhabiting and causes storage rot and decay of the tubers and corms following removal from the field. Careful handling to guard against exposure to contaminated soil or other diseased plant parts should prevent the disease. Sound tubers are probably resistant to infection, but harvesting wounds and exposed surfaces where corms are broken from parents offer avenues of infection. The disease develops rapidly under moist conditions and high temperatures, resulting in a soft, mushy, odorless, colorless decay.

Etiology

The hyaline, branched, septate, and abundant mycelium is white, cottony, and characteristic. There are no vegetative spores. The sclerotia, usually evident on the host tissue or mycelial wefts, are at first white, becoming brown, and finally darker. They are spherical, hard, shiny, resemble radish seed, and are 1 mm or less in diameter. They are easily scattered and remain viable for long periods. The basidial stage is not frequently observed. Under certain conditions, the basidia develop, producing spores that are hyaline, 1-celled, oval to elongate, and $4–9 \times 2–6$ μ. The sterile fungus is *Sclerotium rolfsii.*

Reference

5.

Java black rot, *Diplodia natalensis* Evans

Symptoms

The fungus is a wound parasite and gains entrance to the tubers and corms through mechanical injury or through the point of attachment of corms to the parent tuber. The decay that follows infection is not too distinct externally except for a slight shrinkage and some softening of the internal tissue. There is very little change of color until large portions of the tuber are involved, finally becoming black. The advance of the fungus is slow and complete. The tissue shrinks slightly but does not become mushy and is black throughout. Pycnidia are formed over the outer surface, emerging from the shallow epidermis, and becoming erumpent.

Etiology

The mycelium is brown to black depending on age, septate, branched, and grows through the host cells. Pycnidia are black, scattered or densely clustered, ostiolate, obovate or spherical, usually papillate, and 150–180 μ in diameter. The pycnidiospores are hyaline and 1-celled at first, becoming 2-celled and dark-colored by longitudinal striations. They are oval to elongate, with rounded ends, and measure $20–33 \times 10–18$ μ. The perithecia often associated with the pycnidia are black, subglobose, ostiolate, but larger. The ascospores are hyaline, 1-celled, oval, and measure $30–35 \times 11–14$ μ. The ascospore stage of the fungus is *Physalospora rhodina.*

Reference

5.

Leaf spot, *Phyllosticta colocasiicola* Hoehn.

Symptoms

The spots on the leaves are oval to irregular and vary in size up to 1 in. or more. They are yellowish in outline at first, becoming dark brown with the centers frequently falling away. Spots are visible from both leaf surfaces. Pycnidia are produced on the darkened, dead tissue.

Etiology

The pycnidia are brown to black, ostiolate, subcuticular to emerging, appearing almost superficial, spherical to flattened, and $30–140 \times 50–80$ μ. Pycnidiospores exude in tendrils; the spores are hyaline, oval to elongate, and $4–10 \times 3–5$ μ.

Reference

7.

Leaf spot, *Cercospora colocasiae* (Hoehn.) Chupp

Symptoms

Foliage spots are brown with dark margins, zonate, up to 7 mm in diameter, usually smaller, and show dark stromata on the lower surface.

Etiology

The sporulating structures are dark brown, more or less spherical to ovate, and 20–60 μ in diameter. The conidiophores are in dense fascicles, pale olivaceous, attenuated, usually straight, and $5–30 \times 2–4$ μ. Conidia are pale brown, obclavate, straight, multiseptate, and $25–100 \times 2–4$ μ.

Reference

2.

Powdery dry rot, *Fusarium solani* (Mart.) Appel & Wr.

Symptoms

The decay of tubers and corms is usually dry, hard, and eventually powdery. It is typically a storage rot in which the fungus, which is generally considered saprophytic, penetrates the host through mechanical wounds, largely the exposed areas where corms have been broken from the parent tuber. The decayed tissue does not collapse, but shrinks as it becomes dry. The fungus may be shallow or deep in the flesh, depending on the point of entrance. The tissue finally becomes dull gray and crumbly in consistency.

Etiology

The fungus mycelium is hyaline, branched, septate, and soil-inhabiting. The conidia are hyaline, straight or slightly curved, up to 5-septate, mostly 3, and $30–40 \times 5–7$ μ. Chlamydospores form in the mycelium or terminally; they are 1-celled, smooth to slightly wrinkled, spherical to pyriform, and 8.8×8 μ in diameter.

References

5, 8.

Other fungi associated with dasheen

Cladosporium colocasiae Sawada
Gloeosporium thuemenii Saccardo
Leptosphaeria colocasiae Unamuno
Mycosphaerella colocasiae Hori
Pestalotia vesicolor Spegazzini
Rhizopus nigricans Ehrenberg

References: dasheen

1. Butler, E. J., and G. S. Kulkarni. 1913. Colocasiae blight caused by Phytophthora colocasiae Rac. Mem. Dept. Agr. India 5:233–61.

2. Chupp, C. 1953. A monograph of the fungus genus Cercospora. Published by the author: Ithaca, N.Y., p. 58.
3. Elliott, C. 1951. Manual of bacterial plant pathogens. 2d ed. Chronica Botanica: Waltham, Mass., pp. 39–44.
4. Gomez, E. T. 1925. Leaf blight of gabi. Philippine Agriculturist 14:429–40.
5. Harter, L. L. 1916. Storage-rots of economic aroids. J. Agr. Res. 6:549–71.
6. Matthews, V. D. 1931. Studies on the genus Pythium. University of North Carolina Press: Chapel Hill, pp. 82–87.
7. Parris, G. K. 1941. Diseases of taro in Hawaii and their control. Hawaii Agr. Expt. Sta. Circ. 18:9–17.
8. Roger, L. 1953. Phytopathologie des pays chauds. Encyclopedie Mycologique. 18: 2176–78.
9. Trujillo, E. E. 1965. The effects of humidity and temperature on Phytophthora blight of taro. Phytopathology 55: 183–88.

Date, *Phoenix dactylifera* Linnaeus

Black scorch, *Ceratocystis paradoxa* (Dade) Moreau
Rhizosis, *Ceratostomella radicicola* Bliss
Root rot, *Omphalia pigmentata* Bliss and *O. tralucida* Bliss
Offshoot disease, *Diplodia phoenicum* (Saccardo) Fawcett & Klotz
Leaf spot, *Pestalotia palmarum* Cooke
Brown leaf spot, *Pestalotia phoenicis* Vize
Mold, *Aspergillus phoenicis* (Corda) Thom & Church
Flower blight, *Mauginiella scaettae* Cavara
Pink mold, *Gliocladium roseum* (Link) Thom
Leaf spot, *Exosporium palmivorum* Saccardo
False smut, *Graphiola phoenicis* (Mougeot) Poiteau
Other fungi associated with date

Black scorch, *Ceratocystis paradoxa* (Dade) Moreau

Symptoms

A dark brown to black, rough, irregular, elongate spot or streak along the rachis or leafstalk is the most frequently observed indication of the disease. There is some distortion of the leaf, and the central bud may be eccentric or blackened and killed. A darkening of the florescence and fruitstalk develop within the spathe. The developing fruitstalk and the peduncles become diseased during the growing period; the severity depends on the stage of development at the time of infection.

Etiology

The mycelium is dark-colored, septate, and branched. The conidia or endospores, produced by cutting off sections of the hyaline conidiophores, often appear in chains and measure 5–15×3–7 μ. Black, oval, 1-celled spores, measuring 11–17×7–15 μ, are produced on the dark mycelium. The vegetative stage of the fungus is *Thielaviopsis paradoxa.*

References

2, 6, 8, 11.

Rhizosis, *Ceratostomella radicicola* Bliss

Symptoms

The first visible sign of the disease is a reddish brown discoloration of the pinnae. Leaves begin to wilt and die within a few days. Infection starts from the lower whorls, moves upward, and encompasses the fruitstalks; dates drop from the strands. Before complete wilting of all older leaves, the tightly folded young leaves lose turgor and turn whitish. Affected palms die within 4–6 weeks. Diseased areas on the roots and top of the plant are invaded later by secondary organisms causing rapid decay. It is a severe disease on seedlings. Infections on fruit occur during a later stage.

Etiology

Hyphae are hyaline, septate, branched, and 3–10 μ in diameter. Conidiophores are hyaline, erect, up to 3-septate, and form spores at the tips; the spores are continuous, hyaline, and measure 8–15×6–10 μ. Macroconidiophores are hyaline and septate; conidia are hyaline to dark, exogenous, thick-walled, ovate, and 12–22×11–16 μ. Perithecia are solitary and more or less submerged. There are 1 to many appendages which are dark, branched, and have fimbriate beaks. Asci are deliquescent, and ascospores are hyaline, elliptical, held in a mucoid, pearly bead, and measure 8–15×2–4 μ. The vegetative stage is *Chaloropsis thielavioides.*

References

2, 3.

Root rot, *Omphalia pigmentata* Bliss and *O. tralucida* Bliss

Symptoms

Lesions occur on the leaf bases, main trunk, and roots. Brown discolorations are the first indication of diseased epidermis. Affected areas enlarge and show some depression. Roots are reddish brown externally, dark brown internally, and are often killed when young before penetrating the soil. Offshoots are frequently invaded by the fungus. Rhizomorphic mycelium penetrates the leaf-bases and spaces and spreads out over the surfaces. Diseased trees show some wilting of lower leaves, and, as the infection becomes more severe, additional leaves wilt and die. Trees show an off-color, leaf size is reduced, and stunting is evident. Fruit branches are reduced in number, and the individual dates are undersized, shriveled, and fibrous.

Etiology

Both fungi produce white, silky mycelium with clamp connections. The sporophores of *O. pigmentata* develop stipes, 5–35×1–2 mm, and orange yellow to white pilei, 5–33 mm in diameter. The other fungus shows stipes, measuring 4–23×0.5–1.5 mm, and a white to buff pileus, 3–18 mm in diameter. The basidiospores are hyaline to white, oval, papillate, and measure 6–9×4–7 μ and 11–16×3–6 μ, respectively.

References

1, 4.

Offshoot disease, *Diplodia phoenicum* (Sacc.) Fawc. & Klotz

Symptoms

The outside leaves of offshoots often are the first to die, although when infection occurs in the bud area the inner leaves die first. In offshoots, infection appears to originate at the union with the parent plant. On older plants, there sometimes develops a dark streak up the stem or rachis for a few inches to several feet. The leaf may remain green and unaffected, although cross sections reveal light brown to chocolate-colored tissue involved in the dark streaks.

Etiology

The pycnidia are black, subglobose, ostiolate, submerged to erumpent, and appear on dead stem and leaf tissues. The spores are hyaline to dark, 2-celled, oval to elongate, and 22–24 × 10–12 μ.

References

7, 8.

Leaf spot, *Pestalotia palmarum* Cke.

Symptoms

The fungus develops as a wound parasite. Spots are usually found on leaves that are declining or have been injured. Infections produce pale yellow to brown spots that may be variously scattered and variable in contour. The pustules of the fungus are usually plentiful on dead fronds.

Etiology

The pustules, in the form of acervuli, are black, circular to elliptical, subepidermal becoming erumpent, and exude quantities of colored conidia that measure 16–21 × 4–7 μ. Center cells are colored, large, and constricted, and terminal cells are hyaline and small. The basal cell is short pedicellate, and the apical cell supports 3 setulae that are straight or spreading and measure 8–20 × 2 μ.

Reference

9.

Brown leaf spot, *Pestalotia phoenicis* Vize

Symptoms

Gray to brown, sharply defined, variously scattered spots appear on the pinnae of the leaves, frequently coalescing into extensive, brown, irregular areas.

Etiology

The pustules are variously scattered, mostly on the upper surface, usually somewhat submerged becoming erumpent, and 75–280 × 75–100 μ. The conidia are erect and slightly bent with 3 central colored cells and 2 end cells that are hyaline. The conidia measure 16–22 × 5–7 μ. The terminal cell is conical, supporting mostly 3 setulae that are usually straight, up to 22 μ long, and have characteristically knobbed terminals. The pedicel is 4–6 μ long.

Reference

10.

Mold, *Aspergillus phoenicis* (Cda.) Thom & Church

Symptoms

The fruit is the only part of the host showing the black, powdery mold. The mold usually grows over the mature fruit, causing them to soften and to lose their bright color. The sugary, fleshy parts are destroyed and overgrown by the fungus.

Etiology

The spores of the fungus are black, 1-celled, produced in chains, spherical or nearly so, and 3–5 × 5 μ.

References

5, 14.

Flower blight, *Mauginiella scaettae* Cav.

Symptoms

The disease is found only on the florescence in the spring at blooming time. The spathes show yellow to brown spots with yellow borders. The infection advances from the extreme tips to the main stalk, which becomes weakened by the decay that exudes a brown liquid. The whole florescence falls. Partial invasion sometimes occurs, and some fruit may develop.

Etiology

The mycelium is composed of large cells that support erect conidiophores developing up to 6 cells, mostly 2–3. These are elongate to round, become separated, and function as conidia, measuring 7–9 μ, mostly 8 μ, in diameter.

Reference

14.

Pink mold, *Gliocladium roseum* (Lk.) Thom

Symptoms

This mold is ubiquitous and is a composite of molds that are more or less pink and associated with mature fruit of the date palm as well as many other fruit and plant parts, especially under warm, humid conditions. The disease is observed sometimes in the orchard but is mostly found after the fruit is removed from the tree. It also produces a mold on young leaves.

Etiology

The conidia are hyaline, oval, 1-celled, smooth, apiculate, and 5–7×3–5 μ. They develop in chains from upright conidiophores and often appear in pustular masses in viscid liquid droplets. The penicilli are once- or twice-branched.

Reference

13.

Leaf spot, *Exosporium palmivorum* Sacc.

Symptoms

As a seedling disease the leaf spotting is often severe. The spots are readily detected when about 1 mm in diameter, appearing as cleared areas somewhat transparent as though water-soaked. They enlarge rapidly and form characteristic, somewhat zonate spots resulting from alternating rings of light and dark, depending on proximity to principal veins in the leaf.

Etiology

The spots become speckled with brown to black stromata that are mostly superficial, compact, flat, circular, and 200–400×100–200 μ. The crest is composed of crowded, elongate conidiophores that produce many dark-colored, long, curved, 8–14-septate, obclavate conidia. They are usually attenuated toward the tip and measure 75–120×7–12 μ.

Reference

14.

False smut, *Graphiola phoenicis* (Moug.) Poit.

Symptoms

There appear on the pinnae of the older leaves scattered, crateriform pustules that are brown to black, small, cylindrical, often raised, and sandpapery to the touch. They measure up to 1 mm in diameter and at maturity produce bundles of bright yellow, protruding filaments several millimeters in length and spores as a yellowish dust. There is usually a yellow area surrounding each pustule on green foliage. The leaf petioles and rachises also are attacked and erumpent; lesions up to several inches in extent may develop lengthwise on the stem.

Etiology

The fungus produces a pustule with a prominent peridium that becomes hard and contains a hyaline, septate hypha that bears quantities of spores along its course or in chains or clumps. The spores are yellow, dusty, elliptical, and 3–6 μ.

References

12, 15, 16.

Other fungi associated with date

Alternaria citri Ellis & Pierce
Alternaria stemphylioides Bliss
Armillaria mellea Vahl ex Fries
Aspergillus niger Tieghem
Catenularia fuliginea Saito
Citromyces ramosus Bainier & Sartory
Fusarium albedinis (Killermann & Marignoni) Malencon
Fusarium semitectum Berkeley & Ravenel
Gibberella fujikuros (Sawada) Wollenweber
Haplosporella palmicola Henning
Helminthosporium molle Berkeley & Curtis
Meliola furcata Léveillé
Phomopsis phoenicola Traverso
Phymatotrichum omnivorum (Shear) Duggar
Physalospora rhodina (Berkeley & Curtis) Cooke
Phytophthora palmivora Butler
Poria ravenelae Berkeley & Broome
Zygosaccharomyces cavarae Rodio

References: date

1. Bliss, D. E. 1938. Two new species of Omphalia which cause decline disease in date palms. Mycologia 30:313–26.
2. Bliss, D. E. 1941. A new species of Ceratostomella on the date palm. Mycologia 33:468–82.
3. Bliss, D. E. 1941. Relation of Ceratostomella radicicola to rhizosis of the date palm. Phytopathology 31:1123–29.
4. Bliss, D. E. 1944. Omphalia root rot of the date palm. Hilgardia 16:15–124.
5. Brown, J. G. 1920. Rot of date fruit. Botan. Gaz. 69:521–29.
6. Dade, H. A. 1958. Ceratostomella paradoxa, the perfect stage of Thielaviopsis paradoxa (de Seynes) von Hoehnel. Trans. Brit. Mycol. Soc. 13:184–94.
7. Fawcett, H. S. 1930. An offshoot and leafstalk disease of date palms due to Diplodia. Phytopathology 20:339–44.
8. Fawcett, H. S., and L. J. Klotz. 1932. Diseases of the date palm, Phoenix dactylifera. Calif. Univ. Agr. Expt. Sta. Bull. 522:1–12, 14–25.
9. Guba, E. F. 1929. Monograph of the genus Pestalotia De Notaris. Phytopathology 19:191–232.
10. Guba, E. F. 1961. Monograph of Monochaetia and Pestalotia. Harvard University Press: Cambridge, Mass., pp. 89–91.
11. Klotz, L. J., and H. S. Fawcett. 1932. Black scorch of the date palm caused by Thielaviopsis paradoxa. J. Agr. Res. 44: 155–66.
12. Nixon, R. W. 1957. Differences among varieties of date palm in tolerance of Graphiola leafspot. Plant Disease Reptr. 41:211–13.
13. Raper, K. B., and C. Thom. 1949. A manual of the Penicillia. Williams & Wilkins: Baltimore, Md., pp. 678–80.
14. Roger, L. 1953. Phytopathologie des pays chauds. Encyclopedie Mycologique 18:1471, 1895–97, 2202.
15. Sinha, M. K., R. Singh, and R. Jeyarajan. 1970. Graphiola leaf spot on date palm (Phoenix dactylifera): susceptibility of date varieties and effect on chlorophyll content. Plant Disease Reptr. 54:617–19.
16. Smith, E. F. 1888. A date palm fungus (Graphiola phoenicis Poit.). Botan. Gaz. 13:211–13.

Durian, *Durio zibethinus* Linnaeus

Claret canker, *Pythium complectius* Braun
Patch canker, *Phytophthora palmivora* Butler

Claret canker, *Pythium complectius* Braun

Symptoms

The fungus produces cankers on the lateral roots slightly below the soil surface; the cortical tissues become brown and dry. Trees decline when severely attacked.

Etiology

The mycelium is hyaline, branched, and nonseptate. The sporangia are spherical to oval, and measure 16–22 μ. Chlamydospores develop on the scanty mycelium. The oospores are hyaline, spherical, yellowish, and average 11–21 μ in diameter.

Reference

1.

Patch canker, *Phytophthora palmivora* Butl.

Symptoms

The reddish brown cankers are the most conspicuous stage of the disease. They develop slowly, depending largely on environmental factors of moisture and ventilation, and are most destructive during the rainy season. A general necrosis appears and decomposition of roots continues.

Etiology

The mycelium is hyaline, branched, nonseptate, stout, and produces slender haustoria. On exposed surfaces a whitish, mildewlike mat may develop. Chlamydospores are hyaline, spherical, and measure 20×45 μ. Sporangia are hyaline, papillate, oval, and 32×58 μ. Fifteen to 20 zoospores develop as sporangia germinate. Oospores are hyaline, spherical, smooth, and 18–41 μ. They are rarely observed.

Reference

1.

Other fungi associated with durian

Botryobasidium salmonicolor (Berkeley & Broome) Venkatarayan
Calonectria rigidiuscula (Berkeley & Broome) Saccardo
Ganoderma pseudoferreum (Wakefield) Overeem & Steinmann
Fusarium solani (Martius) Appel & Wollenweber
Glomerella cingulata (Stoneman) Spaulding & Schrenk
Phomopsis durionis Sydow
Phyllosticta durionis Zimmermann
Phytophthora nicotianiae var. *parasitica* (d'Haan) Tucker
Thanatephorus cucumeris (Frank) Donk

Reference: durian

1. Roger, L. 1951. Phytopathologie des pays chauds. Encyclopedie Mycologique 17: 617, 657.

Eggplant, *Solanum melongena* Linnaeus

Brown rot, *Pseudomonas solanacearum* E. F. Smith
Ring rot, *Corynebacterium sepedonicum* (Spieckermann & Kotthoff) Skaptason & Burkholder
Angular leaf spot, *Xanthomonas vesicatoria* (Doidge) Dowson
Leak, *Pythium aphanidermatum* (Edson) Fitzpatrick
Late blight, *Phytophthora infestans* (Montagne) De Bary
Fruit rot, *Phytophthora parasitica* Dastur
Downy mildew, *Peronospora tabacina* Adams
Cottony rot, *Sclerotinia sclerotiorum* (Libert) De Bary
Tipover, *Diaporthe vexans* (Saccardo & Sydow) Gratz
Southern blight, *Botryobasidium rolfsii* (Saccardo) Venkatarayan
Soil rot, *Thanatephorus cucumeris* (Frank) Donk
Rust, *Puccinia angivyi* Bouriquet
Fruit spot, *Ascochyta hortorum* (Spegazzini) C. O. Smith
Wilt, *Verticillium albo-atrum* Reinke & Berthold
Leaf spot, *Cercospora melongenae* Welles
Leaf spot, *Cercospora deightonii* Chupp
Leaf spot, *Alternaria solani* (Ellis & G. Martin) Jones & Grout
Wilt, *Fusarium oxysporum* f. sp. *melongena* Matuo & Ishigami
Other fungi associated with eggplant

Brown rot, *Pseudomonas solanacearum* E. F. Sm.

Symptoms

Bacterial brown rot has been reported to be worldwide in distribution on eggplant and on a large number of other plants. The early symptoms include a drooping of the terminal leaves usually followed within a few days by a general collapse of the plant or sometimes part of the plant. Often one side wilts and becomes brown and dies. Dead plants dry and remain upright in place. The disease may appear on single plants in a field or may be correlated with certain field areas. The organism survives in the soil.

Etiology

The organism is a gram-negative rod, forms no capsules, is motile by a single polar flagellum, and measures $0.5 \times 1.5\ \mu$. Agar colonies are white to dirty white, opalescent, becoming brownish, smooth, shiny, and irregular. Optimum temperature, 35–37C.

References

5, 12.

Ring rot, *Corynebacterium sepedonicum* (Spieck. & Kotth.) Skapt. & Burkh.

Symptoms

Ring rot has not been recognized in nature on this host. Inoculations have shown that a severe wilting and vascular infection resulted in 10–12 days, followed by conspicuous leaf malformation, stunting, bronzing, shriveling, and drying. Most commercial varieties are susceptible, although some Puerto Rican varieties show resistance.

Etiology

The organism is gram-positive, pleomorphic, with globulose to club-shaped cells, and measures 0.8–1×0.4–0.6 μ. Agar colonies are translucent to white, glistening, thin, small, and smooth. Optimum temperature, 20–23C.

References

5, 9.

Angular leaf spot, *Xanthomonas vesicatoria* (Doidge) Dows.

Symptoms

Leaf spot is inconspicuous and unimportant on this host. It is seldom observed and should be verified by culture and inoculation. Spots are water-soaked to brown, small, irregular in outline, and less than 1 mm in diameter.

Etiology

The organism is a gram-negative rod, motile by a single polar flagellum, and measures 1–1.5×0.6–0.7 μ. Agar colonies are yellow, semitranslucent, circular to irregular, and butyrous. Optimum temperature, 30C.

Reference

5.

Leak, *Pythium aphanidermatum* (Eds.) Fitzp.

Symptoms

Leak as found in the field and in transit results from the fungus invasion of the fruit at its blossom-end. The organism invades from the soil, resulting in the production of an abundance of white, cottony mycelium, especially in humid weather. The fruit becomes soft and collapses as it is overgrown by the fungus, resulting in a watery rot and, in transit or storage, a leaking container.

Etiology

The mycelium is hyaline, nonseptate, branched, and 4–6 μ. Sporangia are lobulate. There are few to many reniform, biciliate zoospores that measure 6–12 μ. Oospores are spherical, smooth, and 12–28 μ.

References

4, 12.

Late blight, *Phytophthora infestans* (Mont.) d By.

Symptoms

Blight in the seedling stage does little or no damage with the exception of occasional losses in the seedbed. A brown, rough discoloration may be found on the fruit followed by internal, water-soaked areas containing the fungus mycelium.

Etiology

The hyaline, branched, nonseptate, uniformly smooth mycelium measures 3–8 μ in diameter. The conidiophores are typical, hyaline, with swollen joints, and mostly erect. The conidia are ovate and granular. Diseased fruit when cut open and placed in a moist chamber produce sporangia or conidia in abundance in 12 hours.

References

7, 12.

Fruit rot, *Phytophthora parasitica* Dast.

Symptoms

This decay often results in losses of as much as 50 per cent. The lesions on the fruit are water-soaked at first, circular to oval becoming irregular, and eventually involve the entire fruit. The lesions are not sunken or characteristically marked. The fungus produces an internal softness and sporulates profusely on cut surfaces.

Etiology

The mycelium is hyaline, nonseptate, branched, irregular to smooth, and 3–8 μ in diameter. The conidia are hyaline, ovate, granular, with prominent papillae, and measure 30–34 × 27–30 μ, averaging 34 × 28 μ.

References

2, 3.

Downy mildew, *Peronospora tabacina* Adams

Symptoms

The foliage of seedlings often becomes yellowish, withers, and turns brown and blackish. Plants are not frequently attacked and usually survive infection with some loss of foliage.

Etiology

The mycelium is hyaline, branched, nonseptate, submerged, and produces erect, branched sporangiophores through the stomata. The lemon-shaped conidia measure 17–28 × 13–17 μ and germinate directly. Brown oospores are spherical, rough, and 20–60 μ in diameter.

Reference

12.

Cottony rot, *Sclerotinia sclerotiorum* (Lib.) d By.

Symptoms

This rot causes lesions on the aerial parts of plants in any stage of development, but is most frequent at about blossoming time. The infections may occur on any of the branches and are usually initiated at the nodes. Branches are killed, and white mycelium is often evident on the brown areas. The fungus also invades the main stem at the soil line, causing cankers, an eventual girdling of the stem, and death of the plant.

Etiology

The mycelium is hyaline to white, plentiful, branched, and septate. There are no functional conidia. Sclerotia are black, irregular in shape, up to 1 mm in diameter, and usually located in the stem pith of killed plants. They are the surviving stage of the fungus.

Reference

12.

Tipover, *Diaporthe vexans* (Sacc. & Syd.) Gratz

Symptoms

This disease is so named because at the critical stage of growth the plant often falls over onto the soil because of a stem lesion at or near the soil line. The fungus is seed-borne and may survive in cultivated fields on plant debris for several years. Seedlings may damp-off after emergence or may be killed before that time. Infected plants usually are not killed even though they bear stem infections. They continue to be the source of inoculum as pycnidia appear in the cortex and produce viable pycnidiospores. These infected seedlings may die anytime and often fail to survive transplanting. When set in the field, some plants will respond rapidly and produce large-branched and heavy foliaged plants. However, the stem lesion prevents the main stem from enlarging in diameter sufficiently to support the heavy aerial parts. The disease is found as brown, irregular, scattered, frequently zonate spots on the leaves and as internodal, brown cankers on the aerial branches, petioles, and peduncles. The fruit lesions are readily detected at an early stage of development by areas that are a darker purple than the surrounding healthy fruit surface. However, within a day or two these areas rapidly expand and fade to a tan color. If several infections appear, the entire fruit becomes tan and soft and begins to shrivel and shrink.

Etiology

Pycnidia appear on all of the diseased areas as black, scattered, often zoned specks and are diagnostic of the disease. The pycnidia vary greatly in size, measuring from 60 to 300 μ in diameter. The pycnidiospores are hyaline, 1-celled, oval to elongate, $5–8 \times 2–3$ μ, and are often exuded in undulating tendrils. The stylospores are hyaline, slightly curved or with only a bent tip, nonseptate, and 28×13 μ. This stage is *Phomopsis vexans*. The perithecia have been produced in artificial culture and have not been observed on the host plant in the field. In culture they are generally clustered, partially imbedded in a stroma, and produce elongated, ostiolate beaks. The asci are hyaline, 8-spored, clavate, sessile, and 28–44×5–12 μ. The ascospores are hyaline, biseriate, ellipsoid to oval, 1-septate, with cellular guttulae, and measure 9–12×3–5 μ.

References

2, 6, 12.

Southern blight, *Botryobasidium rolfsii* (Sacc.) Venkat.

Symptoms

This disease is frequently associated with wilted plants in the field. It causes lesions on the main stem near or slightly above the soil line, resulting in a girdling effect.

Etiology

Mycelium is white. Many brown, small, spherical sclerotia may be produced that measure less than 1 mm in diameter.

Reference

12.

Soil rot, *Thanatephorus cucumeris* (Frank) Donk

Symptoms

Rhizoctoniose as associated with the host is frequently important in the damping-off of seedlings and in the occasional stem-girdle at the soil line. The cortex is often killed, and a scaly condition develops on the blossom end of fruit close to or in contact with the soil. This manifestation is usually unimportant unless a soft decay develops following some mechanical injury. The imperfect stage is *Rhizoctonia solani*.

Etiology

The mycelium is hyaline becoming brown, septate, branched, often weblike, and loose around diseased plants. Sclerotia are brown, irregular, usually thin, and superficial. No conidia are formed but basidiospores frequently develop.

References

2, 12.

Rust, *Puccinia angivyi* Bour.

Symptoms

This disease is usually destructive and severe wherever found. All parts of the plant aboveground are susceptible to infection at all stages of their development. The fungus produces reddish orange, raised aecia on the stems and leaves, causing considerable malformation and distortion; the same effect develops on the peduncle, calyx, and fruit. On the stem the lesions are raised; they are the same color as in the aecial stage and gradually become brown as the telia develop in similarly shaped pustules. All the pustules are subepidermal and emerge at maturity, producing the typical peridia which break open, liberating the spores.

Etiology

The mycelium of the fungus is hyaline, septate, branched, and measures about 4 μ in diameter; haustoria are formed. The aecia are globose-elongate, up to 0.5 mm in diameter, and contain chains of orange yellow aeciospores. The aeciospores are subglobose to angular and measure 25×20 μ. The teliospores are brown, 2-celled, attached by a long, slender pedicel, and measure 35–52×21–27 μ. The aecial and telial stages of the fungus are the only spore stages known. Other rust fungi of less importance reported on this host are the following: *Puccinia paspalicola* (P. Henn.) Arth.—pycnia and aecia; *P. substriata* Ell. & Barth.—uredia and telia; and *P. tubulosa* (Pat. & Gaill.) Arth. —pycnia and aecia.

References

2, 12.

Fruit spot, *Ascochyta hortorum* (Speg.) C. O. Sm.

Symptoms

The disease frequently causes cankers of variable dimensions on the fruit.

Etiology

Cankers are characterized by black pycnidia and hyaline, 2-celled pycnidiospores that measure 8–10×4–5 μ.

Reference

14.

Wilt, *Verticillium albo-atrum* Reinke & Berth.

Symptoms

Wilt appears infrequently on this host, resulting in a slow killing of the plant. The aboveground portions develop a droopy, off-color condition. The plant gradually or suddenly deteriorates, becomes brown, and remains standing after dying. The woody stem tissue displays a browning particularly of the vascular system. Cultures are necessary for identification.

Etiology

The mycelium is hyaline to slightly colored, branched, and septate. Conidiophores arise from the mycelium and develop secondary branches terminated by chains of hyaline, 1-celled, oval conidia that measure $5–7 \times 3$ μ.

References

2, 8, 12.

Leaf spot, *Cercospora melongenae* Welles

Symptoms

These spots are circular to irregular and light to dark brown depending on their age and size which is from small points up to 1 cm in diameter.

Etiology

Conidiophores form in fascicles of several to a dozen on the lower leaf surface. The conidia are hyaline, straight or curved, multiseptate, and $12–40 \times 2–5$ μ.

References

2, 12.

Leaf spot, *Cercospora deightonii* Chupp

Symptoms

The fungus forms mostly irregular leaf spots. Viewed from the upper surface they are yellow green to brown. On the undersurface they are smoky to sooty-olivaceous, because of the production of dark, single, conidiophores and colored conidia.

Etiology

The conidia are obclavate to fusiform, multiseptate, and $25–90 \times 2–5$ μ.

Reference

2.

Leaf spot, *Alternaria solani* (Ell. & G. Martin) Jones & Grout

Symptoms

This blight produces brown, irregular, often zonate spots on the foliage of plants of all ages, beginning occasionally on the cotyledons. In the seedling stage infection may be severe, causing extensive killing. However, the disease is not important in the field. There may be brown, scabby areas on the fruit surface.

Etiology

The conidia are olive brown, 5–10-septate, with 3–7 muriform septations, long beaks, and measure $90–140 \times 12–16$ μ.

Reference

12.

Wilt, *Fusarium oxysporum* f. sp. *melongena* Matuo & Ishigami

Symptoms

Seedlings frequently show lack of turgidity followed by a yellowing of the leaf tips. The decline continues in the form of a wilt, resulting in stunted plants and yellowing of the foliage, beginning at the lower areas and developing upward. Severely diseased plants are nonproductive and die prematurely. The vascular tissues become yellow and brown often on only one side of the stem. Cultures are necessary for identification of the causal parasite.

Etiology

The mycelium is hyaline, septate, branched, and various hues of yellow, rose, or blue in culture. The microconidia are hyaline, 1-celled, oval, often in heads, and measure 8–10×3–4 μ. The macroconidia are hyaline, elongate, curved, up to 5-septate, mostly 3-septate, and 25–40×3–5 μ. Sporodochia are frequently produced, and chlamydospores are present. The fungus is specific to eggplant.

Reference

10.

Other fungi associated with eggplant

Botrytis cinerea Persoon ex Fries
Cladosporium fulvum Cooke
Colletotrichum melongenae (Ellis & Halsted) Averna-Saccá
Diplodia natalensis P. Evans
Erysiphe cichoracearum De Candolle
Macrophomina phaseolina (Tassi) Goidanich
Myrothecium roridum Tode ex Fries
Phyllosticta melongenae Sawada
Rhizopus stolonifer (Ehrenberg ex Fries) Lind
Septoria melongenae Lobik
Stemphylium solani Weber

References: eggplant

1. Bouriquet, G. 1946. Les maladies des plantes cultivees a Madagascar. Encyclopedie Mycologique 12:481–90.
2. Chupp, C., and A. F. Sherf. 1960. Vegetable diseases and their control. Ronald Press: N.Y., pp. 341–48, 556–59.
3. Dastur, J. F. 1913. On Phytophthora parasitica nov. spec., a new disease of the castor oil plant. Mem. Dept. Agr. India, Botan. Ser. 5:177–226.
4. Drechsler, C. 1926. The cottony leak of eggplant fruit caused by Pythium aphanidermatum. Phytopathology 16: 47–50.
5. Elliott, C. 1951. Manual of bacterial plant pathogens. 2d ed. Chronica Botanica: Waltham, Mass.
6. Gratz, L. O. 1942. The perfect stage of Phomopsis vexans. Phytopathology 32: 47–50.
7. Harrison, A. L., and D. G. A. Kelbert. 1944. Late blight on eggplant in Florida. Plant Disease Reptr. 28:116.
8. Kamal, M., and C. Saydam. 1970. Verticillium wilt of eggplant in Turkey. Plant Disease Reptr. 54:241–43.
9. Larson, R. H. 1944. The ring rot bacterium in relation to tomato and eggplant. J. Agr. Res. 69:309–25.
10. Matuo, T., and K. Ishigami. 1958. On the wilt of Solanum melongena L. and its causal fungus, Fusarium oxysporum

f. melongena n. f. Ann. Phytopathol. Soc. Japan 23:189–92.
11. Ramsey, G. B., J. S. Wiant, and L. P. McColloch. 1952. Market diseases of tomatoes, peppers and eggplants. U.S. Dept. Agr., Agr. Handbook 28:47–49.
12. Walker, J. C. 1952. Diseases of vegetable crops. McGraw-Hill: N.Y., pp. 297–313.
13. Weber, G. F. 1938. Phomopsis blight of eggplants. Florida Univ. Agr. Expt. Sta. Press Bull. 522:1–2.
14. Wolf, F. A. 1914. Fruit rots of eggplant. Phytopathology 4:38. (Abstr.)

Fig, *Ficus carica* Linnaeus

Fruit decay, *Phytophthora palmivora* Butler
Whiskers, *Rhizopus nigricans* Ehrenberg
Souring, *Saccharomyces apiculatus* Reess
Limb blight, *Sclerotinia sclerotiorum* (Libert) De Bary
Twig blight, *Megalonectria pseudotrichia* (Schweinitz) Spegazzini
Rust, *Physopella fici* (Castagne) Arthur
Felty fungus, *Septobasidium bogoriensis* Patouillard
Southern blight, *Botryobasidium rolfsii* (Saccardo) Venkatarayan
Pink disease, *Botryobasidium salmonicolor* (Berkeley & Broome) Venkatarayan
Thread blight, *Ceratobasidium stevensii* (Burt) Venkatarayan
Web blight, *Botryobasidium microsclerotia* (Matz) Venkatarayan
Root rot, *Ganoderma lucidum* (Leysser ex Fries) Karsten
Mushroom root rot, *Clitocybe tabescens* (Scopoli ex Fries) Bresadola
Trunk canker, *Phomopsis cinerescens* (Saccardo) Traverso
Leaf spot, *Ascochyta caricae* Rabenhorst
Anthracnose, *Colletotrichum carica* F. L. Stevens & Hall
Mold, *Aspergillus niger* Tieghem
Black spot, *Alternaria fici* Farneti
Leaf spot, *Ormadothium fici* Tims & L. Olive
Leaf spot, *Cephalosporium fici* Tims & L. Olive
Limb canker, *Tubercularia fici* Edgerton
Leaf spot, *Cercospora fici* Heald & Wolf
Other fungi associated with fig

Fruit decay, *Phytophthora palmivora* Butl.

Symptoms

The fruit is attacked when it is dark green or purplish. The fungus mycelium covers it more or less superficially with a white florescence. The figs enlarge and shed as they soften. However, if the atmosphere is dry, they remain attached to the peduncle as dry mummies.

Etiology

The mycelium is hyaline, nonseptate, and branched. Penetrating haustoria may be found in the fruit. The chlamydospores are hyaline, oval to spherical, and $35–90 \times 19–41\ \mu$. They germinate by producing 15–25 biciliate zoospores which in turn develop mycelium of the fungus. The oospores are imbedded and spherical.

Reference

19.

Figure 96. Brown rot, *Pseudomonas solanacearum*, on eggplant.
Figure 97. Tipover, *Diaporthe vexans*, causing tan spots on purple eggplant fruit.
Figure 98. Soil rot, *Thanatephorus cucumeris*, on eggplant seedlings. Enlarged 2.5X.
Figure 99. Rust, *Physopella fici*, on fig foliage.
Figure 100. Pink disease, *Botryobasidium salmonicolor*, on fig, showing external mycelium.

Whiskers, *Rhizopus nigricans* Ehr.

Symptoms

Only the ripening fruit frequently become moldy on the tree during wet, humid weather or after removal from the tree. The fruit is softened by the fungus which rapidly penetrates it and causes a wet collapse. The fungus grows abundantly over the outside of the fruit, producing heads of black spores at the terminal end of long, erect hyphae.

Etiology

The arching stolons are hyaline, nonseptate, and give rise to the sporangiophores which are unbranched and fasciculate. The sporangia are terminal, large, globose, and multispored. The columella is prominent; spores are globose to oval, smooth, and 11×14 μ. Zygospores are usually spherical, with a warty wall, and measure 150×200 μ.

Reference

9.

Souring, *Saccharomyces apiculatus* Reess

Symptoms

Spoilage is due largely to fermentation occurring when the fruit begin to ripen and the eye on the end of the fruit is open. The pink pulp becomes colorless, turns watery, and a colored liquid exudes from the eye in drops. Gas bubbles appear, and the fruit loses its shape, begins to disintegrate, and develops an odor of fermentation. As the process continues, the fruit shrivel, may develop the so-called black neck condition, and is shed. The fermentation is caused by the introduction of certain yeasts through the eye by insects.

Etiology

The yeasts are hyaline and 1-celled. Cells that bud apically are vacuolated; no spores are produced, and the vegetative cells are $8\text{–}9 \times 3\text{–}4$ μ.

References

4, 14.

Limb blight, *Sclerotinia sclerotiorum* (Lib.) d By.

Symptoms

Weakened branches and those otherwise injured frequently show a white, cottony fungus growth. Aerial mycelium surrounds and grows over the areas but is usually limited under humid conditions and is most readily observed in early morning. Under drying conditions, the mycelium collapses and shrivels. Black, irregularly shaped sclerotia may develop on the diseased parts associated with the mycelium. The fungus is disseminated by airborne ascospores.

Etiology

The sclerotia, usually found on the twigs or in the pith, are black, shiny, becoming hard, with a white interior. They germinate by producing apothecia containing many asci, each holding 8 ascospores that are hyaline, 1-celled, oval, and 9–12 × 5–6 μ.

Reference

21.

Twig blight, *Megalonectria pseudotrichia* (Schw.) Speg.

Symptoms

The disease is associated with dead twigs and does not occur on foliage or fruit. Young growth dies back several inches, forming a slightly sunken, dry canker. There develop on the oldest killed tissue small, reddish pink, stalked heads that break through the epidermis and expand until they are 1–2 mm high. They are single or joined, and there may be as many as 6 in a group. They are smooth, usually tapering upward, with a stalk up to 200 μ in diameter and heads up to twice that size. The cankers are most frequently found on young twigs but also affect larger branches.

Etiology

The synnemata, or heads, are composed of parallel hyphae that flare at the crest into a somewhat flat, rounding, sporodochial, compact surface. Hyaline, ovate, 1-celled conidia that average 6 × 3 μ in diameter are formed and held together in a gelatinous mass. This is the vegetative stage of the fungus *Stilbum cinnabarinum*. The perithecia develop at the base of the conidial stalks; they are bright red, spherical, resilient, short-stalked, and measure about 0.5 mm in diameter. The asci are up to 130 μ long, averaging 100 × 17 μ. The 8 ascospores are hyaline, muriform, with 5–7 transverse septa, and measure 20–40 × 9–14 μ.

References

6, 26.

Rust, *Physopella fici* (Cast.) Arth.

Symptoms

Small, brown points appear on the lower surface of the leaves in midseason. These points increase in size to about 1 mm and normally become more numerous. Before the end of the growing season there are enough spots to cause the lower leaf surface to become cinnamon brown. The upper leaf surface becomes sunken in spots opposite the brown areas on the lower leaf surface. The leaves are often shed prematurely after curling upward from the margins. A secondary flush of foliage may appear if the regular production of leaves is destroyed. The leaves are the only part of the host plant affected.

Etiology

The pycnial and aerial stages of the fungus are unknown. The uredia appear on the lower leaf surface. They are round, scattered, small, and dehiscent by a central rupture. Paraphyses are pale brown. The uredospores are pale yellow, ovate to ellipsoid, echinulate, and $14–23 \times 18–32\ \mu$. Telia and teliospores are rarely found.

References

1, 3, 16, 32.

Felty fungus, *Septobasidium bogoriensis* Pat.

Symptoms

The fungus forms a thin, felty, resupinate, mycelial web, which may be up to several inches long, surrounding twigs and branches. The resilient layer is light-colored at first, becoming pale violet with gray tints in the older parts. The margins are almost white and show caverns under the felt covering next to the cortex.

Etiology

The fungus produces scattered basidia at the surface and basidiospores which are hyaline, 1-celled, elongate, and $9–16 \times 2–3\ \mu$. The fungus is a parasite of certain stationary parasitic insects on the host.

References

8, 19.

Southern blight, *Botryobasidium rolfsii* (Sacc.) Venkat.

Symptoms

Young plants are girdled at the soil line as the fungus, which is soil-borne, causes infection under highly humid conditions. The diseased area begins on one side and enlarges until the plant is completely girdled. The killed area is generally confined mostly to the soil-line area. Some white, compact mycelium is usually evident at the time the aerial parts of the tree become affected.

Etiology

In advanced stages the tan to brown, spherical, seedlike sclerotia may be found resting on the older hyphae or on the adjacent soil. These sclerotia are the primary means of dissemination of the fungus and measure 0.5–1 mm in diameter. The characteristics of the sclerotia are diagnostic of the fungus. The imperfect stage is *Sclerotium rolfsii.*

Reference

16.

Pink disease, *Botryobasidium salmonicolor* (Berk. & Br.) Venkat.

Symptoms

The disease is usually evident by the wilting and dying of portions of branches. Closer examination reveals a bright, salmon pink, soft, wet fungus mat growing in a resupinate manner over the outer bark. It may extend from a few inches to a foot or two along the branch. The fungus penetrates the cortex and causes a brown discoloration of the cambium. The leaves beyond the diseased zone are shed. The diseased area usually stops abruptly before involving the larger branches. A thin film or layer of brown hyphal threads may be found, preceding the development of the sporulating area. The basidia and basidiospores are produced abundantly during warm, damp, rainy weather. The manifestation of the disease may vary somewhat on cocoa, citrus, and rubber. The fungus varies slightly in color, depending on its maturity.

Etiology

The fungus is a basidiomycete included in the widely distributed group of thread blights. Manifestations in early stages are the superficial, fine, silky, mycelial hyphae. Later whitish to pink, wet, thick incrustations appear, probably depending on environmental factors and the condition of the host tissue. They may appear sparingly as pinkish tufts of the fungus protected in crevices of the cortex. Basidiospores are produced abundantly over the pink areas on closely packed, erect basidia. The abundance of fertile surface of the fungus is related to moisture, temperature, and the source of nutrition from the host. Soft woody twigs and branches of the host are more conducive to the production of extensive sporophores.

References

24, 27, 30.

Thread blight, *Ceratobasidium stevensii* (Burt) Venkat.

Symptoms

The disease manifests itself by causing foliage on the new flush of growth to die and turn brown. Leaves often remain attached to the twig by fungus threads which are brown, of considerable strength, and adhere to the bark of the twig. The growth of these threads is initiated from brown, hard, loaf-like sclerotia located along the twig near or over lenticels or around the terminal bud. Hyphae extend from the twig up the petioles and peduncles and fan out over the leaf blade and fruit. The leaf parenchyma is invaded and is killed in enlarging circles from the point of attachment of the petiole to the leaf blade.

Etiology

The fungus is entirely aerial in its habitat and consequently not frequently found on annuals. The sclerotia survive on the aerial parts of the host. When young they are white, superficial, irregular, and cottony. When mature they are brown, hard, mound-shaped, and 0.5 mm thick and twice as long. The basidia are found on the killed leaf tissue. They are hyaline, short, and produce basidiospores that are hyaline, oval to elongate, 1-celled, and measure 5–9 ×

3–4 μ. The sclerotia are diagnostic of the fungus. The vegetative stage is *Hypochnus stevensii.*

References

13, 23, 24, 31.

Web blight, *Botryobasidium microsclerotia* (Matz) Venkat.

Symptoms

The fungus grows externally on the twigs, petioles, and peduncles. It shows no parasitic tendencies until it comes in contact with foliage and fruit. The fine, brown, silky threads spread over the leaf blades, penetrate the cuticle and parenchyma, kill the cells, and cause the development of brown, circular spots that enlarge rapidly in all directions. Sometimes the centers of the spots become skeletonized or drop out. When well-established, the fungus produces basidiospores in abundance on the killed leaves, petioles, fruit, peduncles, and current woody growth.

Etiology

The mycelium of the fungus is tan to brown and forms a fine, netlike growth of silky hyphae over the host tissue. The basidia and basidiospores develop on the superficial hyphae. The spores are hyaline, oval, thin-walled, and 5–6 × 11 μ. The sclerotia are superficial, light tan when young becoming brown, subglobose to elongate, homogeneous in texture, composed of loosely aggregated hyphae without an outer wall, and measure 200 × 350 μ. More than 50 per cent of them will pass through a 100 mesh screen. These sclerotia are characteristically diagnostic of the disease. The vegetative stage is *Rhizoctonia microsclerotia.*

References

15, 24, 28, 33.

Root rot, *Ganoderma lucidum* (Leyss. ex Fr.) Karst.

Symptoms

The fungus causes a white, soft, spongy decay with scattered areas of black or reddish brown in the lignin, heartwood, and roots. It decays stumps and is essentially a wood parasite. Its natural effect aboveground is a decline indicated by leaf shedding, following extensive yellowing, lack of fruit set, and dieback of smaller twigs. In diagnosis it would be essential to determine that no aboveground disease may be causing some of these symptoms.

Etiology

The fungus produces white mycelium and characteristic brown sporophores that are woody and frequently zonate. They are usually sessile, smooth, and appear slightly varnishedlike with a yellowish margin and a gray lower surface. The tubes are short and the pores small. The basidia are hyaline, globose,

Figure 101. Thread blight, *Ceratobasidium stevensii,* on fig, showing mycelium and killed leaves.

Figure 102. Web blight, *Botryobasidium microsclerotia,* showing sclerotia on fruit and foliage of fig.

Figure 103. Thread blight, *Ceratobasidium stevensii,* showing sclerotia on fig. branches. Enlarged 2X.

Figure 104. Anthracnose, *Colletotrichum caricae,* on fig foliage.

Figure 105. Growth of 14 days, at specific temperature, of *Botryobasidium microsclerotia* (web blight) from fig.

Figure 106. *Cylindrocladium* leaf spot on fig. Left: Inoculated. Right: Noninoculated.

and up to 12 μ in diameter; the basidiospores are hyaline, obovate, 1-celled, and 7–15×5–9 μ.

Reference

19.

Mushroom root rot, *Clitocybe tabescens* (Scop. ex Fr.) Bres.

Symptoms

The aboveground symptom is essentially a gradual decline brought about by the limitation of the proper root functions and indicated by yellowing, dropping of foliage, and dieback. There may develop a rapid wilting of the entire tree, resulting from a severe and extensive canker that more or less girdles the main trunk at or just below the soil line. Plants are killed, and upon removal of some of the outer cortex and bark down to the wood there are usually exposed various amounts of widely scattered white fungus wefts or plates of mycelium. Plants that have been killed produce clusters of honey-colored mushrooms seasonally.

Etiology

The mycelium is white, branched, septate, and in strands or plates. The sporophores are centrally stipitate with a tannish brown pileus supporting fawn-colored, decurrent gills that produce white, oval, 1-celled basidiospores. The cap is several inches in diameter, and the stipe may be up to 6 in. tall and lack an annulus. The basidiospores measure 8–10×4–6 μ.

Reference

18.

Trunk canker, *Phomopsis cinerescens* (Sacc.) Trav.

Symptoms

Cankers appear in the cortex and wood of the trunk and larger branches. The cankers develop slowly and may persist for several years. They are circular to elongate and later cause the area to become necrotic, dried, and cracked, exposing the wood. New callus may develop along the margins near the wood.

Etiology

The pycnidia are black, scattered in the cortex, immersed becoming erumpent, subglobose, ostiolate, and 250–500 μ in diameter. The pycnidiospores are hyaline, oval to elliptical, usually biguttulate, 1-celled, often exuded in tendrils, and measure 6–9×2–3 μ. The stylospores are filiform and produced in fruiting structures that are indistinguishable from the pycnidia producing the pycnidiospores; they are mostly straight except for a terminal, bent-curved section. They measure 20–25×1 μ. The ascospore stage is *Diaporthe cinerescens*.

Reference

11.

Leaf spot, *Ascochyta caricae* Rab.

Symptoms

The leaf spots are reddish brown, circular to irregular, with sunken margins, and usually measure less than 15 mm in diameter. They are often found along the principal veins.

Etiology

The pycnidia appear in the dry central part of the spots and contain oval, 1–2-celled spores, measuring 12×4 μ.

Reference

20.

Anthracnose, *Colletotrichum carica* F. L. Stev. & Hall

Symptoms

Brown leaf spots appear at midseason on older leaves. They are variable in size, sometimes exceeding 1 cm in diameter, circular except irregular along leaf margins, slightly sunken, dull gray brown becoming brown, and harbor numerous, scattered acervuli that may or may not contain setae. The spores are produced in abundance in pink masses within the centers of the leaf or fruit spots.

Etiology

The acervuli are 85–200 μ in diameter, and the conidia measure $8\text{–}20 \times 3\text{–}6$ μ. The ascospore stage of this fungus is *Glomerella rufo-maculans.*

References

9, 16.

Mold, *Aspergillus niger* v. Tiegh.

Symptoms

Only the fruit is affected; the fungus grows over and inside the dried fruit. Fruit show dark, yellowish spots that later become somewhat translucent. The infection takes place following the opening of the eye. Previous to the opening, the interior is sterile. The term "smut" implies the production of many black spores produced abundantly within the ripening fruit.

Etiology

The spores are small, more or less spherical, 1-celled, and produced in chains upon an erect conidiophore that is 1 mm or more high. The conidia are dark, slightly verrucose, and measure 3–5 μ.

References

17, 22.

Black spot, *Alternaria fici* Farneti

Symptoms

Mature figs on the tree show brown to olive spots that gradually darken, are circular, somewhat sunken, and often cause dropping.

Etiology

The conidia are produced on the spots and cause the dark color to deepen. They are dark-colored, muriform septate, multicelled, and measure 46–70 × 12–15 μ.

References

2, 21.

Leaf spot, *Ormadothium fici* Tims & L. Olive

Symptoms

Beginning as dark brown, small, circular spots, these characteristic zonate lesions, measuring up to 8 cm in diameter, develop rapidly into a targetlike lesion on the upper surface of the leaf. Each lesion is composed of as many as a dozen concentric rings uniformly spaced. The zones are demarked by dark lines, and the interim parenchyma tissue bleaches somewhat and frequently falls out. On the lower surface, the centers are brown with a surrounding white mycelium border.

Etiology

Conidiophores typically bear clusters of brown, cylindrical, curved, up to 7-septate, smooth or warty spores, which measure 3.4–5.2 × 3.9–29.6 μ.

Reference

29.

Leaf spot, *Cephalosporium fici* Tims & L. Olive

Symptoms

Brown, small, circular spots on the upper leaf surfaces are the first visible indications of the disease. These spots enlarge by adding a complete concentric ring around the infection points, continuing to as many as a dozen rings. The zones are demarked by dark brown rings between which are light brown bands of dead tissue. The dead tissue falls out leaving a skeletonlike structure. The spots often measure 6 cm in diameter.

Etiology

The lower leaf surface becomes covered with a white mycelial growth of the fungus and ropelike strands of hyphae surmounted by beadlike heads of clustered spores. The conidia are hyaline, oval, usually slightly curved, 1-celled, and measure on an average 3×6 μ.

Reference

29.

Limb canker, *Tubercularia fici* Edg.

Symptoms

The tissue surrounding the fruit peduncle scar turns darker, shrinks, and becomes covered with tufts of pustules. A collapse and drying out of the area around the fruit scar result in a browning of the tissue. Immediately surrounding this sunken area, new healthy cortex is produced, and the older, killed bark drops out. The canker gradually spreads, and often the twig or branch becomes nearly girdled and dies.

Etiology

The mycelium is white, septate, and branched. The sporodochia develop subepidermally and become erumpent; they are pink and up to 4 mm in diameter. Conidiophores are hyaline, oval, and $5\text{–}7 \times 2\text{–}5$ μ.

Reference

9.

Leaf spot, *Cercospora fici* Heald & Wolf

Symptoms

The leaf spots are irregular, circular to angular, depending mostly on the age of the leaf. Most are small but some enlarge up to 6–8 mm in diameter. They are various shades of brown with a darker border. The color is a different shade of brown on the lower leaf surface. The disease is usually not severe enough to cause an important amount of defoliation. The foliage is the only part of the plant affected.

Etiology

The conidia are produced on the top leaf surface on black stromata that are globular and 30–70 μ in diameter. Conidiophores are produced mostly in upright fascicles that appear dark; however, singly they are almost hyaline and measure $2\text{–}4 \times 10\text{–}30$ μ. The conidia are pale-colored, slightly obclavate, usually gently curved, faintly septate, and $2\text{–}5 \times 30\text{–}180$ μ. There are several other *Cercospora* spp. that infect *Ficus* spp.

References

5, 12.

Other fungi associated with fig

Armillaria mellea Vahl ex Fries
Bacterium fici Cavara

Botryosphaeria ribis Grossenbacher & Duggar
Botrytis cinerea Persoon ex Fries
Choanephora cucurbit (Berkeley & Ravenel) Thaxter
Cylindrocladium scoparium Morgan
Diplodia caricae Saccardo
Fusarium moniliforme var. *fici* Caldis
Macrophoma fici de Almeida & Camara
Ophiodothella fici E. A. Bessey
Phoma fici caricae Werwoerd & k'Pleosis
Phyllosticta caricae Massalongo
Phymatotrichum omnivorum (Shear) Duggar
Rosellinia necatrix (Hartig) Berkeley
Septobasidium pseudopedicellatum Burt
Stilbum cinnabarinum Montagne

References: fig

1. Arthur, J. C., and G. B. Cummins. 1962. Manual of the rusts in United States and Canada. Hafner Co.: N.Y., pp. 60–61.
2. Brooks, C., and L. P. McColloch. 1938. Spotting of figs on the market. J. Agr. Res. 56:473–88.
3. Butler, E. J. 1914. Notes on some rusts in India. Ann. Mycologici 12:76–79.
4. Caldis, P. D. 1930. Souring of figs by yeasts and the transmission of the disease by insects. J. Agr. Res. 40:1031–51.
5. Chupp, C. 1953. A monograph of the fungus genus Cercospora. Published by the author: Ithaca, N.Y., p. 396.
6. Condit, I. J. 1947. The fig. Chronica Botanica: Waltham, Mass., pp. 167–76.
7. Condit, I. J., and J. Enderud. 1956. A bibliography of the fig. Hilgardia 25: 520–82.
8. Couch, J. N. 1935. Septobasidium in the U.S. J. Elisha Mitchell Sci. Soc. 51:1–77.
9. Edgerton, C. W. 1911. Diseases of the fig tree and fruit. Louisiana Agr. Expt. Sta. Bull. 126:4–10, 14–16.
10. Edgerton, C. W. 1911. Two new fig diseases. Phytopathology 1:12–17.
11. Grove, W. B. 1935. British stem- and leaf-fungi. Vol. 1. Cambridge University Press: London, pp. 186–87.
12. Heald, F. D., and F. A. Wolf. 1911. New species of Texas fungi. Mycologia 3:5–22.
13. Large, J. R., J. H. Painter, and W. A. Lewis. 1950. Thread blight in tung orchards and its control. Phytopathology 40:453–59.
14. Lodder, J., and N. J. W. Kreger-V. Rij. 1952. The yeasts: a taxonomic study. Interscience: N.Y., pp. 595–99.
15. Matz, J. 1917. A Rhizoctonia of the fig. Phytopathology 7:110–18.
16. Matz, J. 1918. Some diseases of the fig. Florida Univ. Agr. Expt. Sta. Bull. 149: 3–10.
17. Phillips, E., E. Smith, and R. E. Smith. 1925. Fig smut. Calif. Univ. Agr. Expt. Sta. Bull. 387:1–40.
18. Rhoads, A. S. 1956. The occurrence and destructiveness of Clitocybe root rot of woody plants in Florida. Lloydia 19: 193–239.
19. Roger, L. 1951. Phytopathologie des pays chauds. Encyclopedie Mycologique 17:657–58, 912–17, 1042–48.
20. Roger, L. 1953. Phytopathologie des pays chauds. Encyclopedie Mycologique 18:1731.
21. Smith, R. E. 1941. Diseases of fruits and nuts. Calif. Agr. Ext. Serv. Circ. 120: 69–70, 71.
22. Smith, R. E., and H. N. Hansen. 1931. Fruit spoilage diseases of figs. Calif. Univ. Agr. Expt. Sta. Bull. 506:41–49.
23. Stevens, F. L., and J. C. Hall. 1910. Diseases of economic plants. Macmillan: N.Y., pp. 89–91.
24. Talbot, P. H. B. 1965. Studies of 'Pellicularia' and associated genera of Hymenomycetes. Persoonia 3:371–406.
25. Taubenhaus, J. J., and W. N. Ezekiel. 1931. A sclerotinia limb blight of figs. Phytopathology 21:1195–97.
26. Tims, E. C. 1935. A Stilbum disease of fig in Louisiana. Phytopathology 25: 208–22.
27. Tims, E. C. 1963. Corticium salmoni-

color in the United States. Plant Disease Reptr. 47:1055–59.
28. Tims, E. C., and P. J. Mills. 1943. Corticium leaf blights of fig and their control. Louisiana Agr. Expt. Sta. Bull. 367: 1–19.
29. Tims, E. C., and L. S. Olive. 1948. Two interesting leaf spots of fig. Phytopathology 38:707–15.
30. Venkatarayan, S. V. 1950. Notes on some species of Corticium and Pellicularia. Indian Phytopathol. 3:81–86.
31. Weber, G. F. 1927. Thread blight, a fungus disease of plants caused by Corticium stevensii Burt. Florida Univ. Agr. Expt. Sta. Bull. 186:143–62.
32. Weber, G. F. 1931. Fig rust and its control. Florida Univ. Agr. Expt. Sta. Press Bull. 439:1–2.
33. Weber, G. F. 1939. Web-blight, a disease of beans caused by Corticium microsclerotia. Phytopathology 29:559–75.
34. Wolf, F. A., and W. J. Bach. 1927. The thread blight disease caused by Cortecium koleroga (Cooke) Höhn, on citrus and pomaceous plants. Phytopathology 17:689–709.

Ginger, *Zingiber officinale* Roscoe

Wilt, *Pseudomonas solanacearum* E. F. Smith
Rhizome soft rot, *Xanthomonas zingiberi* (Uyeda) Săvulescu
Rot, *Pythium myriothylum* Drechsler
Blight, *Botryobasidium rolfsii* (Saccardo) Venkatarayan
Leaf spot, *Colletotrichum zingiberis* (Sundararaman) Butler & Bisby
Yellows, *Fusarium oxysporum* f. sp. *zingiberia* Trujillo
Rhizome decay, *Fusarium roseum* (Link) Snyder & Hansen
Other fungi associated with ginger

Wilt, *Pseudomonas solanacearum* E. F. Sm.

Symptoms

Wilt is first noticed because of the slight yellowing and flaccid condition of the lower leaves followed by a progressive wilting from the bottom leaves. Finally, a complete yellowing, browning, and death of the shoot occurs followed by a similar collapse of adjacent shoots. Under certain circumstances, yellowing may not accompany the wilting. Detailed examination of the rhizome usually reveals a brownish, soft, wet decay. Soil fumigation, rotation, and clean planting stock should control the disease.

Etiology

The organism is a gram-negative rod, motile by a single polar flagellum, and measuring $1–5 \times 0.5\ \mu$. Agar colonies are opalescent, usually darkening, small, irregular, wet, shiny, and smooth. Optimum temperature, 35–37C.

References

2, 3.

Rhizome soft rot, *Xanthomonas zingiberi* (Uyeda) Săvul.

Symptoms

Diseased plants first show lack of growth and weak turgidity followed by some faint yellowing of older leaves and drooping of foliage. Yellowing and leaf-fall follows rapidly as the rhizome and roots become invaded, softened, and killed.

Etiology

The organism is a gram-negative rod, motile by 1–3 polar flagella, and measures 0.5–1.1 $\times$ 0.7–2 μ. Agar colonies are white. Optimum temperature, 28C.

Reference

2.

Rot, *Pythium myriothylum* Drechs.

Symptoms

The foliage becomes yellow, beginning at the leaf tips and spreading down the leaf, which is killed; the shoot becomes pale and diseased in a soft, watery decay which extends down the main stem. The fungus is readily observed in the underground parts.

Etiology

The mycelium is hyaline, nonseptate, branched, and about 7–9 μ in diameter. It forms clavate appressoria; sporangia are mostly terminal. About 30–40 zoospores are formed, freed in a vesicle, and eventually liberated. They are reniform, biflagellate, and 10–12 μ in diameter. Oogonia are spherical, smooth, thin-walled, mostly terminal, and average 28 μ in diameter. The oospores are yellowish, subspherical, and 12–37 μ in diameter.

References

1, 4, 6, 7.

Blight, *Botryobasidium rolfsii* (Sacc.) Venkat.

Symptoms

A white mycelial growth on rhizomes of ginger is produced by the blight fungus. Inoculation experiments show that it does not invade healthy rhizomes but is apparently a wound-invading fungus.

Etiology

Mycelium is white, floccose, external, and bears pale yellow to clove brown sclerotia that are globose and 0.8–2.5 mm in diameter.

References

5, 8.

Leaf spot, *Colletotrichum zingiberis* (Sund.) Butl. & Bisby

Symptoms

Spots appear on either leaf surface. They are light yellow, circular to oval, small at first, 2–3 mm in diameter, often coalescing as they enlarge, and have numerous black specks which often form concentric rings in the centers. The centers of some leaf spots fall out. The leaf sheath and rhizome scales are also attacked. Plants become stunted and produce only scanty rhizomes.

Etiology

The stromata of the fungus consist of mycelium, setae, and conidia. The areas are black, circular to oval, in dense clusters, and 30–140 μ in diameter. The setae are brown, erect, septate, and 85–168 μ high. Conidia are hyaline, subfusoid, curved, 1-celled, and 17–24×3–5 μ.

Reference

9.

Yellows, *Fusarium oxysporum* f. sp. *zingiberia* Trujillo

Symptoms

Yellowing of lower leaves progresses upward and precedes a wilting of the plant. Rhizome examination shows a discoloration of the vascular system and a decay of the cortical areas. The organism has been found to invade the vascular region in early stages of the disease and to include the cortical tissue in more advanced stages. The fungus survives in the soil and is transmitted to disease-free soil on infected rhizomes used as seed stock.

Etiology

Sporodochia and pionnotes are salmon-colored; conidia are 3-septate and measure 25–40×3–5 μ. Microconidia are hyaline, 1-celled, and 5–12×2–4 μ. The ascigerous stage is not known.

References

10, 11.

Rhizome decay, *Fusarium roseum* Lk.

Symptoms

The fungus was observed growing on the rhizomes of the host in storage where it produced aerial hyphae and an abundance of typical spores. Inoculations showed that it was a weak wound invader and a storage invader of ginger rhizomes.

Etiology

The sporodochia are tan to brown and produce hyaline conidia that are oval to spindle-shaped or sicklelike, 3–5-septate, mostly 5, and measure 30–50× 3–4 μ. Chlamydospores are round, more or less rough, intercalary or terminal, and 6–14 μ in diameter.

Reference

5.

Other fungi associated with ginger

Coniothyrium zingiber F. L. Stevens & Atienza
Hymenula offinis (Fautrey & Lambotte) Wollenweber

Nectriella zingiberi F. L. Stevens & Atienza
Phyllosticta zingiberi Ramakrishnan
Piricularia zingiberi Nisikado
Rosellinia zingiberi F. L. Stevens & Atienza
Thanatephorus cucumeris (Frank) Donk

References: ginger

1. Butler, E. J. 1918. Fungi and diseases in plants. Thacker: Calcutta, India, pp. 348–52.
2. Elliott, C. 1951. Manual of bacterial plant pathogens. 2d ed. Chronica Botanica: Waltham, Mass., p. 152.
3. Ishii, M., and M. Aragaki. 1963. Ginger wilt caused by Pseudomonas solanacearum E. F. Smith. Plant Disease Reptr. 47:710–16.
4. McCarter, S. M., and R. H. Littrell. 1968. Pathogenicity of Pythium myriotylum to several grasses and vegetable crops. Plant Disease Reptr. 52:179–83.
5. Mehrotra, B. S. 1952. Fusarium roseum Link and Sclerotium rolfsii Sacc. on ginger rhizomes. Indian Phytopathol. 5:52–54.
6. Mundkur, B. B. 1949. Fungi and plant disease. Macmillan: London, pp. 56–58.
7. Subramanian, L. S. 1919. A pythium disease of ginger, tobacco, and papaya. Mem. Dept. Agr. India 10:181–94.
8. Sundaram, N. V. 1953. Thread blight of ginger. Indian Phytopathol. 6:80–85.
9. Sundararaman, S. 1922. A new ginger disease in Godavari district. Mem. Dept. Agr. India, Botan. Ser. 11:209–17.
10. Teakle, D. S. 1965. Fusarium rhizome rot of ginger in Queensland. Queensland J. Agr. Sci. 22:263–72.
11. Trujillo, E. E. 1963. Fusarium yellows and rhizome rot of common ginger. Phytopathology 53:1370–71.

Grape, *Vitis vinifera* Linnaeus

Crown gall, *Agrobacterium tumefaciens* (E. F. Smith & Townsend) Conn
Bacterial blight, *Erwinia vitivora* (Baccarini) Duplessis
Downy mildew, *Plasmopara viticola* (Berkeley & Curtis) Berlese & De Toni
Powdery mildew, *Uncinula necator* (Schweinitz) Burrill
Spot anthracnose, *Elsinoë ampelina* (De Bary) Shear
Shoot blight, *Sclerotinia sclerotiorum* (Libert) De Bary
Ripe rot, *Glomerella cingulata* (Stoneman) Spaulding & Schrenk
Black rot, *Guignardia bidwelli* (Ellis) Viala & Ravaz
Dead arm, *Cryptosporella viticola* (Saccardo) Shear
Fruit rot, *Botryosphaeria ribis* Grossenbacher & Duggar
Rust, *Physopella vitis* (Thuemen) Arthur
Shoestring root rot, *Armillaria mellea* Vahl ex Fries
Mushroom root rot, *Clitocybe tabescens* (Scopoli ex Fries) Bresadola
Leaf spot, *Septoria ampelina* Berkeley & Curtis
Stem blight, *Phomopsis viticola* Saccardo
Bitter rot, *Melanconium fuligineum* (Scribner & Viala) Cavara
Gray mold, *Boytrytis cinerea* Persoon ex Fries
Cane spot, *Pestalotia pezizoides* De Notaris
Leaf spot, *Isariopsis clavispora* (Berkeley & Curtis) Saccardo
Leaf spot, *Cercospora viticola* (Cesati) Saccardo
Other fungi associated with grape

Crown gall, *Agrobacterium tumefaciens* (E. F. Sm. & Towns.) Conn

Symptoms

Gall is manifested primarily by the development of tumor or gall-like outgrowths of the parenchyma tissue of the crown and stems. When young, the gall tissue is mostly smooth and spongy, of variable size from small, warty growths to large, knoblike structures or sometimes long, linear protuberances. With age they become dark-colored, hard, hornlike, and rough. Plants are frequently killed or may continue to grow and produce fruit. The severity depends on the location of the gall on the vine and the kind of girdling caused.

Etiology

The organism is a gram-negative rod, forms capsules, is motile by polar flagella, and measures 1–3 × 0.4–0.8 μ. Agar colonies are translucent to white, small, circular, and glistening. Optimum temperature, 25–30C.

References

8, 12.

Bacterial blight, *Erwinia vitivora* (Bacc.) Duplessis

Symptoms

This blight is almost entirely confined to the vinifera grape where pale green to grayish black areas develop on young shoots, and leaves become yellow to brown. Leaf spots are yellowish to reddish brown, circular, and often show exudate. Blossoms may be blackened and result in barren bunches. Shoot infections often continue up several nodes, causing some cracking of the stem and development of adventitious buds on stunted vines.

Etiology

The organism is a gram-negative rod, single or in combinations, produces capsules, is motile by 6–8 peritrichous flagella, and measures 1–2 × 0.4–1.1 μ. Agar colonies are orange yellow to brown, glistening, spreading, raised, and circular. Optimum temperature, 25–30C.

Reference

1.

Downy mildew, *Plasmopara viticola* (Berk. & Curt.) Berl. & de T.

Symptoms

Mildew appears as a white, frostlike mold on any of the aboveground parts of the plant, mostly on leaves, young shoots, and developing fruit clusters. Leaf spots as viewed from above are pale yellow, irregular in outline, and eventually become brown. The sporulating fungus develops on the lower surface of these spots. Young shoots are often distorted by the fungus and frequently killed. Flowers and fruiting clusters are subject to attack, and part or all of the cluster may be killed and turn brown. Some shelling may take place, or berries are killed and retained on the plant as mummies.

Etiology

The fungus mycelium is hyaline, nonseptate, branched, and occupies the intercellular spaces. Haustoria are produced. The conidiophores are initiated in the substomatal spaces and emerge through the stomata singly or in clusters up to 20 directly through the epidermis. They measure 300–500 × 7–9 μ, and the terminal half divides up to three times, giving rise, at the branch tips, to the sterigmata, which bear hyaline, ovoid, papillate, thin-walled sporangia, measuring 15–30 × 11–18 μ. They germinate by zoospores that are pear-shaped, 1-celled, with 2 cilia, and measure 7 × 9 μ. Oospores are formed in the host tissue and measure 25–36 μ in diameter. They are thick-walled and germinate by forming a single conidiophore bearing a sporangium that further develops by forming zoospores. The fungus is an obligate parasite.

References

1, 8, 12.

Powdery mildew, *Uncinula necator* (Schw.) Burr.

Symptoms

Powdery mildew is recognized by the white, dustlike, powdery, superficial mycelium and conidia of the fungus. It is found mostly on the upper surface of leaves and sometimes on shoots, tendrils, stems, flowers, and fruit clusters. Leaves often become yellow and tend to curl upward; some shedding may occur. Flowers fail to set fruit, and late infected berries may show retarded growth, be misshapen, and split. The fruit becomes immune at maturity.

Etiology

The fungus mycelium is hyaline, superficial, septate, produces haustoria, and supports hyaline, erect, septate conidiophores that cut off conidia at their tips in succession, often forming several in a chain. The conidia are hyaline, ovate to oblong, 1-celled, and 25–30 × 15–17 μ. The perithecia are formed later and are variously scattered over the upper surface and associated with the hyaline mycelium. They are black, spherical, 70–128 μ, with 10–30 brown, narrow, tip-curved appendages about 200–300 μ long. There are 4–9 asci within each perithecium. There are usually less than 8 ascospores; they are hyaline, oval, 1-celled, and 18–25 × 10–12 μ. The fungus is an obligate parasite.

References

1, 8, 12.

Spot anthracnose, *Elsinoë ampelina* (d By.) Shear

Symptoms

This disease attacks all parts of the plant but is most destructive on the growing shoots and on the fruit. Dark brown, irregular spots appear between and on the large veins of the leaves and on the petioles, causing distortion. The shoots and tendrils are weakened by the depressed, dark, cankerlike lesions that have a black, raised margin. Shoots are killed. Fruit spots are reddish brown becoming gray in the center with a reddish border and are up to 1/4 in. in diameter. The coloration is very characteristic. The berries seldom shrink and do not drop. The disease is sporadic, occasionally causing serious loss of fruit.

Etiology

The mycelium is hyaline, septate, branched, entirely submerged, and develops acervuli as black specks in the centers of the depressed lesions. The conidia are hyaline, 1-celled, oblong to ovoid, and 5–6×2–3 μ. The sexual stage of the fungus develops on the year-old canes as black, small, inconspicuous ascomata in which are imbedded globular asci containing the hyaline, 3-septate spores that measure 15–16×4–5 μ. The vegetative stage is *Sphaceloma ampelinum*.

References

1, 8, 12.

Shoot blight, *Sclerotinia sclerotiorum* (Lib.) d By.

Symptoms

This blight is mostly unimportant as it may kill the growing shoots only under highly favorable conditions of temperature and humidity. A white, cottony mycelium develops with the drooping and withering of the growing tissue.

Etiology

The mycelium is hyaline, branched, septate, often in cottony wefts, and produces no functioning conidia. Black, shiny sclerotia are usually found on the diseased host parts and are the surviving organs.

Reference

13.

Ripe rot, *Glomerella cingulata* (Ston.) Spauld. & Schrenk

Symptoms

Ripe rot is confined mostly to the fruit where it causes reddish brown spots on light-colored berries and no distinct change of color on purple berries. The berries shrink, and there is usually some shelling. As the disease develops, a series of concentric zones appears.

Etiology

The black, small acervuli emphasize the zoning. They produce masses of hyaline, 1-celled conidia, measuring 10–20×5–7 μ. The ascospore stage is not frequently found. The perithecia are black, flask-shaped, and contain clavate asci. The ascospores are hyaline, elliptical, slightly curved, and 20–28×5–7 μ.

Reference

12.

Black rot, *Guignardia bidwelli* (Ell.) Viala & Ravaz

Symptoms

Black rot is found on any of the aboveground parts of the plant but is most severe on the berries, although foliage and other green parts are generally infected. On the foliage the spots are reddish brown, generally circular, and mostly scattered. Heavy infection may result in larger blotches caused by coalescing. Individual spots may be barely visible to 1 cm in diameter, but they are usually less. Later they become speckled with black, submerged pycnidia that produce the spores of the fungus. The shoots, tendrils, flower clusters, fruit, and stems are characteristically infected, showing a brownish color and the pycnidia. The fruit often shell off and shrink as mummies, covered with pycnidia.

Etiology

The pycnidia are black, submerged at first becoming erumpent, ostiolate, 80–180 μ in diameter. The pycnidiospores develop abundantly and escape, often in long tendrils. They are hyaline, globose, 1-celled, and 8–11 $\times$ 6–8 μ. The perithecia are black, subepidermal to erumpent, globose, leathery, nonparaphysate, and 40–120 μ in diameter. The ascospores are hyaline, elliptic to oblong, 1-celled, and 12–17 $\times$ 4–5 μ. The imperfect stage is *Phyllosticta labrusca*.

References

1, 8, 12.

Dead arm, *Cryptosporella viticola* (Sacc.) Shear

Symptoms

Plants fail to produce normal foliage in the spring on one or more of the main branches or even fail entirely except for basal sprouts. Reddish brown spots develop on the shoots, petioles, and leaves and may cause a splitting of the canes. Fruit infection sometimes takes place. A heart rot slowly progresses in the stems and is instrumental in affecting the branches. Cankers appear on stem lesions and on diseased fruit in the fall and produce mature pycnidiospores in the spring.

Etiology

The pycnidia with labyrinthiform chambers are black and irregular in shape. Pycnidiospores are hyaline, fusoid, 1-celled, and 7.5–15 $\times$ 2–5 μ. The spores are hyaline, long, slender, fusiform, curved, threadlike, and measure 18–40 $\times$ 1–1.5 μ. The perithecia are usually submerged in stromata and exert the ostiolate beaks. The ascospores are hyaline, 1-celled, oval, and 11–15 $\times$ 4–6 μ.

References

1, 6.

Fruit rot, *Botryosphaeria ribis* Gross. & Dug.

Symptoms

Fruit rot appears on the berries, particularly of muscadine varieties, about the time they reach full size. The dark brown spots have tan centers, are circular, flat or sunken, and up to 4 mm in diameter. The pycnidia are imbedded in the centers. Berries showing spots usually shell off but do not form mummies; they dry, leaving the shell-like skin covered with the pycnidia.

Etiology

The pycnidia are black, spherical, and 150–200 μ in diameter. The pycnidiospores are hyaline, ovoid to elliptical, and $14\text{–}26 \times 5\text{–}9\ \mu$. Mature ascocarps produced in the stromata are spherical and 172–315 μ in diameter. The asci are cylindrical, 8-spored, and $102\text{–}157 \times 17\text{–}24\ \mu$; ascospores are hyaline, 1-celled, ovoid to elliptical, and measure $19\text{–}31 \times 8\text{–}12\ \mu$. The vegetative stage is *Macrophoma* sp.

References

6, 7.

Rust, *Physopella vitis* (Thuem.) Arth.

Symptoms

Rust forms orange yellow, small, cushionlike sori on the lower surface of leaves; they are miniature in size. The sori are slightly raised and free the yellow uredospores upon rupturing. The only indication on the upper leaf surface is a very limited yellow speck less than 1 mm in diameter. Some leaf shedding may take place toward the end of the season.

Etiology

Only the uredia and telia stages are known. The uredospores are sessile, globoid, with echinulate walls, and measure $20\text{–}36 \times 12\text{–}25\ \mu$. The teliospores are in chains of 2–7; the cells are oblong with smooth walls, and measure $14\text{–}26 \times 8\text{–}10\ \mu$.

References

3, 12.

Shoestring root rot, *Armillaria mellea* Vahl ex Fr.

Symptoms

Root rot is located at the crown of the plant where the invading hyphae penetrate the cortex, grow through the phloem, and kill the cambium tissue as the stem gradually becomes girdled and the plant dies. The stages of decline are often varied in extent probably depending on ecological conditions. A plant may linger for a season or more, exhibiting symptoms of malnutrition, such as stunting, leaf yellowing, shedding, and lack of growth.

Etiology

The fungus produces hyaline mycelium, black, shiny, heavy, stringlike rhizomorphs usually next to the wood, and, seasonally, honey-colored, gilled, stipitate, annulated, capitate sporophores, 3–4 in. wide and up to 6 in. high.

Reference

12.

Mushroom root rot, *Clitocybe tabescens* (Scop. ex Fr.) Bres.

Symptoms

Mushroom root rot is the source of much concern because the disease may spread extensively, killing plants almost before any external indications are revealed. The fungus is soil-inhabiting and is common on recently cleared oak forest land. The plants are invaded by the mycelium of the fungus in the vicinity of the crown, including the major roots. Plants suffer a girdling effect, the completeness and suddenness of which is reflected by the surrounding factors of moisture, aeration, and temperature. Plants may show a slow decline, entirely or in part, with yellowing of leaves, stunted growth, wilting, and leaf-shedding; leafless canes retain the partially mature fruit. Stem sections show cortex killed to the wood and usually mats of white mycelium growing over the wood or imbedded in the inner cortex.

Etiology

Usually in the fall, during wet periods, the sporophores of the fungus develop around the base of the plant. They are honey-colored, stipitate, without an annulus, and develop gills below a central, spreading cap. The sporophores are 2–5 in. wide and up to 6 in. tall. Basidiospores are hyaline, white en masse, 1-celled, oblong, $8–10 \times 4–6\ \mu$, and are shed in abundance from the gills.

Reference

12.

Leaf spot, *Septoria ampelina* Berk. & Curt.

Symptoms

This disease occurs on the leaf blades, petioles, and canes. It is probably more important on the last, where brown, small, angular spots up to 2 mm in diameter develop that are light-colored at first and rapidly darken.

Etiology

The black, small, globose, ostiolate pycnidia average $100\ \mu$ in diameter and grow in the epidermal tissue. The pycnidiospores are hyaline, somewhat curved, filiform, 3–5-septate, and measure $30–60 \times 2–3\ \mu$.

Reference

13.

Figure 107. Downy mildew, *Plasmopara viticola,* on grape. Left: Lower leaf surface. Right: Upper leaf surfaces. (Photograph by A. S. Rhoads.)

Figure 108. Black rot, *Guignardia bidwelli,* on grape foliage. (Photograph by A. S. Rhoads.)

Figure 109. *Armillaria mellea* (shoestring root rot) sporophores showing diagnostic gills and annulus. (Photograph by A. S. Rhoads.)

Figure 110. Bitter rot, *Melanconium fuligineum,* showing fruit infection on grape. (Photograph by A. S. Rhoads.)

Stem blight, *Phomopsis viticola* Sacc.

Symptoms

Stem blight is readily recognized by the existence of naked branches in the spring and the production in some instances of yellowish green, crinkled, and much dwarfed foliage and shoots. When a foot or two of new shoot has grown, often numerous angular, yellow spots appear with yellow margins and dark centers. Later leaf petioles, canes, and pedicels show brown to black spots and some cane cankers. Fruit decay apparently is associated with peduncle lesions. A wilt may occur in midsummer. A dry rot in the interior part of the trunk extends to the outer cortex and up and down within the trunk.

Etiology

The pycnidial stage of the fungus is very abundant on the old canes of the year before. It is located in the cortex and also on the fruit but is rather scarce on the new canes. Pycnidia are black, submerged becoming erumpent, obovate, often with labyrinthiform interiors, and produce 2 kinds of conidia. The alpha spores are hyaline, 1-celled, oval to slightly elongate, and $6–11 \times 2–5$ μ. The beta spores are hyaline, long, slender, curved, and measure $18–40 \times 1–2$ μ. The ascospore stage is *Cryptosporella viticola.*

References

2, 8.

Bitter rot, *Melanconium fuligineum* (Scribn. & Viala) Cav.

Symptoms

Bitter rot is not conspicuous on the foliage and is difficult to recognize on the fruit clusters. The fungus developing in the stems causes them to become dry and hard, resulting in poorly developed berries which shrivel and begin to shell off. Infection on the berries is indicated by a brown to reddish color. There is some shrinkage, but the berries retain their form. Finally, the whole surface becomes dotted with black, blisterlike structures in which the spores of the fungus are produced. As the spores develop, the berry loses moisture, shrinks, and forms a mummy. The disease continues to develop in transit.

Etiology

The acervuli are scattered uniformly over the surface of the berry; they are mostly brownish black and subepidermal to erumpent. The conidia are hyaline to light olive, ovoid to elliptical, and measure $9–12 \times 4–6$ μ.

Reference

12.

Gray mold, *Botrytis cinerea* Pers. ex Fr.

Symptoms

Mold develops on the fruit, either on the vine or after removal, forming a more or less dense mycelial growth intermingled with myriads of conidia. The berries show a slight discoloration, become watery, and collapse with some package leakage.

Etiology

The fungus mycelium is hyaline, branched, septate, and largely superficial. The sporophores are hyaline, erect, usually stout, with branched terminals bearing clusters of hyaline to gray, 1-celled, oval conidia, measuring 11–15× 8–11 μ. Sometimes black, small, irregularly shaped sclerotia are found.

Reference

6.

Cane spot, *Pestalotia pezizoides* de N.

Symptoms

These spots are usually found on the runners where pseudostromata appear supporting acervuli. Acervuli rise above the epidermis in pezizoid, apothecioid, open fruiting structures, and measure 225–600 × 100–300 μ.

Etiology

The conidia are long, fusiform, curved, tapering, 6-celled, and 27–36 × 6–9 μ. The 4 interior cells are brown, and the 2 exterior cells are hyaline. There are 2–5 setulae; they are hyaline, sometimes branched, and 6–18 μ long. Pedicels are 2–10 μ long.

Reference

5.

Leaf spot, *Isariopsis clavispora* (Berk. & Curt.) Sacc.

Symptoms

Brown spots form on the shaded and older leaves; they are irregular to angular, brittle, have a slightly raised border, vary in size to 0.5 cm, and eventually become black. Later these severely spotted leaves show yellowish blotches and are shed. On the lower surface there are numerous, scattered, erect, hairlike columns of fungus filaments at the crest of which the spores of the fungus are produced.

Etiology

The filaments are brown, septate, loosely aggregated in the synema, terminate in free-spreading tips, and measure 200–300 × 35–60 μ. The conidia borne on the tip ends of these hyphae are olive brown, 3–10-septate, elongate, curved, clavate with a rounded apex tapering to slender base, and measure 50–100 × 7–14 μ. The ascospore stage is recorded as *Mycosphaerella personata.*

References

9, 12.

Leaf spot, *Cercospora viticola* (Ces.) Sacc.

Symptoms

Leaf spot is widely distributed and common. It forms reddish brown to brown black, circular to irregular areas with ashy gray centers; these areas are 2–12 mm in diameter.

Etiology

On the lower surface, there develop olive brown, multiseptate, geniculate, often undulating conidiophores in dense fascicles or coremiumlike clumps measuring 50–300 × 3–6 μ. The conidia are olive brown, obclavate, 3–7-septate, sparingly curved, blunt at the apex, and measure 20–80 × 4–7 μ. The perithecia are scattered, subepidermal becoming erumpent, globose, aparaphysate, and measure 60 × 90 μ. Asci are clavate to elliptic, short, stipitate, and 28–41 × 6–10 μ. The 8 ascospores are hyaline, roughly biseriate, elliptic, 2-celled, and 11–22 × 2.5–4 μ. The perfect stage is *Mycosphaerella personata.*

References

8, 9.

Other fungi associated with grape

Cladosporium herbarum (Persoon) Link
Coniothyrium diplodiella (Spegazzini) Saccardo
Didymosphaeria sarmenti (Cooke & Harkness) Berlese & Voglino
Eutypella vitis (Schweinitz ex Fries) Ellis & Everhart
Hendersonia sarmentorum Westendorp
Macrophoma farlowiana (Viala & Sauvageau) F. Tassi
Micropera ampelina Saccardo & Fairman
Nectria viticola Berkeley & Curtis
Phyllachora picea (Berkeley & Curtis) Saccardo
Phymatotrichum omnivorum (Shear) Duggar
Physalospora obtusa (Schweinitz) Cooke
Pseudovalsa viticola Ellis & Everhart
Rosellinia necatrix (Prillieux) Berlese
Valsa vitis (Schweinitz) Berkeley & Curtis
Verticillium albo-atrum Reinke & Berthold

References: grape

1. Anderson, H. W. 1956. Diseases of fruit crops. McGraw-Hill: N.Y., pp. 355–98.
2. Black, L. L., and D. A. Slack. 1967. Dead-arm disease of grapes in Arkansas and West Virginia. Plant Disease Reptr. 51:1038–39.
3. Clayton, C. N., and W. H. Ridings. 1970. Grape rust, Physopella ampelopsidis, on Vitis rotundifolia in North Carolina. Phytopathology 60:1022–23.
4. Delp, C. J. 1954. Effect of temperature and humidity on the grape powdery mildew fungus. Phytopathology 44:615–26.
5. Guba, E. F. 1961. Monograph of Monochaetia and Pestalotia. Harvard University Press: Cambridge, Mass., pp. 248–51.
6. Harvey, J. M., and W. T. Pentzer. 1960. Market diseases of grapes and other

small fruits. U.S. Dept. Agr., Agr. Handbook 189:6–22.
7. Luttrell, E. S. 1948. Botryosphaeria ribis, perfect stage of the Macrophoma causing ripe rot of muscadine grapes. Phytopathology 38:261–63.
8. McGrew, J. R., G. W. Still, and H. Baker. 1961. Control of grape diseases and insects in the eastern United States. U.S. Dept. Agr. Farmers' Bull. (Rev.) 1893: 1–32.
9. Munjal, R. L., and K. K. Sethi. 1966. Cercospora species causing leafspot of grapes in India. Indian Phytopathol. 19:209–14.
10. Nelson, K. E. 1956. The effect of Botrytis infection on the tissue of Tokay grapes. Phytopathology 46:223–29.
11. Quaintance, A. L., and C. L. Shear. 1922. Insect and fungous enemies of the grape. U.S. Dept. Agr. Farmers' Bull. 1220:1–75.
12. Rhoads, A. S. 1926. Diseases of grapes in Florida. Florida Univ. Agr. Expt. Sta. Bull. 178:75–156.
13. Roger, L. 1953. Phytopathologie des pays chauds. Encyclopedie Mycologique 18: 1178, 1786.
14. Shear, C. L. 1911. The ascogenous form of the fungus causing dead-arm of the grape. Phytopathology 1:116–19.

Guar, *Cyamopsis tetragonoloba* (Linnaeus) Taubenhaus

Bacterial blight, *Xanthomonas cyamopsidis* Patel & Patel

Symptoms

Pale yellow, v-shaped spots extend from the leaf margins down the principal veins toward the petiole and begin as small, yellow areas along the edge of the blade. They increase rapidly, becoming pale brown with a yellow, water-soaked border. The entire margin may become infected, causing the leaf to shrivel, droop, and die. The invasion may continue from the leaf through the petiole into the stem where it becomes systemic and results in a loss of the floral parts.

Etiology

The organism is a rod, single or sometimes in pairs, forming capsules, motile by a polar flagellum, and measuring $1–2 \times 0.4–0.6\ \mu$. Agar colonies are yellow, glistening, entire, circular, smooth, pulvinate, and butyrous.

References

11, 12, 13.

Leaf spot, *Pseudomonas syringae* van Hall

Symptoms

Leaf lesions with light brown centers and dark borders surrounded by chlorotic halos are circular to irregular in outline and 1–5 mm in diameter. Numerous leaf lesions often coalesce, resulting in leaf curling, wilting, and shedding.

Etiology

The organism is a gram-negative rod, single, in pairs, or in short chains, produces capsules, is motile by 1 to several polar flagella, and measures 0.7–1.5 × 1.5–3 μ. Agar colonies are white, transparent, circular to fimbriate, smooth or sometimes slightly wrinkled, and convex. A green fluorescent pigment is produced in culture.

References

5, 10.

Powdery mildew, *Leveillula taurica* (Lév.) Arn.

Symptoms

Light-colored areas appear at various locations on the foliage; they are usually somewhat scattered but often coalesce, involving extensive areas. The coalescing results in a drying out of the invaded tissue, finally causing the leaf to shed.

Etiology

The mycelium is septate, branched, somewhat irregular in diameter, and mostly internal, growing among the parenchyma cells. The conidiophores, single or in small clusters, emerge through the stomata. They are hyaline, septate, branched, and produce a single, terminal conidium. External mycelium may develop from the base of the conidiophores and grow superficially over the surface of the leaf. Mature conidia are shed and secondary ones are produced. The conidia are hyaline, 1-celled, cylindrical to lanceolate, or with pointed or truncated tips, and measure 35–82 × 12–28 μ. The black perithecia are imbedded in the dense surface mycelium. They are spherical to somewhat flattened, surrounded by the hyaline to brown appendages, and measure up to 250 μ in diameter. They contain about 20 large, cylindrical to oval asci, each containing 2 ascospores, and measure 100 × 25–40 μ. The spores are hyaline, 1-celled, elongate, and 25–40 × 14–22 μ.

References

1, 15.

Anthracnose, *Colletotrichum dematium* f. *truncata* (Schw.) Arx

Symptoms

Leaf spots at first are small, dark specks of irregular shape, variously scattered, with a slight yellowish halo. As the spots enlarge, they become more or less circular and dark-colored with no halo effect and often coalesce, measuring up to 1 cm in diameter. Petioles also become diseased.

Etiology

The acervuli develop as black pimples scattered over the centers of the leaf spots. The conidia are hyaline, 1-celled, elongate, with tapering ends, and measure 26–33 × 2–3 μ.

References

4, 16.

Leaf spot, *Alternaria cucumerina* (Ell. & Ev.) Elliott

Symptoms

Leaf spots are light tan to brown with a dark brown border, circular or irregular, with concentric zoning resulting from darker ridges, and up to 6 mm in diameter. Secondary spots may be more angular and coalesce and may also be found on the stems and pods. Under favorable conditions, spots on both leaf surfaces become covered with the dark brown conidiophores and conidia of the fungus.

Etiology

The mycelium is dark, septate, and branched. The conidiophores are dark, septate, erect, usually straight, and 42–78×5–7 μ. The conidia are dark brown, obclavate, septate, restricted, with 1–3 longitudinal septations. The septate beak may be longer than the body of the spore; overall measurements are 81–199×13–24 μ.

References

2, 6, 8, 16.

Leaf spot, *Alternaria cyamopsidis* Rang. & Rao

Symptoms

Dark brown, round to irregular, zonate spots, varying from 2 to 10 mm in diameter develop from small, water-soaked areas appearing on the leaf blades. They are light brown on the lower surfaces. When spots coalesce the leaf usually falls from the plant, and severe defoliation results with poor growth and yield.

Etiology

Inter- and intracellular mycelia develop in diseased areas. Conidia are produced on the spots in groups with chains of up to 4 spores. Conidia are gray olive, straight, obclavate, smooth, and constricted. Beaks are often several times the length of the spore, measuring 50–140×4–6 μ; the muriform conidia measure 62–146×12–19 μ.

Reference

14.

Purple stain, *Cercospora kikuchii* (Matsu. & Tomoyasu) Gardner

Symptoms

Most diagnostic are the extensive discolored areas on the seed. They range from a very faint, delicate lavender tinge to large, dark purple blotches often involving the entire seed. On the foliage, the spots, up to 15 mm, are mostly circular, scattered, dark-colored, with a light center and sometimes a light halo. The stem and pod infections are elongate, blotchy, and tan; more definite lesions are black and distinct in outline.

Etiology

The stromata are small and appear mostly on the lower surface and on the stems and pods. The fascicles of conidiophores are dark brown, fairly well scattered or semidense, multiseptate, not branched, and up to 200 μ long. Conidia are hyaline, circular, several septate, straight or curved, and measure $50\text{–}375 \times 2\text{–}5\ \mu$, mostly less than 100 μ long.

References

3, 7, 9.

Other fungi associated with guar

Ascochyta imperfecta Peck
Botryobasidium rolfsii (Saccardo) Venkatarayan
Erysiphe polygoni De Candolle
Macrophomina phaseolina (Tassi) Goidanich
Phymatotrichum omnivorum (Shear) Duggar
Thanatephorus cucumeris (Frank) Donk

References: guar

1. Butler, E. J. 1918. Fungi and diseases in plants. Thacker: Calcutta, India, pp. 271–73.
2. Chand, J. N., and P. S. Verma. 1968. Occurrence of a new Alternaria leaf spot of clusterbeans (Cyamopsis tetragonoloba) in India. Plant Disease Reptr. 52:145–47.
3. Chupp, C. 1953. A monograph of the fungus genus Cercospora. Published by the author: Ithaca, N.Y., pp. 313, 326.
4. Desai, M. V., and N. Prasad. 1955. A new Colletotrichum from India. Indian Phytopathol. 8:52–57.
5. Elliott, C. 1951. Manual of bacterial plant pathogens. 2d ed. Chronica Botanica: Waltham, Mass., pp. 88–93.
6. Jackson, C. R., and G. F. Weber. 1959. Morphology and taxonomy of Alternaria cucumerina. Mycologia 51:401–8.
7. Johnson, H. W., and J. P. Jones. 1962. Purple stain of guar. Phytopathology 52:269–72.
8. Luttrell, E. S. 1951. Diseases of guar in Georgia. Plant Disease Reptr. 35:166.
9. Murakishi, H. H. 1951. Purple seed stain of soybean. Phytopathology 41:305–18.
10. Orellana, R. G. 1967. Leaf spot of guar caused by Pseudomonas syringae in the United States. Plant Disease Reptr. 51: 182–84.
11. Orellana, R. G., and M. L. Kinman. 1970. A new virulent race of Xanthomonas cyamopsidis, bacterial blight of guar. Plant Disease Reptr. 54:111–13.
12. Orellana, R. G., C. A. Thomas, and M. L. Kinman. 1965. A bacterial blight of guar in the United States. FAO Plant Protect. Bull. 13:9–13.
13. Patel, A. J., and M. K. Patel. 1958. A new bacterial blight of Cyamopsis tetragonoloba (L.) Taub. Current Sci. (India) 27:258–59.
14. Rangaswami, G., and A. V. Rao. 1957. Alternaria blight of clusterbeans. Indian Phytopathol. 10:18–25.
15. Roger, L. 1953. Phytopathologie des pays chauds. Encyclopedie Mycologique 18: 1609–11.
16. Sowell, G. 1965. Anthracnose of guar. Plant Disease Reptr. 49:605–9.

Guava, *Psidium guajava* Linnaeus

Ripe rot, *Glomerella cingulata* (Stoneman) Spaulding & Schrenk
Dieback, *Physalospora psidii* Stevens & Pierce
Rust, *Puccinia psidii* Winter
Thread blight, *Ceratobasidium stevensii* (Burt) Venkatarayan
Mushroom root rot, *Clitocybe tabescens* (Scopoli ex Fries) Bresadola
Fruit canker, *Pestalotia psidii* Patouillard
Wilt, *Fusarium oxysporum* f. sp. *psidii* Prasad, Mehta, & Laly
Leaf spot, *Cercospora sawadae* Yamamoto
Algal spot, *Cephaleuros virescens* Kunze
Other fungi associated with guava

Ripe rot, *Glomerella cingulata* (Ston.) Spauld. & Schrenk

Symptoms

The early indications of the disease are brown, small, sunken, circular, soft, decaying, leaf spots which, as they mature, become covered with acervuli in more or less circular zones containing a heaped up mass of pinkish conidia.

Etiology

The spores are hyaline, 1-celled, cylindrical with rounded ends, and measure 10–15 × 4–5 μ. The perithecia, often scattered but occasionally in clusters, are black, round to ovoid, sunken structures with short necks and no paraphyses. The asci are cylindrical to obclavate and 42–65 × 7–8 μ. The 8 ascospores are hyaline, biseriate, 1-celled, straight or curved, and 14–19 × 3–5 μ. The vegetative stage of the fungus is *Gloeosporium psidii.*

References

1, 3, 12.

Dieback, *Physalospora psidii* Stev. & Pierce

Symptoms

Areas on the branches are discolored, develop longitudinal cracks, become dry, and eventually girdle the twig or branch which becomes brown and dies. In severe cases many cankers develop on the woody parts of the plant and extensive wood-killing results that may involve the entire plant.

Etiology

The mycelium is hyaline to brown, branched, septate, and intercellular. Pycnidia are black, shallowly imbedded in the outer cortex, globose, and ostiolate. The conidia are hyaline to colored, continuous to 1-septate, and 12–15 × 5–8 μ. The perithecia are black and submerged with protruding ostioles. Ascospores are hyaline, oval, and 30–37 × 13–16 μ. The vegetative stage is *Diplodia* sp.

Reference

13.

Rust, *Puccinia psidii* Wint.

Symptoms

The sori are inconspicuous on the lower leaf surfaces. Yellow flecks appear scattered over the affected leaves. The yellow areas become cinnamon brown in the center and are small and circular to irregular in outline.

Etiology

The uredia are submerged becoming erumpent and form spots up to several millimeters in diameter. The uredospores are pale, subglobose to ovate or ellipsoidal, and 16–23 μ in diameter. The telia are intermixed with the uredia, brown, punctiform, gregarious, and irregular in shape. The teliospores are ovate to elongate, 2-celled, constricted, with persistent pedicels, and measure 31–33 × 18 μ.

Reference

11.

Thread blight, *Ceratobasidium stevensii* (Burt) Venkat.

Symptoms

The disease is first detected by the presence of killed foliage on the previous year's wood or on the new flush. The dead leaves are suspended by strong brown mycelial threads growing along the lower side of the twig. As these threads elongate, they continue up the leaf petioles and fruit peduncles where they fan out over the blade, penetrate the epidermis and parenchyma, and kill the tissue. The leaves dehisce but often remain suspended.

Etiology

The fungus produces basidia; the basidiospores are hyaline, oval to elongate, and 8–12 × 4–5 μ. Brown, flat to loaflike sclerotia, mostly 1–3 mm long, are produced by the fungus and are located on a lenticel or aggregated around terminal buds. They survive the dormant season there and generate the mycelium when new growth of the host takes place.

Reference

14.

Mushroom root rot, *Clitocybe tabescens* (Scop. ex Fr.) Bres.

Symptoms

On young trees the decline is very rapid, and a plant may show some lack of vigor followed by yellowing, wilting, and browning of foliage within a few weeks. A certain amount of defoliation may take place, and some dieback of twigs may be observed. Usually such a plant is doomed as the fungus has already developed a girdle and is killing the cambium on the larger roots or on the main trunk at the crown or near the soil line. After exposing the inner

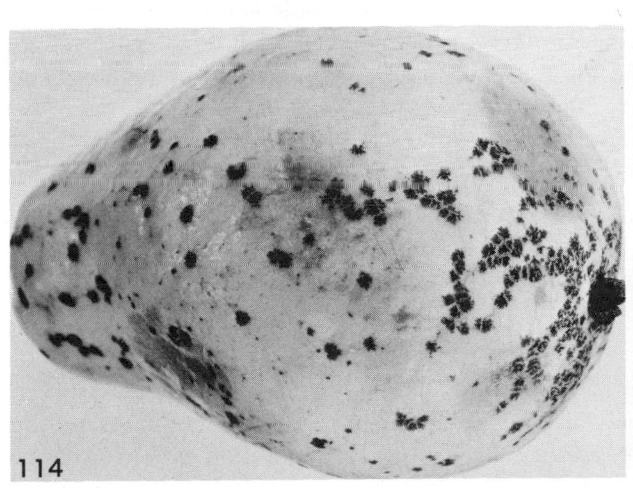

Figure 111. Ripe rot, *Glomerella cingulata,* on mature guava fruit.

Figure 112. Dying guava tree showing sporophores of *Clitocybe tabescens* at base. (Photograph by A. S. Rhoads.)

Figure 113. Algal spot, *Cephaleuros virescens,* on guava foliage and fruit. (Photograph by G. D. Ruehle.)

Figure 114. Algal spot, *Cephaleuros virescens,* on surface of guava fruit. (Photograph by G. D. Ruehle.)

Figure 115. Southern blight, *Botryobasidium rolfsii,* killing jute. (Photograph by J. Gonçalves.)

cortex with a knife cut, the white mycelial strands and plates of the fungus can be observed in the cambium area.

Etiology

Honey-colored, decurrent, gilled, stipitate, nonannulated sporophores, 4–6 in. across and up to 6 in. high, may be found near the soil line attached to the host. Basidiospores are hyaline, 1-celled, oval, and 8–10 × 4–6 μ.

Reference

9.

Fruit canker, *Pestalotia psidii* Pat.

Symptoms

The first indication of fruit infection is a brown, small, circular, necrotic, raised area with a sunken center. The first spots are shallow and cause little damage. Inoculations of green fruit attached to the tree were successful, while detached fruit failed to become diseased. The disease has been mostly confined to the green fruit and is rarely found on foliage or ripe fruit.

Etiology

The fungus produces an abundance of mycelium and conidia in the host and in culture. The pycnidia are black, circular, and usually scattered but may assume a zoning pattern. Pycnidiospores are 5-celled, oblong, clavate, erect, and 13–31 × 5–10 μ. They are brownish except for the 2 end cells which are hyaline. The apical cell supports 3 hyaline, slender, elongate appendages; the basal cell is characterized by a single, short, stout pedicel.

References

5, 6.

Wilt, *Fusarium oxysporum* f. sp. *psidii* Prasad, Mehta, & Laly

Symptoms

A wilting and browning of the foliage is early evidence of the disease and may continue to be detrimental, causing defoliation and eventually killing the plant. Removal of the outer cortex reveals a darker coloration extending down to the cambium where dark streaks are formed.

Etiology

The isolated fungus in culture produces hyaline to white mycelium, sporodochia, and pionnotes. Macroconidia are hyaline, mostly 3–4-septate, curved, and 32–40 × 3–6 μ. Chlamydospores are present.

References

4, 7, 13.

Leaf spot, *Cercospora sawadae* Yamamoto

Symptoms

Dark smoky areas or patches appear on the lower leaf surfaces; these areas are not particularly delimited, and no definite marginal distinctions appear.

Etiology

There are no stromata, and the conidiophores are not fasciculate but arise more or less singly from the subhyaline to smoky brown hyphae. They are septate and usually straight, measuring $10–50 \times 2–4$ μ. The conidia are pale olivaceous, cylindrical, catenulate or not, 3–5-septate, and measure $25–90 \times 2–5$ μ.

References

2, 8.

Algal spot, *Cephaleuros virescens* Kunze

Symptoms

The alga causes reddish brown to purple, small, circular spots, up to 8 mm in diameter, on fruit, leaf tissue, midrib, and veins. Fructifications appear on the spots as minute, erect hairs that are white, topped with yellow. The fungus occurs as a simple epiphyte or parasite on leaves, growing on the surface and penetrating the leaf tissue. Sometimes the penetration of leaf cuticle extends between cuticle and epidermal cells.

Etiology

The organism develops hairs made up of fine filaments, bearing at their ends 3–9 short sterigmata nearly equal in length, and yellow rounded bodies with the sporangia borne at the end of each. Zoospores are produced in the presence of sufficient moisture, escape through a small opening when mature, and swim by means of 2 cilia.

References

1, 10, 15.

Other fungi associated with guava

Alternaria tenuis Nees
Armillaria mellea Vahl ex Fries
Asterina psidii Ryan
Caudella psidii Ryan
Cercosporina psidii (Rangel) Saccardo
Curvularia lunata (Wakker) Boedijn
Gloeosporium psidii Delacroix
Meliola psidii Fries
Pestalotia disseminata Thuemen
Phoma psidii de Camara
Phomopsis psidii de Camara

Phyllachora subcircinans Spegazzini
Phyllosticta guajavae Viégas
Phyllosticta psidii Tassi
Phymatotrichum omnivorum (Shear) Duggar
Phytophthora parasitica Dastur
Rhizopus nigricans Ehrenberg

References: guava

1. Butler, E. J. 1918. Fungi and diseases in plants: Thacker: Calcutta, India, pp. 413–22, 512–15.
2. Chupp, C. 1953. A monograph of the fungus genus Cercospora. Published by the author: Ithaca, N.Y., p. 408.
3. Cook, M. T. 1913. The diseases of tropical plants. Macmillan: London, p. 144.
4. Edward, J. C. 1959. Variation in the guava wilt pathogen, Fusarium oxysporum f. psidii. Indian Phytopathol. 13:30–36.
5. Guba, E. F. 1961. Monograph of Monochaetia and Pestalotia. Harvard University Press: Cambridge, Mass., pp. 139–41.
6. Patel, M. K., M. N. Kamat, and G. H. Hingorani. 1950. Pestalotia psidii Pat. on guava. Indian Phytopathol. 3:165–76.
7. Prasad, N., P. R. Mehta, and S. B. Laly. 1952. Fusarium wilt of guava (Psidium guajava L.) in Uttar Pradesh, India. Nature 169:753.
8. Ragunathan, V., and N. N. Prasad. 1969. Occurrence of Cercospora sawadae on Psidium guajava. Plant Disease Reptr. 53:455.
9. Rhoads, A. S. 1956. The occurrence and destructiveness of clitocybe root rot of woody plants in Florida. Lloydia 19: 193–239.
10. Ruehle, G. D. 1941. Algal leaf and fruit spot of guava. Phytopathology 31:95–96.
11. Saccardo, P. A. 1888. Puccinia psidii Winter. Sylloge Fungorum 7:643.
12. Srivastava, M. P., and R. N. Tandon. 1969. Post harvest diseases of guava in India. Plant Disease Reptr. 53:206–8.
13. Vestal, E. F. 1950. A textbook of plant pathology. Kitabistan Allahabad: India, pp. 495–96.
14. Weber, G. F. 1927. Thread blight, a fungus disease of plants caused by Corticium stevensii Burt. Florida Univ. Agr. Expt. Sta. Bull. 186:143–62.
15. Winston, J. R. 1938. Algal fruit spot of orange. Phytopathology 28:283–86.

Hemp, *Cannabis sativa* Linnaeus

Bacteriosis, *Pseudomonas cannabina* Ŝutić & Dowson var. *italica* Dowson
Twig blight, *Botryosphaeria marconii* (Cavara) Charles & Jenkins
Wilt, *Fusarium oxysporum* f. sp. *cannabis* Noviello & Snyder
Leaf spot, *Cercospora cannabis* Hara & Fukui
Other fungi associated with hemp

Bacteriosis, *Pseudomonas cannabina* Ŝutić & Dows. var. *italica* Dows.

Symptoms

Reddish violet, irregular, scattered or confluent spots of various size, usually less than 2 mm, appear on the leaf blades surrounded by a pale yellow halo. The lesions on the main veins result in rapid leaf shedding. Small, necrotic centers appear on the woody parts, forming elongate cavities filled with bacteria and frequently penetrating to the wood.

Etiology

The organism is a rod, motile by 1–4 polar flagella, and measuring 1.5×0.3 μ.

Agar colonies are yellow gray, smooth, convex, and produce a brown pigment in the medium.

Reference

3.

Twig blight, *Botryosphaeria marconii* (Cav.) Charles & Jenkins

Symptoms

A slight wilting and drooping of leaves rapidly develops into extensive wilt, browning, and death of the plant; the dried leaves remain attached. The tips of branches show the first symptoms, but the lower parts become bleached and bear the fruiting structures of the fungus.

Etiology

The spores of the fungus are produced in two kinds of black, small pycnidia. The microconidia are hyaline, oval, 1-celled, and 4–6 × 1–2 μ. The macroconidia are hyaline, 1-celled, fusiform to elliptical, and 16–18 × 5–6 μ. The asci are developed in the same fruiting stroma. They are clavate, 8-spored, accompanied by filiform paraphyses, and measure 80–90 × 13–15 μ. The ascospores are hyaline to pale green, 1-celled, elongate with rounded ends, and 16–18 × 2–8 μ.

Reference

1.

Wilt, *Fusarium oxysporum* f. sp. *cannabis* Nov. & Snyd.

Symptoms

The disease first appears on 3-month-old plants, producing a yellowish green color and small, dark, irregular spots on the lower leaves. Later a yellowish tan color develops, and the leaf dies and remains pendant. Cortical tissues over the vascular strands become yellow tan and often occur on the side of the stem, causing the foliage symptoms on that side. The plant may bend toward the colored side of the stem into a leaning position. There are no characteristic symptoms of the disease below the soil surface except for a deterioration of cortical tissue in old, long-infected plants.

Etiology

The mycelium is hyaline, septate, branched, and occupies the vascular tissue. Conidiophores are often verticillately branched and bear hyaline, 1-celled microconidia that measure 5–7 × 2.5–3 μ. Macroconidia are hyaline, sickle-shaped, mostly 3-septate, and 22–36 × 4–5 μ. The fungus is soil-borne.

References

4, 5.

Leaf spot, *Cercospora cannabis* Hara & Fukui

Symptoms

Leaf spots are yellow tan to brown, circular at first but becoming irregular, usually small, separate, and distinct.

Etiology

Stromata on the lower leaf surface are brown and inconspicuous. Fascicles are numerous with up to 12 conidiophores that are pale brown, usually curved, septate, and unbranched. Conidia are hyaline, curved to cylindric, multiseptate, and 20–90 × 2–4 μ.

Reference

2.

Other fungi associated with hemp

Botryobasidium rolfsii (Saccardo) Venkatarayan
Botrytis cinerea Persoon ex Fries
Cercospora cinnabina Wakefield
Macrophomina phaseolina (Tassi) Goidanich
Nectria cancri Rutgers
Peronospora cannabina Otth
Phymatotrichum omnivorum (Shear) Duggar
Sclerotinia sclerotiorum (Libert) De Bary
Septoria cannabis (Lasch) Saccardo

References: hemp

1. Charles, V. K., and A. E. Jenkins. 1914. A fungous disease of hemp. J. Agr. Res. 3:81–84.
2. Chupp, C. 1953. A monograph of the fungus genus Cercospora. Published by the author: Ithaca, N.Y., p. 394.
3. Goidanich, G., and F. Ferri. 1960. La batterirosi della canapa da Pseudomonas cannabina Ŝutić et Dowson var. italica Dowson. Phytopathol. Z. 37:21–32.
4. Lee, H. A., and F. B. Serrano. 1923. Banana wilt of the Manila hemp plant. Phytopathology 13:253–56.
5. Noviello, C., and W. C. Snyder. 1962. Fusarium wilt of hemp. Phytopathology 52:1315–17.

Hop, *Humulus lupulus* Linnaeus

Crown gall, *Agrobacterium tumefaciens* (E. F. Smith & Townsend) Conn
Downy mildew, *Pseudoperonospora humuli* (Miyabe & Takahashi) G. W. Wilson
Cottony rot, *Sclerotinia sclerotiorum* (Libert) De Bary
Canker, *Gibberella pulicaris* (Fries) Saccardo
Powdery mildew, *Sphaerotheca humuli* (De Candolle) Burrill
Twig smudge, *Pestalotia truncata* Léveillé
Gray mold, *Botrytis cinerea* Persoon ex Fries
Leaf spot, *Cercospora cantauriensis* Salmon & Wormold
Leaf spot, *Cercospora humuli* Hori
Wilt, *Verticillium albo-atrum* Reinke & Berthold
Other fungi associated with hop

Crown gall, *Agrobacterium tumefaciens* (E. F. Sm. & Towns.) Conn

Symptoms

Tumors or galls from pea-sized protrusions to massive structures, some over 6 in. in diameter, develop on the roots and stems. They usually appear at ground level, but they may be produced on roots or even high on the stems. When young, they are whitish, mostly irregularly spherical, convoluted, soft, and resilient. When old, they are blackish, hard, horny, and rough.

Etiology

The organism is a gram-negative rod, forming capsules, motile by polar flagella, and measuring 0.4–0.8$\times$1–3 μ. Agar colonies are white, translucent, circular, and glistening. Optimum temperature, 25–28C.

Reference

3.

Downy mildew, *Pseudoperonospora humuli* (Miy. & Tak.) G. W. Wils.

Symptoms

The fungus may survive in the perennial root or crown of the plant. The early shoots are a combination of normal, healthy, slender stalks and diseased, stunted, thicker ones that appear pale green to silvery gray. Diseased lateral branches may also appear from any place on the healthy main stem, or a terminal infection may develop. The nodes are close together; the leaves are brittle, gray above, and covered with the fungus as a white to lavender downy mildew on the lower surfaces.

Etiology

The hyaline, nonseptate mycelium is almost entirely submerged. The sporangiophores are 100–400 μ high, dense, divided up to 4 times, and produce sporangia at the subulate tips. The sporangia are oval, thin-walled, papillate, and 22–26$\times$15–18 μ. Upon germination, the sporangia produce 12–20 zoospores that measure 8–12 μ in diameter. They are active for short periods and then settle down and germinate. The oospores found imbedded in the host tissue in late summer are 36–40 μ in diameter and also germinate by producing zoospores.

Reference

1.

Cottony rot, *Sclerotinia sclerotiorum* (Lib.) d By.

Symptoms

The disease is indicated by straw-colored, faded areas on the stems at the soil line and upward. The cortex is devitalized and a certain amount of wilting occurs, depending on the extent of invasion by the fungus in regard to the support of the top growth. Eventually the plant dies. A scanty amount of white, fluffy mycelium may appear externally, but it is plentiful in the pith, where the black, shiny, elongate sclerotia of the fungus are produced. The sclerotia continue to develop until the stem dries out. They remain viable in plant debris or in the soil until the following season.

Etiology

The hyaline, septate, branched mycelium often aggregates into white, cottony tufts. Conidia are not produced. Black, irregular, shiny, sclerotia, measuring up to 1 cm in diameter, are usually formed and are the means of survival of the fungus.

Reference

7.

Canker, *Gibberella pulicaris* (Fr.) Sacc.

Symptoms

The foliage withers, and the lower stem or vine at or near the soil line is discolored, weakened, and easily pulled away from the rootstock. Late-infected vines sometimes show a definite swelling at the base. Whitish pustules of the sporulating fungus develop on the infected stalks from the soil line up to a few feet.

Etiology

The spores of the fungus develop on the hyphal growth. They are hyaline, slightly curved, mostly 3-septate, but up to 5-septate, and $25–35 \times 4–5$ μ. The perithecia appear mostly above the epidermis; they are dark purple, oval, ostiolate, rounded, and 20–30 μ in diameter. The ascospores are hyaline, 4-celled, constricted, with obtuse ends and thicker centers, and measure $17–32 \times 5–7$ μ. The vegetative stage is *Fusarium sambricinum.*

Reference

7.

Powdery mildew, *Sphaerotheca humuli* (DC.) Burr.

Symptoms

This disease is evident by the presence of small, greenish blisters, which later are the focal point of white mycelium that grows in all directions on the leaf surface, petioles, peduncles, and inflorescence. The white hyphae are superficial and produce many conidiophores that are upright and support terminal conidia singly or in chains. The conidia are often so plentiful as to produce a white, dusty coating over the surface.

Etiology

The mycelium is hyaline, septate, branched, and supported by imbedded haustoria. The conidiophores are hyaline, septate, with narrow basal cells widening at the terminal where hyaline, oval, 1-celled conidia are produced, measuring 20–30 × 12–15 μ. The perithecia are scattered to gregarious, vary in color from pink to black depending on maturity, and are 58–120 μ in diameter. The appendages are dark brown, long, undulating to straight, and septate. The asci are elliptical to subglobose and measure 45–90 × 51–72 μ. The ascospores are hyaline, 1-celled, and 20–25 × 12–18 μ.

Reference

1.

Twig smudge, *Pestalotia truncata* Lév.

Symptoms

The fungus develops pustules on the dead petioles, peduncles, and twigs. They are black, globular, immersed or erumpent, and exude myriads of black conidia that are so plentiful as to discolor the immediate host tissue in black, smudgy areas.

Etiology

The conidia are produced in acervuli or pycnidia that lack ostioles. Conidia are 4-celled, fusoid, erect, not constricted, and measure 14–20 × 6–8 μ. The two center cells are olivaceous, cask-shaped, and 11–14 μ. The apical cell is hyaline, subulate, sometimes deciduous or attenuated to a single setula or branched. The pedicel is coarse and long, often up to 50 μ.

Reference

4.

Gray mold, *Botrytis cinerea* Pers. ex Fr.

Symptoms

Brown discolorations appear at the tip of the cone and tend to increase upward toward the peduncle. The spots are largely marginal at first but tend to develop farther up the scale or leaf of the cone. They are almost black, mostly small, angular, and develop into blotches. Cones are often severely discolored, especially in wet weather.

Etiology

The mycelium is hyaline, septate, and branched. It produces stocky conidiophores in abundance. These are divided once or twice at the tips and support conidia that grow in dense clusters. The conidia are hyaline, 1-celled, oval, and 11–15 × 8–11 μ. Under certain conditions black, small, shiny sclerotia may develop on the plant or on the ground.

Reference

7.

Leaf spot, *Cercospora cantauriensis* Salm. & Wormold

Symptoms

Leaf spots are grayish with a dark purple to brown border surrounded by a yellow zone, mostly circular, and up to 5 mm in diameter. There are usually no stromata on the lower leaf surface.

Etiology

The conidiophores are olive brown, short, single, straight, unbranched, 1–7-septate, with obtuse tips, and measure 25–290 × 8–20 μ. The conidia are faintly olive brown, cylindric, mostly straight, 5–19-septate, and 135–500 × 1–2 μ.

Reference

7.

Leaf spot, *Cercospora humuli* Hori

Symptoms

Leaf spots are tan to brown, circular to angular, and up to 5 mm in diameter.

Etiology

The conidiophores are produced from brown stromata on the lower surface of the leaf in dense, spreading fascicles. They are pale olive brown, straight or curved, septate, branched or not, and 55–70 × 3–5 μ. The conidia are pale, olivaceous, mostly straight, 4–11-septate, and 35–120 × 2–4 μ.

Reference

2.

Wilt, *Verticillium albo-atrum* Reinke & Berth.

Symptoms

A yellowing of the older leaves, progressing upward during the season, indicates the presence of the disease. Necrotic black areas occur between the principal veins. The edges of the leaves become dry, curl, finally wither, and may be shed. Associated with the leaf symptoms is a browning of the wood in a sector or uniformly in cross section. There also appears frequently a swelling of the first several feet of the stem. After harvest, black streaks run along the ridges over the primary vascular bundles.

Etiology

The mycelium is hyaline, branched, septate, and bears solitary or loosely aggregated conidiophores that are hyphaelike with verticillate branching. Conidiophores are thick at the base, taper upward, and measure 100–400 × 4–6 μ. The conidia are hyaline, oblong to ovate, 1-celled, produced singly, form chains or wet, ball-like droplets, and measure 4–5 × 2–4 μ. Black, sclerotialike bodies are sometimes present.

References

5, 6.

Other fungi associated with hop

Armillaria mellea Vahl ex Fries
Ascochyta humuli Kabat & Bubak
Cylindrosporium humuli Ellis & Everhart
Fumago vagans Persoon ex Saccardo
Glomerella cingulata (Stoneman) Spaulding & Schrenk
Mycosphaerella erysiphina (Berkeley & Broome) Kirchner
Phoma herbarum Westerdijk
Phyllosticta decidua Ellis & Kellerman
Phyllosticta humuli Saccardo & Spegazzini
Phytophthora cactorum Lebert & Cohn
Septoria humuli Westendorp
Septoria lupulina Ellis & Kellerman
Stagonospora humuli-americani Fairman
Verticillium dahliae Klebahn

References: hop

1. Butler, E. J., and S. G. Jones. 1949. Plant pathology. Macmillan: London, pp. 879–86.
2. Chupp, C. 1953. A monograph of the fungus genus Cercospora. Published by the author: Ithaca, N.Y., pp. 397–98.
3. Elliott, C. 1951. Manual of bacterial plant pathogens. 2d ed. Chronica Botanica: Waltham, Mass., pp. 5–12.
4. Guba, E. F. 1961. Monograph of Monochaetia and Pestalotia. Harvard University Press: Cambridge, Mass., pp. 65–70.
5. Keyworth, W. C. 1942. Verticillium wilt of the hop (Humulus lupulus). Ann. Appl. Biol. 29:346–57.
6. Ludbrook, W. V. 1933. Phytogenicity and environmental studies on Verticillium hydromycosis. Phytopathology 23:117–54.
7. Wormold, H. 1946. Diseases of fruits and hops. Lockwood & Sons: London, pp. 61–62, 250–64.

Jujube, *Zizyphus jujuba* Miller

Anthracnose, *Glomerella cingulata* (Ston.) Spauld. & Schrenk

Symptoms

The premature shedding of fruit when the branches, twigs, and foliage are free of a specific parasite results in a more careful observation of the fruit. Dark, small, circular dots develop on slightly sunken, circular fruit spots that measure 4–6 mm in diameter. In section, the diseased area is a dark olivaceous color. The decay causes the fallen fruit to shrivel and become mummies.

Etiology

The fungus produces masses of salmon pink conidia in acervuli that are hyaline, 1-celled, elongate, round ended, slightly oval, and resemble *Gloeosporium* sp.

References

3, 4, 5.

Leaf spot, *Cercospora jujubae* Chowdhury

Symptoms

A slight yellowing associated with certain leaves develops opposite a dark, moldlike fungus that appears on the lower leaf surfaces.

Etiology

The mycelium is usually colored, septate, branched, and produces smoky conidiophores that emerge through the stomata, are 1–4-septate, constricted with conidial scars, and measure 48–152 × 4–7 μ. Conidia are pale olivaceous, 1–5-septate, obclavate, and 25–45 × 8–10 μ.

References

1, 2.

Other fungi associated with jujube

Phakopsora zizyphi-vulgaris (P. Henning) Dietel
Uredo zizyphi Patouillard

References: jujube

1. Chupp, C. 1953. A monograph of the fungus genus Cercospora. Published by the author: Ithaca, N.Y., p. 469.
2. Golsen, E. W., and H. L. Rubin. 1964. Cercospora jujubae Chowdhury, a new report for the United States. Plant Disease Reptr. 48:914.
3. Srivastava, M. P. 1967. A new fruit rot of Zizyphus jujuba Lamk. Indian Phytopathol. 20:389–90.
4. Taubenhaus, J. J., and W. N. Ezekiel. 1931. An anthracnose of the jujube. Phytopathology 21:1185–89.
5. Thomas, C. C. 1924. The Chinese jujube. U.S. Dept. Agr. Bull. 1215:1–30.

Jute, *Corchorus capsularis* Linnaeus

Leaf spot, *Xanthomonas nakatae* (Okobe) Dows.

Symptoms

The brown, small, circular spots, usually less than 5 mm in diameter, characterize this disease on the leaves and stems, with somewhat sunken areas on the capsules. The leaf spots are often surrounded by a yellow halo; when numerous, the spots coalesce and may cause some defoliation, otherwise only some shot holes appear. The stem lesion may develop into stem girdling, killing the part above that point.

Etiology

The organism is a gram-negative rod, single, in pairs, or in short chains, supporting a single polar flagellum, and measuring 1–2.5$\times$0.2–0.4 μ. Agar colonies are yellow becoming brown, glistening, round, smooth, and have a raised, entire margin. Optimum temperature, 30–32C.

References

5, 9.

Southern blight, *Botryobasidium rolfsii* (Sacc.) Venkat.

Symptoms

The early attacks of the fungus are usually limited to the collar region at the soil line where a soft rot develops. The mycelium invades the epidermis and cortex, resulting in extensive disintegration, tissue shredding, and a final breaking over of the stem. Less severely diseased plants often survive, and the fungus mycelium grows upward above the soil line, forming mats of various densities. At this place, the sporulating stage of the fungus may be observed. Hitherto only the brown, spherical, mustard-seed-like, hard sclerotia have been observed.

Etiology

The clavate, swollen cells branch off from the mycelial mat. Some of these cells form basidia which are stout, clavate, and 13–16$\times$6–8 μ. Each basidium bears 2–4 hyaline, tapering sterigmata and basidiospores which are hyaline, smooth, obliquely pyriform, and 6–8.5$\times$4 μ.

Reference

7.

Brown band, *Macrophoma corchori* Saw.

Symptoms

The stem is attacked by the fungus slightly above the ground line where a reddish brown, shallow canker develops. The color gradually darkens and extends up and down the stem and around it. Plants in all stages are susceptible. In young plants, a typical damping-off stage is most destructive. On mature plants black, small, sclerotialike bodies are formed in the interior of the stem, and black, subepidermal, ostiolate pycnidia develop below the cuticle.

Etiology

The hyphae are hyaline, septate, and branched. The pycnidia are black, submerged, punctiform, with ostioles protruding. Conidiophores are hyaline, numerous, cylindric, tapering, continuous, and $10–14 \times 2–4\ \mu$. Pycnidiospores are hyaline, oval to oblong, round at the apex with an obtuse base, smooth, 1-celled, straight or curved, and $16–32 \times 7–10\ \mu$. The fungus is also known as *Sclerotium bataticola*.

Reference

11.

Stem canker, *Macrophomina phaseolina* (Tassi) Goid.

Symptoms

A severe stem canker develops on month-old seedlings, causing a wilting of the plants and shedding of the foliage. The cankers are slate to black, slightly sunken, and 2–3 in. long.

Etiology

Stem sections show very small, black sclerotia imbedded in the pith, woody tissue, and inner cortex. The exterior is generally smooth. The vegetative stage is *Sclerotium bataticola*.

References

1, 2, 4, 10.

Black band, *Diplodia corchori* Syd.

Symptoms

Diseased plants are often detected in the field by the appearance of wilting, drying, or brown, dead plants among the healthy green ones. Diseased plants shed their leaves, and a dense, discolored to black band forms around the stem a few feet above the soil line. The early stages show the stem band originates at a weak lateral twig and from there spreads up and down the stem. The stem is usually split lengthwise, exposing the internal fibers. The blackened surface eventually becomes crowded with black, subepidermal fruiting bodies of the fungus with ostioles of the pycnidia protruding.

Etiology

The hyphae of the fungus penetrates the cortex and invades the phloem, xylem, and medullary rays. The pycnidia are mostly spherical, imbedded becoming erumpent, and $200–300\ \mu$ in diameter. The pycnidiospores are often exuded in white masses, indicating their immature condition as they appear. They are 1-celled, hyaline, and oval to elongate. The mature spores are dark-colored, 2-celled, and approximately the same size as the immature spores, averaging 24×12 μ. The ascospore stage is *Physalospora rhodina.*

Reference

11.

Stem rot, *Colletotrichum corchorum* Ikata & Yosh.

Symptoms

The first visible indication of the disease is a yellowish brown depressed area, irregular in outline, and up to 1 cm in length. The color gradually changes to brown then black, and necrotic lesions appear. Lesions may coalesce, girdle the stem, and result in breakage. Pods are attacked and do not produce seed except when the disease appears late. The seedlings may not develop normally in weak attacks and are destroyed in severe attacks. The organism is seed-borne.

Etiology

The conidia are hyaline, falcate, guttulate, and measure 17–24×2–3.5 μ, with acervuli including from 10 to 50 setae.

References

6, 8.

Leaf spot, *Cercospora corchori* Saw.

Symptoms

Dark brown, small, irregularly circular spots appear on the leaves and enlarge up to 4–14 mm in diameter. They often coalesce, forming large brown blotches that are followed by a general yellowing and eventual shedding. The disease proceeds from the lower leaves upward. Spots also appear on the stems and seed capsules, and the fibers may become discolored.

Etiology

The mycelium is hyaline becoming olivaceous, septate, branched, and aggregates in the substomatal spaces. There stromatic masses are formed that give rise to the fruiting structures of the fungus. The conidiophores develop on the lower leaf surface in fascicles of 4–10 and are up to 3-septate. The conidia are hyaline, filiform to obclavate, 2–7-septate, and 32–117×3–5 μ.

Reference

3.

Other fungi associated with jute

Ascochyta corchoricola Chochrjazow
Cerotelium corchori Sydow
Epicoccum neglectum Desmazieres

Phoma sabdariffae Saccardo
Pythium perniciosum Gobi
Xanthomonas solanacearum (E. F. Smith) Dowson

References: jute

1. Békési, P., J. Vörös, and O. H. Calvert. 1970. Macrophomina phaseoli in Hungary damaging sunflower. Plant Disease Reptr. 54:286–87.
2. Butler, E. J. 1918. Fungi and diseases in plants. Thacker: Calcutta, India, pp. 272–73.
3. Chowdhury, S. 1947. A Cercospora blight of jute. J. Indian Botan. Soc. 26:227–31.
4. Crandall, B. S. 1954. Jute stem canker in Cuba. Plant Disease Reptr. 38:37–38.
5. Elliott, C. 1951. Manual of bacterial plant pathogens. 2d ed. Chronica Botanica: Waltham, Mass., p. 126.
6. Ghosh, T. 1957. Anthracnose of jute. Indian Phytopathol. 10:63–70.
7. Ghosh, T., and K. V. George. 1955. Sclerotium rolfsii Sacc. on jute and its perfect stage. Indian J. Agr. Sci. 25:171–73.
8. Islam, N., and Q. A. Ahmed. 1964. The cause of tolerance of anthracnose, Colletotrichum corchori, in species and varieties of jute. Trans. Brit. Mycol. Soc. 47:227–33.
9. Sabet, K. A. 1957. Bacterial leaf spot of jute (Corchorus olitorius L.). Ann. Appl. Biol. 45:516–20.
10. Seymour, C. P. 1969. Charcoal rot of nursery-grown pines in Florida. Phytopathology 59:89–92.
11. Shaw, F. J. F. 1921. Studies in diseases of the jute plant. Mem. Dept. Agr. India, Botan. Ser. 11:37–58; 1924, 13: 193–99.

Kenaf, *Hibiscus cannabinus* Linnaeus

Wilt, *Pseudomonas solanacearum* E. F. Smith
Powdery mildew, *Leveillula taurica* f. *hibisci* (Léveillé) Arnaud
Web blight, *Botryobasidium microsclerotia* (Matz) Venkatarayan
Blight, *Phoma sabdariffae* Saccardo
Ashy stem, *Macrophomina phaseolina* (Tassi) Goidanich
Anthracnose, *Colletotrichum hibisci* Pollacci
Gray rot, *Botrytis cinerea* Persoon ex Fries
Leaf spot, *Cercospora abelmoschi* Ellis & Everhart
Leaf spot, *Cercospora hibiscina* Ellis & Everhart
Leaf spot, *Cercospora malayensis* F. L. Stevens & Solheim
Other fungi associated with kenaf

Wilt, *Pseudomonas solanacearum* E. F. Sm.

Symptoms

This is a vascular disease and causes early stunting, some foliage yellowing, general or acute wilting, and a dropping of the leaves. The vascular bundles are darkened, and in extreme instances the pith is invaded. The roots are mostly discolored and many are killed.

Etiology

The organism is a gram-negative rod, forms no capsules, is motile by a single polar flagellum, and measures 0.5×1.5 μ. Agar colonies are small, irregular, smooth, wet, shiny, and opalescent gradually becoming brown. Optimum temperature, 35–37C.

Reference

4.

Powdery mildew, *Leveillula taurica* f. *hibisci* (Lév.) Arn.

Symptoms

A white, chalky, dusty covering develops on the leaf surfaces. The early infections cause very little change in the host. As the disease becomes more severe, a yellowing of the leaves and a droopiness develop. The leaves become functionless, and many are shed. The disease continues to affect leaves to the tops of the plants.

Etiology

The mycelium is hyaline, branched, and submerged into the parenchyma. Conidiophores may develop superficially or protrude through the stomata. The conidia are long, cylindrical, slightly rough, with rounded ends, and measure $40–60 \times 12–16\ \mu$. The cleistothecia measure $125–270\ \mu$ in diameter and are covered with straight-tipped appendages. The asci are numerous, each measuring $60–75 \times 24–30\ \mu$, and usually 2–4-spored. Ascospores are about $19–25 \times 13–15\ \mu$.

References

3, 17.

Web blight, *Botryobasidium microsclerotia* (Matz) Venkat.

Symptoms

First indications of infection are brown, small, circular spots on the leaf blades. They enlarge rapidly and often show a zonate characteristic. The spots have reddish borders and coalesce or enlarge, involving large portions of the leaf. The size of the spots on the upper leaf surface is augmented by the rapid extension of the aerial mycelium that covers the lower surface. This growth advances rapidly from the point of infection, where leaves contact each other, or from the petiole onto the blade if the infection is on the stem. The blade becomes yellow, turns brown, and is shed.

Etiology

The mycelium is yellow to brown, depending on age and rate of growth, septate, branched, and produces basidia on the lower leaf surface in loose, sporodochialike mycelial nets. The basidiospores are hyaline, 1-celled, oval to elongate, and $6–9 \times 3–5\ \mu$. Under certain conditions, brown, small, multicelled sclerotia are produced on the mycelium on the exterior of diseased plant parts. They measure $200–300\ \mu$ and are usually slightly elongate; most of them will pass through a 100 mesh screen.

References

2, 9, 18.

Blight, *Phoma sabdariffae* Sacc.

Symptoms

This disease, found on all aboveground parts, causes small, circular to irregular, water-soaked spots with pale green borders. The spots enlarge rapidly under favorable conditions, up to 5 cm in diameter. Severely spotted leaves are usually shed. Blossoms are often discolored and fail to develop. Shoots are blackened, shrivel, and die back to a node in more hardened tissue.

Etiology

The mycelium is hyaline, white en masse becoming darker, septate, branched, and finely granular. The pycnidia, abundantly produced on the leaf lesions, are at first submerged but become erumpent. They are brown to black, scattered or gregarious, and ostiolate. The conidiophores are hyaline, short, and produce pycnidiospores that are hyaline, ellipsoidal, 1-celled, and 10–13 $\times$ 5–7 μ.

References

6, 12.

Ashy stem, *Macrophomina phaseolina* (Tassi) Goid.

Symptoms

The damping-off of seedlings is important in considering loss of stand. The plants that survive the seedling stage are subject to crown cankers and stem girdling that cause death. Dead plants have ashy gray stems and remain standing. Older plants may show foliage spots that are brown, circular, and often speckled with black pycnidia. Black, small sclerotia develop on the cortex.

Etiology

The mycelium is hyaline, septate, branched, and soil-inhabiting. The sclerotia are black, shiny, irregularly shaped, smooth to angular, hard, plentiful in diseased cortex and wood, and measure 50–150 μ in diameter. The conidia are hyaline, 1-celled, fusiform, slightly curved, with rounded ends, thicker through the centers, and measure 15–30 $\times$ 5–8 μ.

Reference

14.

Anthracnose, *Colletotrichum hibisci* Pollacci

Symptoms

Spots on the stems and branches are tan to brown, oval to elongate, slightly sunken, and sometimes noticeably zonate. A tip blight is very common and destructive in young plants. The tender apices of growing shoots collapse, wither, and die. This is one of the most destructive diseases of the host. It is widespread geographically.

Etiology

The mycelium is hyaline, septate, and branched. It produces acervuli that are oval to elongate or lenticular, scattered or gregarious, and 225 $\times$ 65 μ. The setae are dark to violet, relatively short with acute tips, continuous or septate, and 55×3–4 μ. The conidiophores are short and hyaline, and the conidia are subclavate, elongate, continuous, biguttulate, and 11–25×4–5 μ.

References

7, 11, 15.

Gray rot, *Botrytis cinerea* Pers. ex Fr.

Symptoms

Infection occurs on the aboveground parts of the stem as well as on petioles, flowers, peduncles, and capsules. These areas, which are slightly brownish, become covered with a gray, fluffy accumulation of mycelium and sporulating structures of the fungus. The involved areas tend to become dark, water-soaked, and soft. Under highly humid conditions, plants are killed, but they may survive in a drier environment. More than one infection may be present on a single plant. Individual lesions may extend for 1 foot or more along the stem. Breaking over and lodging may occur.

Etiology

The mycelium is hyaline, branched, septate, and supports erect, dendritic conidiophores bearing clusters of hyaline, oval to oblong, 1-celled conidia, measuring 11–15×8–11 μ. Small, black sclerotia are produced that germinate by producing hyphae.

Reference

10.

Leaf spot, *Cercospora abelmoschi* Ell. & Ev.

Symptoms

The leaf spots are irregular in shape and size and vary from yellow in the centers to yellow green to green along the margins. On the lower surface of the leaves, the conidiophores form brown to sooty patches often limited by the principal veins. As the disease develops, the leaves become yellow then brown.

Etiology

The mycelium is subhyaline, septate, branched, and about 2 μ in diameter. Conidiophores, mostly on the lower surface, emerge from stomata, are compact and tufted, fuliginous, branched, 1–5-septate, and measure 25–140×3–5 μ. The conidia are olivaceous, cylindrical to obclavate, straight or curved, 1–7-septate, constricted, and 25–90×3–5 μ.

Reference

16.

Leaf spot, *Cercospora hibiscina* Ell. & Ev.

Symptoms

The leaf spots are yellowish to green on top view and mostly indefinite in size and shape. The lower surface of the leaf is sooty to dark-colored with the larger veins separating the fuliginous fungus growth. When the disease is severe, the foliage is shed.

Etiology

The mycelium is subhyaline to olivaceous, branched, septate, and about 3 μ in diameter. The conidiophores are on the lower leaf surface, emerging through stomata or growing from surface mycelium. They are dark brown, straight, multiseptate, branched, and 240–1000×3–4 μ. The conidia are subhyaline, olivaceous, cylindrical, 3-septate, and 20–70×3–4 μ.

Reference

1.

Leaf spot, *Cercospora malayensis* F. L. Stev. & Solh.

Symptoms

Leaf spots are brown to gray with purple to reddish borders. They are circular, irregular, or elongate, and appear between the principal veins. The small spots are several millimeters in diameter; those that are elongate may be up to 30 mm. The involved tissues often drop out.

Etiology

The mycelium is light brown, branched, and 3–5 μ in diameter. The conidiophores are sooty to brown, emerging, tufted, straight or flexuous, and 1–8-septate; conidial scars are large and distinct. Conidia are hyaline, acicular to obclavate, multiseptate, and 50–270×2–5 μ.

References

1, 16.

Other fungi associated with kenaf

Botryobasidium rolfsii (Saccardo) Venkatarayan
Choanephora cucurbitarum (Berkeley & Ravenel) Thaxter
Fusarium sarcochroum (Desmazieres) Saccardo
Phytophthora parasitica Dastur
Thanatephorus cucumeris (Frank) Donk
Trichoderma koningi Oudemans
Trichoderma lignorum (Tode) Harz

References: kenaf

1. Chupp, C. 1953. A monograph of the fungus genus Cercospora. Published by the author: Ithaca, N.Y.
2. Crandall, B. S. 1948. Pellicularia target leaf spot disease of kenaf and roselle. Phytopathology 38:503–5.

3. Diehl, W. W. 1952. Powdery mildew of kenaf in Florida. Plant Disease Reptr. 36:52.
4. Elliott, C. 1951. Manual of bacterial plant pathogens. 2d ed. Chronica Botanica: Waltham, Mass.
5. Ghosh, T., and K. V. George. 1953. Brown rot of mesta (Hibiscus cannabinus Linn.). Indian Phytopathol. 6: 106–9.
6. Ghosh, T., and N. Mukherji. 1958. Tip rot of mesta (Hibiscus cannabinus Linn.). Current Sci. (India) 27:67–69.
7. Hartley, C. 1927. Notes on Hibiscus disease in west Java. Phytopathology 17: 25–27.
8. McCaun, L. P. 1952. Kenaf, a biographical survey. U.S. Dept. Agr. Biograph. Bull. 17:1–8.
9. Matz, J. 1917. A rhizoctonia of the fig. Phytopathology 7:110–18.
10. Perez, J. M., and T. E. Summers. 1963. A botrytis disease of kenaf. Plant Disease Reptr. 47:200–201.
11. Presley, J. T. 1952. Colletotrichum tip blight of kenaf: a brief history of the disease. Plant Disease Reptr. 36:333–34.
12. Quiniones, S. S., and F. T. Orillo. 1952. Blight of kenaf. Philippine Agriculturist 36:235–50.
13. Reinking, O. A. 1945. Diseases of roselle fiber plants in El Salvador. Plant Disease Reptr. 29:411–14.
14. Roger, L. 1953. Phytopathologie des pays chauds. Encyclopedie Mycologique 18: 1704–9.
15. Saccardo, P. A. 1898. Sphaeropsidaceae. Sylloge Fungorum 14:1015.
16. Solheim, W. G., and F. L. Stevens. 1931. Cercospora studies. II. Some tropical Cercosporae. Mycologia 23:365–405.
17. Stevenson, J. A. 1945. Powdery mildew of mesquite. Plant Disease Reptr. 29: 214–15.
18. Weber, G. F. 1939. Web-blight, a disease of beans caused by Corticium microsclerotia. Phytopathology 29:559–75.
19. Wilson, F. D., and M. Y. Menzel. 1964. Kenaf (Hibiscus cannabinus) Roselle (Hibiscus sabdariffae). Econ. Botany 18:80–91.

Lang, *Lathyrus sativus* Linnaeus

Wilt, *Fusarium orthoceras* f. sp. *lathyri* Bhide & Uppal
Leaf spot, *Mycosphaerella ontarioensis* Stone

Wilt, *Fusarium orthoceras* f. sp. *lathyri* Bhide & Uppal

Symptoms

Yellowing of leaves is the first visible symptom in the field. It begins on the older leaves, progresses upward, and eventually involves the entire plant. The leaves wilt, droop, and finally dry and die. Discoloration of the stem is found only in the root portions. In fields, patches of wilted plants may appear and these areas gradually enlarge. Ten to 15 per cent losses are common, and 60–70 per cent losses frequently occur.

Etiology

The mycelium is mostly white with occasional green-yellow-orange tints, septate, and branched; it forms no sporodochia, pionnotes, or sclerotia. Microconidia are hyaline, 1-celled, ovoid to elliptical, and 8×4 μ. With age they elongate and become up to 2-septate, measuring at most 31×5 μ. Chlamydospores average about 10 μ in diameter.

Reference

1.

Leaf spot, *Mycosphaerella ontarioensis* Stone

Symptoms

Spots on the foliage, stems, and pods are small, slightly yellow to brown, circular, and slow-growing. They eventually reach 5–6 mm in diameter and become speckled with small, black, spherical pycnidia.

Etiology

Pycnidia become erumpent, measuring 75–160 μ. The pycnidiospores are hyaline, cylindrical, 1-septate, and 8–13 × 2–3 μ. The perithecia are globose, submerged to erumpent, and papillate. The asci are cylindrical, measuring 45–50 × 12–14 μ, and the hyaline ascospores are septate with pointed ends and measure 10–13 × 5–6 μ. The imperfect stage is an *Ascochyta* sp.

Reference

2.

References: lang

1. Bhide, V. P., and B. N. Uppal. 1948. A new fusarium disease of lang (Lathyrus sativus). Phytopathology 38:560–67.
2. Stone, R. E. 1915. The life history of Ascochyta on some leguminous plants. Phytopathology 5:4–10.

Lentil, *Lens esculenta* Moench

Downy mildew, *Peronospora lentis* Gäumann
Rust, *Uromyces fabae* (Persoon) De Bary
Wilt, *Fusarium orthoceras* f. sp. *lentis* Vasudeva & Srinivasan
Other fungi associated with lentil

Downy mildew, *Peronospora lentis* Gäum.

Symptoms

Yellow blotches appear on the upper surface of leaves, which become chlorotic and are rapidly shed. The downy fungus growth appears on the lower surfaces of the yellow blotches.

Etiology

The downy growth is composed of conidiophores of the fungus which protrude through the stomata in an aerial condition, 1 to several at a place. They are hyaline, straight or curved, nonseptate, several times dichotomously branched at the tips, and 200–500 μ high. The conidia are gray to dark, globose to elliptical, papillate, 1-celled, and 13–30 × 12–25 μ. They germinate by producing mycelium. Oospores are brown to dark, imbedded, globose, reticulate, with rough walls, and 20–42 μ in diameter.

References

1, 4.

Rust, *Uromyces fabae* (Pers.) d By.

Symptoms

The first indication of the disease is the appearance of yellowish white, small, circular, slightly raised areas on the upper leaf surface. Later numerous erupting peridia, containing spores of the fungus, develop on the lower surface opposite. Later in the season brown sori appear on all aboveground parts on both sides of the leaves. Teliospores develop in the uredial sori. They are dark to almost black and survive until the following season.

Etiology

The pycnia are small and often inconspicuous in the sorus; they produce hyaline, 1-celled spores. The aecia are prominent on the lower surface, producing ellipsoidal, thin-walled, finely verrucose spores, measuring 18–26 × 15–21 μ. The uredospores are produced in sori on both leaf surfaces and on stems; the spores are golden brown, obovoid, echinulate, and 22–30 × 17–25 μ. The teliospores are produced in the uredia sori and are distinctly dark brown to black, obovoid, pedicellate, 1-celled, and 25–39 × 18–25 μ.

References

1, 3, 5.

Wilt, *Fusarium orthoceras* f. sp. *lentis* Vas. & Srin.

Symptoms

The lower leaves of plants curl, become pale, and may drop; successive leaves become similarly involved, and finally the youngest leaves droop, and the plant dies. The root disease in broadcast fields appears in circular patches, and when in rows it extends to adjoining plants. Frequently, secondary stem roots emerge above those destroyed.

Etiology

The mycelium is hyaline, pink or purple, branched, septate, and produces no pionnotes or sporodochia in culture. The conidiophores are myceliumlike, producing conidia at their tips. The microconidia are hyaline, ovoid to cylindrical, 1-celled, often in wet, false heads, and 2–9 μ; some are curved. The macroconidia are variously fusiform-falcate, mostly 3-septate, and measure 23–48 × 3–5 μ.

Reference

6.

Other fungi associated with lentil

Macrophomina phaseolina (Tassi) Goidanich
Pythium ultimum Trow
Thanatephorus cucumeris (Frank) Donk

References: lentil

1. Arthur, J. C., and G. B. Cummins. 1962. Manual of the rusts in United States and Canada. Hafner Press: N.Y., pp. 242–43.
2. Beniwal, S. P. S., and S. K. Srivastava. 1968. Downy mildew of lentil (Lens esculenta). Plant Disease Reptr. 52: 817–18.
3. Butler, E. J. 1918. Fungi and diseases in plants. Thacker: Calcutta, India, pp. 254–56.
4. Gäumann, E. 1923. Beitr. kryptogamenflora Schweiz 4:198–99.
5. Prasada, R., and U. Nath Verma. 1948. Studies on lentil rust Uromyces fabae (Pers.) De Bary in India. Indian Phytopathol. 1:142–46.
6. Vasudeva, R. S., and K. V. Srinivasan. 1952. Studies on the wilt disease of lentil (Lens esculenta Moench). Indian Phytopathol. 5:23–32.

Lettuce, *Lactuca sativa* Linnaeus

Soft rot, *Erwinia aroideae* (Townsend) Holland
Soft rot, *Erwinia carotovora* (L. R. Jones) Holland
Marginal leaf spot, *Pseudomonas marginalis* (N. A. Brown) F. L. Stevens
Bacterial rosette, *Pseudomonas rhizoctonia* (R. C. Thomas) Burkholder
Bacterial soft rot, *Pseudomonas viridilivida* (N. A. Brown) Holland
Bacterial wilt, *Xanthomonas vitians* (N. A. Brown) Starr & Weiss
Damping-off, *Pythium debaryanum* Hesse
Downy mildew, *Bremia lactucae* Regel
Powdery mildew, *Erysiphe cichoracearum* De Candolle
Drop, *Sclerotinia sclerotiorum* (Libert) De Bary
Rust, *Puccinia extensicola* Plowright var. *hieraciata* (Schweinitz) Arthur
Bottom rot, *Thanatephorus cucumeris* (Frank) Donk
Anthracnose, *Marssonina panattoniana* (Berlese) Magnus
Gray mold, *Botrytis cinerea* Persoon ex Fries
Leaf spot, *Cercospora longissima* Saccardo
Leaf spot, *Septoria lactucae* Passerini
Other fungi associated with lettuce

Soft rot, *Erwinia aroideae* (Towns.) Holland

Symptoms

The disease manifestations result from invasion of the foliage, stem, or roots through wounds. The spots are small at first, appearing as water-soaked points that expand rapidly, resulting in a soft, watery, and usually somewhat slimy decay without much change in color and no odor. The outer foliage and head leaves change rapidly under favorable conditions to a wet decay. The main stem may become involved and disintegrate into a pulpy, wet mass. The disease is found in the field but is most important in transit, storage, and the market.

Etiology

The organism is a gram-negative rod, single, in pairs, or in short chains, produces no capsules, is motile by 2–8 peritrichous flagella, and measures $2\text{–}3 \times 0.5\ \mu$. Agar colonies are white to opalescent, glistening, and round. Optimum temperature, 35C.

Reference

3.

Soft rot, *Erwinia carotovora* (L. R. Jones) Holland

Symptoms

Soft rot is a worldwide decay found on most vegetables and many other kinds of plants. It produces a soft, wet, slimy rot of green parts and a mushy to punk type of decay on more mature or fleshy parts. It develops rapidly under favorable conditions. There is no change in color. Mature plants are usually severely attacked, and in transit, storage, and the market are frequently a total loss.

Etiology

The organism is a gram-negative rod, in short chains, produces no capsules, is motile by 2–5 peritrichous flagella, and measures 1.5–2 × 0.6–1 μ. Agar colonies are whitish, raised, round, and shiny. Optimum temperature, 27C.

References

3, 7, 9.

Marginal leaf spot, *Pseudomonas marginalis* (N. A. Brown) F. L. Stev.

Symptoms

This leaf spot usually originates along the leaf margins and gradually extends from 0.5 to 1.5 cm until the leaf is entirely involved, dries out, becomes brown, and dies. There are some reddish, rust-colored, and some much darker areas.

Etiology

The organism is soil-inhabiting. It is a gram-negative rod, in short chains, produces capsules, is motile by 1–3 polar flagella, and measures 0.8–2.8 × 0.4–1.25 μ. Agar colonies are creamy becoming yellow, thin, round, and smooth. A green fluorescence is produced in culture.

References

1, 3, 8.

Bacterial rosette, *Pseudomonas rhizoctonia* (R. C. Thomas) Burkh.

Symptoms

Rosette frequently occurs in greenhouses. It attacks through the root system, causing plants to become stunted, yellowish, and lose turgor. The main stem becomes discolored in the vascular areas, and most of the smaller roots are killed.

Etiology

The organism is a gram-negative rod, in chains, produces no capsules, is nonmotile, and measures 0.4–0.85 × 0.9–1.9 μ. Agar colonies are greenish yellow darkening later, round, entire, raised, and viscid. A green fluorescence is produced in the medium. Optimum temperature, 25–27C.

References

3, 9.

Bacterial soft rot, *Pseudomonas viridilivida* (N. A. Brown) Holland

Symptoms

Soft rot develops from small, scattered, water-soaked spots on the leaves. As the spots enlarge they coalesce, usually involving most of the leaf in a soft, wet decay that may cause melting away of the outer leaves.

Etiology

The organism is a gram-negative rod, produced in chains, motile by 1–3 polar flagella, and measures 1–3 × 1–1.3 μ. Agar colonies are creamy white, translucent, round, smooth, and entire. A green fluorescent pigment is produced in the medium.

References

3, 8, 10.

Bacterial wilt, *Xanthomonas vitians* (N. A. Brown) Starr & Weiss

Symptoms

This wilt shows up mostly as a leaf spot, causing wilting along the margins and extending into the blade. Stem infections are indicated by greenish brown discolorations of the interior; stems sometimes become hollow and decay. Infected plants in the early stages are a lighter green than nondiseased plants. Roots may also become infected. Diseased plants are easily broken off at the soil line.

Etiology

The organism is a gram-negative rod, in chains, produces capsules, is motile by bipolar flagella, and measures 0.6–1.3 × 0.4–0.9 μ. Agar colonies are creamy to yellow, smooth, thin, round, and sometimes ooze to the surface. Optimum temperature, 25–26C.

References

1, 2, 3, 9.

Damping-off, *Pythium debaryanum* Hesse

Symptoms

Damping-off has been observed wherever the host plant is grown. It is worldwide in distribution and affects seedlings of a large number of different plants under conditions of high humidity and wet soils during periods when sunshine and ventilation are lacking. The seedlings are attacked at or near the soil line by this soil-inhabiting fungus which invades and softens the tender stems, causing them to fall over, wither, and die. The fungus usually grows over the prostrate plants.

Etiology

The mycelium is hyaline, nonseptate, and branched. Sporangia are hyaline, spherical or oval, and 15–26 μ in diameter; they usually germinate by zoospore production. Oospores are hyaline, smooth, spherical, and 12–20 μ in diameter; they germinate by production of germ tubes.

Reference

2.

Downy mildew, *Bremia lactucae* Regel

Symptoms

Yellow green, shadowy areas appear on the foliage as irregular blotches on young and growing plants; on the older leaves of head lettuce these areas become more or less angular, limited by the larger veins. The lower surface of the blotches is usually covered with a white, soft weft of fungus sporangiophores and sporangia growing up from the surface to 1 mm in height. In severe infections some foliage may be killed and become brown, and head leaves show the disease lesions in which secondary organisms that continue to destroy the tissue are often found.

Etiology

The mycelium is hyaline, nonseptate, branched, develops between the cells of the host, and penetrates the cell walls with haustoria. The sporangiophores are irregularly dichotomously branched from an erect, terete stem. The extreme tips of these branches are palmately divided and form a flat, disklike swelling terminated by 3–5 sterigmata each supporting a hyaline, oval, 1-celled, papillate sporangium that measures 18.5 × 17.5 μ. Sporangia germinate by producing germ tubes and mycelium or by zoospore formation. The zoospores are hyaline, globular, biciliate, and 4–5 μ in diameter.

References

2, 9, 10.

Powdery mildew, *Erysiphe cichoracearum* DC.

Symptoms

Powdery mildew is generally recognized by the white, powdery accumulation of mature conidia and the light-colored superficial mycelium on leaf surfaces. The leaves gradually become pale green to yellow, fade, die, and turn brown.

Etiology

The fungus mycelium is hyaline or white, branched, septate, superficial, and produces haustoria. The conidiophores are erect, septate, and produce conidia in chains of up to 6 or more. The conidia are hyaline, 1-celled, barrel-shaped to elongate, and 28–60 × 11–28 μ. The perithecia are black, spherical, with appendages, and measure 80–140 μ; the ascospores measure 13–32 × 12–21 μ.

Reference

2.

Drop, *Sclerotinia sclerotiorum* (Lib.) d By.

Symptoms

The most conspicuous indication of a diseased plant is the flabby or droopy condition of the foliage or head leaves, which continue to decline as a wilt, and the eventual collapse and death of the plant. The damping-off phase of the disease often occurs in seedbeds and field crop plants during seasons favorable for the growth of the fungus. The disease is common in the field, in transit, and storage and is worldwide.

Etiology

The fungus characteristically produces an abundance of cottony mycelium, varying somewhat with specific environmental conditions. The mycelium is hyaline, mostly white, although sometimes pale olive to buff or greenish, branched, and septate. Sclerotia are black outside, whitish within, shiny, rubbery to hard, and highly variable in size up to several centimeters. Germinating sclerotia form few to many cup- or saucer-shaped, funnel to reflexed apothecia that are up to 2 cm in diameter and mostly yellow to pink. The apothecia contain cylindrical to elongate or clavate asci measuring 81–250 × 4–23 μ. The ascospores are hyaline, monoseriate, elliptical to oval, 1-celled, and 12–15 × 6–8 μ. Paraphyses are hyaline, elongate, and narrow. There are many variations in the manifestations of the disease on different host plants under a wide range of environmental conditions. Consequently, there are several described species and forms of the fungus probably not too well defined; nevertheless, the symptoms are usually quite similar.

References

6, 8, 9, 10.

Rust, *Puccinia extensicola* Plowr. var. *hieraciata* (Schw.) Arth.

Symptoms

Rust is not frequently found and is seldom of any importance. It produces some pycnia and aecia on lettuce leaves, appearing as orange yellow sori on either leaf surface.

Etiology

Aeciospores are more or less globoid and measure 13–23 × 12–21 μ. Uredia and telia are on alternate hosts of *Carex* spp.

Reference

2.

Bottom rot, *Thanatephorus cucumeris* (Frank) Donk

Symptoms

Bottom rot results from the invasion of the outer lower foliage leaves and main stem of the plant by the soil-inhabiting fungus. The older leaves adjacent to the soil become overgrown by the mycelium, turn yellowish, shrivel, and die. The fungus causes girdling lesions on the main stem at the soil line, resulting in malnutrition, indicated by a yellow appearance, wilting, and stunting. The fungus mycelium may grow over the head, reducing the outer foliage to a wet, often slimy consistency. The entire head may melt down to a blackish, lifeless mass.

Etiology

The mycelium of the fungus is light brown, septate, branched, and frequently forms sclerotia that adhere to vegetative structures more or less superficially. Sclerotia may survive for long periods; they germinate by producing hyphae. Sometimes the basidiospore stage develops on the lower surface of leaves that are drier, producing spores that are hyaline, 1-celled, oval to elongate, and 5–9 × 3–4 μ.

References

9, 10.

Anthracnose, *Marssonina panattoniana* (Berl.) Magn.

Symptoms

Anthracnose produces small, water-soaked spots on the foliage that rapidly enlarge into irregularly circular, reddish pink areas up to 0.5 cm in diameter. The large veins and seed stalk are affected. Often diseased areas of the foliage drop out, leaving ragged holes. Serious infection causes plants to fade to a sickly yellow color and results in stunting.

Etiology

The fungus mycelium is hyaline, septate, branched, and produces acervuli in the more permanent tissue. The acervuli contain an abundance of conidia that are hyaline, 1-septate, constricted, somewhat curved, and 17 × 4 μ.

References

2, 9.

Gray mold, *Botrytis cinerea* Pers. ex Fr.

Symptoms

The mold is usually considered a saprophyte and is a secondary invader. When environmental conditions are favorable, it becomes of major importance in the field, transit, storage, and on the market. It may kill tissue, as damping-off of seedlings or rotting of stems, petioles, peduncles, pods, flowers, foliage, and fruit. The damping-off stage is commonly found on lettuce. It is of prime importance in cool, humid weather as it rapidly involves foliage and head leaves, causing them to become soft, collapsed, watery, and covered with a superficial gray mold.

Etiology

The mycelium is hyaline, septate, branched, and fast-growing. The conidiophores are erect and divide one or more times, producing conidia that are closely clustered. The conidia are hyaline, oval to elongate, 1-celled, and 11–15 $\times$ 8–11 μ. The fungus produces black, hard, shiny sclerotia, usually not imbedded, of various sizes, and less than 1 cm in diameter. They germinate by the production of hyphae. An apothecial stage has been developed by proper manipulation of cultures from various sources subjected to variable ecological conditions. The sporophores are pinkish, cup- to saucer-shaped, several millimeters in diameter, and short-stipitate.

References

2, 8, 9, 10.

Leaf spot, *Cercospora longissima* Sacc.

Symptoms

Leaf spot has not been recorded as a serious disease although it is widespread and common. Small, water-soaked spots appear variously scattered over the older leaves; they enlarge up to 1 cm and may coalesce, causing the leaves to turn yellow, tan, and finally brown.

Etiology

The fungus produces olive brown conidiophores up to 80 μ high with conidia that are hyaline, septate, straight or curved, and 20–100 $\times$ 3–5 μ.

Reference

2.

Leaf spot, *Septoria lactucae* Pass.

Symptoms

This leaf spot is first indicated by small, water-soaked spots on the leaf blades, irregular in outline and light yellow. These spots enlarge to become elongate to elliptical and olive brown with black, imbedded pycnidia speckled over the surface. Often the central portion of the spot falls out. Sometimes most of the foliage leaves become infected, turn brown, and fall.

Etiology

The fungus mycelium is hyaline, septate, branched, and may aggregate in the stomata and initiate the pycnidia which are black, imbedded, subglobose, ostiolate, and 54×120 μ in diameter. The pycnidiospores are hyaline, curved, 1–3-septate, and 29–40×2–3 μ.

Reference

2.

Other fungi associated with lettuce

Alternaria sonchi J. J. Davis
Botryobasidium rolfsii (Saccardo) Venkatarayan
Phymatotrichum omnivorum (Shear) Duggar
Pleospora herbarum (Persoon ex Fries) Rabenhorst
Pseudomonas cichorii (Swingle) Stapp
Puccinia hieracii (Schumacher) Martius
Puccinia opizii Bubak
Pythium polymastum Drechsler
Pythium ultimum Trow
Sclerotinia intermedia Ramsey
Sclerotinia minor Jagger
Stemphylium botryosum Wallroth f. *lactucum* Padhi & Snyder

References: lettuce

1. Burkholder, W. H. 1954. Three bacteria pathogenic on head lettuce in New York State. Phytopathology 44:592–96.
2. Chupp, C., and A. F. Sherf. 1960. Vegetable diseases and their control. Ronald Press: N.Y., pp. 348–74.
3. Elliott, C. 1951. Manual of bacterial plant pathogens. 2d ed. Chronica Botanica: Waltham, Mass.
4. Groves, J. W., and F. L. Drayton. 1939. The perfect stage of Botrytis cinerea. Mycologia 31:485–89.
5. Padhi, B., and W. C. Snyder. 1954. Stemphylium leaf spot of lettuce. Phytopathology 44:175–80.
6. Purdy, L. H. 1955. A broader concept of the species Sclerotinia sclerotiorum, based on variability. Phytopathology 45:421–27.
7. Ramsey, G. B., and J. S. Wiant. 1944. Market diseases of fruits and vegetables. U.S. Dept. Agr. Misc. Publ. 541:2–40.
8. Ramsey, G. B., B. A. Friedman, and M. A. Smith. 1959. Market diseases of beets, chicory, endive, escarole, lettuce, . . . U.S. Dept. Agr., Agr. Handbook 155:1–42.
9. Walker, J. C. 1952. Diseases of vegetable crops. McGraw-Hill: N.Y., pp. 209–24.
10. Weber, G. F., and A. C. Foster. 1928. Diseases of lettuce, romaine, escarole and endive. Florida Univ. Agr. Expt. Sta. Bull. 195:303–33.

Loquat, *Eriobotrya japonica* Lindley

Fire blight, *Erwinia amylovora* (Burrill) Bergey
Crown rot, *Phytophthora cactorum* (Lebert & Cohn) Schroeter
Leaf blotch, *Fabraea maculata* Atkinson
Canker, *Physalospora rhodina* (Berkeley & Curtis) Cooke

Anthracnose, *Glomerella cingulata* (Stoneman) Spaulding & Schrenk
Collar rot, *Botryobasidium rolfsii* (Saccardo) Venkatarayan
Mushroom root rot, *Clitocybe tabescens* (Scopoli ex Fries) Bresadola
Leaf spot, *Phyllosticta eriobotryae* Thuemen
Leaf spot, *Pestalotia versicolor* Spegazzini
Scab, *Fusicladium eriobotryae* (Cavara) Saccardo
Other fungi associated with loquat

Fire blight, *Erwinia amylovora* (Burr.) Bergey

Symptoms

Infection in the growing tips of branches is evident by the change in color from typical green to a faded, scalded, lighter color, then to brown. They decline slowly, stiffen, become dark brown, and adhere to the twig. The top group of leaves is involved, and they succumb successively until the previous year's wood is encountered. Blossoms in the panicle become infected and are killed along with accompanying foliage of the fruit-spur to old wood. Cankers develop in the base of the new flush or fruiting panicle. Infection may appear on the older branches where wounds occur. Severely diseased trees reveal many blighted branches and twigs.

Etiology

The organism is a gram-negative rod, growing singly or in chains, forming no capsules, motile by peritrichous flagella, and measuring $0.9–1.5 \times 0.7–1\ \mu$. Culture colonies are white, opalescent, glistening, small, circular, and butyrous. Optimum temperature, 30C.

References

2, 3.

Crown rot, *Phytophthora cactorum* (Leb. & Cohn) Schroet.

Symptoms

Usually the first indication of the disease is an off-color of the foliage of a specific branch or larger portion of the tree. This change of color precedes a general yellowing and even an almost sudden wilting and death of the tree. A common situation is a general thinning of the foliage and dying back of the twigs and small branches in the growing perimeter of the tree. Closer observation of the trunk and larger branches generally reveals a definite cankered area of various dimensions, located variously but often in association with an old branch scar. These cankers involve and include all tissue down to the wood, which often becomes darkened. The outer cortex becomes dry, cracked, blackened, and breaks loose and curls outward. Cankers resulting from infection by this fungus may develop on any of the aerial parts of the tree; however, they are usually found on the main trunk or any of the larger branches, often 10–15 feet above the soil surface.

Etiology

The mycelium is hyaline, nonseptate, intercellular, and branched; it sup-

Figure 116. Downy mildew, *Bremia lactucae,* on lettuce.
Figure 117. Drop, *Sclerotinia sclerotiorum,* showing collapsed lettuce plants in a field.
Figure 118. Gray mold, *Botrytis cinerea,* sporulating on decaying head of lettuce.
Figure 119. Mushroom root rot, *Clitocybe tabescens,* on loquat, showing dead branches.
Figure 120. Fire blight, *Erwinia amylovora,* on loquat leaf and twigs.
Figure 121. Crown rot, *Phytophthora cactorum,* canker on loquat tree.
Figure 122. Leaf blotch, *Fabraea maculata,* on loquat, caused by the imperfect stage of the fungus *Entomosporium maculatum.*

ports pale yellow, spherical to elongate, granular, papillate sporangia that measure 20–46 × 18–35 μ. Zoospores are about 10 μ in diameter. Chlamydospores are spherical and 20–40 μ in diameter; and oospores are very similar except with thicker walls.

References

5, 11.

Leaf blotch, *Fabraea maculata* Atk.

Symptoms

Circular spots may be found on the stems, leaves, and fruit, beginning as small lavender dots that later enlarge up to 4–6 mm and become brown. Black, small, fruiting acervuli appear in the center of the leaf spots and are subcuticular, wide, and shallow. On the fruit the spots are black and sunken, causing them to shrink and eventually drop or form mummies. The stem infections are not common, although lesions are sometimes produced on the peduncle and flowers.

Etiology

The fungus mycelium is hyaline becoming dark, septate, branched, and produces a subcuticular stroma particularly on the upper leaf surface. The spores are hyaline, 4-celled, and 18–20 × 12–14 μ. The 2 larger unequal cells are accompanied by 2 smaller lateral cells attached cruciately at the sides of the larger pair. Each of 3 cells has long setae, and the basal cell has a stipe that measures 20 × 1.0 μ. The sexual stage develops on fallen leaves in the form of disklike apothecia. The ascospores are hyaline, 2-celled, unequal, slightly curved, constricted, and about 12–20 × 5–7 μ. The imperfect stage is *Entomosporium maculatum.*

References

1, 6.

Canker, *Physalospora rhodina* (Berk. & Curt.) Cke.

Symptoms

The fungus develops primarily from mechanical wounds anywhere on the trunk or larger branches. The outer cortex begins to loosen and shrink away from the healthy margins; the lesion elongates from the focal point up and down the branch, becoming dark-colored and deepening to the wood. The cankers are several times longer than wide. Often the infection takes place at the scar of a cut off branch. The cortical tissue is rapidly invaded. Frequently the branch is girdled, and the foliage turns brown and remains attached.

Etiology

The mycelium is dark, branched, septate, and forms a black stromata in which the pycnidia of the fungus are shallowly imbedded. Pycnidia are globose to slightly flattened, subepidermal becoming erumpent, and ostiolate. The pycnidiospores are dark, 2-celled, striate, oval, and 18–24 × 10–14 μ. The perithecia are black, flask-shaped, imbedded in the dead cortical tissue, ostiolate, sometimes in groups, and 150–180 μ in diameter. The ascospores are hyaline, 1-celled, oval to elongate, and 20–33 × 10–18 μ. The vegetative stage of the fungus is *Diplodia natalensis.*

Reference

8.

Anthracnose, *Glomerella cingulata* (Ston.) Spauld. & Schrenk

Symptoms

Scattered spots appear on the older leaves. They are light green becoming tan and finally brown. Marginal blotches are irregular, measuring 4–12 mm long and about 5–6 mm wide. On the fruit, pale yellow to brown sunken areas develop that are soft with no definite margins. The fruit seldom drop but remain, shriveled, dry, and brown. The fruit spurs occasionally are killed and become dark-colored and dry after the fruit ripen, resulting in a dieback condition. The fungus sporulates abundantly on these parts in wet weather.

Etiology

The mycelium is hyaline, septate, branched, and usually scanty on the host. The acervuli are not conspicuous, about 40 μ in diameter, and 4–8 μ high. There may or may not be black setae. The conidia are hyaline, cylindrical, with round ends, mostly straight, and 12–17 × 5–6 μ. The perithecia, not frequently found on the host, are brown to black, gregarious, flask-shaped, immersed becoming erumpent, and about 300 μ in diameter. The asci are 8-spored, and the ascospores are hyaline, 1-celled, oval to oblong, and 20–28 × 5–8 μ. The vegetative stage is *Colletotrichum gloeosporioides.*

References

8, 9.

Collar rot, *Botryobasidium rolfsii* (Sacc.) Venkat.

Symptoms

The tan to brown diseased areas develop at the crown of the plant or at the soil line. The discoloration is faint at first, gradually changing to tan. The area becomes sunken and frequently rough and scaly. The margins advance until girdling of the stem is complete. Yellowing of the foliage may develop prior to death of the plant.

Etiology

The fungus mycelium is hyaline or white in exposed areas. It is usually scanty but may develop abundantly and become mixed with light-colored to tan or brown, spherical, hard, superficial sclerotia measuring less than 1 mm in diameter. No conidia are produced. The vegetative stage of the fungus is *Sclerotium rolfsii.*

Reference

10.

Mushroom root rot, *Clitocybe tabescens* (Scop. ex Fr.) Bres.

Symptoms

Trees infected with this fungus show the early symptoms of poor nutrition. The leaves appear stunted, and there is little or no new growth. Considerable shedding of leaves continues throughout the year, and there is a light production of fruit which is usually small and lacks turgor. Some dying back of terminal twigs takes place. This is followed by a slow decline until the tree dies. A more rapid development of the disease shows an off-color of the foliage, its rapid wilting and browning, and death. All of these may occur within 2 weeks. Trees thus affected harbor a crown rot or a canker at the soil line or on the main roots in various stages of girdling of the tree. The cortex is invaded by the white mycelium that penetrates the cambium to the wood. In dry seasons the bark cracks and peels off, and girdling is rapid. In wet seasons the tree frequently produces callous to regenerate new tissue around the canker. However, trees do not recover.

Etiology

Sporophores of the fungus usually develop at ground level. They are honey-colored, with decurrent gills, centrally stipitate, and up to 14 cm tall; the cap is up to 10 cm in diameter. Basidiospores are white and 8–10 $\times$ 4–6 μ.

Reference

8.

Leaf spot, *Phyllosticta eriobotryae* Thuem.

Symptoms

On the older leaves, brown, circular spots develop, 2–4 mm in diameter and variously scattered. Frequently, they are so numerous as to cause the leaf to shed. The spots are quite uniform in size and seldom coalesce.

Etiology

The pycnidia of the fungus appear as a few, black, scattered specks at first submerged in the brown, killed tissue of the leaf. The beak protrudes through the epidermis. The pycnidia are mostly globose, measuring 60–120 μ in diameter. The pycnidiospores are hyaline, small, 1-celled, ovate to elongate, and 6–10 $\times$ 2–3 μ.

Reference

8.

Leaf spot, *Pestalotia versicolor* Speg.

Symptoms

Brown, circular, sunken spots up to 3 mm in diameter develop scattered over the upper surfaces of the leaves. Similar spots usually appear on the lower surfaces. The sporulating pustules are mostly on the upper surface; they are black, scattered, globular, imbedded becoming erumpent, retain some epidermal shreds, and measure up to 200 μ in diameter.

Etiology

The conidia are 5-celled, erect with some curvature, clavate, mostly short, and measure 22–27 $\times$ 7–10 μ. The 3 central cells are fuliginous, 1 cell of the 3 paler and smaller than the other 2. The basal cell is hyaline and small with a pedicel 4–7 μ long. The terminal cell is hyaline, small, with 3 or 4 flexuous, diverging setulae that measure 17–27 $\times$ 1 μ.

Reference

4.

Scab, *Fusicladium eriobotryae* (Cav.) Sacc.

Symptoms

Tan to brown spots appear on the fruit about the time they assume full size and show some yellow shades. The spots expand rapidly and cause the fruit to become malformed and disfigured, sometimes shrunken, hard, and dry. The leaves sometimes show similar circular to irregular spots on the upper surface, but they are mostly scattered and inconspicuous. The disease rarely appears on the branches.

Etiology

The mycelium is dark, septate, branched, and develops in circular areas beneath the cuticle. The conidiophores are hyaline, 1-celled, erect to undulating, stocky, and 50–60 $\times$ 6–8 μ. The conidia are hyaline, solitary, terminal, obclavate, 1-celled, and 12–22 $\times$ 6–9 μ.

References

2, 8.

Other fungi associated with loquat

Armillaria mellea Vahl ex Fries
Ascochyta eriobotryae Voglino
Botryobasidium salmonicolor (Berkeley & Broome) Venkatarayan
Clasterosporium eriobotryae Hara
Fusarium udium Butler
Leptosphaeria puttemansii Maublanc
Microdiplodia miyake Saccardo
Omphalia flavida Maublanc & Rangel

Phaeosphaeria eriobotryae Miyake
Phoma eriobotryae Bongard
Phymatotrichum omnivorum (Shear) Duggar
Sclerotinia fuckeliana (De Bary) Fuckel
Septoria eriobotryae Maffei
Sphaeropsis eriobotryae Peglion

References: loquat

1. Anderson, H. W. 1956. Diseases of fruit crops. McGraw-Hill: N.Y., pp. 149–53.
2. Condit, I. J. 1915. The loquat. Calif. Univ. Expt. Sta. Bull. 250:251–84.
3. Elliott, C. 1951. Manual of bacterial plant pathogens. 2d ed. Chronica Botanica: Waltham, Mass., pp. 30–35.
4. Guba, E. F. 1961. Monograph of Monochaetia and Pestalotia. Harvard University Press: Cambridge, Mass., pp. 227–34.
5. Miller, P. A. 1942. Phytophthora crown rot of loquat. Phytopathology 32:404–9.
6. Piehl, A., and E. M. Hildebrand. 1936. Growth relations and stages in the life history of Fabraea maculata in pure culture. Am. J. Botany 23:663–68.
7. Rhoads, A. S. 1942. Notes on Clitocybe root rot of bananas and other plants in Florida. Phytopathology 32:487–96.
8. Roger, L. 1953. Phytopathologie des pays chauds. Encyclopedie Mycologique 18: 1376–87, 1404–17, 2004–6.
9. Shear, C. L., and A. Wood. 1913. Studies of fungus parasites belonging to the genus Glomerella. U.S. Dept. Agr. Bull. 252:1–110.
10. Tandon, I. N. 1965. Collar-rot of loquat by Sclerotium sp. and its control. Indian Phytopathol. 18:240–45.
11. Weltzien, H. C., and F. J. Schwinn. 1966. Phytophthora trunk rot on loquat trees, Eriobotrya japonica, in Lebanon. Phytopathol. Z. 56:331–39.

Macadamia, *Macadamia integrifolia* Mueller

Raceme blight, *Botrytis cinerea* Persoon ex Fries
Canker, *Phytophthora cinnamomi* Rands

Raceme blight, *Botrytis cinerea* Pers. ex Fr.

Symptoms

The disease appears on the rachis, florets, and buds as small, necrotic flecks that become brownish black. The flowers often remain attached to the rachis, and during early stages of disease development these parts become overgrown with a gray mold composed of the mycelium, conidiophores, and conidia of the fungus. The disease develops abundantly at relatively cool temperatures under highly humid conditions.

Etiology

The mycelium is hyaline to olive drab, branched, and septate. The conidia are borne in compact clusters on branched conidiophores. Conidia are hyaline, globose, smooth, and 6–12 μ in diameter. Black, irregularly shaped sclerotia, 3–9 μ in diameter, are often formed.

Reference

1.

Canker, *Phytophthora cinnamomi* Rands

Symptoms

Trunk cankers and infection of branches tend to girdle, kill the bark, and cause a brown discoloration of the wood immediately beneath. These cankers are present on both stock and grafts of nursery plants. No external symptom is visible except for a slightly sunken area. The disease is apparently a wood parasite on this host.

Etiology

The mycelium is hyaline, nonseptate, branched, and mostly submerged; chlamydospores are hyaline, terminal, globose to pyriform, and 28–60 μ in diameter. Sporangia are seldom produced in nature. Oospores are spherical and 25–27 μ in diameter.

References

2, 3.

References: macadamia

1. Halgmann, O. V. 1963. Raceme blight of macadamia in Hawaii. Plant Disease Reptr. 47:416–17.
2. Zentmyer, G. A. 1960. Phytophthora canker of macadamia trees in California. Plant Disease Reptr. 44:819.
3. Zentmyer, G. A., and W. B. Stoney. 1961. Phytophthora canker of macadamia trees. Calif. Avocado Soc. Yearbook, 1961, pp. 107–9.

Mango, *Mangifera indica* Linnaeus

Bacterial leaf spot, *Pseudomonas mangifera-indicae* Patel, Kulkarni, & Moriz
Soft rot, *Erwinia carotovora* (Jones) Holland
Scab, *Elsinoë mangiferae* Bitancourt & Jenkins
Fruit rot, *Physalospora rhodina* (Berkeley & Curtis) Cooke
Anthracnose, *Glomerella cingulata* (Stoneman) Spaulding & Schrenk
Sooty mold, *Capnodium mangiferae* P. Henning
Black mildew, *Meliola mangiferum* Earle
Collar rot, *Ustulina zonata* Léveillé
Brown felt, *Septobasidium pilosum* Boedijn & Steinmann
Thread blight, *Ceratobasidium stevensii* (Burt) Venkatarayan
Pink disease, *Botryobasidium salmonicolor* (Berkeley & Broome) Venkatarayan
Shoestring root rot, *Armillaria mellea* Vahl ex Fries
Leaf spot, *Phyllosticta mortoni* Fairman
Blight, *Macrophoma mangiferae* Hingorani & Shiarma
Leaf spot, *Pestalotia mangiferae* P. Henning
Powdery mildew, *Oidium mangiferae* Berthold
Leaf spot, *Cercospora mangiferae* Koorders
Spotting, *Alternaria tenuissima* (Nees ex Fries) Wiltshire
Algal leaf spot, *Cephaleuros virescens* Kunze
Other fungi associated with mango

Bacterial leaf spot, *Pseudomonas mangifera-indicae* Patel, Kulkarni, & Moriz

Symptoms

Small, angular, water-soaked spots, varying from 1 to 4 mm in diameter, are early indications of infection. They are first light yellow but become brown and retain a distinct yellow halo. The surface of the leaf spots is rough and is usually covered with an exudate. Spots appear on the fruit and sometimes on young stems.

Etiology

The organism is a gram-negative rod, produces no capsules, is motile by 2–4 polar flagella, and measures $0.3–0.6 \times 0.4–1.5\ \mu$. Agar colonies are creamy white, shiny, round, flat, smooth, and entire. Optimum temperature, 27C.

Reference

20.

Soft rot, *Erwinia carotovora* (Jones) Holland

Symptoms

A soft rot of the fruit, caused by bacteria that enter through natural openings on wounds, causes premature dropping. The decay advances rapidly during warm, humid weather. No other part of the host is affected by the bacteria.

Etiology

The organism is a gram-negative rod, in chains, forms no capsules, is motile by 2–5 peritrichous flagella, and measures $0.6–0.9 \times 1.5–2\ \mu$. Colonies are dirty white, opalescent, shining, smooth, and have irregular margins. Optimum temperature, 27C.

Reference

19.

Scab, *Elsinoë mangiferae* Bitanc. & Jenkins

Symptoms

Irregular, raised, corky lesions form on fruit, twigs, and leaves. Corky projections may cause distinct distortions of the fruit or irregular, slightly raised scabs without much final distortion. On ripened fruit, scabby areas remain green. On young fruit, lesions are the same type as on leaves. On unfolded leaves, there appear small, translucent dots with defined papillary elevations. In older lesions, a fine velvety surface is formed by the fruiting structure of the causal fungus. Affected leaves are distorted and wrinkled or stunted. Lesions are usually seen on one surface only. On rapidly growing nursery plants, characteristics are the same as on older leaves and fruit.

Etiology

The fungus in its perfect stage, which is not common, develops a pulvinate ascostroma above the epidermis. One to 20 irregularly imbedded, scattered, circular to elliptical asci develop in the ascomata. Ascospores are hyaline and 2–4-celled. The imperfect stage is *Sphaceloma mangiferae*.

References

2, 9, 23.

Fruit rot, *Physalospora rhodina* (Berk. & Curt.) Cke.

Symptoms

Discoloration first starts around the button and is followed by a leathery and pliable condition. Decay becomes darker both externally and internally, showing wide bands on the surface. Usually decay is accompanied by the exudation of a considerable amount of amber-colored, sticky juice. In a dry atmosphere, affected fruit becomes mummified. In a moist atmosphere, a dark, feltlike mycelial growth and soft decay develop.

Etiology

Mycelium is gray to dark in culture, not evident in nature. Pycnidia are black, clustered or separate, submerged in the outer cortex, ostiolate, and emerging. Pycnidia are borne on dead twigs, bark, and fruit and are 150–180 μ in diameter. Pycnidiospores are hyaline, dark, striate, elliptical, at first 1-celled then septate, and measure 26×14 μ. Imperfect stages are *Diplodia mangiferae*, *D. natalensis*, or *Botryodiplodia theobromeae*.

References

9, 21, 23.

Anthracnose, *Glomerella cingulata* (Ston.) Spauld. & Schrenk

Symptoms

On either young or quite mature leaves light green spots appear that soon turn brown. Spots are usually located on the margins or tip of the leaf. The disease is characterized by dying back of mature twigs. In some cases, advance is so rapid as to cause leaves not to drop but to wither and dry up. On the surface of the apparently unbroken rind of the fruit there are markings referred to as tear staining. Stains are reddish to dull reddish green, depending on the maturity of the fruit.

Etiology

This weak parasite attacks only extremely young tissue or old twigs and branches of trees weakened through injuries. Conidia are hyaline, elongate, 1-celled, sometimes become septate upon germination, and measure 12–20×3–7 μ. The ascospore stage is not frequently associated with incipient decay. The vegetative stages are *Gloeosporium mangiferae* and *Colletotrichum gloeosporioides*.

References

1, 14, 16, 18, 23.

Sooty mold, *Capnodium mangiferae* P. Henn.

Symptoms

A more or less dense, black film or crust on leaves and twigs is characteristic. A black, thin net, mainly on the upper leaf surface, appears first, later increases in density, and becomes granular. Sometimes it peels off in papery flakes.

Etiology

The mycelium is black, more or less dense, branched, and septate. Pycnidia are more or less cylindrical and contain many hyaline, small, elliptical spores, measuring 5×2 μ. Perithecia measure 35×60 μ and contain asci, which produce smoky green, 3-septate ascospores, measuring $13–16 \times 5$ μ.

References

6, 23.

Black mildew, *Meliola mangiferum* Earle

Symptoms

A black, sooty covering of both leaf surfaces, stems, petioles, peduncles, and fruit develops in association with certain insects and is made up of the dark brown to black mycelium of the fungus.

Etiology

The mycelium is opaque, branched, septate, composed of many cells, and measures $100–120 \times 10$ μ. The perithecia are black, globose, and 130–160 μ; they contain 4-celled ascospores that are cylindrical, constricted, rounded at the ends, and $35–42 \times 14–18$ μ.

Reference

23.

Collar rot, *Ustulina zonata* Lév.

Symptoms

The collar and upper parts of the roots are affected, causing dry rot of tissue that is not apparent on other organs. On surrounding areas, a few black spots or lesions rupture the surface. Internally, very clear, snaky, black lines are present. The bark is blackened, and lesions are evident on the surface only at later stages of the disease, especially on dead stumps.

Etiology

The sporophores appearing at the crown of the plant are grayish black, thick, rough, irregular, and resupinate or rarely stipitate. Perithecia are imbedded, oval, with ostioles protruding. Asci are elongate, cylindrical, and contain 8 ascospores that are hyaline, 1-celled, oval, and $30–40 \times 9–12$ μ.

Reference

23.

Brown felt, *Septobasidium pilosum* Boed. & Steinm.

Symptoms

A soft, deep brown, resupinate, feltlike, almost leathery covering up to 10 mm wide and 1.5 mm thick completely surrounds the twigs and often grows on the petiole and blades of leaves. The affected surface is smooth and compact, whereas the area beneath is soft, spongy, and made up of a minute mycelial thread closely woven and packed together pillarlike. The mycelium is closely applied but never penetrates the bark and harms the twig only indirectly.

Etiology

The mycelium is brown, septate, 5–8 μ thick, and produces clusters of conidia that are hyaline, round, and thick-walled. Probasidia are spherical and 12–17 μ thick.

References

8, 22.

Thread blight, *Ceratobasidium stevensii* (Burt) Venkat.

Symptoms

Thread blight is characterized by fungal threads on the lower side of twigs. It also forms a thick web, covering the undersurface of the leaves. The threads are white at first and later become brown. A further stage is represented by development of the web into a parchmentlike membrane completely covering the lower side of leaves. Internal hyphae serve to absorb food material from the leaves.

Etiology

Superficial hyphae of the fungus are brown, septate, branched, and form threadlike strands on twigs. On foliage the fungus fans out in a thin, weblike covering of the blade. Basidia and basidiospores are formed in a loose net of hyphae. The basidia are 8–12 μ high; the sterigmata are 9–12 μ long; and the basidiospores are hyaline, oval, and $8\text{–}10 \times 4\text{–}5$ μ. The fungus is also known as *Pellicularia, Corticium,* or *Hypochnus* spp.

References

6, 21, 25.

Pink disease, *Botryobasidium salmonicolor* (Berk. & Br.) Venkat.

Symptoms

This disease chiefly affects the twigs or smaller branches. The first noticeable symptom is the loss of green color in leaves, which wither. The fungus appears in different forms, but the most easily recognized and destructive is the rose-colored or whitish crust. When dried this crust cracks somewhat like badly dried whitewash. At later stages, the bark splits and peels away from the wood. The fungus spreads rapidly over the surface after killing the bark. A girdle develops, and all foliage beyond it dies.

Etiology

The hyphae are 10–15 μ in diameter; a hymenial area produces the basidia which are short, cylindrical, and 16–33×5–8 μ. They each produce 4 sterigmata, 4–6 μ long, which support hyaline, pyriform, 1-celled basidiospores, measuring 9–12×6–7 μ. The fungus is also known as a ***Corticium*** sp.

References

6, 22, 26.

Shoestring root rot, *Armillaria mellea* Vahl ex Fr.

Symptoms

Aboveground indication of this disease is not apparent until a sudden wilting and collapse occur. Beginning as a puffiing or swelling of the bark at the crown and on the roots, the fungus later shows white, felty, fan-shaped growth under and in the bark. Cordlike, purplish brown to black rhizomorphs appear on the root surface, and light brown mushrooms show aboveground during late fall and early winter.

Etiology

The fungus produces white mycelium, black rhizomorphs, and honey-colored, annulated mushrooms 4–6 in. tall, which shed white basidiospores.

References

3, 5, 9, 15.

Leaf spot, *Phyllosticta mortoni* Fairm.

Symptoms

The organism appears principally on the leaves where it produces numerous carbonaceous pycnidia that are imbedded in white lesions on the leaf.

Etiology

Pycnidia are prominently erumpent, spherical or elongated between the veins, and 60–100 μ in diameter. Pycnidiospores are hyaline, oblong to ovoid, continuous, and 5–7×2.5 μ.

Reference

23.

Blight, *Macrophoma mangiferae* Hing. & Shir.

Symptoms

This blight is indicated by the production on the foliage of irregular brown spots with purplish borders, speckled with black pycnidia. The disease is present throughout the year, but heaviest infection occurs during the monsoon season. The spots are circular to elongate, often involving more than half the leaf, which dies and becomes brown.

Etiology

Pycnidia appear on the lower leaf surface, producing an abundance of pycnidiospores that are hyaline, 1-celled, elliptical, and 19.8 × 6.5 μ.

References

12, 13.

Leaf spot, *Pestalotia mangiferae* P. Henn.

Symptoms

The disease appears on the leaf as brown, round spots, 1 cm or more in diameter. Spots most frequently occur on the edges of the leaf where rainwater droplets accumulate. Lesions with slightly raised borders and concentric striations are usually surrounded by ill-defined halos. Lesions are mostly on the upper surface. Later black dots arranged concentrically appear on the spots. Fungus fructifications arise from beneath the epidermis.

Etiology

Pycnidia usually appear on the lower leaf surface. They are numerous, globose to elongate, subepidermal then erumpent, and 75–175 μ in diameter. Conidia are 5-celled, oblong to clavate, and 22–26 × 8–10 μ. The 3 central cells are brown and the terminal cells hyaline. The 3 setulae are widely divergent and are about 19–26 μ long; the pedicel is short.

References

11, 23, 24.

Powdery mildew, *Oidium mangiferae* Berth.

Symptoms

This disease first attacks the young developing foliage, later the inflorescences, then the fruit. Young foliage is usually distorted, curled, stunted, and often shed.

Etiology

Fungus hyphae are superficial, branched, and septate, with haustoria. The conidiophores are simple, erect, septate, 64–163 μ long, with 2 or several basal cells supporting the conidia. Conidia are oval to elliptic, often in chains, and measure 42 × 18–21 μ.

References

10, 23.

Leaf spot, *Cercospora mangiferae* Koord.

Symptoms

Leaf spots are brown to almost black and usually surrounded by a wide, green to yellow red border. They are generally small, seldom over 2 mm in diameter, and more or less circular. Black stroma in dark tissue supports fascicles of the fungus.

Etiology

The conidiophores are formed in clusters, dark-colored at the base, almost hyaline at the tips, and 3–5×5–20 μ. Conidia are olivaceous, cylindric to obclavate, 3–7-septate, and 4–5.5×20–65 μ.

Reference

7.

Spotting, *Alternaria tenuissima* (Nees ex Fr.) Wiltsh.

Symptoms

Small, brown, circular spots on leaves, fruit, and twigs are indications of the disease. These spots enlarge and become more or less irregular in outline and darker colored. A soft decay may follow infection of the fruit.

Etiology

The mycelium is dark-colored, intercellular, and septate. Conidiophores arise from the host epidermis in twos and threes; they are olivaceous, branched, and septate. The conidia are dark-colored, beaked, 2–9-septate, with or without longitudinal divisions, linear or obclavate, muriform, and average 56.8×13.4 μ.

Reference

17.

Algal leaf spot, *Cephaleuros virescens* Kunze

Symptoms

The alga causes small, purple, circular to elongate spots on leaf tissue, midrib, veins, and twigs. Fructifications are produced on the spots as minute, erect, greenish yellow hairs, topped with yellow. The fungus occurs as a simple epiphyte on leaves, growing on the surface and penetrating the cuticles of the leaf tissue. Sometimes it penetrates the leaf and extends between cuticle and epidermal cells.

Etiology

The organism develops vegetatively throughout most of the year. In reproduction, certain cells produce vertical stalks which support 3–6 terminal sporangia that are oval to spherical, thick-walled, about 40–50 μ in diameter, with orange-colored contents. The sporangia liberate a number of zoospores which generate, forming new plants.

References

4, 6.

Other fungi associated with mango

Botryosphaeria ribis (Tode ex Fries) Grossenbacher & Duggar
Chaetothyrium mangiferae Mendoza
Colletotrichum nigrum Ellis & Halsted
Curvularia lunata (Wakker) Boedijn
Dimerosporium mangiferum (Cooke) Saccardo
Ganoderma mangiferae (Léveillé) Patouillard
Leptothyrium circumcissum Sydow
Lophodermium mangiferae Koorders
Omphalia flavida (Cooke) Maublanc & Rangel
Phaeosphaerella mangiferae F. L. Stevens & Weedon
Phymatotrichum omnivorum (Shear) Duggar
Phytophthora palmivora Butler
Polyporus gilvus (Schweinitz) Fries
Rosellinia echinata Massee
Sphaerostilbe repens Berkeley & Broome
Thielaviopsis paradoxa (De Seynes) Hoehnel

References: mango

1. Anderson, H. W. 1956. Diseases of fruit crops. McGraw-Hill: N.Y., pp. 64–72.
2. Barnes, H. 1946. The mango: a list of references. U.S. Dept. Agr. Library List 29.
3. Bouriquet, G. 1940. Les maladies des plantes cultivees a Madagascar. Encyclopedie Mycologique 12:277–79.
4. Boyce, J. S. 1961. Forest pathology. 3d ed. McGraw-Hill: N.Y., pp. 104–8.
5. Brooks, F. T. 1953. Plant diseases. 2d ed. Oxford University Press: London, pp. 334–35.
6. Butler, E. J. 1918. Fungi and diseases of plants. Thacker: Calcutta, India, pp. 413–22, 477–79, 487–89, 499–506.
7. Chupp, C. 1953. A monograph of the fungus genus Cercospora. Published by the author: Ithaca, N.Y., pp. 34–40.
8. Couch, J. N. 1938. The genus Septobasidium. University of North Carolina Press: Chapel Hill.
9. Fawcett, H. S. 1936. Citrus diseases and their control. 2d ed. McGraw-Hill: N.Y., pp. 119–29, 449–53, 529–49.
10. Gangolly, S. R., et al. 1957. The mango. Indian Council of Agricultural Research: New Delhi, pp. 498–99.
11. Guba, E. F. 1961. Monograph of Monochaetia and Pestalotia. Harvard University Press: Cambridge, Mass., pp. 238–39.
12. Hingorani, M. K., and O. P. Sharma. 1956. Blight disease of mango. Indian Phytopathol. 9:195–96.
13. Hingorani, M. K., O. P. Sharma, and H. S. Sohi. 1960. Studies on blight disease of mango caused by Macrophoma mangiferae. Indian Phytopathol. 13:137–43.
14. Knorr, L. C., R. F. Suit, and E. P. DuCharme. 1957. Handbook of citrus diseases in Florida. Florida Univ. Agr. Expt. Sta. Bull. 587:11–17.
15. Konrad, P., and A. Maublanc. 1948. Les agaricales. Encyclopedie Mycologique 14:383–84.
16. McMurran, S. M. 1914. The anthracnose of the mango in Florida. U.S. Dept. Agr. BPI Bull. 52:1–15.
17. Mukherji, S. K., and S. K. Bhattacharya. 1965. A new disease of Mangifera indica in India, caused by Alternaria tenuissima. Plant Disease Reptr. 49:405–7.
18. Nowell, W. 1923. Diseases of crop-plants

in the Lesser Antilles. West India Committee: London, pp. 237–38.
19. Patel, M. K., and Y. A. Padhye. 1948. Bacterial soft rot of mango in Bombay. Indian Phytopathol. 1:127–28.
20. Patel, M. K., Y. S. Kulkarni, and L. Moniz. 1948. Pseudomonas mangiferae-indicae pathogenic on mango. Indian Phytopathol. 1:147–52.
21. Pathak, V. N., and D. N. Srivastava. 1967. Mode of infection and prevention of diplodia stem-end rot of mango fruits (Mangifera indica). Plant Disease Reptr. 51:744–46.
22. Roger, L. 1951. Phytopathologie des pays chauds. Encyclopedie Mycologique 17:917–21, 941–58.
23. Roger, L. 1953. Phytopathologie des pays chauds. Encyclopedie Mycologique 18:1202–12, 1377–95, 1404–17, 1457–58, 1598–99, 1625–29, 1646–52, 1663–72, 1819, 1866–79.
24. Sarkar, A. 1960. Leaf spot disease of Mangifera indica L. caused by Pestalotia mangiferae Butl. Lloydia 23:1–7.
25. Venkatarayan, S. V. 1949. The validity of the name Pellicularia koleroga Cooke. Indian Phytopathol. 2:186–89.
26. Venkatarayan, S. V. 1950. Notes on some species of Corticium and Pellicularia. Indian Phytopathol. 3:81–86.

Mangosteen, *Garcinia mangostana* Linnaeus

Fruit rot, *Physalospora rhodina* (Berkeley & Curtis) Cooke
Thread blight, *Ceratobasidium stevensii* (Burt) Venkatarayan
Leaf spot, *Pestalotia espaillatii* Ciferri & Gonzáles
Leaf spot, *Leptostroma garciniae* Fragoso & Ciferri
Other fungi associated with mangosteen

Fruit rot, *Physalospora rhodina* (Berk. & Curt.) Cke.

Symptoms

The mature fruit rots after removal from the tree. A dark discoloration begins around the peduncle attachment then advances rapidly through the entire fruit. The outside skin turns black and shiny and later becomes dull with some eruptions caused by the production of pycnidia.

Etiology

Pycnidia are black, oval, ostiolate, and 150–180 μ in diameter. The pycnidiospores are oval, 2-celled, striate, and 24×15 μ. Perithecia sometimes appear on the outside of fruit or on the stems or branches; they are ostiolate, contain 8-spored asci, and are up to 300 μ in diameter. The ascospores are hyaline, 1-celled, oval, and $24–42 \times 7–17$ μ. The vegetative state is *Diplodia natalensis*.

Reference

4.

Thread blight, *Ceratobasidium stevensii* (Burt) Venkat.

Symptoms

The tan to brown, fine, silky mycelium grows from the main stems, where it persists during the nongrowing part of the year as sclerotia, onto the petioles and spreads out in a fan-shape over the lower leaf surface, kills the parenchyma tissue, and causes the leaf to become brown, die, and shed. However, the threadlike mycelium may cause the leaf to remain suspended. The peduncles are likewise attached and suffer some decay and discoloration.

Etiology

The sclerotia form along the stems and are usually located near lenticels or around the terminal buds. Basidia are produced on the lower leaf surfaces. Basidiospores are hyaline, 1-celled, oval to elongate, and 8–9×3–6 μ. The fungus is also known as *Pellicularia, Corticium,* and *Hypochnus* spp.

References

1, 3.

Leaf spot, *Pestalotia espaillatii* Cif. & Gonz.

Symptoms

This fungus is found associated with dead leaf tissue near the tips, along margins, and sometimes in pale spots on the leaf blade. The pustules occur on the upper leaf surface; they are black, numerous, somewhat flattened, circular, and up to 200 μ in diameter.

Etiology

The conidia are 5-celled. The 3 interior cells are dark olivaceous. The end cells are hyaline; the basal cell has a short pedicel; and the terminal cell has 2–3 divergent setulae that are hyaline, nonseptate, and up to 16 μ long. The conidia measure 12–17×5–6 μ.

References

2, 5.

Leaf spot, *Leptostroma garciniae* Fragoso & Cif.

Symptoms

Irregular leaf spots are red to chestnut brown with a reddish black border, measure 5–10 mm in diameter, appear more or less scattered, and cause minor damage.

Etiology

The black pycnidia located on the substratum are submerged becoming erumpent, oblong to elongate, and ostiolate. Pycnidiospores are ovoid, oblong to bean-shaped, mostly hyaline, and 1-celled.

Reference

3.

Other fungi associated with mangosteen

Botrytis cinerea Persoon ex Fries
Cephaleuros virescens Kunze
Fomes noxius Corner
Glomerella cingulata (Stoneman) Spaulding & Schrenk
Helminthosporium garciniae Petch

Macrophomina phaseolina (Tassi) Goidanich
Pythium debaryanum Hesse
Rhizopus nigricans Ehrenberg

References: mangosteen

1. Feldkamp, C. L. 1946. The mangosteen, a list of references. U.S. Dept. Agr. Library List 32:1–29.
2. Guba, E. F. 1961. Monograph of Monochaetia and Pestalotia. Harvard University Press: Cambridge, Mass., p. 105.
3. Roger, L. 1951. Phytopathologie des pays chauds. Encyclopedie Mycologique 17: 948–58, 18:1796.
4. Su, M. T. 1933. Report of the mycologist. Burma, Mandalay, p. 12. (R.A.M. 13: 78.)
5. Su, M. T. 1938. Report of the mycologist. Burma, Mandalay, pp. 1–12. (R.A.M. 18:155–56.)

Maté, *Ilex paraguariensis* Saint Hilaire

Leaf spot, *Pestalotia paraguariensis* Maublanc
Leaf spot, *Cercospora ilicicola* Maublanc
Other fungi associated with maté

Leaf spot, *Pestalotia paraguariensis* Maubl.

Symptoms

The fungus is mostly associated with injured or vegetatively declining foliage, appearing in the centers of leaf spots, where the tissues are brown and dead, or in larger, deteriorating blotches.

Etiology

The fruiting pustules are black, more or less scattered on the upper leaf surface, and up to 250 μ in diameter. The conidia are pyriform to fusoid, often curved, not equilateral, 5-celled, and 23–28×8–10 μ; the 2 terminal cells are hyaline, and the 3 remaining cells are dark to black. Usually the bottommost of the 3 is lighter-colored than the rest. The apical cell is hyaline and usually produces 3 setulae; the basal hyaline cell has a pedicel 6–7 μ long.

Reference

2.

Leaf spot, *Cercospora ilicicola* Maubl.

Symptoms

The circular leaf spots are most prominent on the upper surface and somewhat indistinct on the lower surface. They are dull gray to white with a purple border that fades into the surrounding green. The spots may be up to 3 mm in diameter. The disease is seldom severe enough to cause extensive defoliation. A similar leaf spot is caused by another species of *Cercospora*.

Etiology

The fruiting structures are produced on the lower leaf surface. The stromata are dark brown to black, rounded, and slightly raised. The conidiophores are colored, septate, straight, geniculate, and 15–100×4–6 μ. The conidia are hyaline, obclavate, straight or curved, septate, and 35–75×2–4 μ. The second fungus, *Cercospora maté*, is similar except that the conidiophores are not geniculate and measure 5–40×3–4 μ; the conidia measure 20–60×3–4 μ. Probably the conidial stage of the fungus is *Leptosphaeria paraguariensis*.

Reference

1.

Other fungi associated with maté

Asterina maté Spegazzini
Cercospora maté (Spegazzini) Marchionatto
Cercospora yerbae Spegazzini
Cercosporina maté Spegazzini
Colletotrichum yerbae Spegazzini
Leptosphaeria paraguariensis Maublanc
Mycosphaerella ilicola Maublanc
Phyllosticta maté Spegazzini
Phyllosticta yerbae Spegazzini

References: maté

1. Chupp, C. 1953. A monograph of the fungus genus Cercospora. Published by the author: Ithaca, N.Y., pp. 51–54.
2. Guba, E. F. 1961. Monograph of Monochaetia and Pestalotia. Harvard University Press. Cambridge, Mass., p. 213.

Millet (ragi), *Eleusine coracana* Gaertner

Bacterial blight, *Xanthomonas coracanae* Desai, Thirumalachar, & Patel
Downy mildew, *Sclerophthora macrospora* (Saccardo) Thirumalachar, Shaw, & Narasimhan
Smut, *Melanopsichium eleusinis* (Kulkarni) Mundkur & Thirumalachar
Leaf spot, *Helminthosporium nodulosum* Berkeley & Curtis
Other fungi associated with millet (ragi)

Bacterial blight, *Xanthomonas coracanae* Desai, Thirum., & Patel

Symptoms

Water-soaked, translucent, hyaline, linear to elongate spots are the first indication of infection. The streaks are pale yellow to brown, parallel with the midrib, and 5–10 mm long. The brown areas often cover the leaf and cause it to wither.

Etiology

The organism is a rod, single or in pairs, produces capsules, is motile by a polar flagellum, and measures 1.1–1.8×0.5–0.7 μ. Agar colonies are yellow, glistening, circular, entire, smooth, pulvinate, and butyrous.

References

2, 7.

Downy mildew, *Sclerophthora macrospora* (Sacc.) Thirum., Shaw, & Naras.

Symptoms

Several healthy spikes, 1–3 in. long, diverge from the culm composing the florescence. Diseased plants show a proliferation of the first glumes of the lower spikelets, gradually extending and involving the florescence, into leafy shoots producing a brushlike structure at the top of the culm. The plants are usually pale green, and there is no shedding.

Etiology

The mycelium is thin-walled, branched, nonseptate, and contorted; none appears above the epidermis. Hyphal branches emerge through the stomata and function as sporangiophores, bearing large, oval sporangia that are delicately attached and easily separated. Upon germination, an apical pore develops through which 24–32 zoospores emerge. The sporangia measure 60–100 $\times$ 43–64 μ and germinate a few hours after formation. The oogonia are pale to yellow, spherical to subglobose, and 43–61 $\times$ 40–54 μ. The antheridia, 2–5 in number, persist. The oospore is hyaline, spherical, and 38–50 μ in diameter.

References

3, 11, 12.

Smut, *Melanopsichium eleusinis* (Kulkarni) Mund. & Thirum.

Symptoms

The smut is first evident during the flowering period and is in florets scattered at random in the florescence. The ovules are invaded and transformed into galls several times larger than normally developed seed. The diseased ovules are slightly greenish, 2–3 mm in diameter, and extend slightly beyond the glumes. At maturity the galls may grow to a diameter of 16 mm. Infection takes place only in the floral parts and only by disseminating chlamydospores.

Etiology

Mycelium is not plentiful and is rapidly utilized in the formation of chlamydospores, which are black, subglobose to spherical, minutely verrucose, and 7–11 μ in diameter. The sorus is locular, forming several cavities. The chlamydospores germinate by forming a 4-celled basidium with hyaline, lateral, terminal, ovate sporidia.

References

4, 10.

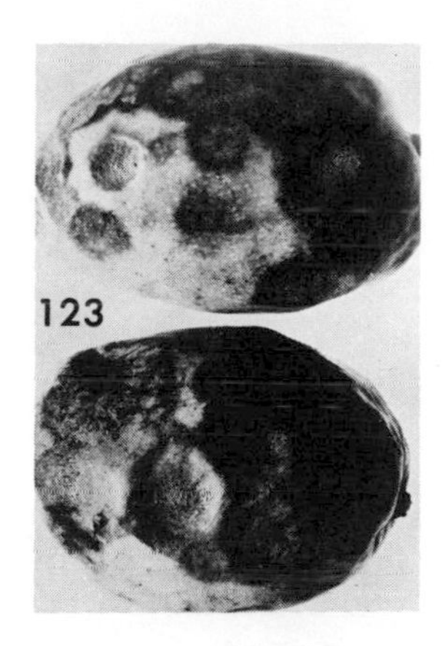

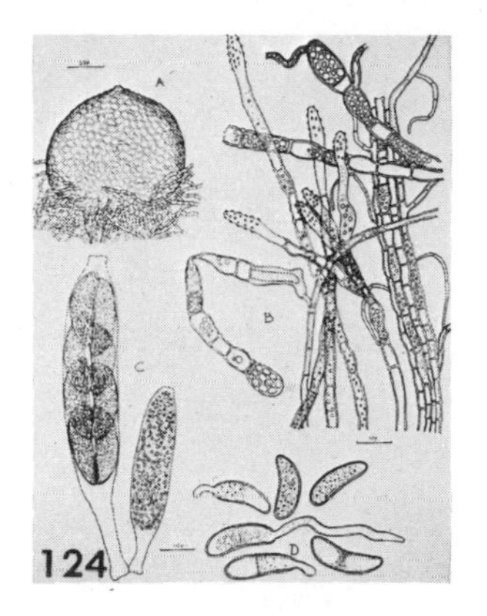

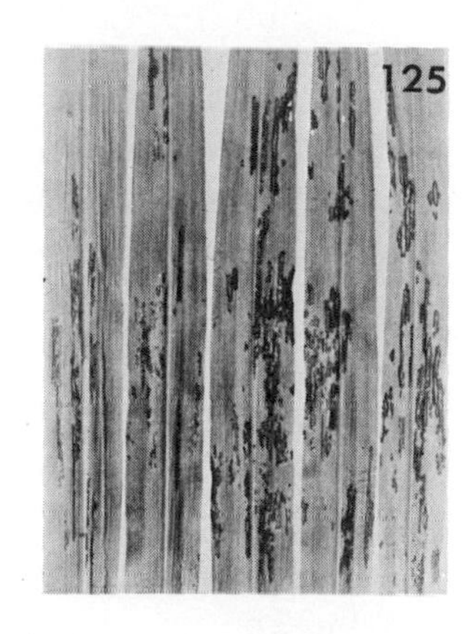

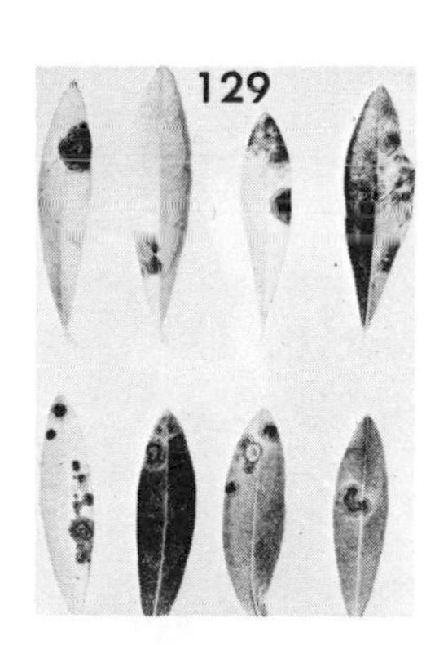

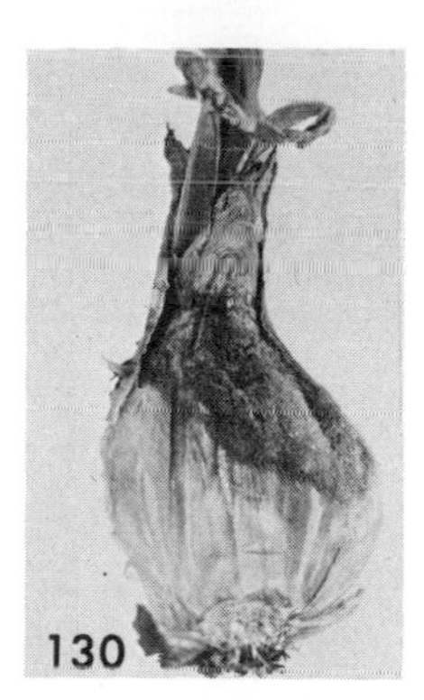

Figure 123. Anthracnose, *Glomerella cingulata*, on mango.

Figure 124. Anthracnose, *Glomerella cingulata*. A. Perithecium. B. Mycelium. C. Asci. D. Ascospores. (Photograph by S. T. Doeung.)

Figure 125. Downy mildew, *Sclerophthora macrospora*, on millet.

Figure 126. Leaf spot, *Pestalotia espaillatii*, on inoculated mangosteen leaf. (Photograph by W. Changsri.)

Figure 127. Popcorn disease, *Ciboria carunculoides*, on mulberry fruit. (Photograph by D. G. Kelbert.)

Figure 128. Leaf spot, *Cercospora abelmoschi*, on okra.

Figure 129. Peacock spot, *Cyclonium oleaginum*, on olive. (Photograph by E. E. Wilson.)

Figure 130. Gray mold neck rot, *Botrytis allii*, on onion.

Leaf spot, *Helminthosporium nodulosum* Berk. & Curt.

Symptoms

The diseased areas appear on the stem, leaves, and sometimes on the florescence. The spots are pale yellow to black, long and narrow, and usually measure 10×1 mm. They cause a decay of the sheaths and lower parts. Severe infection results in stunting, drying of the leaves, and poor production of seed heads.

Etiology

The conidiophores appear mostly on the lower leaf surface. They are brown, septate, erect, and 105×5 μ. The conidia are brown, 4–8-septate, and 48×10 μ. The ascospore stage is *Cochliobolus nodulosus.*

References

1, 6, 8, 9.

Other fungi associated with millet

Balansia claviceps Spegazzini
Botryobasidium microsclerotia (Matz) Venkatarayan
Botryobasidium rolfsii (Saccardo) Donk
Botryodiplodia theobromae Patouillard
Colletotrichum graminicola (Cesati) G. W. Wilson
Helminthosporium giganteum Heald & Wolf
Helminthosporium leucostylum Drechsler
Helminthosporium oryzae B. de Haan
Macrophomina phaseolina (Tassi) Goidanich
Ophiobolus miyabeanus Ito & Kuribayashi
Piricularia grisea (Cooke) Saccardo
Piricularia oryzae Briosi & Cavara
Ustilago eleusinis Kulkarni
Ustilago rabenhorstiana Kuehn

References: millet

1. Butler, E. J. 1918. Fungi and diseases in plants. Thacker: Calcutta, India, p. 241.
2. Desai, S. G., M. J. Thirumalachar, and M. K. Patel. 1965. Bacterial blight disease of Eleusine coracana Gaertn. Indian Phytopathol. 18:384–86.
3. Dickson, J. G. 1956. Diseases of field crops. 2d ed. McGraw-Hill: N.Y., pp. 115–21.
4. Fischer, G. W. 1951. The smut fungi. Ronald Press: N.Y., p. 25.
5. Kulkarni, G. S. 1922. The smut of nachani or ragi (Eleusine coracana Gaertn.). Ann. Appl. Biol. 9:184–86.
6. Luttrell, E. S. 1957. Helminthosporium nodulosum and related species. Phytopathology 47:540–48.
7. Rangawami, G., N. N. Prasad, and K. S. S. Eswaran. 1961. Bacterial leafspot diseases of Eleusine coracana and Setaria italica in Madras State. Indian Phytopathol. 14:105–7.
8. Roger, L. 1953. Phytopathologie des pays chauds. Encyclopedie Mycologique 18: 2033.
9. Sprague, R. 1950. Diseases of cereals and grasses in North America. Ronald Press: N.Y., pp. 369–70.
10. Thirumalachar, M. J., and B. B. Mundkur. 1947. Morphology and the mode of transmission of the ragi smut. Phytopathology 37:481–86.

11. Thirumalachar, M. J., and M. J. Narasimhan. 1949. Downy mildew on Eleusine coracana and Iseilema laxum in Mysore. Indian Phytopathol. 2:46–51.
12. Thirumalachar, M. J., C. G. Shaw, and M. J. Narasimhan. 1953. The sporangial phase of the downy mildew on Eleusine coracana with a discussion of the identity of Sclerospora macrospora Sacc. Bull. Torrey Botan. Club 80:299–307.
13. Venkatarayan, S. V. 1945. Diseases of ragi (Eleusine coracana). Mysore Agr. J. 24:50–57.
14. Weber, G. F. 1939. Web-blight, a disease of beans caused by Corticium microsclerotia. Phytopathology 29:559–75.

Mulberry, *Morus* spp.

Bacterial blight, *Pseudomonas mori* (Boyer & Lambert) F. L. Stevens
Popcorn disease, *Ciboria carunculoides* (Siegler & Jenkins) Whetzel
Powdery mildew, *Phyllactinia moricola* Sawada
Leaf spot, *Mycosphaerella mori* (Fuckel) Lindau
Coral canker, *Nectria cinnabarina* Tode ex Fries
Blight, *Coryneum mori* Homuri
Leaf spot, *Cercospora missouriensis* Winter
Leaf spot, *Cercospora moricola* Cooke
Other fungi associated with mulberry

Bacterial blight, *Pseudomonas mori* (Boyer & Lambert) F. L. Stev.

Symptoms

Water-soaked lesions develop on the shoots and foliage of black and white mulberry, which are the most susceptible. These spots later enlarge, coalesce, and become brown to black, surrounded by a yellow area. Leaves often become misshapen and distorted, showing dark, sunken areas on the principal veins. Stripes appear on the young shoots as long, narrow, sunken, translucent-bordered areas, often producing a yellowish exudate. Wood is invaded and twigs are killed. Cankers develop on twigs.

Etiology

The organism is a gram-negative rod, single, in pairs, or in chains, forms no capsules, is motile by 1–7 polar flagella, and measures $0.9–1.3 \times 1.8–4.5\ \mu$. Agar colonies are white, translucent, circular, smooth, flat, entire to undulatory, and slow-growing.

References

3, 4.

Popcorn disease, *Ciboria carunculoides* (Siegler & Jenkins) Whet.

Symptoms

Infection takes place at flowering time, and evidence of the disease is manifested several weeks later when some of the drupelets of the syncarp are much larger than others, resembling small popcorn kernels. The sepals become blighted. As the fruit matures, the diseased drupelets continue to be larger than the remaining healthy ones. At ripening time, the diseased fruit are shed, and the fungus survives in the soil. These popcornlike drupelets are the sclerotia of the fungus and remain viable, but quiescent, until the approximate time of the next flowering of the host or about 10 or 11 months later. Ascospores are produced, causing primary infection.

Etiology

The sclerotia produce hyaline, oval, 1-celled spermatia that measure 2–4 × 2–2.5 μ. These are followed later by the apothecia that are brown, disk-form to cupulate, 1 to several per sclerotium, and measure 4–12 mm in diameter on a flexuous stipe 15–42 mm high. The asci average 117 × 7 μ; each contains 8 ascospores that are hyaline, uniseriate, reniform, 7–8 × 3–4 μ, and composed of 2 cells of different sizes. The paraphyses are hyaline, branched, and septate. Spermatia and ascospores are the only spore forms of the fungus. The fungus is also known as *Sclerotinia* sp.

References

6, 7.

Powdery mildew, *Phyllactinia moricola* Saw.

Symptoms

The early stage of the disease is characterized by the appearance of fine, whitish, weblike mycelium, producing a chalky white powder of spores, and developing mostly on the lower surface of leaves. The host tissue is not immediately killed by the fungus so that well-defined leaf spots are not formed, but the leaves gradually become light green, fading to yellow. Some brown spots develop in the region of the heavy concentration of mycelium, and the leaf may become brown and die.

Etiology

The hyaline, branched, septate mycelium is mostly superficial, growing over the epidermal cell in which haustoria are formed. The conidiophores of the fungus, which arise vertically from the mycelium, are multicelled. The hyaline, 1-celled, oval conidia, measuring mostly 53–95 × 12–24 μ, are produced on the conidiophores one after another, sometimes forming in chains. The perithecia are partially imbedded in the mycelium. They are black, spherical, adorned with stiff appendages with long, attenuated tips and bulbous bases, and measure about 200 μ in diameter. The asci, several per cleistothecium, are generally oval to elongate, measure 74 × 40 μ, and contain 2 hyaline, oval ascospores that measure 33 × 22 μ.

Reference

5.

Leaf spot, *Mycosphaerella mori* (Fckl.) Lindau

Symptoms

The spots have light brown centers and reddish brown borders. They are circular at first but later become irregular to angular and measure up to 10 mm in diameter. The fruiting structures of the fungus develop on these spots.

Etiology

The conidia-producing stroma are light brown and develop under the epidermis, mostly on the upper leaf surface. They appear in concentric circles in the centers of the spots. The conidia are hyaline to faintly colored, elongate to cylindrical, with 3–10 cross walls, and measure 20–60 × 5–8 μ. The perithecia appear later on the fallen foliage. They develop after a resting period and are 60–80 μ in diameter. The asci are not accompanied by paraphyses, measure 35–40 × 5–7 μ, and contain 8 ascospores that are 2-celled, slightly curved, and 12–14 × 3–4 μ. The imperfect stage is *Cercosporella maculans.*

Reference

5.

Coral canker, *Nectria cinnabarina* Tode ex Fr.

Symptoms

The mycelium enters the cortex and the wood primarily through wounds or through dieback tissues weakened from other causes. The cortex may shrink and gum exudes, indicating the infection that later is usually definitely demarked by the brownish bark color and shrinkage. Eventually the spore-producing structures that are submerged break through the cortex and form small, pink eruptions.

Etiology

The mycelium is hyaline to brownish, septate, branched, and slowly penetrates the wood. It produces small, firm pads that appear rose-colored becoming brown, are mostly superficial, and up to 2 mm in diameter. The conidia are oblong to cylindrical, 1-celled, and 4–6 × 2 μ. The macroconidia are falcate, 4–11-septate, and 37–60 × 3–5 μ. The perithecia form in groups in scattered areas and are red, rough, protruding, and about 375–400 μ in diameter. The asci are 50–90 × 7–12 μ, and the 8, 2-celled ascospores are hyaline and 14–16 × 5–7 μ.

Reference

5.

Blight, *Coryneum mori* Homuri

Symptoms

Small twigs, less than 1 in. in diameter, are usually most severely attacked by the fungus. Infection takes place near the base and extends up the branch, often one-sided at first, but later involving the whole circumference. The involved area may be somewhat sunken or depressed. Young trees often show lesions on the main stem. The diseased area shrinks as it dries, becomes brown, and dies. Black, fruiting pustules appear on the killed areas usually in clusters or short lines at the twig bases and on the lower twig areas above the crotch. The fungus sometimes develops as a wound parasite.

Etiology

The stromata are brownish black and give rise to a palisade of short, stout conidiophores, each terminated with a hyaline to brown, multiseptate conidium measuring 25–40 × 10–18 μ. The fungus remains viable on fallen, dead host branches and on the soil surface.

Reference

1.

Leaf spot, *Cercospora missouriensis* Wint.

Symptoms

The leaf spots are more or less indefinite in outline, although when only several millimeters in diameter, they can be discerned by the dark, smoky shade of the lower surface of the leaf. The spots usually average more than 10 mm in diameter and often coalesce, covering a larger part of the leaf surface.

Etiology

The stromata appear on the lower surface in dense, medium brown fascicles and are up to 80 μ in diameter. The conidiophores are pale brown, branching, and 5–25 × 1–3 μ. Conidia are yellowish, linear, obclavate, straight or curved, and 25–70 × 2–4 μ.

Reference

2.

Leaf spot *Cercospora moricola* Cke.

Symptoms

The leaf spots are gray to tan with a definite purple to black border, mostly circular, and usually small, seldom over 8 mm in diameter.

Etiology

Stromata are almost lacking and are medium in size. Conidiophores are pale brown, undulate, curved or bent, not branched, and 5–30 × 2–4 μ. Conidia are hyaline, obclavate to cylindrical, straight or curved, and 25–125 × 2–4 μ.

Reference

2.

Other fungi associated with mulberry

Armillaria mellea (Vahl) Patouillard
Ascochyta mori Miyake
Botryodiplodia anceps Saccardo & Sydow
Cercospora mori March & Steyaert

Cercosporella mori Peck
Diaporthe mori Berlese
Dimerosporium mori Endo
Diplodia mori Westendorp
Gibberella moricola (Cesati & De Notaris) Saccardo
Massaria mori Miyake
Mycosphaerella arachnoidea Wolf
Myriangium duriaei Montagne & Berkeley
Pestolotia mori (Castagne) Montagne
Phoma mororum Saccardo
Phomopsis moricola (Saccardo) Grove
Physalospora rhodina (Berkeley & Curtis) Cooke
Physopella fici (Castagne) Arthur
Rosellinia aquila (Fries) De Notaris
Rosellinia necatrix (Hartig) Berkeley
Septoria mori Léveillé
Sphaeropsis mori Berlese
Stagonospora mori Miyake
Uncinula mori Miyake

References: mulberry

1. Butler, E. J. 1909. The mulberry disease caused by Coryneum mori Nom. in Kashmir, with notes on other mulberry diseases. Mem. Dept. Agr. India, Botan. Ser. 2:1–18.
2. Chupp, C. 1953. A monograph of the fungus genus Cercospora. Published by the author: Ithaca, N.Y., pp. 399–400.
3. Doidge, E. M. 1915. The South African mulberry blight Bacterium mori (Boy. & Lamb.) Smith. Ann. Appl. Biol. 2: 113–24.
4. Elliott, C. 1951. Manual of bacterial plant pathogens. 2d ed. Chronica Botanica: Waltham, Mass., pp. 74–75.
5. Roger, L. 1953. Phytopathologie des pays chauds. Encyclopedie Mycologique 18: 1340–42, 1475–79, 1600–1604.
6. Siegler, E. A., and A. E. Jenkins. 1923. Sclerotinia carunculoides, the cause of a serious disease of the mulberry (Morus alba). J. Agr. Res. 23:833–36.
7. Whetzel, H. H., and F. A. Wolf. 1945. The cup fungus, Ciboria carunculoides, pathogenic on mulberry fruits. Mycologia 37:476–91.

Natal plum, *Carissa carandas* Linnaeus

Anthracnose, *Colletotrichum inamdarii* Lal & Singh
Leaf spot, *Pestalotia carissae* Guba
Other fungi associated with natal plum

Anthracnose, *Colletotrichum inamdarii* Lal & Singh

Symptoms

Small, scattered, pinkish red spots appear on the leaves and as they enlarge, a single one may spread over large portions of a leaf. The spots become irregular in outline and turn brown, surrounded by a red margin and a yellow halo. The central part bleaches somewhat and becomes speckled with the acervuli of the fungus.

Etiology

The mycelium is hyaline usually darkening, septate, branched, and 2–5 μ in diameter. The acervuli appear as black specks, stromatic, grooved, and bearing black, septate setae. The conidia are hyaline, 1-celled, oblong with rounded ends, straight, and 11–26×3–6 μ.

Reference

2.

Leaf spot, *Pestalotia carissae* Guba

Symptoms

Black pustules of the fungus appear on the upper surface of brown leaf spots that are surrounded with a red border and a yellow halo. With age, these spots fade to a lighter color and become speckled with the fruiting structures of the fungus.

Etiology

The acervuli average about 175 μ in diameter. The conidia are 5-celled, long, fusiform, usually curved, and measure 21–26×5–6 μ. The 3 interior cells are olivaceous; the 2 terminal cells are hyaline; the basal cell has a pedicel 4–7 μ long; and the terminal cell has 3 widely divergent setulae that are hyaline, curved, nonseptate, and 14–28 μ long.

Reference

1.

Other fungi associated with natal plum

Botryobasidium microsclerotia (Matz) Venkatarayan
Colletotrichum gloeosporioides Penzig
Diplodia natalensis P. Evans
Phymatotrichum omnivorum (Shear) Duggar
Physalospora obtusa (Schweinitz) Cooke

References: natal plum

1. Guba, E. F. 1961. Monograph of Monochaetia and Pestalotia. Harvard University Press: Cambridge, Mass., pp. 141–42.
2. Lal, A., and R. P. Singh. 1953. Anthracnose disease of Carissa carandas Linn. caused by Colletotrichum inamdarii. J. Indian Botan. Soc. 32:54–63.

Niger seed, *Guizotia abyssinica* (Linnaeus) Cassini

Leaf spot, *Xanthomonas guizotia* Yirgou

Symptoms

Small, brown leaf spots surrounded by halos are produced in a week following infection. Leaf distortion develops as the spots enlarge. Marginal infections are most common, primarily located at the hydathodes. The leaves are slowly killed.

Etiology

The organism is gram-negative and motile by a single polar flagellum. Agar colonies are yellow, smooth, circular, and entire. Optimum temperature, 28–30C.

Reference

4.

Powdery mildew, *Erysiphe cichoracearum* DC.

Symptoms

Areas on the stems show a purplish tinge at points of infection. The powdery mildew spreads to the foliage from the stems. As the disease becomes more advanced, the leaves fade to a lighter green and finally become yellow.

Etiology

The mycelium is superficial, growing over the leaf surfaces from central points. It is hyaline, branched, septate, and develops haustoria that invade the epidermal cells. Septate conidiophores may be up to 190 μ tall. The conidia, produced successively on the conidiophores, are hyaline, 1-celled, somewhat barrel-shaped, and 30–50 × 4–5 μ. The ascospore stage is a cleistothecium, with flexuous appendages, that contains 10–15, 2-spored asci. The hyaline, 1-celled ascospores measure 20–25 × 12–15 μ.

Reference

3.

Rust, *Puccinia guizotiae* Cumm.

Symptoms

Brown spots appear on the lower leaf surface and are of various dimensions up to 7 mm in diameter. These lesions include groups of cinnamon brown telia. The upper surface of these spots is yellowish and chlorotic.

Etiology

The pycnia and aecia are unknown. The telia are subepidermal, pulvinate, and up to 0.3 mm in diameter. The teliospores are oblong to elliptical, with rounded tips and attenuated bases, 2-celled with a restricted median, stipitate, and 40–55 × 18–19 μ.

Reference

2.

Leaf spot, *Cercospora guizotiae* Siem.

Symptoms

The leaf spots are yellow to brown, distinct, circular to irregular, often coalesce, and appear first on the lower surface.

Etiology

Conidiophores produced on the stromata on the lower leaf surface are mostly in dense fascicles, brown, frequently septate, not branched, and straight or slightly curved. The conidia are hyaline to light brown, circular, faintly multiseptate, and 20–160×5–7 μ.

Reference

1.

Other fungi associated with niger seed

Alternaria porri (Ellis) Ciferri f. *dauci* (Kuehn) Neergaard
Cercospora guizotia Govindu & Thirumalachar
Macrophomina phaseolina (Tassi) Goidanich

References: niger seed

1. Chupp, C. 1953. A monograph of the fungus genus Cercospora. Published by the author: Ithaca, N.Y., p. 140.
2. Cummins, G. B. 1952. Uredinales from various regions. Bull. Torrey Botan. Club 79:212–34.
3. Yirgou, D. 1964. Some diseases of Guizotia abyssinica in Ethiopia. Plant Disease Reptr. 48:672.
4. Yirgou, D. 1964. Xanthomonas guizotiae sp. nov. on Guizotia abyssinica. Phytopathology 54:1490–91.

Nutmeg, *Myristica fragrans* Houttuyn

Root rot, *Rosellinia pepo* Patouillard
Thread blight, *Ceratobasidium stevensii* (Burt) Venkatarayan
Pink disease, *Botryobasidium salmonicolor* (Berkeley & Broome) Venkatarayan
Thread blight, *Marasmius pulcher* (Berkeley & Broome) Petch
Fruit rot, *Diplodia natalensis* Evans
Nut rot, *Coryneum myristicae* Stein
Other fungi associated with nutmeg

Root rot, *Rosellinia pepo* Pat.

Symptoms

The aboveground parts of the plant tend to decline because of malfunctions of the root system or the prevention of movement of moisture to the upper plant parts. The foliage becomes off-color, yellowish, and begins to shed. The fruit becomes dry and shrivels, followed by dieback. The fungus invades the roots and overgrows them with a black layer. The wood is penetrated and killed, and as the fungus develops the plant declines.

Etiology

The fungus develops black, upright, hairlike structures which produce conidia that are hyaline, oval, 1-celled, and 5×2 μ. The perithecia are black, imbedded, globose, and contain ascospores that are brown, straight, fusiform, pointed at each end, and $67–72 \times 8–9$ μ. White, star-shaped mats of mycelium develop in the cambium.

Reference

3.

Thread blight, *Ceratobasidium stevensii* (Burt) Venkat.

Symptoms

The dead, brown leaves either remain in place or are pendant along a growing branch or twig. In these areas the threads of the fungus grow more or less superficially along the lower surfaces of twigs and branches. The threads grow rapidly, and upon encountering a petiole, a portion extends up to the blade where the larger threads fan out in a fine, light brown, silky, netlike covering of the lower surface. The parenchyma is invaded and the cells killed. The leaves die, the older ones first, and often remain hanging from the twigs by the threads of the fungus.

Etiology

The brown, superficial, meandering, cordlike threads of hyphae compose the fungus. The basidia, which develop on the netlike mycelium that covers the leaf blade, are hyaline, oval, 1-celled, and $8–12 \times 4–5$ μ. The basidiospores are hyaline, oval, 1-celled, and $8–9 \times 3–6$ μ.

Reference

2.

Pink disease, *Botryobasidium salmonicolor* (Berk. & Br.) Venkat.

Symptoms

The disease kills foliage, branches, and fruit. The external appearance is of a rosy pink, resupinate, relatively thin overgrowth that may extend many inches along a branch. The fungus is at first internal, invading the cambium and cortex.

Etiology

The mycelium is hyaline to tan, branched, and multiseptate. The conidia are hyaline, oval to spherical, 1-celled, and quite variable in size, averaging $14–18 \times 9–13$ μ. The pinkish hymenium is composed of intertwined hyphae, producing short, upright branches. The basidia are short, cylindrical, narrow, and 16–34×5–8 μ. They produce 4 sterigmata, 4–6 μ long, that develop hyaline, globose to pyriform, apiculate, 1-celled basidiospores that measure 9–12×6–7 μ. The conidial stage is *Necator decretus*.

Reference

2.

Thread blight, *Marasmius pulcher* (Berk. & Br.) Petch

Symptoms

The fungus produces strings or threads, composed of fine, parallel, mycelial filaments, up to 3 μ in diameter, that grow more or less superficially along the lower surfaces of branches and twigs. As they advance up the petiole and fan out over the leaf blade, the blade is killed and becomes brown. The process is continued up the leafy branch to the peduncles and fruits. The diseased parts are shed, and sporophores develop on the soil.

Etiology

The mycelium survives in the aerial parts of the host. The sporophores of the fungus develop on the fallen, diseased plant parts. The fruiting structures may be sessile, resupinate, or have short stipes attached directly to the mycelium. The pileus is white, circular or disk-shaped, convex, up to 2.5 mm in diameter, with a very small lamella. The basidiospores are white, boat-shaped, and 6–8×4 μ.

Reference

2.

Fruit rot, *Diplodia natalensis* Evans

Symptoms

The disease is confined to the fruit of nutmeg, which are more susceptible when half mature. The entire fruit becomes rotted, beginning at the peduncle. The tissues become dull green to brown and appear water-soaked. The pericarp prematurely splits. The rotted seed become overgrown with dark mycelium.

Etiology

Pycnidia are formed over the surface of the rind. They are black, erumpent, globose, and imbedded in the stroma. The pycnidiospores are hyaline, 1-celled, and oval when young, later becoming dark, 2-celled, and not constricted; they measure 2×13 μ. The ascospore stage of this fungus is *Physalospora rhodina*.

Reference

1.

Nut rot, *Coryneum myristicae* Stein

Symptoms

This fungus causes damage to the nuts when they break open after falling. Sometimes they become covered with greenish black spots up to 2 cm in diameter, and the tissue on the lower side is brown. Sometimes the damage is light, but the quality and yield of the product is reduced.

Etiology

The acervulus, which is submerged in the host tissue, becomes erumpent. The oblong conidia are pale olive, claviform to cylindric, straight or slightly curved, 4–8-celled, without constrictions, and measure 30–80 × 3–5 μ.

Reference

3.

Other fungi associated with nutmeg

Cladosporium lauri Rayband
Phytophthora palmivora Butler
Pseudomonas tonelliana (Ferraris) Burkholder

References: nutmeg

1. Ramakrishnan, T. S., and A. P. S. Damodaran. 1954. Fruit rot of nutmeg. Indian Phytopathol. 7:7–17.
2. Roger, L. 1951. Phytopathologie des pays chauds. Encyclopedie Mycologique 17: 941–54, 1089–90.
3. Roger, L. 1953. Phytopathologie des pays chauds. Encyclopedie Mycologique 18: 1294–95.

Okra, *Hibiscus esculentus* Linnaeus

Pod spot, *Ascochyta abelmoschi* Harter
Leaf spot, *Cercospora abelmoschi* Ellis & Everhart
Wilt, *Fusarium oxysporum* f. sp. *vasinfectum* (Atkinson) Snyder & Hansen
Wilt, *Verticillium albo-atrum* Reinke & Berthold
Other fungi associated with okra

Pod spot, *Ascochyta abelmoschi* Harter

Symptoms

The disease is found on the pods, peduncles, and stem but not on the foilage. Pod spots increase from 1 to several centimeters in very short periods. They are somewhat circular or oval to oblong with a brown to black margin, usually zonate. The fungus advances internally in the pod surrounding the seed. Numerous pycnidia appear scattered or somewhat gregarious on the pod and also on the individual seed coat.

Etiology

The pycnidia are brown to black, lenticular, pyriform to globose, thick-walled, ostiolate, submerged becoming erumpent, and 65–225 μ in diameter.

Small pycnidiospores often extended in sinuous tendrils, 3–4 mm in length, are oval to cylindrical, straight or curved, and $4–14 \times 2–5\ \mu$. They are hyaline, single-celled at first and 1-septate when mature, with slightly curved sides and rounded ends.

References

2, 5, 6.

Leaf spot, *Cercospora abelmoschi* Ell. & Ev.

Symptoms

Sooty to dark olivaceous blotches of fungus hyphae grow extensively over the undersurface of the leaves.

Etiology

The mycelium is dark, septate, branched, forming small specks or no stromata, and often dense fascicles. The conidiophores are pale olivaceous, uniform in color, multiseptate, branched, with a small spore scar, and measure $20–140 \times 3–6\ \mu$. The conidia are pale brown, obclavate, usually curved, up to 8-septate, and $20–90 \times 3–7\ \mu$.

Reference

4.

Wilt, *Fusarium oxysporum* f. sp. *vasinfectum* (Atk.) Snyd. & Hans.

Symptoms

Seedlings frequently damp-off under adverse growing environments. Yellow discoloration of older leaves or even a yellowing of leaves on one side of the plant occurs. They deteriorate rapidly and are usually shed. The main stem becomes more or less naked, eventually turns brown, and dies. The root system becomes slightly discolored, and the discoloration extends into the vascular tissues up the main stem. The light brown tissue readily yields the fungus in pure culture.

Etiology

The mycelium is hyaline, septate, branched, and usually abundant. The sporodochia and pionnotes form yellow to pink masses. The conidia are hyaline, 3–5-septate, sickle-shaped, with the tip attenuated to a blunt apex and the base showing the place of attachment, and $23–56 \times 3–6\ \mu$. The 3-septate spores are smaller than the 5-septate spores. Microconidia are hyaline, oval to elongate, 1-celled, and 1-septate. Chlamydospores are hyaline, 1-celled, terminal or intercalary, and $8–15\ \mu$.

References

1, 3.

Wilt, *Verticillium albo-atrum* Reinke & Berth.

Symptoms

The lower leaves are the first to show decline and wilt followed by a general stunting or lack of growth and rapid collapse of the entire plant. The vascular tissues are invaded, causing some discoloration. The black, narrow, threadlike streaks in the wood are most diagnostic. They are more or less scattered and not too abundant.

Etiology

The mycelium is hyaline becoming brown, branched, septate, and supports long, narrow conidiophores that are verticillately branched at the septations. The conidia are hyaline, oval, 1-celled, and 5–7×3 μ. Chlamydospores or sclerotia may develop.

References

3, 7.

Other fungi associated with okra

Alternaria hibiscinum Thuemen
Botryobasidium rolfsii (Saccardo) Venkatarayan
Botrytis cinerea Persoon ex Fries
Cercospora malayensis F. L. Stevens & Solheim
Choanephora cucurbitarum (Berkeley & Ravenel) Thaxter
Colletotrichum gloeosporioides Penzig
Diaporthe arctii (Lasch) Nitschke
Erysiphe cichoracearum De Candolle
Macrophomina phaseolina (Tassi) Goidanich
Meliola hibisci (Sprengel) Fries
Mycosphaerella hibisci Gutner
Oidium abelmoschi Thuemen
Phoma okra Cooke
Phyllosticta hibiscina Ellis & Everhart
Phymatotrichum omnivorum (Shear) Duggar
Sclerotinia sclerotiorum (Libert) De Bary
Thanatephorus cucumeris (Frank) Donk
Thielaviopsis basicola (Berkeley & Broome) Ferraris

References: okra

1. Armstrong, G. M., B. S. Hawkins, and C. C. Bennett. 1942. Cross inoculations with isolates of fusaria from cotton, tobacco and certain other plants subject to wilt. Phytopathology 32:685–98.
2. Bond, T. E. T. 1943. Pod spot of okra (Hibiscus esculentus L.) and a leaf spot of Hibiscus rosa-sinensis L. in Ceylon. Trop. Agr. (Trinidad) 20:67–70.
3. Carpenter, C. W. 1918. Wilt diseases of okra and the verticillium-wilt problem. J. Agr. Res. 12:529–46.
4. Chupp, C. 1953. A monograph of the fungus genus Cercospora. Published by the author: Ithaca, N.Y., pp. 367–68.
5. Ellis, D. E. 1950. Ascochyta blight of okra in western North Carolina. Phytopathology 40:1056–58.

6. Harter, L. L. 1918. A hitherto unreported disease of okra. J. Agr. Res. 14:207–12.

7. Rudolph, B. A. 1931. Verticillium hydromycosis. Hilgardia 5:197–353.

Olive, *Olea europaea* Linnaeus

Knot, *Pseudomonas savastanoi* (E. F. Smith) F. L. Stevens
Galls, *Pseudomonas tonelliana* (Ferraris) Burkholder
Gelatina, *Omphalotus olearius* Delacroix ex Fries
Anthracnose, *Gloeosporium olivarum* de Almeida
Leaf spot, *Ascochyta oleae* Scalia
Leaf spot, *Phyllosticta oleae* Ellis & G. Martin
Fruit spot, *Macrophoma dalmatica* (Thuemen) Berlese & Voglino
Fruit spot, *Cylindrosporium olivae* Petri
Peacock spot, *Cycloconium oleaginum* Castagne
Leaf spot, *Cercospora cladosporioides* Saccardo
Other fungi associated with olive

Knot, *Pseudomonas savastanoi* (E. F. Sm.) F. L. Stev.

Symptoms

The gall or knot is most often initiated by the invasion of wounds by the bacterial pathogen, resulting in the extensive development of host tissues stimulated to rapid proliferation. The growing galls are soft and spongy at first but later become hard, loaflike, hemispherical, or elongate. The knots are perennial and, when wet, exude the organism, which is disseminated by rain and dew. The knots may appear anywhere on roots, trunk, branches, and leaves. Terminal shoots are killed when infected; often large branches and entire trees die. The knots develop in the wood.

Etiology

The organism is a gram-negative rod, forms in groups or short chains, is motile by 1–4 polar flagella, and measures $0.4–0.8 \times 1.2–3.3\ \mu$. The colonies are white, glistening, flat, entire, and smooth. Optimum temperature, 23–24C.

References

3, 10, 12.

Galls, *Pseudomonas tonelliana* (Ferr.) Burkh.

Symptoms

The organism produces swelling, galls, or tubercles on all parts of the plant except the root system. The swellings are more or less spherical to elongate, soft and spongy when fresh, becoming dark and woody with age. Some splitting of the stems or shoots may take place. Infected leaves show small swellings on the veins and are often curled and thickened.

Etiology

The organism is a gram-negative rod, motile by 1–3 polar flagella, produces no capsules, and measures $1–3 \times 0.5–0.6\ \mu$. Colonies are gray, circular, flat, glistening, with irregular margins. This disease and its causal agent are probably the same as the preceding knot disease and are considered synonymous with *P. savastanoi.*

References

3, 11.

Gelatina, *Omphalotus olearius* Del. ex Fr.

Symptoms

The foliage of affected trees is yellowish and stunted; the leaves are smaller and drop; and the twigs die back. Entire trees or portions of the branches may show these symptoms. Single trees or usually several in proximity may show the effects. Cortical areas of the trunk and roots split lengthwise, developing fissures. Necrotic areas appear on the roots. These areas become soaked with gum, absorb water, and become gelatinous.

Etiology

Rhizomorphs or mycelial mats have not been observed in nature. Sporophores have developed in culture and have been observed on the host. The mycelium is orange to brown in culture, septate, branched, and shows clamp connections. The sporophore is small, variously colored, and is very similar to *Clitocybe olearius.*

Reference

6.

Anthracnose, *Gloeosporium olivarum* Alm.

Symptoms

Small, brown spots appear on the surface of green or ripe fruit. These spots gradually enlarge and become irregular in shape and depressed. The surface of the spots gradually becomes brick red, and tends to become darker, due to the sporulation of the fungus in closely clustered acervuli.

Etiology

The conidia are hyaline, with parallel sides and round ends; they are 1-celled, typical of the genus, and $12\text{–}20 \times 3\text{–}7\ \mu$. The ascigerous stage is *Glomerella cingulata.*

References

4, 8, 13.

Leaf spot, *Ascochyta oleae* Scalia

Symptoms

The fungus produces watery spots on the foliage; the pycnidia are formed on the upper surface of the spots.

Etiology

Pycnidia are globose to slightly flattened, measuring 140–160×180–195 μ. The pycnidiospores are hyaline to lightly colored, oblong, 2-celled, and 7–11× 3–5 μ.

Reference

9.

Leaf spot, *Phyllosticta oleae* Ell. & G. Martin

Symptoms

Small, round spots, pale yellow to brown with a white border, and up to 1 mm in diameter, appear on the leaves.

Etiology

The pycnidia develop on the upper leaf surface, and the pycnidiospores are hyaline, 1-celled, oval, and 2–3×3–4 μ.

Reference

9.

Fruit spot, *Macrophoma dalmatica* (Thuem.) Berl. & Vogl.

Symptoms

Round to oval, depressed, tan to brown spots surrounded by a distinct darker margin appear on the green fruit. The fungus develops on the fruit as a wound parasite.

Etiology

Pycnidia are produced in more or less circular fashion. The pycnidiospores are hyaline, somewhat elongate, 1-celled, and 22×6–7 μ.

Reference

9.

• Fruit spot, *Cylindrosporium olivae* Petri

Symptoms

Faintly purple to brown spots surrounded by a dark violet band appear at the base of fruit. They are large, round to oval, and have depressed centers. The fruit become wrinkled, dry, and often drop or become mummified on the tree.

Etiology

The acervuli appear subepidermally as small cavities that enlarge and support hyaline, elongate, cylindrical, straight or curved conidia that measure 13–15×1–3 μ.

Reference

9.

Peacock spot, *Cycloconium oleaginum* Cast.

Symptoms

A spot develops on olive leaves, petioles, peduncles, and fruit during any season of the year. Young spots are circular, sooty brown, later becoming dark brown, and measure up to 10 mm. These spots often show a faint yellow halo; the remainder of the leaf becomes yellow and is usually shed. The sooty brown color is caused by the abundance of conidia. Conidia are produced on the leaves on the tree and also on fallen leaves. No other spore stage of the fungus has been found.

Etiology

The subcuticular mycelium, which seldom penetrates the epidermal layer, is scarce in the host tissue. It produces greenish to yellowish brown, elongate-ovoid, pyriform, 1–2-celled conidia on short, globose conidiophores. Conidia are usually straight and measure 14–27 × 9–15 μ.

References

1, 7, 13.

Leaf spot, *Cercospora cladosporioides* Sacc.

Symptoms

Exceedingly small, olive brown to dark purplish spots appear on the lower leaf surfaces. On the green fruit, purplish blotches appear variously scattered, only slightly sunken, and may coalesce, involving most of the surface. Ripe olives show no differentiation between the normal dark colors and fungus spots.

Etiology

The fungus grows slowly in culture, forming dark, compact turbinate colonies. On the host, conidiophores form in groups, effuse on the lower leaf surface, with small stromata. They are olivaceous to black in fascicles, pale olivaceous singly, sparingly septate, dense, rarely branched, usually curved, and 20–60 × 3–5 μ. Conidia are very pale olivaceous, cylindrical, straight or curved, up to 5-septate, and 25–65 × 4–6 μ.

References

2, 5.

Other fungi associated with olive

Armillaria mellea Vahl ex Fries
Asterina oleina Cooke
Botryosphaeria ribis (Tode ex Fries) Grossenbacher & Duggar
Capnodium elaeophilum (Montagne) Prillieux

Gloeosporium fructigenum Berkeley
Macrosporium oleae Reichert
Phoma olivarium Thuemen
Rosellinia necatrix (Hartig) Berlese
Septobasidium bagliettoanum (Fries) Bresadola
Septoria oleae Durieu & Montagne
Sphaeropsis delmatica (Thuemen) Gigante
Verticillium albo-atrum Reinke & Berthold

References: olive

1. Arnaud, G., and M. Arnaud. 1931. Maladies de l'olivier. Encyclopedie Mycologique 4:1573–78.
2. Chupp, C. 1953. A monograph of the fungus genus Cercospora. Published by the author: Ithaca, N.Y., p. 414.
3. Elliott, C. 1951. Manual of bacterial plant pathogens. 2d ed. Chronica Botanica: Waltham, Mass.
4. Gorter, G. J. M. 1961. Investigations on Gloesporium fructigenum . . . anthracnose of olive. Agr. Tech. Serv., Rep. So. Africa, p. 347.
5. Hansen, H. N., and T. E. Rawlins. 1944. Cercospora fruit and leaf spot of olive. Phytopathology 34:257–59.
6. Kouyeas, V. 1964. On the etiology of gelatina, a disease causing decline of olive trees. Ann. Inst. Phytopathol., Benaki, Greece 6:107–16.
7. Miller, H. N. 1949. Development of the leaf spot fungus on the olive leaf. Phytopathology 39:403–10.
8. Pontis, R. E., and H. N. Hansen. 1942. Olive anthracnose in the United States. Phytopathology 32:642–44.
9. Roger, L. 1953. Phytopathologie des pays chauds. Encyclopedie Mycologique 18: 1670, 1686, 1731, 1880.
10. Wilson, E. E. 1935. The olive knot disease: its inception, development, and control. Hilgardia 9:233–64.
11. Wilson, E. E., and A. R. Magie. 1963. Physiological, serological, and pathological evidence that Pseudomonas tonelliana is identical with Pseudomonas savastanoi. Phytopathology 53:653–59.
12. Wilson, E. E., and H. N. Miller. 1949. Olive leaf spot and its control with fungicides. Hilgardia 19:1–24.
13. Zachos, D. G., and S. A. Makris. 1963. Researches on Gloesporium olivarum Alm. in Greece. Ann. Inst. Phytopathol., Benaki, Greece 5:128–30, 238–59.

Onion, *Allium cepa* Linnaeus

Soft rot, *Erwinia carotovora* (L. R. Jones) Holland
Slippery skin, *Pseudomonas alliicola* Starr & Burkholder
Sour skin, *Pseudomonas cepacia* Burkholder
Damping-off, *Pythium debaryanum* Hesse
White tip, *Phytophthora porri* Foister
Downy mildew, *Peronospora destructor* (Berkeley) Caspary
Smut, *Urocystis cepulae* Frost
Rust, *Puccinia porri* (Sowerby) Winter
Rust, *Puccinia asparagi* De Candolle
Southern blight, *Botryobasidium rolfsii* (Saccardo) Venkatarayan
Soil rot, *Thanatephorus cucumeris* (Frank) Donk
Pink root, *Pyrenochaeta terrestris* (Hansen) Gorenz, J. C. Walker, & Larson
Dry rot, *Diplodia natalensis* P. Evans
Smudge, *Colletotrichum circinans* (Berkeley) Voglino
Black mold, *Aspergillus niger* Tieghem
Blue mold, *Penicillium expansum* Link ex Thom
Gray mold neck rot, *Botrytis allii* Munn
Mycelial neck rot, *Botrytis byssoidea* J. C. Walker
Small sclerotial neck rot, *Botrytis squamosa* Walker
Leaf spot, *Heterosporium allii* Ellis & G. Martin
Leaf spot, *Cercospora duddiae* Welles

Purple blotch, *Alternaria porri* (Ellis) Ciferri
Basal rot, *Fusarium oxysporum* f. sp. *cepae* (Hanzawa) Snyder & Hansen
White rot, *Sclerotium cepivorum* Berkeley
Other fungi associated with onion

Soft rot, *Erwinia carotovora* (L. R. Jones) Holland

Symptoms

Soft rot, often destructive in storage and on the market, is not frequently found in the field and then mostly when the plants are mature. The decay develops in the neck region and progresses downward, causing a soft, wet, odoriferous condition.

Etiology

The organism is a gram-negative rod, motile by 2–5 peritrichous flagella, and measuring 1.5–2 $\times$ 0.6–0.9 μ. Agar colonies are dirty white, glistening, round, smooth, and raised. Optimum temperature, 27C.

References

4, 8.

Slippery skin, *Pseudomonas alliicola* Starr & Burkh.

Symptoms

The central scales in the bulb may be readily exuded by slight pressure on the root crown. The organism enters the bulb through the neck and usually penetrates a single scale which becomes colored and softened. After reaching the root crown, the invaded area expands to other scales, softening them at their bases. The entire bulb begins to dry, shrivel, and collapse.

Etiology

The organism is a gram-negative rod, motile by 1 to several polar or bipolar flagella, and measuring 1–2.8 $\times$ 0.7–1.4 μ. Agar colonies are white becoming grayish or creamy, viscid, with waxy margins. The medium shows stages of brown. Optimum temperature, 30C.

References

3, 4, 8.

Sour skin, *Pseudomonas cepacia* Burkh.

Symptoms

The disease originates in the neck area where it causes a wet, yellow discoloration on the outer, fleshy bulb scales. In handling, the outer, dry scales slip away from the infected scales when some pressure is applied. The inner, fleshy scales of the bulb remain healthy and firm until secondary organisms become prominent and result in complete decay.

Etiology

The organism is a gram-negative rod, single or in pairs, forms no capsules, is motile by 1–3 polar flagella, and measures 1–2.8×0.8 μ. Culture colonies are sulfur yellow, slightly viscid, and rough. A yellow green pigment forms in certain culture media. Optimum temperature, 30C.

References

2, 3.

Damping-off, *Pythium debaryanum* Hesse

Symptoms

Seedlings are killed before and after emergence. The germinating seed is invaded by the fungus and the young seedling killed. After emergence the plant is softened at or near the soil line by the invading fungus, resulting in the collapse and rapid disintegration of the tender tissues. The fungus grows through the soil, affecting seedlings as they are encountered.

Etiology

The mycelium is hyaline, nonseptate, branched, and produces spherical to oval sporangia, measuring 16–26 μ in diameter. Up to 20 zoospores emerge upon germination, each measuring 7–12 μ. Oospores are spherical, smooth, and 12–20 μ in diameter.

Reference

8.

White tip, *Phytophthora porri* Foister

Symptoms

The disease is generally found on certain *Allium* spp., but not on onion. It causes the tip portion of leaves and the area toward the bulb to become yellow, die, and bleach out to white. Plants may be stunted, and in severe infections they collapse.

Etiology

The fungus mycelium is hyaline, nonseptate, branched, and produces hyaline, pyriform sporangia, measuring 37–75×31–48 μ. The sporangia germinate directly or by producing 20 or more zoospores. Oogonia measure 44×29 μ, and oospores are yellowish, spherical, and 36×19 μ.

References

3, 8.

Downy mildew, *Peronospora destructor* (Berk.) Casp.

Symptoms

The leaves show a white to lavender mold that covers paler green areas. Frequently, lesions develop that cause the leaf to break over and die. The disease is also found on the seed stalks, which develop long, elliptical lesions that become yellow to brown. The stalks usually remain standing. Young plants are not frequently affected. On the older plants the outer leaves are attacked first and killed. They die down to the bulbs, resulting in reduced foliage, stunted bulbs, and death of the plant. The fungus survives on diseased bulbs and sets and on over-wintering plants.

Etiology

The fungus mycelium is hyaline, nonseptate, branched, and produces nonseptate sporangiophores that may be 1 mm high, with an arborescent, branched, aerial portion with many pointed, acute sterigmata, bearing single, hyaline, pyriform sporangia, measuring 40–72 × 18–29 μ. Sporangia germinate directly. The oogonia are hyaline, spherical, and 43–54 μ in diameter. The oospores are 40–44 μ in diameter. The fungus survives as perennial mycelium in bulbs or as oospores in the soil. Infections may be systemic.

References

3, 8, 9.

Smut, *Urocystis cepulae* Frost

Symptoms

The fungus lives in the soil almost indefinitely, infecting seedlings in the very early pre-emerging stage and shortly thereafter. Seedlings become immune several days after emergence. The disease spreads from the cotyledons to the true leaves, causing extensive linear, black, subepidermal, blisterlike lesions.

Etiology

The chlamydospores of the fungus are black, spherical, and opaque. Around the perimeter of these spores are numerous hyaline, sterile, empty, blisterlike cells. The black spores germinate by producing mycelium, measuring 12–15 μ in diameter. The peripheral sterile cells add considerably to the size of the spore ball.

References

3, 8, 9.

Rust, *Puccinia porri* (Sow.) Wint.

Symptoms

Rust is generally found on the various onion relatives. The uredo stage of the fungus produces orange red sori that free the dusty spores. This stage of the fungus survives from one season to the next.

Etiology

The telial stage is confined to the host tissue until the plant is mature and dies. The sori are black, and the teliospores germinate mostly in place. The pycnial and aecial stages of the rust are not frequently found. The uredospores are mostly globose, echinulate, and 23–29 × 20–24 μ. The teliospores are chestnut brown, oval-ellipsoidal, 2-celled, constricted, pedicellate, and 32–37 × 17–21 μ.

References

3, 8.

Rust, *Puccinia asparagi* DC.

Symptoms

This rust, usually producing all the spore stages on asparagus, has been observed in the pycnial and aecial stages on onions. Onions apparently become infected by uredospores from neighboring asparagus plants. Uredinia and telia are not common on onion.

Etiology

The pycnia and aecia are caulicolous and cupulate. Pycniospores are hyaline, 1-celled, oval, and small. The aeciospores are faintly colored, globoid, finely verrucose, and 13–21 × 17–26 μ. Uredia are brown, and uredospores are golden yellow, echinulate, globoid to elliptical, and 22–30 × 18–25 μ. The telia are dark brown, bearing oblong to ellipsoidal, 2-celled, constricted teliospores measuring 30–48 × 18–25 μ.

References

1, 8.

Southern blight, *Botryobasidium rolfsii* (Sacc.) Venkat.

Symptoms

The disease is important usually in late, mature stages of plants that have been neglected in harvest. The fungus produces white mycelium that grows over the outer scales of the bulb and sometimes penetrates the fleshy scales, causing a softening and leaking.

Etiology

The mycelium is septate, branched, and produces tan to brown, spherical, hard sclerotia, measuring less than 0.5 mm in diameter. The fungus is also known as *Sclerotium rolfsii.*

Reference

8.

Soil rot, *Thanatephorus cucumeris* (Frank) Donk

Symptoms

Soil rot may be definitely associated with dying seedlings in a damping-off condition. The mycelium is found around growing and mature plants but is not important except in rare instances where environmental conditions are most favorable for its development. It is also known as *Rhizoctonia solani.*

Etiology

The weblike mycelium appears on the exterior of the host and on the surrounding soil surface. It is tan to brown, septate, branched, and 4–15 μ in diameter. It survives in the soil or by forming sclerotia or basidiospores.

Reference

8.

Pink root, *Pyrenochaeta terrestris* (Hans.) Gorenz, J. C. Walker, & Larson

Symptoms

Infection takes place on the roots causing them to become yellow or pink; some die and disintegrate. The onion plant does not grow well, although the plants are seldom killed.

Etiology

The mycelium is hyaline, branched, septate, and produces setose pycnidia on the diseased roots and leaf scales. The pycnidia are brown to carbonous, immersed, globose, ostiolate, papillate, and 120–450 μ in diameter. The setae are brown, more or less scattered, septate, and may be more than 100 μ high. Pycnidiospores are hyaline, 1-celled, oblong, and 3.7–6×1.8–2.4 μ.

References

3, 8.

Dry rot, *Diplodia natalensis* P. Evans

Symptoms

Dry rot occurs in transit and storage on bulbs that have apparently been injured in harvest operations. The upper portions of the bulbs are discolored gray to black in the outer scales.

Etiology

The mycelium is dark, septate, branched, and produces black, globose, ostiolate pycnidia, which measure 20–29×10–18 μ. The pycnidiospores are dark, oval, 2-celled, and 22–32×12–18 μ. The ascigerous stage is probably ***Physalospora rhodina.***

Reference

8.

Smudge, *Colletotrichum circinans* (Berk.) Vogl.

Symptoms

Smudge is confined almost entirely to the outer scales of the bulbs of white onions, although most colored ones are partially diseased. Infection takes place on the lower leaf scales, where white, thin, superficial mycelium develops. This expands to the inner surface of the outside scale and produces dark, circular to irregular areas adorned with many black, thick setae. The stromata bear acervuli in which masses of pink spores develop.

Etiology

The mycelium is septate, branched, and hyaline becoming darker to black. The conidia are hyaline to pale yellow, 1-celled, curved, fusiform, and 14–30 × 3–6 μ.

References

3, 7.

Black mold, *Aspergillus niger* v. Tiegh.

Symptoms

Mold develops on the bulbs, originating in bruises and cuts at points of injury. The parts are overgrown by the dark-colored mycelium with minute, black, sporulating heads. Moldy bulbs gradually soften and shrivel.

Etiology

The fungus produces colored, septate, branched mycelium from which erect conidiophores arise, terminating in a swollen vesicle adorned with sterigmata that in turn develop chains of dark, oval to spherical, 1-celled conidia that measure 2–5 μ in diameter.

References

3, 8.

Blue mold, *Penicillium expansum* Lk. ex Thom

Symptoms

Blue mold grows on bulbs that have been injured or bruised, covering the areas with a fine mycelium that produces a fine, velvety, blue green mold composed of conidiophores and conidia of the fungus.

Etiology

The mycelium is hyaline, branched, and septate. The conidiophores are hyaline, erect, produce penicilli at their tips and chains of yellow-blue-green conidia that are 1-celled, oval to spherical, smooth, and 3–3.5 μ in diameter.

References

3, 8.

Gray mold neck rot, *Botrytis allii* Munn

Symptoms

This mold is initiated at the neck region, where the leaves have been removed, and other places on the bulb that have suffered injury. The early impression is of a pale, lifeless area of the outer scale. It gradually becomes overgrown with a pale mycelial mold, turning gray when conidiophores and conidia are produced. The top of the bulb shows a water-soaked margin followed by the advancing mycelium, after which a darker, wet, shrinking condition involves the entire bulb.

Etiology

The invaded area becomes overgrown externally by a dense mass of gray mycelium, producing an abundance of spores and a crusty covering including flat, concave sclerotia, measuring up to 4 mm in diameter. They germinate by the production of mycelium. The mycelium is hyaline, branched, and septate. Erect conidiophores with branched, swollen tips develop clusters of conidia that are hyaline, 1-celled, elliptical to oblong, and $6–16 \times 4–8\ \mu$.

References

6, 8.

Mycelial neck rot, *Botrytis byssoidea* J. C. Walker

Symptoms

This is the most destructive disease of the neck rot type. It also may cause infection on leaves and stalks. The bulbs become infected through the dead leaf bases and mechanical injury to the outer scales. A softening of the bulb takes place as the mycelium grows into the inner tissues of the scales. A dark-colored, moldy, wet, and shriveling condition finally evolves into a mummy. The decaying area becomes overgrown by dense masses of hyaline, septate, branched mycelium, and a rather scanty amount of spores.

Etiology

The conidiophores are often colored, branched and rebranched, producing round tips covered with minute sterigmata and gray, oval, 1-celled, smooth conidia, measuring $8–10 \times 5–11\ \mu$.

References

3, 7, 8.

Small sclerotial neck rot, *Botrytis squamosa* Walker

Symptoms

The disease can be readily distinguished by the presence of black sclerotia that measure up to 4 mm in diameter. They most frequently appear on the dry scales of the white onion bulbs shortly after harvest. A typical neck rot develops at the top of the bulbs or where there may be mechanical injury. The disease develops slowly with a less pronounced decay and a scanty amount of mycelium.

Etiology

The mycelium is hyaline, septate, and branched. Conidiophores are erect, hyaline to colored, sparse, septate, and branched. Conidia develop on sterigmata in clusters. They are gray, hyaline, 1-celled, elliptical, smooth, and 13–22×10–17 μ.

References

3, 7, 8.

Leaf spot, *Heterosporium allii* Ell. & G. Martin

Symptoms

This disease appears on plants that are grown in marginal areas. The spots on the leaves are small, sunken, elliptical, with gray brown centers and a darker margin; they are up to 1.5 in. long and 1/4 in. wide. These spots may coalesce, involving large leaf areas.

Etiology

The mycelium is hyaline, septate, and branched, and the conidiophores emerge through the stomata as one, or in fascicles of several. They are brown, septate, rigid, with swollen tips, and measure 90–120×5–6 μ. The conidia are hyaline, slender, straight or slightly curved, obclavate, 3–15-septate, and 48–100×5.3–8 μ.

References

3, 10.

Leaf spot, *Cercospora duddiae* Welles

Symptoms

Leaf spot appears on young leaves in circular yellow spots up to 5 mm in diameter. They are most numerous at the leaf ends and decrease toward the bases. The leaf ends turn grayish brown and dry. Often all leaves on a plant become diseased.

Etiology

The mycelium is hyaline, septate, and branched. The conidiophores are light brown, filiform, septate, narrow, and fasciculate. The conidia are hyaline, slender, straight or slightly curved, obclavate, 3–15-septate, and 3–5×50–150μ.

References

3, 10.

Purple blotch, *Alternaria porri* (Ell.) Cif.

Symptoms

The leaf spots are white with purplish borders, small, sunken, and, under favorable conditions, weaken the parts attacked so that they fall over; however, seed stalks remain erect. At maturity, the neck region is often invaded, producing a wet, conspicuously yellow to red decay, which results in a dry, shrunken mummy.

Etiology

The fungus mycelium is hyaline or colored, branched, septate, and produces brown, erect, septate conidiophores, singly or in fascicles, and measuring 20–180 × 4–18 μ. The conidia, borne singly or in short chains at the apices, are elongate, slightly constricted, obclavate with a long beak, muriform septate, 105–200 × 12–24 μ, with 6–12 cross walls and up to 3 longitudinal septations.

References

3, 8, 9.

Basal rot, *Fusarium oxysporum* f. sp. *cepae* (Hanz.) Snyd. & Hans.

Symptoms

Fusarium usually is found on plants nearing maturity. The foliage may die back from the tips, and the whole plant appears stunted. Such plants exhibit extensive root damage. This decay advances from the root crown to the lower scales and slowly involves the bulb in a shrunken, rapidly drying decay. It continues in storage where losses are often extensive.

Etiology

The mycelium is hyaline, septate, branched, and produces short conidiophores in loose sporodochia. Conidia are hyaline, typically 3-septate, slightly curved, attenuated toward the apex, distinctly pedicellate, and 21–47 × 3–5 μ. Microconidia are not plentiful. Chlamydospores are intercalary or terminal.

References

3, 6.

White rot, *Sclerotium cepivorum* Berk.

Symptoms

The fungus lives in the soil for several years and may attack plants at any time during their growing period. The disease produces a general yellowing, gradual dying back, stunting, and final collapse of the leaves. The bulbs are rapidly involved, and the fine, white mycelium causes a wet decay of the invaded areas.

Etiology

Small, black, spherical sclerotia often develop. The decay continues in storage. The ascospore stage was designated as *Stromatinia cepivorum.*

References

3, 6, 8.

Other fungi associated with onion

Alternaria allii Nolla
Aspergillus alliaceus Thom & Church
Colletotrichum chardonianum Nolla
Curvularia lunata (Wakker) Boedijn
Helminthosporium allii Campanile
Macrophomina phaseolina (Tassi) Goidanich
Mycosphaerella allicina (Fries) Vestergren
Phyllosticta allii Tehon & Daniels
Phytophthora allii Sawada
Phytophthora drechsleri Tucker
Rhizopus stolonifer (Ehrenberg ex Fries) Lind
Sclerotinia sclerotiorum (Libert) De Bary
Stemphylium botryosum Wallroth
Thielaviopsis basicola (Berkeley & Broome) Ferraris

References: onion

1. Arthur, J. C., and G. B. Cummins. 1962. Manual of the rusts in United States and Canada. Hafner Press: N.Y.
2. Burkholder, W. H. 1950. Sour skin, a bacterial rot of onion bulbs. Phytopathology 40:115–17.
3. Chupp, C., and A. F. Sherf. 1960. Vegetable diseases and their control. Ronald Press: N.Y.
4. Elliott, C. 1951. Manual of bacterial plant pathogens. 2d ed. Chronica Botanica: Waltham, Mass.
5. Nolla, J. A. B. 1927. A new alternaria disease of onions (Allium cepa L.). Phytopathology 17:115–32.
6. Ramsey, G. B., and J. S. Wiant. 1941. Market diseases of fruits and vegetables. U.S. Dept. Agr. Misc. Publ. 440:2–60.
7. Walker, J. C. 1944. Onion diseases and their control. U.S. Dept. Agr. Farmers' Bull. (Rev.) 1060:1–26.
8. Walker, J. C. 1952. Diseases of vegetable crops. McGraw-Hill: N.Y.
9. Walker, J. C., and R. H. Larson. 1961. Onion diseases and their control. U.S. Dept. Agr., Agr. Handbook 208:1–27.
10. Welles, C. B. 1923. A new leaf spot disease of onion and garlic. Phytopathology 13:362–65.

Palm, oil, *Elaeis guineensis* Jacquin

Dry basal rot, *Ceratocystis paradoxa* (Dade) Moreau
Leaf spot, *Cochliobolus heterostrophus* (Drechsler) Drechsler
Anthracnose, *Leptosphaeria elaeidis* Booth & Robertson
Stem rot, *Ustulina vulgaris* Tulasne
Basal rot, *Ganoderma lucidum* (Leysser ex Fries) Karsten
Shoestring root rot, *Armillaria mellea* Vahl ex Fries
Dieback, *Botryodiplodia theobromae* Patouillard
Eyespot, *Helminthosporium carbonum* Ullstrup
Leaf spot, *Helminthosporium halodes* var. *elaeicola* Kovachich
Leaf spot, *Pestalotia mayumbensis* Steyaert
Seedling blight, *Curvularia maculans* (Bancroft) Boedijn
Leaf spot, *Cercospora elaeidis* Steyaert
Vascular wilt, *Fusarium oxysporum* Schlechtendahl
Other fungi associated with oil palm

Dry basal rot, *Ceratocystis paradoxa* (Dade) Moreau

Symptoms

This disease is most severe on young, bearing palms, although older trees that are partially recovered show stunted foliage and produce little or no fruit. The first symptom is an infection and rotting of the developing fruit panicle. Foliage symptoms follow in which the rachis of lower leaves breaks, yellows, and eventually dies. Further development successively involves the older leaves until the emerging leaf is included, and the palm dies. Internal symptoms show a decay, usually in the base of the plant and often on one side or near a leaf sheath. Invasion may be through the petiole or roots.

Etiology

The microconidia, produced endogenously, are hyaline, cylindrical, with flat ends, and measure 9.6×5.2 μ. The macroconidia are black, thick, oval, and 15.0×9.2 μ. Both spore forms are produced abundantly. The perithecia are black, submerged, and long-necked. Ascospores are hyaline, ellipsoid with curved sides, produced in a gelatinous matrix, and measure $7–10 \times 2–4$ μ.

References

3, 11.

Leaf spot, *Cochliobolus heterostrophus* (Drechs.) Drechs.

Symptoms

The spots are brown with very limited necrotic halos. They coalesce, often involving large portions of the leaf in a general rot. A shredding may begin at the tip and advance toward the base.

Etiology

The conidiophores are pale brown, appear singly, and have an enlarged basal cell. The conidia are pale olivaceous, cylindrical to obclavate, 6–12-septate, and $60–120 \times 11–14$ μ. The vegetative stage is *Helminthosporium maydis.*

Reference

8.

Anthracnose, *Leptosphaeria elaeidis* Booth & Robertson

Symptoms

Lesions and a necrosis of the foliage of oil palm seedlings develop at transplanting time. The lesions measure 3.2×1.5 cm and become papery in texture. They are surrounded by a dark brown zone and a yellow halo gradually merging into green. Acervuli develop on the papery tissue. Perithecia have been collected on diseased host parts.

Etiology

The mycelium is sparse in culture, white, and later light brown. The acervuli imbedded in the mycelium produce black droplets of conidia, which form on the host, erupting through the epidermis. They are fusiform to clavate, 4-septate, and 27–34 × 6–7 μ; the middle cells are brown, the others are hyaline, and there are 3 setae. Perithecia are black, scattered, suberumpent, ostiolate, and pseudoparaphysate. Ascospores are brown, uniseriate, cylindric to ellipsoidal, 2-septate, constricted, and 12–21 × 4.5–6 μ. The vegetative stage is *Pestalotiopsis* sp.

Reference

1.

Stem rot, *Ustulina vulgaris* Tul.

Symptoms

A dark-colored canker develops at the base of the petioles and the trunk following invasion through wounds. The tissue becomes dark-colored, and the diseased area slowly extends, resulting eventually in the weakening of the plant and a noticeable off-color of the foliage. Production is reduced, and such trees often become totally nonproductive.

Etiology

The fruiting structure consists of a black, rough, irregular stroma in which is formed the perithecia with protruding ostioles. The asci are long, cylindric, 250–260 × 8–10 μ, and contain 8 uniseriate ascospores measuring 30–40 × 8–13 μ.

Reference

12.

Basal rot, *Ganoderma lucidum* (Leyss. ex Fr.) Karst.

Symptoms

Trees show an aerial deterioration and decline which eventually and surely results in death. Top deterioration has been associated with diseased roots. Infection is initiated in the tree by the advance of the fungus through the interior of the root to the inner areas of the palm base. Coconut palm stumps are considered the source of inoculum when left during replanting with oil palm.

Etiology

The fungus is generally considered to be saprophytic in its habitat and is essentially a wound parasite. The sporophores are dark to reddish brown, developing on killed trees. They may be stipitate or sessile and enlarge to 30–50 cm in diameter. The upper surface is reddish with a white margin and yellow zone. At maturity, the surface appears varnished. The hymenial surface is whitish, crowded with short tubes with small pores. The basidia that line the tubes are hyaline, globose, and about 12 μ in diameter. The basidiospores are pale brown, obovate, 1-celled, and 11×7–8 μ.

References

10, 12, 14, 15.

Shoestring root rot, *Armillaria mellea* Vahl ex Fr.

Symptoms

External symptoms are chlorosis, wilting of the leaves, and breaking of the petioles. The lowermost leaf bases become rotten, and soft rot frequently develops within the trunk. The decay may be general or local in relation to the trunk. In the former, the decay spreads rapidly, and the plant finally reaches a stage of complete collapse. Local rot areas may develop quite extensively, become inactive, and be more or less sealed off with very little effect on the palm. The disease has been observed on trees from 4 to 10 years of age.

Etiology

The fungus produces abundant mycelium and black rhizomorphs, the latter being more conspicuous after the late development of the disease. At certain times of the year the mushroomlike sporophores appear on the soil around the diseased palm or over the stump after it is dead. The sporophores are honey-colored, centrally stipitate, and 2–6 in. tall with a cap up to several inches in diameter. The basidiospores are white. A conspicuous annulus develops around the upper part of the stipe of the sporophore.

References

13, 16.

Dieback, *Botryodiplodia theobromae* Pat.

Symptoms

The fungus develops rapidly in injured leaf and petiole tissue, forming cankers, producing decay, and often causing exudation of thickened sap. The deterioration is slow in most instances, but eventually the plant becomes weak and dies. The fungus may be found on any aboveground parts.

Etiology

The fungus produces dark, septate, branched mycelium that penetrates the parenchyma tissue. The stromata are raised, black, and carbonous. Pycnidia are imbedded in the stroma with ostioles protruding. They vary in size and number and measure 140–240×100–140 μ. The pycnidiospores are 1-celled and hyaline when young; at maturity they are 2-celled, dark, oval to elongate, and 23–31×11–17 μ. The ascospore stage is *Physalospora rhodina* (Berk. & Curt.) Cke.

Reference

12.

Eyespot, *Helminthosporium carbonum* Ullstrup

Symptoms

Although not fatal, the disease is highly detrimental to young palms. The spear leaf is most susceptible. Small, yellow spots appear, surrounded by a pale green halo, and enlarge up to 0.5 mm. Heavy infection results in coalescence of spots, causing a distinct chlorosis of the entire leaf.

Etiology

Conidia are hyaline to pale olive green, cylindrical, slightly curved, 3–8-septate, show broad scars, and measure 36–56 × 11–14 μ. Cross inoculation made on the weed "nagundo" (*Sarcophrynium* sp.) and oil palm from respective isolations of the fungus proved that both were susceptible.

References

8, 9.

Leaf spot, *Helminthosporium halodes* var. *elaeicola* Kovac.

Symptoms

Leaf spot is first found on either of the 2 youngest, open leaves or on the closed spear leaf as small, pale green spots scattered over the laminae. Later, enlarging spots become yellow with a pinpoint brown spot at the center. Finally, spots are usually oval with the upper surface slightly sunken and a pale brown color on the lower surface. With a heavy infestation, large areas of the lamina dry out. The spear leaf becomes highly chlorotic, with ill-defined spots, and extensively rotted.

Etiology

The mycelium is immersed in the substratum. Conidia are produced singly or in small groups, emerging through the outer epidermal wall with one or more well-defined scars and a swollen basal cell. Conidia are olive brown, straight or slightly curved, elliptical or obclavate, and 33–125 × 10–18 μ with a protruding basal hilum.

Reference

7.

Leaf spot, *Pestalotia mayumbensis* Stey.

Symptoms

Elliptical, straw-colored spots, 3–4 cm long, usually surrounded by a reddish purple zone, appear on the foliage. Spots are usually scattered except on young plants.

Etiology

Pustules forming on the spots are circular to elongate, slightly erumpent, and 150–300 μ in diameter. The conidia are 5-celled, straight or slightly curved, fusiform, tapering to their base, slightly constricted, and $22–28 \times 6–9\ \mu$. The 3 center cells are olivaceous, and the end cells are hyaline. The basal cell has a $2–8\ \mu$-long pedicel, and the terminal cell has 3 spreading, $8–15\ \mu$ long, cylindrical setulae.

Reference

4.

Seedling blight, *Curvularia maculans* (Bancroft) Boed.

Symptoms

Small, translucent yellow, circular spots visible from both leaf surfaces first appear; as they become older they enlarge and become irregular to elongate in shape, turn brown, and show sunken centers. A well-defined, narrow, yellow halo surrounds each spot. The center leaves are the most susceptible, and they become more resistant with age. Plants may be defoliated and killed.

Etiology

The conidiophores are colored, septate, single, and produce conidia in whorls. The conidia are 3-septate, ellipsoidal, straight, and $20–27 \times 12–14\ \mu$. The 2 middle cells are larger, thick-walled, and darker than the end cells, which are small and subhyaline with obtuse ends.

Reference

5.

Leaf spot, *Cercospora elaeidis* Stey.

Symptoms

The disease is first observed as a pinpoint chlorotic spot that rapidly becomes dull brown, slightly sunken, and seldom exceeds 0.5 mm in diameter. Conidia develop on these spots, and secondary infection occurs on the same leaf. The spike leaf is usually free of spots, with each succeeding leaf showing more spotting. Older leaves show a progressive desiccation.

Etiology

Hyaline, septate, intercellular mycelium may be observed in the host tissue. Septate, dark brown conidiophores grow at stromatal openings from hyphal clusters and appear in groups of 1–8. Conidia are hyaline to slightly olivaceous and later brown, 3–9-septate, obclavate, with a basal scar, and measure $75–160 \times 6–7\ \mu$. Germ tubes from conidia usually enter the host through stomata.

References

2, 6.

Vascular wilt, *Fusarium oxysporum* Schlecht.

Symptoms

A wilting of the leaves is the early symptom. They bend downward, show a yellow coloration, appear flaccid, and break near the leaf base. The broken petiole allows the entire leaf to hang downward along the trunk. Occasionally the disease develops rapidly, resulting in the death of the tree within a few months. The root system shows brown to reddish streaks in the vascular tissues. The streaks extend upward through the wood vessels in the trunk, darken, and become almost black. The presence of the fungus causes a profuse exudation of gum which blocks the vessels.

Etiology

Isolations of the fungus have been readily obtained from planted root sections on artificial synthetic media. The mycelium develops luxuriantly, giving an indigo blue color to the medium. Microconidia develop plentifully. They are hyaline, 1-celled, and 3×1–$5\ \mu$. The macroconidia are hyaline, mostly 3-septate, rather straight, and 13–33×3–$45\ \mu$. Chlamydospores are abundant and $4.5 \times 9\ \mu$.

Reference

17.

Other fungi associated with oil palm

Botryobasidium rolfsii (Saccardo) Venkatarayan
Fomes applanatus (Persoon) Wallroth
Hysterotomella elaeicola Maublanc
Marasmius palmivorus Sharples
Meliola elaeis F. L. Stevens
Mycosphaerella elaeidis (Beeli) Hendrickx
Pestalotia palmarum Cooke & Greville
Phytophthora palmivora Butler
Poradiella gloeosporidia Steyaert
Poria ravenelae Berkeley & Broome
Venturia elaeidis Marchal & Steyaert

References: oil palm

1. Booth, C., and J. S. Robertson. 1961. Leptosphaeria elaeidis sp. nov. isolated from anthracnosed tissue of oil palm seedlings. Trans. Brit. Mycol. Soc. 44: 24–26.
2. Chupp, C. 1953. A monograph of the fungus genus Cercospora. Published by the author: Ithaca, N.Y.
3. Dade, H. A. 1928. Ceratostomella paradoxa, the perfect stage of Thielaviopsis paradoxa (De Seynes) von Höhnel. Trans. Brit. Mycol. Soc. 13:184–94.
4. Guba, E. F. 1961. Monograph of Monochaetia and Pestalotia. Harvard University Press: Cambridge, Mass., p. 148.
5. Johnston, A. 1959. Oil palm seedling blight. Malayan Agr. J. 42:14–20.
6. Kovachich, W. G. 1954. Cercospora elaeidis leaf spot of the oil palm. Trans. Brit. Mycol. Soc. 37:209–12.
7. Kovachich, W. G. 1954. A leafspot disease of the oil palm (Elaeis guineensis) caused by Helminthosporium halodes Drechsler var. elaeicola var. nov. Trans. Brit. Mycol. Soc. 37:422–25.
8. Kovachich, W. G. 1957. Three leaf diseases of young oil palms associated with Helminthosporium spp. Trans. Brit. Mycol. Soc. 40:90–94.
9. Martinez, A. P. 1958. Helminthosporium leafspot of palms. Principes 2:105–6.
10. Navaratnam, S. J. 1961. Successful inoc-

ulation of oil palms with a pure culture of Ganoderma lucidum. Malayan Agr. J. 43:233–38.
11. Robertson, J. S. 1962. Dry basal rot, a new disease of oil palms caused by Ceratocystis paradoxa (Dade) Moreau. Trans. Brit. Mycol. Soc. 45:475–78.
12. Roger, L. 1951. Phytopathologie des pays chauds. Encyclopedie Mycologique 17: 1046–48; 1953, 18:1454–58, 1758–1805.
13. Thomas, H. E. 1934. Studies on Armillaria mellea (Vahl) Quel., infection, parasitism and host resistance. J. Agr. Res. 48:187–218.
14. Turner, P. D. 1965. The incidence of ganoderma disease of oil palms in Malaya and its relation to previous crop. Ann. Appl. Biol. 55:417–23.
15. Turner, P. D. 1965. Infection of oil palms by Ganoderma. Phytopathology 55:937.
16. Wardlaw, C. W. 1950. Armillaria root and trunk rot of oil palms in the Belgian Congo. Trop. Agr. (Trinidad) 27: 95–97.
17. Wardlaw, C. W. 1950. Vascular wilt disease of the oil palm caused by Fusarium oxysporum Schl. Trop. Agr. (Trinidad) 27:42–47.

Palms, ornamental (also see coconut, date, oil palm)

Leaf spot, *Pseudomonas washingtoniae* (Pine) Elliott, of palm *Washingtonia filifera* (Linden) Wendland
Bud rot, *Phytophthora palmivora* Butler, of betel-nut palm *Areca cathecu* L.
Trunk rot, *Phytophthora parasitica* Dastur, of palm *Washingtonia filifera* (Linden) Wendland
Leaf spot, *Catacauma sabal* Chardon, of palmetto *Sabal* spp.
Leaf spot, *Myrianginella sabaleos* (Weedon) Limber & Jenkins, of palmetto *Sabal* spp.
Mushroom root rot, *Clitocybe tabescens* (Scopoli ex Fries) Bresadola, of palm *Butea capitata* Beccari
Butt rot, *Ganoderma sulcatum* Murrill, of palm *Arecastrum romanzoffianum* (Chamisso) Beccari
Top rot, *Penicillium vermoeseni* Biourge, of palm *Washingtonia robusta* Wendland
Leaf spot, *Pestalotia phoenicis* Vize, of palm *Ptychosperma elegans* Blume
Leaf spot, *Cylindrocladium macrosporium* Sherbakoff, of palm *Washingtonia robusta* Wendland
Leaf spot, *Exosporium arecae* (Berkeley & Broome) Petch, of betel-nut palm *Areca cathecu* L. Chabaud
Leaf spot, *Exosporium palmivorum* Saccardo, of canary palm *Phoenix canariensis* Chabaud
False smut, *Graphiola phoenicis* (Mougeot) Poiteau, of canary palm *Phoenix canariensis* Chabaud
Other fungi associated with ornamental palms

Leaf spot, *Pseudomonas washingtoniae* (Pine) Elliott

Symptoms

Small, pinpoint, water-soaked dots appear on the petiole and blade. They enlarge, forming a lesion up to 15 mm long between the veins; the central portion dies and becomes yellow green, while the zones at each end of the lesion advance under water-soaked conditions. The spots on the petioles are somewhat similar in color to those on the blade but are roughly circular and when numerous may fuse and involve most of the surface. Extensive infection is more common on older leaves.

Etiology

The organism is a gram-negative rod, single or in chains, motile by 1–3 polar flagella, and measuring 1.6–7 μ. Colonies are light, glistening, creamy, smooth, convex, entire, circular, and butyrous. A green fluorescent pigment is produced in culture.

Reference

5.

Bud rot, *Phytophthora palmivora* Butl.

Symptoms

This disease is frequently determined by the infection of the bud, resulting in a wet decay of the base of the most recent unfurled leaf as it projects from the sheath. This leaf may fall over due to the softening at the base. Sometimes one or more older leaves show symptoms of deterioration. The tree does not recover but may linger for several months before dying. In addition to the top disease, the nuts are readily attacked resulting in their rotting and falling. From the bunch of diseased nuts the fungus may grow down the peduncle to the main stalk, into the trunk, and to the growing tip of the plant.

Etiology

The mycelium is hyaline, branched, and nonseptate; the sporangia are hyaline, sparse, spherical, and 42–55 μ. Chlamydospores are not always produced. Oospores are spherical, granular, and 24–42 μ.

References

6, 23.

Trunk rot, *Phytophthora parasitica* Dast.

Symptoms

Leaves wilt and die from no local symptom except malnutrition resulting from trunk cankers located near the soil line. Cankers are soft, spongy, and enlarge sufficiently within a few months to cause top wilting and death of the tree. Infection apparently develops through wounds.

Etiology

The mycelium is hyaline, nonseptate, branched, and produces sporangia that measure 25–50 × 20–40 μ. About 20–30 zoospores are produced upon germination, each measuring 8–12 × 5–8 μ. Oospores are spherical, hyaline to granular, and 15–20 μ in diameter.

References

4, 10.

Leaf spot, *Catacauma sabal* Chardon

Symptoms

A brownish color begins at the apex of the leaf blade and advances toward the petiole. Leaf margins become affected, fade to a pale yellow brown, and die. Conspicuous, diamond-shaped spots are formed on the diseased areas. Clustered around the centers of these spots are black, scattered, subepidermal, conspicuous stromata.

Etiology

The stromata are usually on the upper leaf surface, raised, brown to black or purplish, and show a yellow area on the opposite leaf surface. They measure 200–400 × 93–132 μ in diameter and consist of several locules. The asci are clavate with short pedicels and measure 68–85 × 35–43 μ. The ascospores are hyaline, 1-celled, elongate with rounded ends, and 27–30 × 8–10 μ. Paraphyses are scarce. Conidia are in acervuli and are not frequently produced. They are hyaline and obclavate, measuring 4.5–6 × 2–3 μ.

Reference

8.

Leaf spot, *Myrianginella sabaleos* (Weedon) Limber & Jenkins

Symptoms

The leaf spots are pale to yellow, wrinkled, with a definite margin, small, inconspicuous, elongate up to 5 mm, and 200 μ wide.

Etiology

Ascomata develop in the subepidermal area, raising the surface that ruptures the clypeus. The stromata are circular to irregular, thick, and measure 225–750 × 200–500 μ. The asci are spherical to ovate, solitary, variously distributed, 8-spored, and 21–29 × 17–23 μ. Ascospores are hyaline, clavate, muriform with 3–6 and 2–4 septations, and measure 18–22 × 6–8 μ. Conidia are hyaline, 1-celled, ovoid to clavate, and measure 3–5 × 1–2 μ.

Reference

9.

Mushroom root rot, *Clitocybe tabescens* (Scop. ex Fr.) Bres.

Symptoms

A yellow color and some drooping of one or more of the older leaves is the earliest indication of the disease. This is followed by a browning of the lower leaflets or pinnae. A leaf becomes brown and dead, followed by others in more rapid succession; the youngest leaves or those on the opposite side of the plant from the first affected leaf are the last ones to show disease symptoms before the plant is killed. An examination of the trunk at the crown reveals white strands or plates of the fungus thallus distributed throughout the host tissue.

Etiology

Sometimes the honey-colored mushrooms of the fungus are present. The stipe of the sporophore does not show an annulus. The spores are white. The hyaline to white mycelium is septate, branched, and is found in parenchyma stem tissue. Sporophores usually appear in clusters; each cluster is 4–8 in. tall and half as wide. Basidiospores are hyaline, 1-celled, oval, and $8–10 \times 4–6\ \mu$.

References

16, 17, 21.

Butt rot, *Ganoderma sulcatum* Murr.

Symptoms

The loss of foliage is indicative of a malfunction of the trunk or roots of these palms. In most cases, closer examination reveals sporophores of a shelving basidiomycete. Sections of the stem near the sporophores show an advanced stage of decay several feet in length. The advancing margins are reddish with a brown border. Farther back, the trunk tissues are yellow and soft.

Etiology

The fruiting structures may vary in width, depth, and thickness, but do not often exceed $10 \times 8 \times 3$ in., respectively, at the point of attachment, naturally being thinner at the margins. They are brown, varying from light to dark on the upper surface, which is also rough, striated, or mildly zonate. The lower surface is a dull, cream color and is composed of the hymenium of closely crowded pores. The texture is woody, soft during growth and hard at maturity.

References

3, 15.

Top rot, *Penicillium vermoeseni* Biourge

Symptoms

Wilted, yellowing, and drooping fronds and decay of terminal buds constitute visible symptoms of this palm disease. Large decaying areas of the trunk, extending a foot or more vertically, may be observed. The fungus sporulates abundantly in these decaying areas and at the base of the fronds.

Etiology

In about 10 days in artificial culture, a floccose growth covers a petri dish with mycelium that is white to salmon to rose pink, septate, branched, sinuous, coarse, vacuolate, and $3–6\ \mu$ in diameter. Conidiophores are $100–200 \times 4–5\ \mu$. Fertile branches are penicillate; sterigmata are $8–12\ \mu$. Conidia are 1-celled, elliptical, colorless, $4–6 \times 1–2\ \mu$, and often in chains 1–2 mm long.

References

1, 2, 14.

Leaf spot, *Pestalotia phoenicis* Vize

Symptoms

Ashy gray or variously brown discolorations of the foliage usually result from mechanical injury or malnutrition. The tissues, which are materially weakened or dying, frequently become more or less speckled with variously scattered black pustules of the fungus.

Etiology

The subepidermal fruiting structures become erumpent. They are usually on the upper leaf surface, more or less circular, and 75–200 μ in diameter. The conidia are 5-celled, often slightly curved, and 16–22×5–7 μ. The 2 end cells are hyaline; the basal cell has a short pedicel, and the terminal cell has 3 hyaline, mostly straight setulae that are 15–22 μ long. The 3 intermediate cells are olive brown.

Reference

7.

Leaf spot, *Cylindrocladium macrosporium* Sherb.

Symptoms

Leaf spots are brown with a greenish border, circular or elongate, and up to 2 mm in diameter. They appear on the leaves during humid weather.

Etiology

Under favorable conditions the fungus develops hyaline, septate mycelium, conidiophores, and conidia on the surface of the lesion. The mycelium is branched; the conidiophores are erect; and the conidia are long, cylindric, straight, 1-septate, with rounded ends, and measure 102×5–6 μ.

Reference

20.

Leaf spot, *Exosporium arecae* (Berk. & Br.) Petch

Symptoms

The spots are produced on the pinnae of the leaves and on the petiole, stalk, and rachis of the inflorescence. They are oily in appearance when young and average less than 2 mm in diameter. As they become older, a zonate condition frequently develops. The blade is usually killed when the infections are numerous, and entire leaves may die.

Etiology

The mycelium is more or less irregular and forms large strands up to 2 mm in diameter. These spread out over the foliage and form a central stroma that is black, mostly superficial, compact, and 0.5 mm in diameter and half as high. It is covered with conidiophores that are short, erect, and 200×6 μ. The conidia are mostly clavate to obclavate with the terminal end attenuated, multiseptate, and 32–48×14–26 μ.

Reference

19.

Leaf spot, *Exosporium palmivorum* Sacc.

Symptoms

On both surfaces of the leaf, there appear numerous minute, circular, black spots, which enlarge to 15 mm and become irregular to semizonate. At this stage the lesions have an oily appearance, are semitransparent with a brown center, and surrounded by a pale green border zone that contains fructifications. Upon sporulation, lesions appear to be covered with a brownish, velvety layer. The disease is often severe on plants up to several years old.

Etiology

Mycelium is brown, limited, branched, and septate. It supports a superficial sporodochialike apex, rises in thick clusters, and bears elongated, cylindrical, straight or curved, 8–10-septate, nonconstricted, olive brown conidia, measuring $75–120 \times 7–12$ μ.

References

19, 22.

False smut, *Graphiola phoenicis* (Moug.) Poit.

Symptoms

On both sides of the leaflets and on the rachis, there are small yellow or brown spots, consisting of single or grouped black cups of carbonaceous texture, up to 1 mm in diameter. From the interior of these cups project conspicuous, buff-colored, cottony tufts, thickly powdered with light yellow pollenlike spores. Each infection appears very narrowly localized, but the number of spots progresses with the age of the leaf, so that old leaves are heavily diseased.

Etiology

Very fine mycelium becomes united in a stroma consisting of 2 peridia of which the internal one is hyaline and the external dark. Yellow, sterile hyphae are produced in columns. Fertile hyphae are colorless, septate, and bear globose or elliptical spores or probasidia, either laterally or in chains. Spores are 3–6 μ in diameter. Clouds of spores are released as yellow dust. Upon germination, probasidia produce pseudobasidia that bear spores. The fungus penetrates directly into host tissue.

References

11, 12, 18.

Other fungi associated with ornamental palms

Armillaria mellea Vahl ex Fries
Asterina sabalicola Earle
Auerswaldia palmicola Spegazzini
Colletotrichum gloeosporioides Penzig

Figure 131. Sporophores of *Clitocybe tabescens* (mushroom root rot) at base of killed *Butea* palm. (Photograph by A. S. Rhoads.)

Figure 132. Butt rot, *Ganoderma sulcatum*, on Queen palm. (Photograph by A. S. Rhoads.)

Figure 133. False smut, *Graphiola* sp., on Phoenix palm. Enlarged 2X.

Figure 134. Phoenix palm killed by *Clitocybe tabescens*. (Photograph by A. S. Rhoads.)

Figure 135. Black leaf spot, *Asperisporium caricae*, on young papaya fruit. (Photograph by H. G. McMillan.)

Figure 136. Powdery mildew, *Oidium caricae*, on papaya foliage. (Photograph by H. G. McMillan.)

Figure 137. Leaf spot, *Cercospora papayae*, on papaya. (Photograph by H. G. McMillan.)

Cytospora palmarum Cooke
Didymella phacidiomorpha (Cesati) Saccardo
Diplodia natalensis P. Evans
Ellisiodothis inquinans (Ellis & Everhart) Theissen
Glomerella cingulata (Stoneman) Spaulding & Schrenk
Helminthosporium molle Berkeley & Curtis
Melanconium palmarum Cooke
Meliola palmicola Winter
Mycosphaerella palmae Miles
Phyllosticta palmetto Ellis & Everhart
Phymatotrichum omnivorum (Shear) Duggar
Physalospora rhodina (Berkeley & Curtis) Cooke
Thielaviopsis paradoxa (De Seynes) Hoehnel
Ustulina vulgaris Tulasne

References: palm

1. Bliss, D. E. 1934. The relation of Penicillium vermoeseni to a disease of ornamental palms. Phytopathology 25: 896. (Abstr.)
2. Carpenter, T. R., W. D. Wilbur, and W. Young. 1962. Mexican palm trees apparently killed by Penicillium vermoeseni. Plant Disease Reptr. 46:750.
3. Childs, J. F. L., and E. West. 1953. Butt rot of queen palms in Florida associated with Ganoderma sulcatum. Plant Disease Reptr. 37:632–33.
4. Darley, E. F., and W. D. Wilbur. 1953. Phytophthora trunk rot of Washingtonia palms. Phytopathology 43:469–70. (Abstr.)
5. Elliott, C. 1951. Manual of bacterial plant pathogens. 2d ed. Chronica Botanica: Waltham, Mass., p. 100.
6. Gadd, C. H. 1927. The relationship between the Phytophthorae associated with the bud-rot diseases of palms. Ann. Botany (London) 41:253–79.
7. Guba, E. F. 1961. Monograph of Monochaetia and Pestalotia. Harvard University Press: Cambridge, Mass., pp. 89–91.
8. Limber, D. P., and A. E. Jenkins. 1949. Catacauma sabal Chardon identified in the United States and Mexico. Mycologia 41:537–44.
9. Limber, D. P., and A. E. Jenkins. 1949. Weedon's myriangium on Sabal. Mycologia 41:545–52.
10. McFadden, L. A. 1959. Palm diseases. Principes 3:69–75.
11. Nixon, R. W. 1957. Differences among varieties of the date palm in tolerance of graphiola leaf spot. Plant Disease Reptr. 41:1026–28.
12. Nowell, W. 1923. Diseases of crop-plants in the Lesser Antilles. West India Committee: London, p. 236.
13. Pine, L. 1943. A hitherto unreported disease of the Washington palm. Phytopathology 33:1201–4.
14. Raper, K. B., and C. Thom. 1949. A manual of the Penicillia. Williams & Wilkins: Baltimore, Md., pp. 680–82.
15. Rhoads, A. S. 1944. Some observations on diseases of woody plants in Florida. Plant Disease Reptr. 28:260–61.
16. Rhoads, A. S. 1950. Clitocybe root rot of woody plants in the southeastern United States. U.S. Dept. Agr. Circ. 853:1–25.
17. Rhoads, A. S. 1956. The occurrence and destructiveness of clitocybe root rot of woody plants in Florida. Lloydia 19: 193–239.
18. Roger, L. 1951. Phytopathologie des pays chauds. Encyclopedie Mycologique 17:793–96.
19. Roger, L. 1953. Phytopathologie des pays chauds. Encyclopedie Mycologique 18:2201–2.
20. Sherbakoff, C. D. 1928. Washingtonia palm leaf spot due to Cylindrocladium macrosporum n. sp. Phytopathology 18: 219–25.
21. Simonson, L. M. 1957. Some palms in central Florida. Principes 1:86–89.
22. Trelease, W. 1898. A new disease of cultivated palms. Missouri Botan. Garden 9th Rept. 1899, p. 159.
23. Venkatarayan, S. V. 1932. Phytophthora arecae, parasitic on areca tops, and a strain of P. palmivora Butl. (P. faberi Maubl.) on a new host, Aleurites fordi. Phytopathology 22:217–27.

Pan, *Piper betle* Linnaeus

Bacterial leaf spot, *Pseudomonas betle* (Ragunathan) Burkholder
Foot rot, *Phytophthora parasitica* Dastur
Anthracnose, *Glomerella cingulata* (Stoneman) Spaulding & Schrenk
Basal wilt, *Botryobasidium rolfsii* (Saccardo) Venkatarayan
Anthracnose, *Colletotrichum dasturii* Roy
Leaf spot, *Fusarium semitectum* Berkeley & Ravenel
Other fungi associated with pan

Bacterial leaf spot, *Pseudomonas betle* (Ragun.) Burkh.

Symptoms

The early manifestation of the disease is the appearance of small, water-soaked spots on the underside of the leaves, somewhat limited by the veins. As the spots mature they become visible on the upper surface as a yellowish halo surrounding a tan to brown area, angular in shape, and up to 1/2 in. in diameter. If numerous they may coalesce and cause leaves to become yellow and shed. The disease is often so severe that eradication is necessary.

Etiology

The organism is a gram-negative rod, single or in short chains, produces no capsules, is nonmotile, and measures 0.5×1–$2.5\ \mu$. Agar colonies are yellowish tan, glistening, circular, and raised. A green pigment is produced in certain media.

Reference

4.

Foot rot, *Phytophthora parasitica* Dast.

Symptoms

A distinct pale green to faintly yellow color first indicates the diseased condition. This is followed by a drooping of the new growth and a darkening of the main stem at the soil line. The stem discoloration may extend upward for several inches, and the stem often breaks off at the ground level. The submerged roots are usually dark-colored and rotted, leaving mostly vascular tissue.

Etiology

The mycelium is hyaline, branched, nonseptate, and supports hyaline conidiophores that emerge from the stomata, usually in groups. The conidia or sporangia are hyaline, terminal or lateral, papillate, lemon-shaped, and 30–34×27–$30\ \mu$. They germinate, producing 20–30 zoospores that measure 8–12×5–$8\ \mu$. Oospores are hyaline, spherical, and 15–20 μ in diameter.

References

2, 7.

Anthracnose, *Glomerella cingulata* (Ston.) Spauld. & Schrenk

Symptoms

This fungus causes pale yellow, irregular spots on the branches and twigs but is not reported on other parts of the host. Infection apparently depends on predisposing wounds or physiological conditions.

Etiology

The mycelium is hyaline, septate, branched, and confined to a rather small area. The conidiophores are short and densely clustered in the acervuli that erupt the epidermis and give rise to hyaline, abundant, 1-celled, elongate to oval conidia that measure $10\text{–}20 \times 5\text{–}8\ \mu$. An abundance of spores produced under high humidity are pink en masse. There are no setae. The imperfect fungus is *Gloeosporium* spp.

Reference

5.

Basal wilt, *Botryobasidium rolfsii* (Sacc.) Venkat.

Symptoms

A definite wilt, following a partial yellowing and stunting of the plant are the most prominent indications of the disease. Closer examination reveals a crown lesion at the soil line where invasion of the host tissue has occurred. The area is dark-colored, sunken, irregular in extent, and usually girdles the plant stem, killing it. White mycelium and brown sclerotia are often present.

Etiology

The fungus mycelium is white, septate, branched, and produces an abundance of sclerotia on the host tissue, mycelium, and soil surface. The sclerotia are white when young and dark brown when mature; they vary in diameter and are mostly less than 1 mm. The basidiospores are not frequently found. They are hyaline, 1-celled, oval to elongate, and $4\text{–}10 \times 2\text{–}6\ \mu$.

Reference

5.

Anthracnose, *Colletotrichum dasturii* Roy

Symptoms

Leaf spots are circular and brownish, with a yellow halo and a black center. These spots may coalesce, forming blotches that cause drying of the tissue, curling, shrinkage, and premature shedding. Infection takes place anywhere along the green stem and is indicated by the black, small, circular areas that often expand to girdle the stem, producing some shredding.

Etiology

Acervuli develop as black spots on stems and branches and sparsely on the foliage. They are brown, erumpent, and show marginal, septate setae. The conidia are hyaline, falcate, 1-celled, tapered toward the ends, and 19–26× 2–4 μ.

Reference

6.

Leaf spot, *Fusarium semitectum* Berk. & Rav.

Symptoms

Leaf spots form on central portions of the blade. They are of various sizes, often including large areas composed of a number of broad, concentric zones alternating light and dark brown in color.

Etiology

Often a pink or white mycelium develops on the spots, producing hyaline, oval, 1-celled, or septate spores. Intercalary chlamydospores are formed, and conidia that are hyaline, slightly curved, and mostly 3–4-septate are produced.

Reference

1.

Other fungi associated with pan

Botryobasidium salmonicolor (Berkeley & Broome) Venkatarayan
Capnodium betel Sydow & Butler
Cercospora piperis Patouillard
Colletotrichum necator Massee
Colletotrichum stevensii Roy
Marasmius scandens Massee
Oospora piperis, Uppal, Kamat, & Patel
Pestalotia piperis Petch
Phyllosticta piperis Henning
Pythium piperinum Dastur
Rosellinia bunodes Berkeley & Broome
Synchytrium piperi Mhatre & Mundkur
Xanthomonas betlicola Patel, Kulkarni, & Dhande

References: pan

1. Chattopadhyay, S. B., and S. K. Sen Gupta. 1955. A new leaf spot disease of Piper betle in West Bengal. Indian Phytopathol. 8:105–11.
2. Dastur, J. F. 1957. A short note on the footrot disease of pan in central provinces. Agr. J. India 22:105–8.
3. Elliott, C. 1951. Manual of bacterial plant pathogens. 2d ed. Chronica Botanica: Waltham, Mass., pp. 58–59.
4. Park, M. 1934. Bacterial leaf-spot of betel. Trop. Agriculturist 82:393–94.
5. Roger, L. 1951. Phytopathologie des pays chauds. Encyclopedie Mycologique 17: 974–75; 18:1414–15.
6. Roy, T. C. 1948. Anthracnose disease of Piper betle L. caused by Colletotrichum dasturii Roy sp. nov. in Bengal. J. Indian Botan. Soc. 27:96–102.
7. Vestal, E. F. 1950. A textbook of plant pathology. Allahabad Agricultural Institute: Kitabistan, India, pp. 498–506.

Papaya, *Carica papaya* Linnaeus

Soft rot, *Erwinia papayae* (Rant) Magrou

Symptoms

Wet lesions appear on the stems and leaves of young plants and extend rapidly to the petioles and main stem. The leaves become yellow and develop many dry spots, and at the same time all the upper parts of the plant wither and dry. The leaf blades remain attached to the petioles. The petioles hang pendant and finally drop. The fruits deteriorate and become brown from the blossom end but remain attached to the trunk.

Etiology

The organism is a gram-negative rod, in chains, and is 1×1.5 μ long. Agar colonies are gray white, glistening, thin, and circular.

References

5, 10.

Fruit rot, *Phytophthora palmivora* Butl.

Symptoms

Root and crown infections cause decline in the upper parts of the plant. The foliage becomes yellow and gradually sheds. The severity of the infection determines the longevity of the plant. Cankers frequently girdle the upper stem, causing the top part to die. Fruit infection is common, resulting in a soft decay.

Etiology

The mycelium is hyaline, cottony white in the host parts, and nonseptate. Sporangia are hyaline, oval, and $40\text{–}60 \times 27\text{–}35$ μ. Zoospores, 15–30 per sporangium, measure 7–12 μ.

References

8, 11, 17.

Canker, *Phytophthora parasitica* Dast.

Symptoms

Disease conditions are first evidenced by the drooping, wilting, yellowing, and eventual premature shedding of the foliage. The fruit of such plants are also shed. In this respect the symptoms are similar to those of several root diseases. Plants affected by this fungus, however, are usually free from root troubles, but instead they harbor stem cankers usually located in the top one-third of the plant. At first these cankers are small, up to 1 in. in diameter, ovate, darker than normal tissue, and often partly covered with a white crust. As the canker enlarges, the fruit are attacked and drop. These cankers may enlarge up to 6–8 × 2–4 in., girdle the stem, or seriously weaken it.

Etiology

The vegetative stage of the fungus on the diseased surfaces is a hyaline, nonseptate, branched mycelium that supports hyaline, ovate, papillate sporangia that measure 30–38 × 22–25 μ. Sporangia germinate by zoospore production. Inoculations produce stem cankers, decayed fruit, and root infections.

References

6, 8, 18.

Anthracnose, *Glomerella cingulata* (Ston.) Spauld. & Schrenk

Symptoms

The disease is found on leaves and especially on young fruit where the penetration is accomplished through the stigma of the flower or at the level of the scars left by the falling petals. Young, infected fruit cease to develop and mummify or decay. In certain cases, natural inoculation is general on plant parts; infection remains latent during a part of the growth and resumes activity on petioles and foliage that have dehisced and on fruit only at maturity.

Etiology

Foliage spots on mature fruit are usually covered with concentric circles of acervuli with or without setae, measuring 90–270 μ in diameter. Spores, produced abundantly, are hyaline, oval to oblong, 1-celled usually with 1 or 2 guttulae, and measure 10–18 × 3–6 μ. The vegetative stage is *Gloeosporium papayae* or *Colletotrichum papayae*, depending on the presence of setae. The conidial forms are polymorphic. Both morphological races and host specialization have been recognized. This pathogen is polyphagous and ubiquitous. The perithecia are brown, scattered or in groups, imbedded in black stroma, ostiolate, without paraphyses, and 125–250 μ in diameter. The asci are oblong to claviform, 55–70 × 8–10 μ, and contain 8 hyaline, semibiseriate, 1-celled, crescent-shaped ascospores, with rounded ends, measuring 12–22×3–5 μ. The ascospore stage usually develops on nonliving media.

References

6, 12, 14, 15, 19.

Leaf spot, *Mycosphaerella caricae* Syd.

Symptoms

Small, circular or irregular spots, up to 4 mm in diameter, are produced on the limbs. They also appear on both leaf surfaces, mostly on the lower side and sometimes on the fruit. At first they are yellowish with a brown border and concentric lines. The centers may become gray or whitish. The spots are speckled with pycnidia that change from brown to black and measure 2–3 mm in diameter. The perithecia are visible on the upper leaf surface.

Etiology

The mycelium forms a subepidermal stroma from which emerge numerous brown, septate conidiophores in clusters of few to many. They measure 25–40 ×7–10 μ and produce conidia successively. The conidia are light gray, oblong to oval, 1–2-septate, verrucose, and 10–20×7–10 μ. The black, globular perithecia appear on the leaves and occasionally on the fruit. The asci are sessile, cylindrical, and 40–50×10–12 μ; the ascospores are hyaline, fusoid, straight or slightly angular, 2-celled, and 15–18×3–4 μ. The conidial stage is probably *Fusicladium caricae.*

Reference

12.

Root rot, *Sphaerostilbe repens* Berk. & Br.

Symptoms

Decline of aerial parts of the plant is indicated by the off-color foliage, yellowing, stunting, and drooping of lower leaves, followed by withering and death of the plant. Root decay is usually initiated by mechanical wounds.

Etiology

The mycelium is mostly submerged in the cortex and wood where rhizomorphic mats form, with branches and septations, up to 5 mm in diameter. The mycelium is hyaline, becoming brown and dark, and invaded roots become yellow, purple, or brown. The fructifications compose a cylindrical column 2–6 mm in diameter and 8 mm high and are made up of hyphae or conidiophores. The spores are hyaline, ovoid, 1-celled, and 10–20×6–9 μ. The perithecia usually appear near the base of the synnemata. They are reddish to brown and 1/3–2/3 mm. The ascospores are pale brown, oval, 2-celled, constricted, and 16–21×7–8 μ.

Reference

12.

Leaf spot, *Phoma caricina* Hopk.

Symptoms

The fruit may decay on the tree or after removal. It develops spots under humid conditions, causing a dark area.

Etiology

The pycnidia that develop on these dark areas contain many pycnidiospores that are hyaline, 1-celled, and 4–5 × 12 μ.

Reference

12.

Leaf spot, *Phyllosticta caricae-papayae* Allesch.

Symptoms

Leaf spots are whitish, yellow, or brown, with a halo, circular to angular, often coalesce, and measure 2–5 mm in diameter. The spots dry and portions often fall out.

Etiology

The mycelium in the host tissue produces black pycnidia that are submerged at first then crumpent. They are globose and 80–125 μ in diameter. The pycnidiospores are hyaline, straight or curved, and 4–6 × 1–2 μ.

Reference

12.

Black leaf spot, *Asperisporium caricae* (Speg.) Maubl.

Symptoms

Small spots appear on laminae; they are few or numerous, round, slightly angular, and occur on both sides of leaves but mostly on the lower side. The spots turn ashy gray or white in the center surrounded by dark brown margins of concentric lines. Infection begins mostly from lower leaves, sometimes on random leaves, buds, and fruit. It causes yellowing, wilting, and premature shedding of leaves.

Etiology

Mycelium aggregates in subepidermal stroma. Conidiophores bear many conidia successively. Conidia are cylindrical to pear-shaped, 2-celled, verrucose, and 10–20 × 7–10 μ. Perithecia on leaves and fruit are black, globose, and have an ostiole. Ascospores are hyaline, fusoid, 2-celled, straight, and 15–18 × 1–3 μ. The ascospore stage is *Mycosphaerella caricae.*

References

12, 20, 21.

Black fruit spot, *Ascochyta caricae* Pat.

Symptoms

Brown areas on flowers and young fruit become conspicuous as dark, sunken spots that often extend through the peduncle to the main trunk of the tree. These young flowers and fruit shrivel and drop. Infections on older fruit take the form of black, circular spots, not exceeding 1.5 in. in diameter. On mature fruit, the disease appears as numerous brown, circular spots, separate or coalescing on the exposed surface.

Etiology

Within the spots there appear black, minute, raised pustules that exude whitish masses of oval pycnidiospores that are 1- or 2-celled and measure $9–15 \times 3–5\ \mu$, mostly $12 \times 4\ \mu$.

References

3, 6, 16.

Fruit rot, *Botryodiplodia caricae* (Sacc.) Petri

Symptoms

The spots on the petioles are brown with a yellow border, more or less cankerous, sunken, and elongate. Following infection, mature fruits develop a black, soft, leathery rot.

Etiology

Pycnidia are globose, surrounded by thick walls, imbeded in the tissue, rarely erumpent, with underdeveloped papillae, but without voluminous stroma. Spores are dark, oval, 2-celled, $22–26 \times 10–13\ \mu$, and borne mingled with filiform paraphyses. The ascospore stage is *Physalospora rhodina*.

Reference

12.

Powdery mildew, *Oidium caricae* Noack

Symptoms

The disease causes wilting of young plants, beginning with the lowest leaves which turn yellow and shed. On young crown leaves of old plants, it causes an irregular, yellow, marble pattern. Infections occur on peduncles of male flowers and on young fruit. In many instances, the plant may outgrow the disease, but, when severe, dead areas appear between the larger veins, resulting in some stunting and even death of the plant.

Etiology

White mycelium, often faintly visible, develops on the stem, petiole, fruit, and large veins on the lower face of the lamina. The conidiophores are hyaline, septate, erect, and produce successively hyaline, cylindrical to oval, 1-celled conidia, measuring 28×18 μ.

References

6, 7, 16.

Mildew, *Ovulariopsis papayae* v. Bijl

Symptoms

A slight yellowing of the foliage is produced in irregular patches; on the lower surface of these patches there is a powdery, frosty production of conidia and conidiophores, supported by internal and external mycelium.

Etiology

The mycelium grows over the leaf surface, producing hyaline, simple, erect, septate conidiophores, each producing a single, 1-celled, subclavate conidium, measuring 72×14 μ. The ascospore stage is *Phyllactinia* sp.

Reference

12.

Leaf spot, *Helminthosporium papayae* Syd.

Symptoms

Small, yellow to brown spots, with gray centers, and measuring up to 8–10 mm appear on both surfaces of the leaves, petioles, and branches.

Etiology

The conidiophores are brown, varying up to 600 μ in length, and produce brown, 4–9-septate conidia, measuring 17–123×11–21 μ.

References

9, 13.

Leaf spot, *Cercospora papayae* Hansf.

Symptoms

The leaf spots are whitish to ashen in color, depending on the amount of sporulation present, and are subcircular to irregular in outline, prominent on the upper leaf surface but indistinct on the lower surface. The fruit spots appear as small, black dots, enlarging up to 3 mm; they are mostly shallow and do not cause fruit decay. The stromata are produced on the upper leaf surface, are faintly distinct, and support few to many fascicles.

Etiology

The conidiophores are brown with paler tips, multiseptate, unbranched, and

$50–200 \times 3–6\ \mu$. The conidia are hyaline, variously curved, septate, and $20–75 \times 3–5\ \mu$.

References

4, 6.

Algal spot, *Cephaleuros virescens* Kunze

Symptoms

Lamina spots are orange or purple red, becoming green or gray with age, circular or elongate, slightly raised, and velvety. Spots may be surrounded by gray zones, and the infected leaf appears chlorotic. In severe infection, defoliation occurs.

Etiology

The vegetative thallus is chlorophyll-bearing, cellular, branched, irregularly fan-shaped, and subcuticular. It produces yellow, stipitate, sporangia that appear knoblike and contain zoospores that escape and generate to form a new plant.

References

2, 13.

Other fungi associated with papaya

Alternaria tenuis Nees
Armillaria mellea Vahl ex Fries
Asterina caricarum Rehm
Asterinella papayae Fragoso & Ciferri
Asterostomella caricae Henning
Blakeslea trispora Thaxter
Botryobasidium rolfsii (Saccardo) Venkatarayan
Chaetostroma papaya Patouillard
Choanephora americana Moeller
Corynespora cassiicola (Berkeley & Curtis) Wei
Curvularia lunata (Wakker) Boedijn
Didymella caricae Tassi
Erysiphe cichoracearum De Candolle
Fusarium roseum (Link) Snyder & Hansen
Guignardia caricae (Marchal & Steyaert) Hendrick
Macrophomina bataticola Taubenhaus & Butler
Phomopsis caricae-papayae Petri & Ciferri
Phymatotrichum omnivorum (Shear) Duggar
Pythium butleri Subramaniam
Rhizopus nigricans Ehrenberg
Rosellinia bunodes (Berkeley & Broome) Saccardo
Thanatephorus cucumeris (Frank) Donk

References: papaya

1. Aragaki, M., R. D. Mobley, and R. B. Hine. 1967. Sporangial germination of phytophthora from papaya. Mycologia 59:93–102.
2. Butler, E. J. 1918. Fungi and diseases in plants. Thacker: Calcutta, India.
3. Chowdhury, S. 1950. A fruit rot of papaya (Carica papaya L.) caused by Ascochyta caricae Pat. Proc. Brit. Mycol. Soc. 33:317–22.
4. Chupp, C. 1953. A monograph of the fungus genus Cercospora. Published by the author: Ithaca, N.Y.
5. Elliott, C. 1951. Manual of bacterial plant pathogens. 2d ed. Chronica Botanica: Waltham, Mass., p. 158.
6. Hine, R. B., et al. 1965. Diseases of papaya (Carica papaya L.) in Hawaii. Hawaii Agr. Expt. Sta. Bull. 136:5–25.
7. Parris, G. K. 1941. Diseases of papaya in Hawaii and their control. Hawaii Agr. Expt. Sta. Bull. 87:32–44.
8. Parris, G. K. 1942. Phytophthora parasitica on papaya (Carica papaya) in Hawaii. Phytopathology 32:314–20.
9. Patel, M. K., and L. Moniz. 1968. A new leaf blight disease of papaya in India. Plant Disease Reptr. 52:784–85.
10. Rant, von A. 1931. Wereine bakterienkrankheit bei dem melonenhaume (Carica papaya Linn.) auf Java. Zentr. Bakteriol Parasitenk. Abt. 84:481–87.
11. Roger, L. 1951. Phytopathologie des pays chauds. Encyclopedie Mycologique 17:639–44.
12. Roger, L. 1953. Phytopathologie des pays chauds. Encyclopedie Mycologique 18:1335–36, 1377, 1404–17, 1483–86, 1491–95, 1521–24, 1617–18, 1667–68, 1678–79, 1758–76, 2046.
13. Roger, L. 1954. Phytopathologie des pays chauds. Encyclopedie Mycologique 19:2260–67.
14. Shear, C. L., and A. K. Wood. 1913. Studies of fungous parasites belonging to the genus Glomerella. U.S. Dept. Agr. Bur. Plant Ind. Bull. 252:1–110.
15. Simmonds, J. H. 1937. Diseases of the papaw. Queensland Agr. J. 48:544–52.
16. Simmonds, J. H. 1938. Plant diseases and their control. D. Whyte: Brisbane, Australia.
17. Teakle, D. S. 1957. Papaw root rot caused by Phytophthora palmivora Butl. Queensland J. Agr. Sci. 14:81–91.
18. Trujillo, E. E., and R. B. Hine. 1965. The role of papaya residues in papaya root rot caused by Pythium aphanidermatum and Phytophthora parasitica. Phytopathology 55:1293–98.
19. Trujillo, E. E., and F. P. Obrero. 1969. Anthracnose of papaya leaves caused by Colletotrichum gloeosporioides. Plant Disease Reptr. 53:323–25. (Abstr.)
20. Uphot, J. C. 1925. Das verhalten von Pucciniopsis caricae Earle auf der papaya (Carica papaya) in Florida. Z. Pflanzenkrankh. 35:118–22.
21. West, E. 1943. Papaya leaf spot. Florida Univ. Agr. Expt. Sta., Press Bull. 584: 1–2.
22. Yee, W. 1970. Papayas in Hawaii. Univ. of Hawaii, Ext. Serv. Circ. 436:1–53.

Passionflower, *Passiflora edulis* Sims

Bacterial spot, *Pseudomonas passiflorae* (Reid) Burkholder
Anthracnose, *Glomerella cingulata* (Stoneman) Spaulding & Schrenk
Leaf spot, *Septoria passiflorae* Louw
Anthracnose, *Gloeosporium fructigenum* Berkeley
Scab, *Cladosporium herbarum* (Persoon) Link
Brown spot, *Alternaria passiflorae* Simmonds
Wilt, *Fusarium oxysporum* (Schlechtendahl) Snyder & Hansen f. sp. *passiflorae* Purss
Other fungi associated with passionflower

Bacterial spot, *Pseudomonas passiflorae* (Reid) Burkh.

Symptoms

Well defined, greasy, circular to irregular patches develop on the fruit. Brown spots surrounded by a faint yellow halo appear on the leaves. Light brown, depressed areas, surrounded by irregular swellings, develop on the stems and gradually become darker brown, dry, spongy, and split, exposing the wood. The brown discoloration forms streaks in the wood.

Etiology

The organism is a gram-negative rod, usually single, forms capsules, is motile by 1–5 polar flagella, and measures 0.4×1–$3\ \mu$. Agar colonies are gray, translucent, shining, flat, and butyrous. A fluorescent pigment forms in culture.

Reference

2.

Anthracnose, *Glomerella cingulata* (Ston.) Spauld. & Schrenk

Symptoms

This disease is characterized by brown spots on the foliage and cankers on the principal branches and twigs. The leaf spots are at first surrounded by a green border that later becomes dark brown. The stem cankers appear as brownish areas, and the vine beyond them is usually killed.

Etiology

The acervuli are scattered on the leaf spots and often clustered on the stems. They produce an abundance of hyaline, oval, 1-celled conidia that measure 16–18×4–$6\ \mu$. The setae are scattered over the surface of the acervulus homogeneously. The perithecia are usually dark brown to almost black, gregarious, grow in tufts, are membranous, rostrate, flask-shaped, hairy, immersed in a stroma but becoming erumpent, and measure 250–$350 \times 150\ \mu$. Asci are sessile, clavate, and 8-spored. Ascospores are hyaline, 1-celled, slightly curved to allantoid, and 20–28×5–$7\ \mu$. The vegetative stage is *Colletotrichum gloeosporioides*.

Reference

5.

Leaf spot, *Septoria passiflorae* Louw

Symptoms

The fungus causes lesions on the foliage, flowers, and fruit. The spots on the leaves are circular to irregular, well-defined, 5–10 mm in diameter, with a light brown border and a yellow margin. They may be so plentiful as to completely cover the surface; the tissue becomes necrotic and turns dark. On the fruit the spots are small, circular, and superficial; when plentiful, they cause the tissue to become necrotic and dry. Attacks on the flowers cause them to drop.

Etiology

The pycnidia are plentiful on the leaf spots, appearing first in the stomatal areas. They are black, ostiolate, submerged becoming erumpent, and exude pycnidiospores in tendrils. The conidia are hyaline or lightly colored, filiform, septate, and 30–$60 \times 2\ \mu$.

Reference

5.

Anthracnose, *Gloeosporium fructigenum* Berk.

Symptoms

Spots develop first on the leaves, then on the fruit. A dieback of the vine also occurs. The leaf spots are circular and brown surrounded by a green border which finally becomes dark brown. The centers of these spots bleach to a lighter color, dry, and often crack. The spots on the stems elongate and develop into a canker, often causing the branch beyond to die back to the canker area. Small, brown, concentrically zoned spots that become dry are produced on the fruit and fruit are shed.

Etiology

The acervuli develop on the leaves, fruit, and stems and are devoid of setae. The conidiophores are short, narrow, and pointed, producing conidia at their tips. The conidia are hyaline, oval, 1-celled, often guttulate, and $10\text{–}15 \times 5\text{–}7\ \mu$. The ascospore stage is *Glomerella cingulata.*

Reference

5.

Scab, *Cladosporium herbarum* (Pers.) Lk.

Symptoms

Small, circular, translucent spots appear on the leaves and shoots, which gradually become overgrown with a dark grayish, powdery mass of spores. Affected leaves are rapidly shed. On the fruit the spots are at first small and brown; they enlarge up to 3–6 mm and develop a hard, raised, scablike area that is somewhat craterlike. These spots are shallow and do not cause any decay unless secondary infection occurs.

Etiology

The sporulating clusters are dense, forming an olive black, velvety, effuse coating. The conidiophores are erect, septate, branched, and colored. Conidia are produced on tips in chains of 2–3; they are pale brown, oval to cylindrical, 1–3-septate, and $10\text{–}15 \times 4\text{–}7\ \mu$.

Reference

7.

Brown spot, *Alternaria passiflorae* Simm.

Symptoms

This disease is found on all aerial parts of the host plant. Leaf spots are brown, more or less translucent, small, circular, and enlarge to 3 mm in diameter. With maturity the centers may become light brown, the margins dark brown, and concentric zones develop. On the fruit the spots are dark green with a light central speck, small, and water-soaked. The larger spots, up to 2 cm, show a uniform brown color and possibly some shrinkage, wrinkling, and sinking. On the branches the brown lesions often exceed 4 mm and are usually associated with the nodal area, affecting leaf and tendril.

Etiology

In culture the mycelium is light gray, cottony, septate, and branched. The conidia, usually produced singly or in chains, are dark, obclavate, with a hyaline, septate, often branched beak, which is 44–120 μ long. The body of the conidium averages 85×20 μ, with 5–13 cross septations and up to 3 longitudinal septations.

References

1, 8.

Wilt, *Fusarium oxysporum* (Schlecht.) Snyd. & Hans. f. sp. *passiflorae* Purss

Symptoms

Diseased plants may show a partial wilt, involving one side, or a general wilted condition, which is usually severe and sudden. Plants usually show the disease when less than one year old. Root examination reveals brown decay, beginning with the fine smaller roots and gradually involving the larger ones and even the crown of the plant. The vascular system of the roots and upper parts is characteristically discolored.

Etiology

The mycelium is light-colored, red or purplish in culture. The conidiophores are produced in sporodochia; the microconidia average 2–3.5×5–12 μ. Hyaline, 3–4-septate, sickle-shaped macroconidia are plentiful and measure 3–4.5 ×40–50 μ. Chlamydospores are formed. The ascospore stage is not known.

References

3, 4.

Other fungi associated with passionflower

Armillaria mellea Vahl ex Fries
Ceratobasidium stevensii (Burt) Venkatarayan
Colletotrichum passiflorae Stevens & Young
Diplodia passifloricola Henning
Gloeosporium passifloricolum Sawada
Melampsora passiflorae Hariot
Sclerotinia sclerotiorum (Libert) Massee

References: passionflower

1. Aragaki, M. S., S. Nakata, and C. Long. 1969. The etiology, including nutritional considerations, of passion fruit brown spot in Hawaii. Plant Disease Reptr. 52: 789–92.
2. Elliott, C. 1951. Manual of bacterial plant pathogens. 2d ed. Chronica Botanica: Waltham, Mass., p. 78.
3. McKnight, T. 1951. A wilt disease of the passion vine (Passiflora edulis) caused by a species of Fusarium. Queensland J. Agr. Sci. 8:1–4.
4. Purss, G. S. 1954. Identification of the species of Fusarium causing wilt in passion vines in Queensland. Queensland J. Agr. Sci. 11:79–81.
5. Roger, L. 1953. Phytopathologie des pays chauds. Encyclopedie Mycologique 18: 1404–17, 1781–85, 1826.
6. Simmonds, J. H. 1930. Brown spot of the passion vine. Queensland Dept. Agr. Bull. 6:1–15.
7. Simmonds, J. H. 1932. Powdery spot and fruit scales of passion vine. Queensland Dept. Agr. Pamphlet 1:1–7.
8. Simmonds, J. H. 1938. Alternaria passiflorae n. sp., the causal organism of the brown spot of the passion vine. Proc. Roy. Soc. Queensland 49:150–51.

Peanut, *Arachis hypogaea* Linnaeus

Brown rot, *Pseudomonas solanacearum* E. F. Smith
Stunt, *Olpidium trifolii* (Passerini) Schroeter
Root rot, *Pythium ultimum* Trow
Cottony rot, *Sclerotinia sclerotiorum* (Libert) De Bary
Anthracnose, *Glomerella cingulata* (Stoneman) Spaulding & Schrenk
Stem blight, *Diaporthe phaseolorum* var. *sojae* (Lehman) Wehmeyer
Leaf spot, *Mycosphaerella arachidicola* W. A. Jenkins
Leaf spot, *Mycosphaerella berkeleyi* W. A. Jenkins
Necrosis, *Calonectria crotalariae* Bell & Sobers
Rust, *Puccinia arachidis* Spegazzini
Soil rot, *Thanatephorus cucumeris* (Frank) Donk
Southern blight, *Botryobasidium rolfsii* (Saccardo) Venkatarayan
Stem rot, *Diplodia natalensis* Evans
Charcoal rot, *Macrophomina phaseolina* (Tassi) Goidanich
Scab, *Sphaceloma arachidis* Bitancourt & Jenkins
Crown rot, *Aspergillus niger* Tieghem
Wilt, *Fusarium martii* Appel & Wollenweber var. *phaseoli* Burkholder
Wilt, *Verticillium dahliae* Klebahn
Texas root rot, *Phymatotrichum omnivorum* (Shear) Duggar
Black rot, *Thielaviopsis basicola* (Berkeley & Broome) Ferraris
Concealed damage, Fungus spp.
Other fungi associated with peanut

Brown rot, *Pseudomonas solanacearum* E. F. Sm.

Symptoms

The organism invades the root system and eventually the main stem, causing the vascular tissue to become darkened, slate to brown. Diseased plants wilt, the youngest leaves showing the effects first. The wilting becomes increasingly more pronounced, and the plant eventually collapses and dies. The discolored vascular tissue is filled with bacteria that may exude in white droplets from cut ends of stems under pressure by squeezing.

Etiology

The organism is a gram-negative rod, motile by a single polar flagellum, and measuring 0.5×1.5 μ. Agar colonies are opalescent becoming brown, wet, shiny, smooth, small, and irregular. Optimum temperature, 35–37C.

Reference

8.

Stunt, *Olpidium trifolii* (Pass.) Schroet.

Symptoms

The disease is of little more than passing interest; it seldom reflects on the average health of the host and, in most instances, is probably overlooked. The organism parasitizes the roots and root hairs and causes some root discoloration that is seldom noticed without careful search.

Etiology

The vegetative thallus invades the epidermal and cortical cells. Within the host cells it forms resting spores that are thick-walled, spherical, smooth or verrucose, 8–25 μ, and germinate by forming swarm spores that emerge through exit tubes. Sporangia are globose, thin-walled, elongate, 12–20 μ or more, and germinate by swarm spore production. The zoospores are hyaline, uniciliate, and about 3 μ in diameter.

Reference

7.

Root rot, *Pythium ultimum* Trow

Symptoms

The disease is frequently associated with the damping-off of seedlings and the soft rot of many succulent plant parts. Diseased plants collapse if attacked when young, but when more mature they wilt and die, remaining more or less erect.

Etiology

White to hyaline, branched, nonseptate mycelium is produced in abundance under cool, humid conditions. Spherical sporangia are 16–22 μ. Oospores are spherical and about 20 μ in diameter. Both germinate directly. The fungus is soil-inhabiting.

References

5, 6, 24.

Cottony rot, *Sclerotinia sclerotiorum* (Lib.) d By.

Symptoms

This disease may be observed on any part of the host plant aboveground and at the soil line on the main stem of young plants. Infection consists of stem cankers often limited to 1 in. in length but usually deep enough to partially or completely girdle the stem. The diseased area becomes sunken and dry and when severe the seedling or branch is killed. Infection on the aerial parts is usually scarce. Collapsed tissue may occur on any part—leaves, stipules, peduncles, pegs, and stems.

Etiology

Usually a fluffy, white mycelium grows superficially in association with the diseased areas, and often the black, large, irregular sclerotia develop in association with the mycelium, either on the surface or within the succulent stem.

Reference

7.

Anthracnose, *Glomerella cingulata* (Ston.) Spauld. & Schrenk

Symptoms

This disease occurs only sparingly in the imperfect stage on stems, where discolored spots develop on the cortex. The spots are shallow, mostly limited in extent, elongate, and usually scattered.

Etiology

Only during humid weather do the acervuli appear as pinkish dots in the discolored areas; they represent the *Colletotrichum* stage. The disease is not common or destructive.

References

7, 10.

Stem blight, *Diaporthe phaseolorum* var. *sojae* (Lehman) Wehm.

Symptoms

Frequently destructive on many legumes, this disease is not important on peanuts even though it is often found fruiting on stems and foliage.

Etiology

Ostiolate pycnidia develop in the epidermal and cortical tissue. The pycnidiospores are hyaline, oval, 1-celled, and $6–7 \times 2–3\ \mu$; stylospores are also present, representing the *Phomopsis* stage. The spherical perithecia form in black stromata. The asci are hyaline, clavate, and contain hyaline, elongate, 1-septate ascospores, measuring $10–18 \times 2–5\ \mu$.

Reference

16.

Leaf spot, *Mycosphaerella arachidicola* W. A. Jenkins

Symptoms

This disease is widely distributed and frequently destructive to the extent that there is almost complete defoliation in midseason. Circular, brown spots appear on the lower leaves, and secondary spots gradually develop over the entire plant; as they grow older the spots darken and show a yellow margin. The fungus develops abundant conidia on the lower surface, causing a darker brown color.

Etiology

The conidia are hyaline to greenish, clavate, more or less curved, 3–12-septate, and 37–100 × 2–5.5 μ. The vegetative stage is *Cercospora arachidicola*.

Reference

15.

Leaf spot, *Mycosphaerella berkeleyi* W. A. Jenkins

Symptoms

This disease is often not distinguished from a similar leaf spot of peanut caused by another fungus. The spots are dark brown to almost black, particularly on the lower leaf surfaces; they usually appear when the plants are past the critical stage. The yellow margins are mostly very slight or nonexistent.

Etiology

The fungus develops conidia on the lower surface on brown, coarse conidiophores. The conidia are olive brown, more or less cylindrical, straight, up to 4-septate, and 18–60 × 5–11 μ. This stage is known as *Cercospora personata*.

References

5, 15.

Necrosis, *Calonectria crotolariae* Bell & Sobers

Symptoms

A wilting accompanied by partial blighting and chlorosis of leaf tips and margins is associated with scattered field plants involving the primary branches. The taproots and hypocotyls develop a necrosis and distinctive blackening; killing of the root tips results in stubby ends. The pegs are discolored by dark brown lesions that are also found on the pods.

Etiology

The mycelium is hyaline, tending to become orange. The conidiophores that arise from the mycelium are hyaline at the tips with olivaceous bases, septate, and become branched at the tips; they form phialides that are nonseptate and measure 6.5–17 × 3.5–5 μ. The conidia are hyaline, 3-septate, cylindrical, and 58–107 × 4.8–7.1 μ. The perithecia are orange to red, subglobose, and 320–465 × 270–290 μ. The asci are hyaline, clavate, and long-stalked. Ascospores are hyaline, elongate fusoid, 1–3-septate, and 34–58 × 6.3–7.8 μ. The vegetative stage is *Cylindrocladium crotolariae*.

Reference

4.

Rust, *Puccinia arachidis* Speg.

Symptoms

Rust, known only in the uredo- and teliospore stages, causes some shedding of foliage but is not usually severe. Leaves appear brownish.

Etiology

The small, brown, dusty uredia pustules appear pimplelike, mostly on the lower surface of the leaf. Uredospores are obovate to elliptical, measuring $23–29 \times 16–22$ μ. Teliospores are not common, dark brown, pedicellate, often 3- or 4-celled, and $38–42 \times 14–16$ μ.

Reference

1.

Soil rot, *Thanatephorus cucumeris* (Frank) Donk

Symptoms

Soil rot manifests itself in several ways on the host plant: as a damping-off fungus, as a stem canker parasite, and as an agent responsible for stunting or killing the gynophores from the time they come in contact with the soil until the pods are mature. The damping-off disease causes the seedling plumules to become weakened at the soil line, fall over, and die.

Etiology

Older plants often show stem lesions above the soil line accompanied by some dryness and some strands of brown, septate, branched, superficial mycelium. Sometimes brown, oval to loaf-shaped sclerotia develop on killed stems or other plant debris in the vicinity. Other names are *Pellicularia filamentosa* and *Rhizoctonia solani.*

Reference

7.

Southern blight, *Botryobasidium rolfsii* (Sacc.) Venkat.

Symptoms

This blight is most frequently observed as a disease of peanut, affecting the parts in contact with the soil in the seedling stage and the nuts at maturity. So-called damping-off of the newly germinated seed and young seedlings is not common. It develops on plants a month or two old as a lethal infection or canker around the main stem at the soil line or immediately below. It occurs below, under dry conditions, or above in wet weather. The plants at first show drooping foliage, soon die, and become brown. The fungus also attacks the pegs as they contact the soil; when the nuts become more mature, the fungus softens the stolons so that they become weak and are easily broken from the nuts.

Etiology

Some white mycelium is often observed around the stems, growing into the surrounding soil. White, spherical sclerotia about 1 mm in diameter form rapidly, harden, turn brown, and when observed are diagnostic of the disease. The vegetative stage is *Sclerotium rolfsii.*

Reference

17.

Stem rot, *Diplodia natalensis* Evans

Symptoms

Stem rot is first indicated by a browning of the invaded stem and branch tissues, which later become necrotic and change rapidly to black. The fungus invades the stem most often at or slightly above the soil line and continues to grow up and down the central stem or on main branches. Leaves cut off by the disease from the root system wilt and die but remain attached to the main stem. The disease rapidly spreads from the main stem to the several branches, and eventually the entire plant collapses. The darkened stem tissue extends outward from the main stem.

Etiology

Numerous closely associated pycnidia are black, submerged becoming erumpent, subglobose, papillate, and $150 \times 180\ \mu$ in diameter. They produce dark, 2-celled, oval to elongate, striate pycnidiospores that measure $19\text{–}31 \times 11\text{–}17\ \mu$. This fungus is frequently associated with the so-called concealed damage to the seed. The ascospore stage of the fungus is *Physalospora rhodina.*

References

14, 18, 26.

Charcoal rot, *Macrophomina phaseolina* (Tassi) Goid.

Symptoms

Charcoal rot causes discoloration and often girdling and killing of plants from invasions of root stems and branches. The decay is soft and spongy but not wet except at first. Drying takes place rapidly, and the cortex usually becomes soft and disintegrates, leaving the exposed woody parts.

Etiology

On killed plants many black, imbedded, hard, small sclerotia develop in the wood. The pycnidia, not frequently found, are black, multiostiolate, and subepidermal. The pycnidiospores are hyaline, 1-celled, ovate to elongate, and $14\text{–}31 \times 11\text{–}12\ \mu$. The sterile stage of the fungus is *Sclerotium bataticola.*

References

7, 25.

Scab, *Sphaceloma arachidis* Bitanc. & Jenkins

Symptoms

Scab is encountered on all the aerial parts of the host. On the leaves the small scab spots are circular to irregular, scattered or often confluent, with raised borders and depressed centers. On the branches the surface is light-colored with a darker, yellow to dark brown border.

Etiology

The scabby spots are numerous on the petioles, oval, up to 3 mm in diameter, and often cause a malformation of the affected parts. The acervulus is variable, 20–50 μ in diameter, and produces 2–4-celled conidia that are hyaline, oval, and 8 × 4 μ.

Reference

23.

Crown rot, *Aspergillus niger* v. Tiegh.

Symptoms

Crown rot of the hypocotyl results in yellowing, wilting, and death of the seedlings. A yellow brown lesion appears near the soil line or below, often correlated with the seed, and darkens to almost black, causing the tissue to collapse and disintegrate.

Etiology

The disease is not important except under poor growing conditions. The mycelium is colored, septate, branched, and produces conidiophores terminated by chains of dark, oval, 1-celled conidia, measuring 2–5 μ in diameter.

References

2, 12.

Wilt, *Fusarium martii* Appel & Wr. var. *phaseoli* Burkh.

Symptoms

Wilt shows up on plants at about blossoming time or 8–10 weeks after planting. The youngest leaves become yellow, degenerate rapidly through a drooping or wilt stage, and finally die because of the stem lesion at or near the soil line. Eventually the entire stem is completely destroyed. The disease is usually variously scattered in a field with no indication that it spreads from plant to plant.

Etiology

The mycelium is dull white to reddish orange or brown, septate, and branched. Chlamydospores are produced singly or in chains, are intercalary or terminal, globose, verrucose, and measure 8–12 μ in diameter. Microconidia develop on conidiophores on the aerial mycelium in chains or are confined to wet droplets in heads, mostly 1-celled but often 1-septate, oval to elongate, and 3–22×4–6 μ. The macroconidia are hyaline, mostly 3-septate, constricted, and 24–40×4–7 μ.

Reference

20.

Wilt, *Verticillium dahliae* Kleb.

Symptoms

Wilt is indicated by leaf chlorosis about blossoming time and the yellowing, wilting, browning, and death of the plant begin on the lower leaves and advance rapidly through the indicated stages. Diseased plants become stunted and often mature early. The wilt phase of the disease is usually not present except in rather dry periods. The fungus invades the root system and is not evident except in late stages of the disease. Plants finally develop a brown discoloration in the vascular tissues.

Etiology

The mycelium is hyaline, septate, and branched. Microsclerotia are dark, ovoid, and 80×40 μ. Conidiophores are verticillately branched. Conidia are hyaline, oblong, and 4–5×2–4 μ.

References

7, 9, 23.

Texas root rot, *Phymatotrichum omnivorum* (Shear) Dug.

Symptoms

Texas root rot causes a root decay of peanuts in alkaline soil. Diseased plants show symptoms of malnutrition as indicated by yellow foliage, stunted growth, and death. The roots are invaded, and often a white, moldy growth of the fungus is evident. The sterile stage of the fungus is *Ozonium omnivorum* Ov.

Etiology

The fungus survives as sclerotia that develop at or near the soil surface. The mycelium is hyaline, septate, and branched. Conidia are mostly nonfunctional. The vegetative stage is dirty white to brown.

Reference

7.

Black rot, *Thielaviopsis basicola* (Berk. & Br.) Ferr.

Symptoms

Black rot is usually considered a wound parasite and is widespread. The fungus is mostly observed on the roots where it causes the infected tissue to darken and finally to disintegrate and die. The aerial parts of the plant suffer some stunting, a definite yellowing, and, during dry periods, some wilting. As a rule, plants do not die quickly.

Etiology

The mycelium is hyaline becoming brown, branched, septate, and 3–7 μ in diameter. The conidia are hyaline, usually becoming dark-colored, thin-walled, and $8–30 \times 3–5$ μ.

Reference

7.

Concealed damage, Fungus spp.

Symptoms

This disease is caused by the invasion of the seed or the shell and the development of the fungus in the shell cavities. The space within the shell surrounding the seed becomes invaded by and often packed full of mycelium of any of several of the fungi mentioned below. They are wound parasites and usually invade the shell cavity through the attachment end. The effects on the seed are variable, depending on the fungus and surrounding conditions. In many instances, the seed show variable resistance to invasion. However, in most cases the seed is invaded and decay results with no particular indication of the condition from the external coat of the shell.

Etiology

Fungi causing concealed damage to peanuts: *Alternaria* spp., *Aspergillus* spp., *Botrytis* spp., *Diplodia natalensis* Evans, *Fusarium* spp., *Mucor* spp., *Neocosmospora vasinfectum* E. F. Smith, *Penicillium* spp., *Rhizoctonia solani* Kuehn, *Rhizopus* spp., *Sclerotium bataticola* Taubenhaus, *Sclerotium rolfsii* Saccardo, and *Trichoderma* spp.

References

9, 11, 24.

Other fungi associated with peanut

Ascochyta arachidis Woronichin
Aspergillus flavus Link
Botrytis cinerea Persoon ex Fries
Cercosporella arachidid Heim
Cladosporium herbarum Persoon ex Link
Coniothecium arachideum Lucks
Fusarium oxysporum f. *vasinfectum* (Atkinson) Snyder & Hansen
Penicillium glaucum Link
Rhizopus stolonifer (Ehrenberg ex Fries) Link
Trichoderma viride Persoon ex Fries

References: peanut

1. Arthur, J. C., and G. B. Cummins. 1962. Manual of the rusts in United States and Canada. Hafner Co.: N.Y., p. 244.
2. Ashworth, L. J., et al. 1964. Epidemiology of a seedling disease of Spanish peanut caused by Aspergillus niger. Phytopathology 54:1161–66.
3. Atkinson, R. E. 1944. Report on diseases of peanut. Plant Disease Reptr. 28: 1096–97.
4. Bell, D. K., and E. K. Sobers. 1966. A peg, pod, and root necrosis of peanuts caused by a species of Calonectria. Phytopathology 56:1361–64.
5. Butler, E. J. 1918. Fungi and diseases in plants. Thacker: Calcutta, India, pp. 319–23.
6. Crawford, R. F. 1952. Seed treatment for control of damping-off in peanuts. New Mexico Agr. Expt. Sta. Bull. 370:1–11.
7. Dickson, J. G. 1956. Diseases of field crops. 2d ed. McGraw-Hill: N.Y.
8. Dowson, W. J. 1949. Manual of bacterial plant diseases. Macmillan: N.Y., pp. 86–103.
9. Frank, Z. R., and J. Krikun. 1969. Evaluation of peanut (Arachis hypogaea) varieties for verticillium wilt resistance. Plant Disease Reptr. 53:744–46.
10. Garren, K. H. 1966. Peanut (groundnut) microfloras and pathogenesis in peanut pod rot. Phytopathol. Z. 55:359–67.
11. Garren, K. H., and B. B. Higgins. 1947. Fungi associated with runner peanut seeds and their relation to concealed damage. Phytopathology 37:512–22.
12. Jackson, C. R. 1965. Peanut kernel infection and growth in vitro by four fungi at various temperatures. Phytopathology 55:46–48.
13. Jackson, C. R., and D. K. Bell. 1969. Diseases of peanut (groundnut) caused by fungi. Georgia Agr. Expt. Sta. Res. Bull. 56:4–136.
14. Jacoway, T. H. 1951. Stemrot, Diplodia natalensis Pole-Evans, of peanut, Arachis hypogaea L. Master's thesis, University of Florida, pp. 1–42.
15. Jenkins, W. A. 1938. Two fungi causing leaf spot of peanut. J. Agr. Res. 56: 317–32.
16. Luttrell, E. S. 1947. Diaporthe phaseolorum var. sojae on crop plants. Phytopathology 37:445–65.
17. McClintock, J. A. 1917. Peanut-wilt caused by Sclerotium rolfsii. J. Agr. Res. 8:441–48.
18. McGuire, J. M., and W. E. Cooper. 1965. Interaction of heat injury and Diplodia gossypina and other etiological aspects of collar rot of peanut. Phytopathology 55:231–36.
19. Matthews, V. D. 1931. Studies on the genus Pythium. University of North Carolina Press: Chapel Hill.
20. Miller, J. H., and H. W. Harvey. 1932. Peanut wilt in Georgia. Phytopathology 22:371–83.
21. Miller, L. I. 1946. Peanut leafspot control. Virginia Agr. Expt. Sta. Bull. 104: 1–85.
22. Porter, D. M. 1970. Peanut wilt caused by Pythium myriotylum. Phytopathology 60:393–94.
23. Roger, L. 1953. Phytopathologie des pays chauds. Encyclopedie Mycologique 18:1928; 19:2739–40.
24. Shurtleff, M. C. 1966. How to control plant diseases in home and garden. 2d ed. Iowa State University Press: Ames, pp. 394–96.
25. Sprague, R. 1950. Diseases of cereals and grasses in North America. Ronald Press: N.Y.
26. Voorhees, R. K. 1942. Life history and taxonomy of the fungus Physalospora rhodina. Florida Univ. Agr. Expt. Sta. Bull. 371:1–91.

Pepper, black, *Piper nigrum* Linnaeus

Collar rot, *Phytophthora parasitica* var. *piperina* Dastur
Anthracnose, *Glomerella cingulata* (Stoneman) Spaulding & Schrenk
Stump rot, *Rosellinia bunodes* (Berkeley & Broome) Saccardo
Wilt, *Botryobasidium rolfsii* (Saccardo) Venkatarayan
White thread, *Marasmius scandens* Massee
Leaf spot, *Pestalotia pipericola* Mundkur & Kheswalla
Leaf spot, *Cercospora piperis* Patouillard
Other fungi associated with black pepper

Figure 138. Crown rot, *Aspergillus niger,* on peanut. (Photograph by C. R. Jackson.)

Figure 139. Root rot, *Phytophthora parasitica,* on black pepper seedling. (Photograph by M. A. Porres.)

Figure 140. Root rot of black pepper initiated by *Phytophthora parasitica.* (Photograph by J. Gonçalves.)

Figure 141. *Fusarium* wilt and decline of black pepper. (Photograph by J. Gonçalves.)

Collar rot, *Phytophthora parasitica* var. *piperina* Dast.

Symptoms

This disease consists of wet, cortical, crown rot preceded by yellowing and wilting. The condition develops gradually, and the plant eventually dies back to the crown. New sprouts arise below the affected area of the crown.

Etiology

The mycelium is hyaline, branched, nonseptate, and up to 9 μ in diameter. Sporangiophores are 100–300 μ long, emerging from stomata. The sporangia are hyaline, ovoid to pyriform, and 16–60 × 10–45 μ. There are about 30 ovoid, biciliate zoospores, each measuring 8–12 × 5–8 μ. Oogonia are terminal, and oospores are hyaline, spherical, and 15–20 μ in diameter.

References

3, 5, 6.

Anthracnose, *Glomerella cingulata* (Ston.) Spauld. & Schrenk

Symptoms

Young leaf lesions are small, circular to irregular, and yellow to light brown, becoming darker-colored to almost black as they develop, and retaining a narrow, water-soaked border. Often they coalesce, involving large portions of the blade and causing the leaf to drop.

Etiology

The mycelium is hyaline, coarse, septate, branched, and becomes compact and brown. Conidiophores arise from the hyphae and are hyaline, simple, more slender than the hyphae, and produce conidia at their apex one after another by abscission. The conidia are hyaline, cylindrical to oblong, nonseptate, and 12–20 × 3–7 μ. The vegetative stage of the fungus is *Gloeosporium* sp.

Reference

7.

Stump rot, *Rosellinia bunodes* (Berk. & Br.) Sacc.

Symptoms

The common name is derived from the fact that the fungus apparently develops on tree stumps left neglected in planted areas. The fungus spreads through the soil by strands of sterile mycelium and contacts roots of different plants, killing them. Aboveground symptoms are often obscure, becoming evident as an unhealthy condition appears or as the plant suddenly dies as the cambium is invaded and the stem girdled.

Etiology

Brown strands of the fungus surround the roots and flatten out under the cortex. Some hyphae near the surface of the soil produce purplish black, velvety patches of the conidial stage of the fungus. The conidiophores are short, black stalks, bearing hyaline, small, oval, 1-celled conidia. The perithecia are black, spherical, warty, and 1–1.5 mm in diameter. The asci are up to 300 μ long; ascospores are brown, long, spindle-shaped, with threadlike terminals, and 40–110 × 10–12 μ. The parasitic stage of the fungus is sterile.

Reference

1.

Wilt, *Botryobasidium rolfsii* (Sacc.) Venkat.

Symptoms

Early indications of the disease may be a slight off-color of the foliage, a paler green or partial yellowing, accompanied by a wilt. The plant is stunted and often lingers or collapses suddenly, because of root decay or the development of a stem lesion at the soil line. Under dry conditions, there is very little evidence of the fungus, whereas under humid conditions a white mycelium becomes evident, surrounding the stem just above the soil line.

Etiology

The sclerotia of the fungus are brown, spherical, less than 1 mm in diameter, and sometimes can be observed on the plant stems.

Reference

5.

White thread, *Marasmius scandens* Mass.

Symptoms

Leaves dry along the stems, die, become brown, and eventually hang pendant. The mycelium is white and is usually located on the lower leaf surfaces and on the lower side of the branches. It is mostly threadlike, branches profusely, and grows out toward the terminals, involving the leaf blades by extending up the petioles and to the fruit by way of the peduncles.

Etiology

The mycelial strands are more or less rhizomorphic, extend indefinitely, and are mostly sterile. However, under highly favorable humidity small sporophores develop that are mushroomlike with stipe, cap, and 3–5 gills. The cap is creamy white, delicate, directly attached to the rhizomorphs, and measures 3–6 mm in diameter. The basidiospores are hyaline, 1-celled, ellipsoid, and 6–8 × 4 μ.

Reference

5.

Leaf spot, *Pestalotia pipericola* Mund. & Kheswalla

Symptoms

The leaf spots are gray and irregular to indefinite in shape and size.

Etiology

The fruiting structures are black, submerged to erumpent, and scattered on the upper leaf surface. The conidia are 5-celled, mostly straight, thicker in the center, and 16–30 μ. The 3 middle cells are dark brown. The terminal cells are hyaline; the basal pedicel is hyaline and short, and the upper cell supports mostly 3 diverging, hyaline setulae.

Reference

4.

Leaf spot, *Cercospora piperis* Pat.

Symptoms

The leaves become spotted with yellow, small, circular lesions which gradually become brown and may enlarge to 1 cm in diameter. As they become older, they turn light gray with a yellow margin.

Etiology

Conidiophores, developing on the lower surface, are nonfasciculate, olivaceous, multiseptate, constricted, branched, and show spore scars. Conidia are obclavate, mostly straight, multiseptate, and 25–130 × 3–5 μ.

Reference

2.

Other fungi associated with black pepper

Botryobasidium salmonicolor (Berkeley & Broome) Venkatarayan
Capnodium bettle Sydow & Butler
Oidium piperis Uppal, Kamat, & Patel
Phyllosticta piperis Henning
Pythium piperinum Dastur
Thanatephorus cucumeris (Frank) Donk

References: black pepper

1. Butler, E. J. 1918. Fungi and diseases in plants. Thacker: Calcutta, India, pp. 357–59.
2. Chupp, C. 1953. A monograph of the fungus genus Cercospora. Published by the author: Ithaca, N.Y., pp. 441–42.
3. Leather, R. I. 1967. The occurrence of a phytophthora root and leaf disease of black pepper in Jamaica. FAO Plant Protect. Bull. 15:15–16.
4. Mundkur, B. B., and K. F. Kheswalla. 1942. Indian and Burman species of the genera Pestalotia and Monochaetia. Mycologia 34:308–17.
5. Roger, L. 1951. Phytopathologie des pays chauds. Encyclopedie Mycologique 17: 668, 774–95, 1087–89.
6. Ruppel, E. G., and N. Almeyda. 1952. Susceptibility of native Piper species to the collar rot pathogen of black pep-

per in Puerto Rico. Plant Disease Reptr. 49:550–51.
7. Vimuktanandana, V. Y., and M. S. Celino. 1940. Anthracnose of black pepper (Piper nigrum Linn.). Philippine Agriculturist 29:124–41.

Pepper, red, *Capsicum annuum* Linnaeus

Soft rot, *Erwinia aroideae* (Townsend) Holland
Soft rot, *Erwinia carotovora* (Jones) Holland
Brown rot, *Pseudomonas solanacearum* E. F. Smith
Bacterial spot, *Xanthomonas vesicatoria* (Doidge) Dowson
Damping-off, *Pythium debaryanum* Hesse
Leak, *Pythium aphanidermatum* (Edson) Fitzpatrick
Blight, *Phytophthora capsici* Leonian
Downy mildew, *Peronospora tabacina* Adam
Blossom mold, *Choanephora cucurbitarum* (Berkeley & Ravenel) Thaxter
Mold, *Blakeslea trispora* Thaxter
Yeast spot, *Nematospora coryli* Peglion
Pink joint, *Sclerotinia sclerotiorum* (Libert) De Bary
Anthracnose, *Glomerella cingulata* (Stoneman) Spaulding & Schrenk
Soil rot, *Thanatephorus cucumeris* (Frank) Donk
Southern blight, *Botryobasidium rolfsii* (Saccardo) Venkatarayan
Leaf spot, *Phyllosticta capsici* Spegazzini
Pod rot, *Phoma destructiva* Plowright
Charcoal rot, *Macrophomina phaseolina* (Tassi) Goidanich
Leaf spot, *Alternaria solani* (Ellis & Martin) Jones & Grout
Leaf spot, *Stemphylium floridanum* Hannon & Weber
Leaf spot, *Cercospora capsici* Heald & Wolf
Leaf spot, *Cercospora unamunoi* Castellani
Fruit spot, *Bipolaris tetrameral* (McKinney) Schoen
Wilt, *Fusarium oxysporum* f. sp. *vasinfectum* (Atkinson) Snyder & Hansen
Wilt, *Verticillium albo-atrum* Reinke & Berthold
Other fungi associated with red pepper

Soft rot, *Erwinia aroideae* (Towns.) Holland

Symptoms

Soft rot develops on pods, attached to the plant, that become infected through mechanical injury. The infection develops from a small, sunken, circular spot. It involves the entire contents of the pod. A watery breakdown of the internal tissue is held temporarily by the epidermis and evolves into a bag of liquid, hanging pendant. The epidermis or peduncle usually breaks, the liquid escapes, and the bacteria may infect other pods. The disease is most prominent after midseason.

Etiology

The organism is a gram-negative rod, forms no capsules, is motile by 2–8 peritrichous flagella, and measures $2–3 \times 0.5\ \mu$. Agar colonies are white to opalescent, glistening, and circular to irregular. Optimum temperature, 35C.

References

4, 12, 13.

Soft rot, *Erwinia carotovora* (Jones) Holland

Symptoms

This decay results from infection through mechanical injury to pods, causing a rapid breakdown after the pod has been removed from the plant. The decay develops in the picking and packing processes and often is very serious when the pods are washed without a disinfectant. It continues to develop in transit and on the market.

Etiology

The organism is a gram-negative rod, which forms in chains, produces no capsules, is motile by 2–5 peritrichous flagella, and measures $1.5–2 \times 0.6–0.9\ \mu$. Colonies on agar are dirty white, glistening, raised, circular, entire, and smooth. Optimum temperature, 27C.

References

4, 12, 13.

Brown rot, *Pseudomonas solanacearum* E. F. Sm.

Symptoms

This disease, also known as southern bacterial wilt, is characterized by the brown discoloration of the vascular tissue of the stems of wilting plants. Early symptoms include dull color, slowed growth, and faint drooping of the youngest leaves. The symptoms become more pronounced as the disease develops, and finally the wilting condition becomes permanent, and browning and death ensue. The organism survives in the soil from season to season and enters the plants through the roots. It causes the vascular tissue to become slate brown, often showing dark streaks.

Etiology

The organism is a gram-negative rod, forms no capsules, is motile by a single polar flagellum, and measures $1–5 \times 0.5\ \mu$. Agar colonies are opalescent becoming darker, small, smooth, wet, shiny, and circular to irregular. Optimum temperature, 35–37C.

References

4, 12.

Bacterial spot, *Xanthomonas vesicatoria* (Doidge) Dows.

Symptoms

This disease, which is seed-transmitted, appears in seed beds as black, marginal or scattered, irregular, small spots. As the plants enlarge, infections are spread from cotyledons to the leaves and often are so numerous as to cause considerable defoliation. On mature leaves the spots are yellowish, slightly raised, water-soaked, and blisterlike. When numerous, they cause the leaves to become yellow and shed. If scattered, they enlarge and the leaves remain attached longer. The pod infections are usually scattered, raised, rough, and penetrate deeply. No soft decay is produced, and pods are not shed because of it.

Etiology

The organism is a gram-negative rod, single, in pairs, or in chains, produces capsules, is motile by a single polar flagellum, and measures $1\text{–}1.5 \times 0.6\text{–}0.7\ \mu$. Agar colonies are yellow, semitranslucent, circular to spreading, butyrous, and viscid. Optimum temperature, 30C.

References

3, 12, 13.

Damping-off, *Pythium debaryanum* Hesse

Symptoms

Damping-off causes losses of seedlings and is usually detected in the early stages by the slight drooping of the cotyledon tips in midday. The disease develops rapidly. Wilting becomes permanent, and the seedlings die without much change in color. If they are several days old when attacked, they may remain somewhat erect rather than falling over.

Etiology

The fungus produces hyaline to white mycelium that is branched, nonseptate, and often supports sporangia, oospores, zoospores, and chlamydospores.

References

6, 12.

Leak, *Pythium aphanidermatum* (Edson) Fitzp.

Symptoms

Leak develops largely from field inoculation through wounds or under highly favorable humid conditions. The pods become soft and watery in the field and disintegrate; the decay may develop in storage or transit.

Etiology

The mycelium is hyaline or white, branched, nonseptate, and produces sporangia, zoospores, and oospores. Field infection occurs on pods that touch or are close to the soil surface.

References

3, 6.

Blight, *Phytophthora capsici* Leonian

Symptoms

Blight results from the invasion of the stem through the root system. Stem lesions blacken the tissue up the stem, resulting in complete girdling, stem shrinkage, and death of the aerial growth. Infection takes place on the branches, foliage, and fruit. Soil surface girdling under favorable moisture conditions kills the top growth. It causes large, irregular, scalded leaf lesions and an invasion of the pods. The pods are invaded through the peduncle, and the fungus overgrows the entire pod, which remains attached. If the plant remains erect, it dries and produces sporangia over the outer surface.

Etiology

The mycelium is hyaline, branched, nonseptate, and produces sporangia in abundance. The sporangia have prominent papillae, measure 35–105 × 21–56 μ, and germinate by zoospore formation or directly.

References

3, 11, 12, 13.

Downy mildew, *Peronospora tabacina* Adam

Symptoms

This mildew produces a faint yellowing of the leaf tissue and a rapid invasion of the parenchyma, causing the leaves to collapse and blacken. The sporangiophores, bearing abundant sporangia, are produced as a fine, thin, fuzzy, slightly colored growth on the lower leaf surfaces. This disease is rather rare on pepper.

Etiology

The fungus does not infect the pods. The organism produces hyaline, branched, nonseptate mycelium, as well as sporangia, zoospores, oospores, and chlamydospores.

References

3, 12.

Blossom mold, *Choanephora cucurbitarum* (Berk. & Rav.) Thaxt.

Symptoms

Under humid, warm conditions mold develops on flowers that are beginning to decline. It envelops the floral part, which is shed at the first node.

Etiology

The fungus produces many hairlike structures bearing numerous purple to brown spores in loose clusters. The spores are dark-colored, 1-celled, oval to elongate, straight, and possess polar tufts of delicate hairs. Zygospores are produced in host tissue.

References

12, 13.

Mold, *Blakeslea trispora* Thaxt.

Symptoms

This mold, a weak parasite, develops on stamens, pistils, and petals of declining flowers and also in pods that have been injured. The flowers are shed, and pods showing the fungus deteriorate more rapidly due to invasion of other, more vigorous organisms.

Etiology

The spores are oval to elongate, striate, produced on sporangiola, and show indistinctly polar clusters of fine hairs. Zygospores are produced by mating plus and minus strains.

References

13, 14.

Yeast spot, *Nematospora coryli* Pegl.

Symptoms

This disease, also known as cloudy spot, has been observed most frequently on pods in midsummer after the major production period. It appears as light, submerged, blotchy areas of variable size. External injury may be detected as small, black specks found immediately above the diseased areas. The larger areas often show some brownish to slate-colored zones. Opening of the pods sometimes exposes the fungus in wet, glistening beads or mounds.

Etiology

The vegetative cells are produced by budding and are spherical, measuring up to 20 μ in diameter. The asci are elongate and contain 8 needle-shaped ascospores with terminal appendages.

References

3, 13, 14.

Pink joint, *Sclerotinia sclerotiorum* (Lib.) d By.

Symptoms

Pink joint kills the aerial branches of half-grown plants and causes a girdling of the main stem at the soil line. Lesions begin at the leaf axils or in the growing tips. The portion beyond the lesion wilts and blackens. The fungus invades the branch downward, usually stopping at a prominent node. The invaded branch is often overgrown with the white mycelium of the fungus. The larger stems sometimes dry out and bleach to a light color. Sclerotia develop in the pith cavities of the stems. The diseased nodes are often pink to rose-colored. Cankers at the soil line result in a general wilting.

Etiology

The sclerotia produce apothecia upon germination, forming many asci and hyaline, 1-celled, oval, small ascospores.

References

12, 13.

Anthracnose, *Glomerella cingulata* (Ston.) Spauld. & Schrenk

Symptoms

Anthracnose occurs on pepper pods and becomes more severe as the pods ripen. On green pods, the infection appears as colorless, circular, slightly sunken, water-soaked spots that mature with the pods and frequently involve much of the surface. They become elongate and deeply sunken with darkened centers. Secondary invaders may aid in the deterioration of the pod.

Etiology

The fungus produces numerous acervuli clustered in the spots, and black, septate, bristlelike setae are scattered among them. The vegetative spores are hyaline, pink en masse, 1-celled, cylindrical, with rounded ends, and measure 15–19 $\times$ 5–7 μ. The vegetative stages are *Colletotrichum* and *Gloeosporium* spp.

References

12, 13.

Soil rot, *Thanatephorus cucumeris* (Frank) Donk

Symptoms

Soil rot develops mostly in warm, humid weather on pods that contact the soil. The pod becomes softened at the contact point, and the breaking down of the tissue continues. This fungus also causes damping-off. Soil rot is manifested in the form of stem cankers, developing at the soil line and extending to form a stem lesion or girdle.

Etiology

The parasitic stage of the fungus produces brown sclerotia and is *Rhizoctonia solani* Kuehn. The basidiospores are frequently produced by the mycelium on plant stems associated with the cankers.

Reference

13.

Southern blight, *Botryobasidium rolfsii* (Sacc.) Venkat.

Symptoms

This blight consists of a slight off-color of the green parts and a mild drooping of the growing tips and young leaves. There is a lack of recovery from the wilted condition, or general yellowing of the older leaves, and final death. The stem is girdled, and whitish mycelium that grows out into the surrounding soil for 1 in. or more is usually found around this area.

Etiology

This mycelium supports many brown, spherical, hard, superficial sclerotia, mostly less than 1/2 mm in diameter. They are the principal means of dissemination.

References

12, 13.

Leaf spot, *Phyllosticta capsici* Speg.

Symptoms

This disease consists of small, circular, sometimes raised, scattered leaf spots. They are bordered by a darker band, separating the whitish central areas from the healthy green blade. Some yellowing develops, and sometimes during humid weather the centers fall away.

Etiology

The spots become speckled with black, ostiolate, pycnidia, visible on both leaf surfaces. The pycnidiospores are hyaline, small, oval, 1-celled, and measure $3–6 \times 1.5–3\ \mu$.

Reference

3.

Pod rot, *Phoma destructiva* Plowr.

Symptoms

This disease frequently develops on the peduncles, calyx, and pod tissue, causing blackened areas on stem tissue and a decay of the pods.

Etiology

The diseased areas are characterized by the light-colored pycnidia that are found on the pods. The pycnidiospores are hyaline, 1-celled, oval, and $6–10 \times 3–6\ \mu$.

References

12, 13.

Charcoal rot, *Macrophomina phaseolina* (Tassi) Goid.

Symptoms

This disease, also known as ashy stem rot, girdles the stem of young preblooming plants, resulting in a wilt followed by death. The plant sheds its leaves, and the stem usually remains erect, black except for the area near the soil line that is a lighter color. Black, minute, hard, irregular, imbedded sclerotia, measuring $75–190 \times 60–150\ \mu$, speckle the cortex and wood.

Etiology

The pycnidia, not often observed, are black, ostiolate, up to 150 μ in diameter, and contain conidia that are hyaline, 1-celled, oval to elongate, and 17–32×5–8 μ.

References

3, 12.

Leaf spot, *Alternaria solani* (Ell. & Martin) Jones & Grout

Symptoms

This disease, also known as early blight, is associated with the seedling stage of the plants and causes defoliation. On older plants, the spots are usually scattered, circular, up to 8–10 mm in diameter, frequently zonate, with a light center surrounded by a wide, black band and a slight yellow halo.

Etiology

The spores of the fungus appear on either surface on short conidiophores. The conidia are dark, muriform, with 5–12 cells and 3–7 longitudinal septations. The spores are variable, measuring 90–140×12–16 μ.

References

12, 13.

Leaf spot, *Stemphylium floridanum* Hannon & Weber

Symptoms

Very small, water-soaked spots appear on the leaves. They soon become black as the central area is killed and later bleach to almost white, retaining the black border. The spots are seldom more than 1 mm in diameter, appear on both surfaces, and are mostly sunken.

Etiology

The hyphae and conidiophores are dark-colored, and the conidia are dark, muriform septate, 5–8 celled, restricted, and 19–63×7–23 μ.

References

1, 3, 5, 12.

Leaf spot, *Cercospora capsici* Heald & Wolf

Symptoms

Leaf spot is conspicuous and destructive, producing large, scattered, circular, leaf lesions, often measuring 1 cm in diameter. Spots are almost white in the centers, surrounded by a slate green border. The center areas are sometimes finely zonate on the upper surface. Leaves may be heavily diseased before shedding takes place. The fungus causes spots on the leaf petioles, peduncles, and calyx, but very seldom on the pods.

Figure 142. Bacterial spot, *Xanthomonas vesicatoria*, on red pepper pod.

Figure 143. Blight, *Phytophthora capsici*, on red pepper pod, stem infection.

Figure 144. Leaf spot, *Stemphylium floridanum*, on red pepper. (Photograph by C. I. Hannon.)

Figure 145. Leaf spot, *Cercospora capsici*, on pepper.

Figure 146. Wilt, *Cephalosporium diospyri*, on persimmon tree. (Photograph by A. S. Rhoads.)

Figure 147. Sporophore of *Polyporus rhoadsii* found on killed persimmon tree.

Etiology

The fungus produces conidia in abundance on the upper surfaces. The septate and olivaceous brown conidiophores arise in fascicles from superficial stroma of 4–10 in a cluster. The conidia are hyaline, acicular, straight or curved, septate, and 30–200 × 2–4 μ.

References

2, 3, 12, 13.

Leaf spot, *Cercospora unamunoi* Castell.

Symptoms

This disease causes a yellowing of the foliage but seldom produces a definite lesion. The diseased areas are blotches, lacking definite margins. The leaf tissue is ordinarily not killed.

Etiology

The fungus sporulates profusely on the lower surface, corresponding with the blotches that may be small or cover the entire lower leaf surface with an olivaceous to almost black, compact, moldy growth. Conidia are 15–100 × 3–7 μ.

References

2, 3, 7.

Fruit spot, *Bipolaris tetrameral* (McK.) Schoen

Symptoms

Spots develop on the pods with very little deep penetration. They are gray to reddish brown, oval to irregular, with a definite margin and slightly depressed centers. Sporulation occurs on the surface of these scattered spots.

Etiology

The mycelium is olive brown, septate, branched, and 4–9 μ wide. The conidiophores are brown, septate, rather stout, and produce clusters of conidia. The conidia are brown, 4-celled, cylindrical with rounded ends, and 22–38 × 9–13 μ. The conidia germinate by forming mycelium.

Reference

9.

Wilt, *Fusarium oxysporum* f. sp. *vasinfectum* (Atk.) Snyd. & Hans.

Symptoms

A slight drooping of the lower leaves is followed by their becoming limp, wilted, and finally brown. On the lower stem, some sunken areas may be found where the main lateral roots join the central stem. The stem may become cankerous, and several such lesions cause girdling and a quick killing of the plant. The vascular tissue is often darkened.

Etiology

The fungus produces hyaline, small, 1-celled microconidia, measuring 4–10 $\times$ 2–5 μ; hyaline, large, sickle-shaped, 2–6-celled macroconidia, measuring 27–40 $\times$ 3–6 μ; and thick-walled, 1-celled, oval or oblong chlamydospores, measuring 10–12 μ in diameter.

Reference

3.

Wilt, *Verticillium albo-atrum* Reinke & Berth.

Symptoms

Wilt generally results from the disorganization of the functions of the vascular tissue. The fungus invades the roots and the main stem, where tissue is killed and darkened. Stem cankers appear at the soil line or rather at the moisture line, and the plant succumbs. The plant first shows some wilting, stunting, and an off-color as the only aboveground symptoms.

Etiology

The mycelium is hyaline, branched, septate, and produces hyaline, oval, 1-celled conidia on whorls of branches from the main hyphae. The spores measure 3–12 $\times$ 1.5–3 μ.

References

3, 12.

Other fungi associated with red pepper

Ascochyta capsici Bondartzeva-Montverde
Botrytis cinerea Persoon ex Fries
Brachysporium capsici Hiroë & Watson
Cladosporium herbarum (Persoon ex Fries) Link
Curvularia lunata (Wakker) Boedijn
Diaporthe phaseolorum (Cooke & Ellis) Saccardo
Helminthosporium curvulum Saccardo
Meliola capsicola F. L. Stevens
Microdiplodia capsici Sarejanni
Oidiopsis capsici Sawada
Phymatotrichum omnivorum (Shear) Duggar
Puccinia capsicola Kern
Puccinia paulensis Rangel
Stemphylium solani Weber

References: red pepper

1. Braverman, S. W. 1968. A new leaf spot of pepper incited by Stemphylium botryosum f. sp. capsicum. Phytopathology 58:1164–67.
2. Chupp, C. 1953. A monograph of the fungus genus Cercospora. Published by the author: Ithaca, N.Y.
3. Chupp, C., and A. F. Sherf. 1960. Vegetable diseases and their control. Ronald Press: N.Y., pp. 454–68.
4. Elliott, C. 1951. Manual of bacterial plant pathogens. 2d ed. Chronica Botanica: Waltham, Mass.
5. Hannon, C. I., and G. F. Weber. 1955. A leaf spot of tomato caused by Stemphylium floridanum sp. nov. Phytopathology 45:11–16.
6. Matthews, V. D. 1931. Studies on the genus Pythium. University of North Carolina Press: Chapel Hill, pp. 1–135.
7. Muntañola, M. 1954. A study of a newly identified pepper disease in the Americas. Phytopathology 44:233–39.
8. Ramsey, G. B., J. S. Wiant, and L. P. McColloch. 1952. Market diseases of tomatoes, peppers and eggplants. U.S. Dept. Agr., Agr. Handbook 28:42–47.
9. Rao, V. G. 1967. A new storage disease of chillies (Capsicum annuum). Phytopathol. Z. 58:277–80.
10. Snyder, W. C., and B. A. Rudolph. 1939. Verticillium wilt of pepper, Capsicum annuum. Phytopathology 29: 359–62.
11. Tompkin, C. M., and C. M. Tucker. 1941. Root rot of pepper and pumpkin caused by Phytophthora capsici. J. Agr. Res. 53:417–26.
12. Walker, J. C. 1952. Diseases of vegetable crops. McGraw-Hill: N.Y., pp. 297–313.
13. Weber, G. F. 1932. Diseases of peppers in Florida. Florida Univ. Agr. Expt. Sta. Bull. 244:5–46.
14. Weber, G. F., and F. A. Wolf. 1927. Heterothallism in Blakeslea trispora. Mycologia 19:302–7.

Persimmon, *Diospyros kaki* Linnaeus

Wilt, *Cephalosporium diospyri* Crandall
Other fungi associated with persimmon

Wilt, *Cephalosporium diospyri* Crandall

Symptoms

The host develops a vascular wilt, causing discolored foliage to appear more or less suddenly in any branch or the tops of trees. This condition is often followed by extensive wilting, shedding of variously colored leaves, and finally the death of the tree. Internally the current wood ring and sometimes older rings show dark brown to black streaks. This streaking is also evident in the cambial area of branches and twigs. As the bark loosens on dead trees, orange or pink masses of spores of the fungus may be found.

Etiology

In culture the mycelium is hyaline, vesiculose, branched, septate, and 3–8 $\times$ 6–7 μ in diameter. It may be more or less wet and flat or pinkish white and fluffy. Often the aerial mycelium appears in clumps or in ropelike strands. The conidiophores are hyaline, nonseptate, tapering, and 1–5 $\times$ 3–5 μ wide. Conidia are hyaline, orange pink en masse, continuous, ovate to cylindrical, and 2.7–11.7 $\times$ 1.8–5.4 μ, averaging 2.7 $\times$ 4.5 μ. They are produced acrogenously and form globose heads which often adhere to others, forming a conglomerate spore mass. The spores are spread in water or less easily under dry conditions.

References

3, 4.

Other fungi associated with persimmon

Armillaria mellea Vahl ex Fries
Botryosphaeria ribis (Tode ex Fries) Grossenbacher & Duggar
Capnodium citri Berkeley & Desmazieres
Ceptothyrium pomi (Montagne & Fries) Saccardo
Ceratobasidium stevensii (Burt) Venkatarayan
Ceratostomella pilifera (Fries) Winter
Cercospora kaki Ellis & Everhart
Clitocybe tabescens (Scopoli) Bresadola
Gloeosporium diospyri Ellis & Everhart
Isariopsis linderae (Ellis & Everhart) Saccardo
Lasiosphaeria pezicula (Berkeley & Curtis) Saccardo
Monochaetia diospyri Yoshii
Pestalotia diospyri Sydow
Phoma diospyri Saccardo
Phyllosticta biformis Heald & Wolf
Physalospora rhodina (Berkeley & Curtis) Cooke
Phytophthora cactorum (Lebert & Cohn) Schroeter
Phytophthora citrophthora (Smith & Smith) Leonian
Polyporus rhoadsii Murrill
Polyporus spraguei Berkeley & Curtis
Verticillium albo-atrum Reinke & Berthold

References: persimmon

1. Camp, A. F., and H. Mowry. 1929. The Japanese persimmon in Florida. Florida Univ. Agr. Expt. Sta. Bull. 205:527–62.
2. Chupp, C. 1953. A monograph of the fungus genus Cercospora. Published by the author: Ithaca, N.Y.
3. Crandall, B. S. 1945. A new species of Cephalosporium causing persimmon wilt. Mycologia 37:495–98.
4. Crandall, B. S., and W. L. Baker. 1950. The wilt disease of American persimmon, caused by Cephalosporium diospyri. Phytopathology 40:307–25.
5. Fletcher, W. F. 1942. The native persimmon. U.S. Dept. Agr. Farmers' Bull. (Rev.) 685.1–22.
6. Hubert, E. E. 1921. Notes on sap stain fungi. Phytopathology 11:214–24.
7. Hume, H. H., and F. C. Reimer. 1904. Japanese persimmons. Florida Univ. Agr. Expt. Sta. Bull. 71:69–105.
8. Smith, C. O. 1937. Crown gall on incense cedar, Libocedrus decurrens. Phytopathology 27:844–49.
9. Wolf, F. A. 1928. Further observations on Corticium koleroga (Cke.) v. Höhn. Phytopathology 18:147–48. (Abstr.)

Pineapple, *Ananas comosus* (Linnaeus) Merrill

Fruitlet black rot, *Pseudomonas ananas* Serrano
Fruitlet brown rot, *Erwinia ananas* Serrano
Bacterial heart rot, *Erwinia carotovora* (Jones) Holland
Wilt, *Rhizidiocystis ananasi* Sideris
Heart rot, *Phytophthora parasitica* Dastur

Root rot wilt, *Phytophthora cinnamomi* Rands
Water blister, *Ceratocystis paradoxa* (Dade) Moreau
Fruit rot, *Diplodia natalensis* Evans
Black mold, *Aspergillus niger* Tieghem
Fruitlet core rot, *Penicillium funiculosum* Thom
Fruitlet core rot, *Fusarium moniliforme* Sheldon
White leaf spot, *Thielaviopsis paradoxa* (De Seynes) Hoehnel
Other fungi associated with pineapple

Fruitlet black rot, *Pseudomonas ananas* Serrano

Symptoms

Externally the pineapple fruit appears natural. A cross section through the middle part of the fruit shows various fruitlets with dark brown to black spots. A portion of all of the fruitlets may show some discoloration. The discoloration appears when the ripening stage begins but does not continue in storage. It is difficult to diagnose without cutting the fruit.

Etiology

The organism is a gram-negative rod, single or in short chains, produces no capsules, is motile by 1–4 polar flagella, and measures $0.6 \times 1.8\ \mu$. On beef extract agar, colonies are white, glistening, smooth, wet, and have entire or lobate margins. A green fluorescent pigment is produced in culture. Optimum temperature, 31–33C.

References

2, 16.

Fruitlet brown rot, *Erwinia ananas* Serrano

Symptoms

This disease usually shows no external symptoms and is only detected by cutting the fruit. It develops in the ripening period and is indicated by a definite dull color, frequently purplish dots, some green mottling, and an unnatural hardness. When cut, the fruitlets show various stages of faintly brown to dark brown discoloration. The entire fruitlet may eventually be blackened, and the tissue becomes dry and hard. Infection takes place through the floral stamens, extending to the placenta in the eye cavity.

Etiology

The organism is gram-negative, single or in short chains, produces capsules, is motile by 4–8 peritrichous flagella, and measures $0.6 \times 0.9\ \mu$. On potato glucose agar, colonies are yellow, glistening, usually convex, and circular with a spreading tendency. Optimum temperature, 30–35C.

References

3, 15, 18.

Bacterial heart rot, *Erwinia carotovora* (Jones) Holland

Symptoms

The basal parts of the youngest leaves assume a water-soaked, dull green color, developing into a yellowish area frequently bounded by a brown or purple band. Further progress results in a core decay affecting the leaf bases. Bearing plants may become diseased. The decay usually involves the growing point and the plant dies, or, in certain instances, some healthy sprouts are produced. A collapse of ripening fruits in a very few days following infection results in a soft, oozing, odoriferous mass and total disintegration.

Etiology

The organism is a gram-negative rod, forms in chains, produces no capsules, is motile by 2–5 peritrichous flagella, and measures 1.5–5 × 0.6–0.9 μ. Agar colonies are dirty white, glistening, raised, entire, circular, and smooth.

References

5, 6.

Wilt, *Rhizidiocystis ananasi* Sideris

Symptoms

Diseased plants show decline, resulting from malfunction of the roots. The fungus may destroy all of the root hairs as rapidly as they are formed. The main root tissue ceases to function, slowly becomes dry, and dies. This results in the wilt symptom. However, the fungus must be observed for a correct diagnosis. The parasite is restricted to pineapple roots in Hawaii.

Etiology

The mycelium is white and evanescent, the hyphae are 1 μ in diameter and extend from one root hair to another. Turbinate cells produce hyphal branches. Kidney-shaped sporangia, measuring 16 × 8 μ, are produced on the hyphae and become attached to the root hairs. They discharge their contents into the root hair in a plasmodic state. There the plasmodia kill the host cell and produce smooth or echinulate hypnospores that are about 20 μ in diameter.

Reference

17.

Heart rot, *Phytophthora parasitica* Dast.

Symptoms

This disease results in rotting of the central tissues near the base of the plant. Young leaves are easily detached by pulling. They exhibit a soft, brownish base at point of attachment. Diseased plants become yellow in advanced stages of the root and stem decay.

Etiology

Mycelium in culture is both aerial and submerged, measuring 3–6 μ in diameter; the submerged, coarser chlamydospores are yellowish, spherical, granular, and germinate directly. They average 27 μ in diameter, varying from 10 to 44 μ. Conidia or sporangia, depending on the manner of germination, are lemon-shaped with prominent papillae borne on long conidiophores. They are yellowish brown, thin-walled, granular, and measure 13×54 μ, averaging 28×35 μ. Upon germination they often produce secondary fruiting structures, mycelium, or zoospores. Zoospores, produced 15–50 per sporangium, are amoeboid to spherical and germinate by developing a germ tube. Oospores are yellow brown, spherical, thick-walled, and 19–26 μ, averaging 23 μ.

References

8, 12.

Root rot wilt, *Phytophthora cinnamomi* Rands

Symptoms

Root rot causes a yellow brown or reddish color of the heart leaves on bearing plants after flowering. The leaves lose their turgidity, droop, and are easily pulled out. The base of the plant is often water-soaked, brown, and distinctly demarked from healthy tissue. The fruit colors prematurely. The root system is completely invaded and mostly destroyed, and plants are not firmly anchored. The development of the symptoms is variable, depending often upon the environmental conditions.

Etiology

The hyphae are hyaline, branched, nonseptate, and 4–6.5 μ in diameter. Chlamydosporelike structures are mostly spherical, measuring 29 μ in diameter. The sporangia measure 55×28 μ and upon germination produce zoospores that measure 8–14 μ and germinate to form mycelium.

References

8, 10.

Water blister, *Ceratocystis paradoxa* (Dade) Moreau

Symptoms

A soft rot develops in fruit, exhibiting no external indication except that the skin appears to be wet and water-soaked until the decay is well advanced. The decay develops from wounds, particularly common on the fruit after they are removed from the plant. It invades the fruit from the cut stem end and advances rapidly through the core, reducing the entire fruit to a soft, watery mass with a characteristic sweet odor. Base rot and white leaf spot are symptoms resulting from infection by this fungus.

Etiology

The mycelium is hyaline, septate, branched, and abundant. Conidia are produced endogenously by conidiophores that are myceliumlike in structure, septation, and diameter. The conidia are cut off at the terminal points and emerge one after another, often in chains. They are hyaline, 1-celled, cylindric at first but may become oval, and $10–15 \times 3–5\ \mu$. The second type of conidia are produced singly or in short chains; they are black, thick-walled, oval, and $10–18 \times 7–10\ \mu$. They are produced in abundance and appear as a sooty coat over the well-developed, diseased areas. Perithecia are black, submerged, spherical, long-necked, and ostiolate; hyphae are hyaline, and ascospores are hyaline, elongate, in a matrix, and $7–10 \times 2–4\ \mu$. The vegetative stage is *Thielaviopsis paradoxa.*

References

2, 7, 10.

Fruit rot, *Diplodia natalensis* Evans

Symptoms

The decay produced by this wound parasite is distinguished with difficulty from other diseases of this fruit. The texture of the diseased area is rather firm and has a water-soaked appearance, but there is little or no wetness evident. The fungus usually sporulates, forming black pycnidia slightly imbedded in the carbonous mycelial covering of diseased parts.

Etiology

The mycelium is dirty white to black. The pycnidia are black, single or clustered, submerged at first, then erumpent, ostiolate, and $140–175\ \mu$ in diameter. Pycnidiospores are hyaline when young, later colored, oval to elliptical, 1-celled in the hyaline stage, and 1-septate in the colored stage, often finely striated, and $26 \times 14\ \mu$. The ascospore stage is *Physalospora rhodina.*

Reference

14.

Black mold, *Aspergillus niger* v. Tiegh.

Symptoms

This fungus is worldwide in distribution and develops as a wound parasite. Infection is usually confined to ripe fruit at harvest time or in handling. The mycelium penetrates the host, causing a soft, colorless decay except for the areas that support the black sporulating structures, which are superficial and sooty in appearance.

Etiology

The conidia are produced in more or less spherical heads supported by hyaline, upright conidiophores. The heads with conidia in place average about

300–500 μ in diameter but vary greatly. The terminal swelling may be 100 μ in diameter with many primary penicilli mostly 20 μ long, each of which develops 2–3 secondary sterigmata which bear the spores in chains. The conidia are black, finely echinulate, spherical, 1-celled, and 2.5–4 μ in diameter.

References

9, 14, 19.

Fruitlet core rot, *Penicillium funiculosum* Thom

Symptoms

There is often no external symptom unless certain of the fruitlets fail to color as the fruit matures. Some badly affected eyes may become brown and sunken. Cutting the fruit crosswise reveals the darkening of the fruitlets. The dark areas appear near the margins, converging toward the center, and often involve the core. The fungus develops from mere specks to cover several fruitlets and enters through injuries of the cavity lining.

Etiology

The fungus grows on the remnants of decaying floral parts. The mycelium develops ropes of hyphae that are shades of red, yellow, and green, sporulating irregularly. Conidiophores arise at right angles to the funicular hyphae and range in size from 100 to 300 × 2.5 to 3 μ. Penicilli are biverticillate, symmetrical, with 5–8 metulae and 5–7 sterigmata. Conidia are smooth, elliptical to subglobose, and 2.5–3.5 × 2–2.5 μ.

References

4, 9, 11, 13.

Fruitlet core rot, *Fusarium moniliforme* Sheldon

Symptoms

External symptoms are seldom expressed before the fruit is ripening, and then only certain of the fruitlets fail to color. Sectioning the fruit reveals the brown to black decay that is usually scattered among the fruitlets as each becomes infected individually. The diseased areas are somewhat cone-shaped, wide at the periphery, converging toward the core.

Etiology

The hyphae are hyaline, septate, and branched. The conidiophores are loose sporodochia, producing macroconidia sparingly. Macroconidia are 3–5-septate, curved, and 40–60 × 3–4 μ. Microconidia are hyaline, mostly 1-celled, oval to elongate, produced in heads, and measure 6–10 × 2 μ. The perithecia are black, smooth, superficial, and globose. The 8-spored asci are clavate with the ascospores arranged biseriately. The ascospores are 1–3-septate, straight or curved, and taper toward the tip. The ascospore stage is *Giberella fujikuroi*.

References

11, 20.

White leaf spot, *Thielaviopsis paradoxa* (De Seyn.) Hoehn.

Symptoms

Small, brown to yellowish spots appear on leaves that may be broken over or injured. The color gradually fades and up to several inches of the leaf become involved, reaching the tip and across the blade, causing a girdling effect and a withering of the part beyond. The affected areas become straw-colored to almost white and of a papery consistency.

Etiology

The fungus does not penetrate sound surfaces. The mycelium is hyaline to gray, septate, branched, and abundant. The conidiophores develop conidia endogenously one after another. They are hyaline, cylindrical, and 8–12 $\times$ 3.5–5 μ. A second conidial spore is produced acrogenously in short chains by hyaline conidiophores arising from the mycelium. These spores are olive to dark, 1-celled, oval, thick-walled, and 14–17 $\times$ 11 μ. The ascospore stage of the fungus is *Ceratocystis paradoxa*.

Reference

9.

Other fungi associated with pineapple

Asterinella stuhlmanni (P. Henning) Theissen
Colletotrichum ananas Garud
Nematosporangium rhizophthoron Sideris
Pseudopythium phytophthoron Sideris & Paxton
Pythium butleri Subramaniam
Wallrothiella bromeliae Rehm

References: pineapple

1. Barker, H. D. 1926. Fruitlet black rot disease of pineapple. Phytopathology 16:359–63.
2. Bratley, C. O., and A. S. Mason. 1939. Control of black rot of pineapples in transit. U.S. Dept. Agr. Circ. 511:1–12.
3. Elliott, C. 1951. Manual of bacterial plant pathogens. 2d ed. Chronica Botanica: Waltham, Mass.
4. Hepton, A., and E. J. Anderson. 1968. Interfruitlet corking of pineapple fruit, a new disease in Hawaii. Phytopathology 58:74–78.
5. Johnston, A. 1957. A bacterial heart rot of the pineapple. Malayan Agr. J. 40:2–8.
6. Johnston, A. 1957. Pineapple fruit collapse. Malayan Agr. J. 40:253–63.
7. Lewcock, H. K. 1945. Softrot (water blister) disease of pineapples. Queensland Agr. J. 60:42–45.
8. Mehrlich, F. P. 1936. Pathogenicity and variation in Phytophthora species causing heart rot of pineapple plants. Phytopathology 26:23–43.
9. Oxenham, B. L. 1953. Notes on two pineapple diseases in Queensland. Queensland J. Agr. Sci. 10:237–45.
10. Oxenham, B. L. 1957. Diseases of the pineapple. Queensland Agr. J. 83:13–26.
11. Oxenham, B. L. 1962. Etiology of fruitlet core rot of pineapple in Queensland. Queensland J. Agr. Sci. 19:27–31.
12. Quebral, F. C., et al. 1962. Heart rot of pineapple in the Philippines. Philippine Agriculturist 46:432–50.
13. Raper, K. B., and C. Thom. 1949. A

manual of the Penicillia. Williams & Wilkins: Baltimore, Md.
14. Roger, L. 1953. Phytopathologie des pays chauds. Encyclopedie Mycologique 18:1568–71, 1743–57.
15. Serrano, F. B. 1928. Bacterial fruitlet brownrot of pineapple in the Philippines. Philippine J. Sci. 36:271–300.
16. Serrano, F. B. 1934. Fruitlet blackrot of pineapple in the Philippines. Philippine J. Sci. 55:337–59.
17. Sideris, C. P. 1929. Rhizidiocystis ananasi Sideris, nov. gen. & sp., a root hair parasite of pineapples. Phytopathology 19:367–82.
18. Smith, M. A., and G. B. Ramsey. 1950. Bacterial fruitlet brownrot of Mexican pineapple. Phytopathology 40:1132–35.
19. Thom, C., and K. B. Raper. 1945. A manual of the Aspergilli. Williams & Wilkins: Baltimore, Md.
20. Wineland, G. O. 1924. An ascigerous stage and symptoms for Fusarium moniliforme. J. Agr. Sci. 28:909–22.

Pomegranate, *Punica granatum* Linnaeus

Bacterial leaf spot, *Xanthomonas punicae* Hingorani & Singh
Dry rot, *Nematospora coryli* Peglion
Fruit rot, *Coniella granati* (Saccardo) Petrak & Sydow
Blotch, *Mycosphaerella lythracearum* Wolf
Other fungi associated with pomegranate

Bacterial leaf spot, *Xanthomonas punicae* Hing. & Singh

Symptoms

Small, irregular, water-soaked spots that measure up to 5 mm appear on the leaves. These appear translucent when viewed with transmitted light and become yellow to tan and finally brown surrounded by a wet margin. When plentiful they may coalesce, causing the leaf to become yellow and to shed. An exudate associated with the leaf spots may be observed during humid periods.

Etiology

The organism is a rod, with rounded ends, developing singly, in pairs, or in chains, motile by a single polar flagellum, producing capsules, and measuring $1–2.5 \times 0.5\ \mu$. Agar colonies are colorless to pale yellow, circular, raised, wet, shiny, entire, and are pathogenic only on pomegranate.

Reference

2.

Dry rot, *Nematospora coryli* Pegl.

Symptoms

It is generally known that this yeast is associated with mechanical punctures produced by certain plant bugs as they feed. The organism enters the host plants through siphonaceous activities of the insect and develops in the tender juicy contents of the fruit. The fungus breaks down the cell walls, freeing the liquids; the immediate host tissues collapse, shrivel, dry, and become pale brown. As the tissue dries, it causes an external depression of various dimensions and of a dark brown color. Severe infestations may cause loss of the fruit.

Etiology

The vegetative cells are hyaline, spherical, 1-celled at maturity, and about 20 μ in diameter. They increase by budding, and some mycelial hyphae may develop. The asci are hyaline, cylindrical, and contain 8 hyaline, elongate, 2-celled, appendaged ascospores that measure 60–85 × 10–12.4 μ.

Reference

3.

Fruit rot, *Coniella granati* (Sacc.) Petr. & Syd.

Symptoms

A dark, soft to leathery rot of the fruit penetrates into the interior and causes a conspicuous shrinking and undulation of the skin. The fruit become involved rapidly, and pycnidia are produced over the outer surface.

Etiology

The brownish fruiting structures, usually scattered, produce hyaline pycnidiospores that appear brownish en masse, are 1-celled, elongate, and measure 12–19 × 2–5 μ. This stage has been known as *Phoma granati*.

Reference

1.

Blotch, *Mycosphaerella lythracearum* Wolf

Symptoms

Circular to partially vein-limited brown spots up to 5 mm in diameter appear on the leaves. Such diseased leaves are usually pale green, show yellowish streaks, and are shed prematurely. Fruiting structures of the fungus occur on the lower surfaces of the spots. Small, dark brown spots, developing on the maturing fruits, are somewhat circular at first but soon become angular and extensive, often covering large portions of the surface.

Etiology

Dense clumps of spores develop under favorable moisture and temperature conditions on the surface of the leaf spots and less frequently on the fruit. They measure 20–30 × 3 μ and are slightly colored. The conidia are hyaline, filiform, clavate, up to 5-septate, and 20–56 × 3–4 μ. The spermagonial stage develops in the leaf lesions. Spermagonia are hyaline, rod-shaped, and 3–5 × 1–2 μ. The perithecia are black and develop in the leaf spot tissue, usually on fallen leaves. They are aparaphysate, globose, 75–95 μ in diameter, and immersed becoming erumpent. Asci are cylindrical, measuring 42–50 × 6–8 μ. The ascospores are hyaline, biseriately arranged, curved, unequal, 2-celled, and 11–14 × 2–4 μ. The conidial stage is *Cercospora lythracearum*.

Reference

4.

Other fungi associated with pomegranate

Aspergillus niger van Tieghem
Botrytis cinerea Persoon ex Fries
Ceratobasidium stevensii (Burt) Venkatarayan
Clitocybe tabescens (Scopoli ex Fries) Bresadola
Colletotrichum gloeosporioides Penzig
Curvularia pallescens Boedijn
Diplodia natalensis Evans
Dothiorella sanninii Ciferri
Ganoderma lucidum (Leysser) Karsten
Nectriella versoniana Saccardo & Penzig
Nigrospora oryzae (Berkeley & Broome) Petch
Penicillium expansum Link & Thomas
Sphaceloma punicsi Bitancourt & Jenkins

References: pomegranate

1. Hebert, T. T., and C. N. Clayton. 1963. Pomegranate fruit rot caused by Coniella granati. Plant Disease Reptr. 47: 222–23.
2. Hingorani, M. K., and N. J. Singh. 1959. Xanthomonas punicae sp. nov., on Punica granatum L. Indian J. Agr. Sci. 29: 45–48.
3. Roger, L. 1953. Phytopathologie des pays chauds. Encyclopedie Mycologique 18: 1150–59.
4. Wolf, F. A. 1927. Pomegranate blotch. J. Agr. Res. 35:465–69.

Potato, *Solanum tuberosum* Linnaeus

Slimy soft rot, *Erwinia aroideae* (Townsend) Holland
Soft rot, *Erwinia carotovora* (L. R. Jones) Holland
Black leg, *Erwinia astroseptica* (van Hall) Jennison
Brown rot, *Pseudomonas solanacearum* E. F. Smith
Ring rot, *Corynebacterium sepedonicum* (Spieckermann & Kotthoff) Skaptason & Burkholder
Scab, *Streptomyces scabies* (Thaxter) Waksman & Henrici
Powdery scab, *Spongospora subterranea* (Wallroth) Lagerheim
Wart, *Synchytrium endobioticum* (Schilbersky) Percival
Leak, *Pythium debaryanum* Hesse
Tuber rot, *Phytophthora drechsleri* Tucker
Pink rot, *Phytophthora erythroseptica* Pethybridge
Late blight, *Phytophthora infestans* (Montagne) De Bary
Soft rot, *Rhizopus nigricans* Ehrenberg
Powdery mildew, *Erysiphe cichoracearum* De Candolle
Cottony rot, *Sclerotinia sclerotiorum* (Libert) De Bary
Dry rot, *Xylaria apiculata* Cooke
Smut, *Polysaccopsis hieronymi* (Schroeter) P. Henning
Smut, *Thecaphora solani* Barrus
Rust, *Puccinia pittieriana* P. Henning
Black scurf, *Thanatephorus cucumeris* (Frank) Donk
Southern blight, *Botryobasidium rolfsii* (Saccardo) Venkatarayan
Violet root rot, *Helicobasidium purpureum* (Tulasne) Patouillard

Shoestring rot, *Armillaria mellea* Vahl ex Fries
Dry rot, *Phoma tuberosa* Melhus, Rosenbaum, & Schultz
Hard rot, *Phomopsis tuberivora* Güssow & W. R. Foster
Anthracnose, *Colletotrichum atramentarium* (Berkeley & Broome) Taubenhaus
Gray mold, *Botrytis cinerea* Persoon ex Fries
Silver scurf, *Spondylocladium atrovirens* Harzer
Skin spot, *Oospora pustulans* Owen & Wakefield
Charcoal rot, *Macrophomina phaseolina* (Tassi) Goidanich
Leaf blotch, *Cercospora concors* (Caspary) Saccardo
Early blight, *Alternaria solani* (Ellis & Martin) Jones & Grout
Wilt, *Verticillium albo-atrum* Reinke & Berthold
Wilt, *Fusarium solani* f. *eumartii* (Carpenter) Wollenweber
Tuber rot, *Fusarium solani* f. *radicicola* (Wollenweber) Snyder & Hansen
Other fungi associated with potato

Slimy soft rot, *Erwinia aroideae* (Towns.) Holland

Symptoms

Slimy rot is usually considered a secondary decay following mechanical injury. It is not important except in transit and storage.

Etiology

The organism is a gram-negative rod, produces no capsules, is motile by 2–8 peritrichous flagella, and measures $2–3 \times 0.5\ \mu$. Agar colonies are white, glistening, partially opalescent, and circular to irregular.

Reference

17.

Soft rot, *Erwinia carotovora* (L. R. Jones) Holland

Symptoms

Soft rot follows some mechanical injury to the tubers in the field, in transit, or in storage. It is distinguishable only through laboratory culturing.

Etiology

The organism is a gram-negative rod, forms chains, produces no capsules, is motile by 2–5 peritrichous flagella, and measures $1.5–5 \times 0.6–0.9\ \mu$. Agar colonies are dirty white, glistening, smooth, entire, circular, and raised. Optimum temperature, 27C.

References

7, 8, 17.

Black leg, *Erwinia astroseptica* (van Hall) Jennison

Symptoms

Black leg occurs wherever the host plant is grown and is disseminated by vegetative seed. Seed pieces usually appear free of the organism but if overlooked and present, develop a slowly advancing decay. Young diseased plants become pale green to yellowish, and the older leaves on the main stem become yellow and may shed. The uppermost leaves tend to become erect and curl inward. The main stem darkens at the seed piece, and discoloration may extend up the stem several inches to above the soil line. Plants die before maturing.

Etiology

The organism is a gram-negative rod, in chains, forms capsules, is motile by peritrichous flagella, and measures 1–2 × 6–8 μ. Agar colonies are white, glistening, round, smooth, and butyrous. Optimum temperature, 25C.

References

7, 17.

Brown rot, *Pseudomonas solanacearum* E. F. Sm.

Symptoms

Brown rot is recognized in the field by the appearance of droopy terminal foliage that rapidly develops into a daily wilt followed by permanent wilt, stunt, fading of color, and death of the plant. The tubers may show slight discoloration of the vascular area at the point of union with the stolon, or the discoloration may extend completely through the vascular area and in extreme development cause an exudate from the tuber eyes sufficient to cause the soil to become wet and adhere. Such tubers continue to decay in the soil, in transit, and in storage.

Etiology

The bacterium is a gram-negative rod, motile by a single polar flagellum, produces no capsules, and measures 1.5 × 0.5 μ. Agar colonies are opalescent becoming darker, smooth, wet, shiny, small, and irregular. Optimum temperature, 35–37C.

References

8, 11, 19.

Ring rot, *Corynebacterium sepedonicum* (Spieck. & Kotth.) Skapt. & Burkh.

Symptoms

Early symptoms are pale yellow chlorosis, wilting, stunting, and marginal burning of the foliage. There is a general browning of the vascular tissue in the main stem and larger branches. In the tubers the extent of the disease is variable. Early infection may not be easily detected, as the discoloration is very faint, sometimes only a pale yellow. This gradually darkens to a chocolate brown centered in the vascular ring. The dark color spreads toward the epidermis and into the interior of the tuber, finally forming a slimy wet mass or, when dryer, a mass of chalky consistency. Often the central area is completely destroyed.

Etiology

The organism is gram-positive and may vary from coccus types to rods, globular bodies, and club-shaped cells. It is nonmotile and usually measures $0.8–1.2 \times 0.4–0.6\ \mu$. Agar colonies are white, translucent, glistening, thin, smooth, and small.

References

7, 17.

Scab, *Streptomyces scabies* (Thaxt.) Waks. & Henrici

Symptoms

Scab shows no prominent symptoms aboveground. Lesions of various sizes up to 1 cm in diameter and depths from superficial to several millimeters develop on the tubers. Infection occasionally occurs on some roots and stolons. On the tubers the lesions are brownish, usually scattered, slightly raised at first, and corky. They may develop superficially or cause deep pits containing loose, brown, corky tissue. Young tubers are more susceptible, and lesions do not continue to develop after maturity.

Etiology

The parasite produces hyaline mycelium that forms septations that separate into hyaline, 1-celled spores that measure $1–2 \times 0.6–0.7\ \mu$.

References

7, 8, 17.

Powdery scab, *Spongospora subterranea* (Wallr.) Lagh.

Symptoms

This scab produces no aboveground symptoms. On the tubers the lesions begin as very small, raised, brownish specks that may enlarge to several millimeters. At first spots are subepidermal; as they mature they become chocolate brown, and the epidermis is ruptured and curls back. The contents of the lesion consist of brown host-cell tissue intermingled with spore balls of the fungus.

Etiology

The fungus as a plasmodium enters the potato epidermis and migrates to the host cells where it expands to adjoining cells. There it forms zoosporangia and zoospores. Resting spores are formed in the host cells and disseminated into the soil. Upon germination, swarm spores are formed that measure $2–4\ \mu$ in diameter and produce infection.

References

7, 11, 17.

Wart, *Synchytrium endobioticum* (Schilb.) Perc.

Symptoms

Wart develops on all underground parts of the plant. Some secondary decline may occur. The tubers are most conspicuously attacked; they harbor a soft, warty, irregular, spongy series of protuberances that are variously attached, white to brownish black, of different sizes, and mostly in loose, roundish masses.

Etiology

The resting spores develop in the diseased tissue and are disseminated into the soil. Upon germination, they produce zoospores which penetrate the host epidermis, increase in size, and expand to adjoining cells. When mature, the thallus changes to form resting spores that are variable in size, spherical, and thick-walled.

References

7, 11, 17.

Leak, *Pythium debaryanum* Hesse

Symptoms

Leak, a wet, soft decay, results from infection through wounds of tubers following removal from the field and develops in transit and storage. The interior of the tuber is invaded and more or less dissolved.

Etiology

The fungus mycelium is hyaline, nonseptate, branched, and produces spherical to oval sporangia, measuring 15–26 μ in diameter. They germinate by mycelium or zoospore formation. Zoospores are biciliate, and measure 8–12 μ. Oospores are smooth, spherical, and 12–20 μ.

References

7, 11.

Tuber rot, *Phytophthora drechsleri* Tucker

Symptoms

This rot invades the roots of the young plants soon after the seed piece sprouts. It often causes root discoloration and continues into the lower portion of the main stem. The fungus advances from the main stem to the new tubers. The tubers become discolored at the stolon end, and the fungus proceeds to the bud end, causing a resilient softening best described as lack of turgidness. The outer tuber cortex becomes faintly colored, and, when cut open, the starchy white interior shows a pinkish color often deepening to rose and later black.

Etiology

The fungus mycelium is hyaline, branched, nonseptate, and produces under certain conditions nonpapillate, ovate to obpyriform sporangia, measuring 24–38 × 15–24 μ and averaging 31 × 21 μ. No chlamydospores develop, and oogonia are hyaline, spherical to clavate, smooth, and 21–54 μ in diameter. Oospores are hyaline to yellowish, spherical, smooth, and 16–45 μ in diameter. Antheridia are amphigynous.

Reference

16.

Pink rot, *Phytophthora erythroseptica* Pethyb.

Symptoms

Pink rot of tubers causes little or no aboveground symptoms except a mild form of wilt when the plant infection originates with the decaying seed piece. The tubers become infected through the stolon when in the soil and through the eyes afterward. They become dull brown in color and somewhat resilient in texture with the eyes being somewhat darker. The decay is pale white to pink as it progresses in the tuber.

Etiology

The fungus mycelium is hyaline, nonseptate, branched, and produces oval to lemon-shaped sporangia, measuring 16–30 × 24–40 μ, that germinate by zoospore production. Zoospores are 8–12 μ in diameter. The oospores contained in decayed tubers and other plant parts are spherical, measuring 20–38 μ in diameter.

References

11, 12, 18.

Late blight, *Phytophthora infestans* (Mont.) d By.

Symptoms

The disease appears on the foliage as water-soaked spots. They are up to 1/2 in. or more in diameter with a frosty band of hyphae around the margin of the spot on the lower surface. The band is composed of sporangiophores and sporangia of the fungus. Infection is spread to other leaves and stems. The disease is favored by cool weather. The tuber infection takes place in the soil by zoospores produced by the sporangia on the leaves. They show a dark vascular ring and therein the fungus survives until the following planting season. Planting of diseased seed pieces may cause the fungus to infect the new sprout and be carried above the soil surface; newly formed sporangia cause infection of the foliage.

Etiology

The mycelium is hyaline, branched, and nonseptate. The sporangia are hyaline, oval, papillate, and measure 12–24×21–40 μ. They germinate by forming 8–10 zoospores that measure 6–12 μ. The zoospores, in turn, cause infection. Oospores are rarely found.

References

6, 7, 8, 17, 19.

Soft rot, *Rhizopus nigricans* Ehr.

Symptoms

Soft rot develops in storage and transit on tubers mechanically injured after removal from the soil. It usually begins as a brown discoloration at the point of injury and gradually involves the entire tuber, which melts down in a soft, squashy mass usually overgrown by the mycelium and frequently producing many erect, black heads of conidia.

Etiology

Spores of the fungus are widely scattered. They are dark, 1-celled, oval, and 6×17 μ. Zygospores are dark brown, spherical, and 160×220 μ in diameter.

Reference

11.

Powdery mildew, *Erysiphe cichoracearum* DC.

Symptoms

Mildew is not a frequently observed disease of potato, although it affects many other hosts. The white mycelium develops on the upper leaf surface, small and webby at first, but more compact as the area enlarges to cover the leaves, petioles, and stems.

Etiology

Conidia, produced consecutively on the erect conidiophores, are hyaline, 1-celled, oval to barrel-shaped, and 30–35×16–19 μ. The fungus produces appendaged cleistothecia that contain asci and ascospores.

Reference

17.

Cottony rot, *Sclerotinia sclerotiorum* (Lib.) d By.

Symptoms

This disease usually appears under favorable moisture conditions when ascospore infection takes place in mechanically injured aboveground parts of the plant. Once within the plant, the fungus rapidly invades healthy tissue and causes foliage, petioles, and stems to soften and collapse. The fungus mycelium often grows on the outside of infected tissue in white, cottony wefts. The main

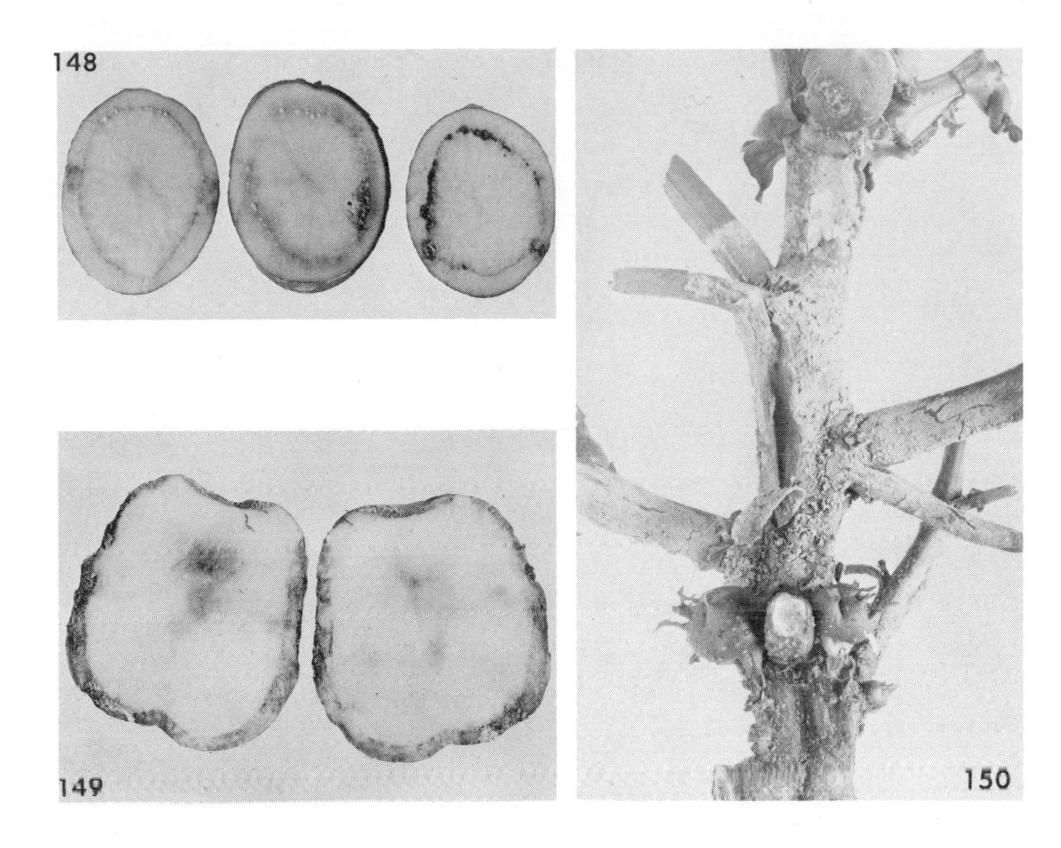

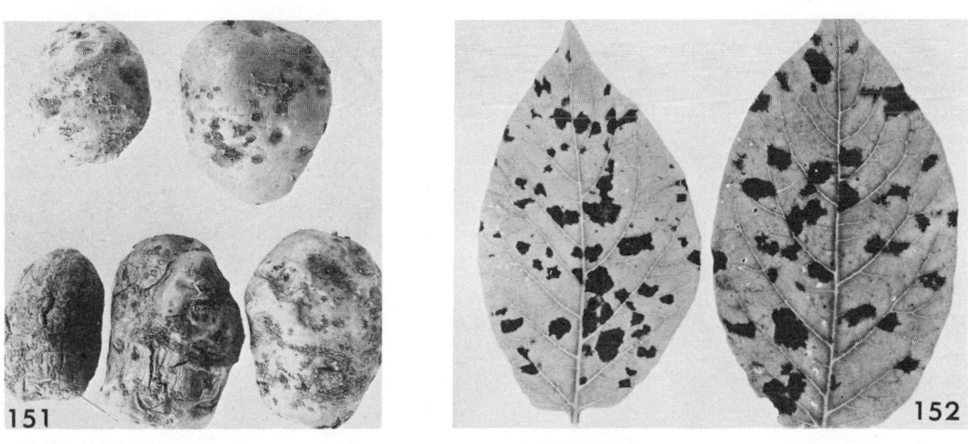

Figure 148. Brown rot, *Pseudomonas solanacearum,* oozes from vascular area of cut diseased potato.

Figure 149. Late blight, *Phytophthora infestans,* on potato, showing cortical invasion.

Figure 150. Basidial stage of *Thanatephorus cucumeris* on potato stem.

Figure 151. Powdery scab, *Spongospora subterranea,* on potato.

Figure 152. Early blight, *Alternaria solani,* on potato foliage. (Photograph by D. G. Kelbert.)

stem may become invaded at the soil line by the fungus. The tubers are seldom infected.

Etiology

The fungus forms black, irregularly shaped, shiny sclerotia, up to 1 cm long, on or within the diseased plant parts. These sclerotia enable the fungus to survive nongrowing periods. They germinate to form apothecia containing many asci and ascospores. The ascospores are hyaline, 1-celled, oval, and 11–15 × 5–8 μ. The microconidia are formed in chains. True conidia are not known.

References

8, 17.

Dry rot, *Xylaria apiculata* Cke.

Symptoms

This is a rather rare disease of tubers that develops on lands that have not been thoroughly cleared of tree roots prior to potato planting. The rhizomorphs of the fungus extending from infected woody tree roots come in contact with the tubers, and the mycelium invades the tuber, causing a light-colored, soft, cheesy decay, which rapidly involves the whole tuber.

Etiology

The mycelium is dark, branched, septate, and forms black to cream-colored stromata that may be branched and up to 8 cm × 3 mm. The conidia are light brown, 1-celled, oval to elliptical, and 6–7 × 2–3 μ. As the stromata mature, perithecia are formed containing uniseriate, 8-spored asci that contain brown, elliptical, 1- or 2-celled ascospores that measure 16–25 × 4–7 μ.

References

8, 14.

Smut, *Polysaccopsis hieronymi* (Schroet.) P. Henn.

Symptoms

This rare disease of the aerial parts of the plant involves leaves, stems, and berries.

Etiology

The sori, covered by a yellow membrane, may be 3–5 × 1–3 cm in size, including spore masses. The spore balls, globose to angular, include a central fertile spore and a peripheral layer of sterile, almost hyaline cells. The spores are mostly chestnut brown, smooth, and 22 × 26 μ.

Reference

3.

Smut, *Thecaphora solani* Barrus

Symptoms

This smut is characterized by warts or lumps on the surface of leaves and internal brown specks. Rusty brown spore balls occupy the cavities of these specks.

Etiology

The spore balls, consisting of 1–8 spores, are covered with verrucose markings on the exposed surfaces. Each spore may have a thin inner wall and a thick, rough outer wall. Spore balls are yellow to brown, ovoid, and 12–48 × 12–35 μ. The spores, measuring 7.5–20 × 8–18 μ, are dusty when dry.

References

2, 9.

Rust, *Puccinia pittieriana* P. Henn.

Symptoms

Rust causes some defoliation in Mexico and Peru. It is not widespread or frequently observed.

Etiology

Not studied, described, or published.

Reference

1.

Black scurf, *Thanatephorus cucumeris* (Frank) Donk

Symptoms

Black scurf is generally recognized on potato tubers by the black, very small, up to 1 cm wide, thin, irregularly shaped sclerotia adhering superficially to the tubers. Seed pieces contain the sclerotia, and as the potato sprout grows so does the fungus. Young sprouts are often diseased at the growing tips or along the sides and frequently account for barren hills in a field. Plants surviving initial injury may grow to the critical period, at which time the foliage yellows, and the leaves seem to be more upright and erect. Older plants may show axillary tuber formation aboveground because of inadequate translocation to the stolons. Stem cankers girdling the plant at near maturity result in premature killing.

Etiology

The fungus is soil-inhabiting but usually survives as sclerotia on the tubers. No conidia are produced. The basidial spore stage may develop in a thin, tan, mycelial mat or net on the lower areas of the main stem and branches. The basidiospores are ellipsoid to oblong and 7–13 × 4–7 μ.

References

7, 8, 17, 19.

Southern blight, *Botryobasidium rolfsii* (Sacc.) Venkat.

Symptoms

This fungus most often causes a girdling of the main stem; as the lesion develops the stems and foliage become yellow, stunted, wilted, and finally die. The tubers sometimes rot in the field when too close to the main stem and shallowly located.

Etiology

The fungus produces white mycelium that surrounds the stem at the soil line and gives rise to brown, spherical sclerotia, measuring 1 mm or less in diameter. No conidia are produced. Basidiospores, seldom observed, are hyaline, 1-celled, oval to elongate, and $5–10 \times 3–6\ \mu$. The imperfect fungus is *Sclerotium rolfsii.*

References

7, 8, 19.

Violet root rot, *Helicobasidium purpureum* (Tul.) Pat.

Symptoms

This underground disease causes the aerial parts to appear stunted, chlorotic, frequently wilted and brown, and dead. The tubers are attacked and become surrounded by and covered with a thick mycelial mat that acts as a cushion for the penetrating, absorbing hyphae.

Etiology

The mycelium is purple to tan brown, often thickened strands, interwoven, septate, and branched. Aggregates of mycelium form microsclerotia. This vegetative stage is known as *Rhizoctonia crocorum.* The basidial stage may be found on any herbaceous material as a violet to purple, compact felt, producing scattered basidia and hyaline, oval to oblong, 1-celled basidiospores, measuring $10–12 \times 6–7\ \mu$.

References

11, 17.

Shoestring rot, *Armillaria mellea* Vahl ex Fr.

Symptoms

This rot has been found on the tubers of plants growing on recently cleared land. Tree stumps and roots showed rhizomorphs similar to those attached to the potatoes, causing areas of soft, cheesy, light-colored decay. The disease is considered accidental on this host.

Etiology

In contacting the host, the extending rhizomorphs decentralize, producing mycelial strands that penetrate the cortex. The mycelium is hyaline, branched, and septate. When mature, it forms honey-colored, capitate, mushroomlike sporophores. The caps are 4–6 in. high, the stipes are annulated, and the basidiospores measure $9 \times 6\ \mu$.

References

11, 17.

Dry rot, *Phoma tuberosa* Melhus, Rosenb., & Schultz

Symptoms

Only tuber infections occur, forming brown to blackish, sunken, irregular but distinctive, membraneous spots, variously scattered over the surface, associated with wounds, and measuring 6–25 mm in diameter.

Etiology

The mycelium is brown, septate, and branched. The pycnidia are black, submerged to erumpent, scattered, and up to 160 μ in diameter. The pycnidiospores are hyaline, 1-celled, and up to $3.7–6 \times 1.8–3.7\ \mu$.

Reference

13.

Hard rot, *Phomopsis tuberivora* Güssow & W. R. Foster

Symptoms

Hard rot causes infrequent damage to leaves and tubers. Tubers are infected at the stolon end.

Etiology

Pycnidia are black, globose, and submerged becoming erumpent; pycnidiospores are hyaline, 1-celled, spindle-shaped, and $10–12 \times 4–6\ \mu$. Stylospores are filiform, curved, and $8–30 \times 0.5–1.5\ \mu$. The ascospore stage probably is *Diaporthe tulasnei.*

Reference

10.

Anthracnose, *Colletotrichum atramentarium* (Berk. & Br.) Taub.

Symptoms

Anthracnose, also known as black dot, causes a tip yellowing of the foliage, gradually involving the entire leaf along with a leaf rolling. Eventually the entire plant becomes involved, the yellow leaves become brown, and the plant appears to ripen prematurely. Sometimes aerial tubers appear. Stem infections show a purple red color in the vascular regions, and stolons are likewise affected; the roots are invaded and decay. The tubers may be stunted and undersized because of stolon infections and often are covered with pseudosclerotia.

Etiology

The sclerotia are black, with or without setae, and up to 2 mm in diameter. The conidiophores form a palisade layer, supporting conidia that are hyaline, pink en masse, 1-celled, curved, and 17–22×3–8 μ.

Reference

17.

Gray mold, *Botrytis cinerea* Pers. ex Fr.

Symptoms

Mold is mostly a storage and transit disease. The fungus enters through wounds or under other favorable conditions and produces a brown, odorless decay, resulting in a rubbery, shrunken tuber.

Etiology

The conidia of the fungus are hyaline, 1-celled, oval, and 11–15×8–11 μ. Sclerotia may be produced.

Reference

17.

Silver scurf, *Spondylocladium atrovirens* Harz.

Symptoms

Silver scurf develops on the tubers in the soil. It forms extensive, irregular, almost superficial blotches; deeper, better defined lesions develop under favorable conditions on certain varieties. The silvery appearance is evident on new infections.

Etiology

The fungus mycelium is hyaline to brown, septate, and branched. The conidiophores are erect, septate, and 150–350×7–8 μ. Conidia are produced in whorls and are brown, up to 8 septate, with rounded base and pointed tips, and measure 18–64×7–8 μ.

References

7, 11, 13, 17.

Skin spot, *Oospora pustulans* Owen & Wakef.

Symptoms

This disease is recognized on tubers by the appearance of olive brown, round, raised skin pustules. The specks in slightly sunken areas develop in storage and are not ordinarily recognized in the field. The spots may involve the eye-buds, preventing their further development and resulting in poor seed pieces.

Etiology

The fungus mycelium is hyaline to pale brown and 2–4 μ in diameter. It produces branches that function as conidiophores and bear chains of conidia that are hyaline, small, ellipsoidal to cylindrical, and 6–12×2–3 μ.

Reference

17.

Charcoal rot, *Macrophomina phaseolina* (Tassi) Goid.

Symptoms

This rot is usually recognized on many plants by the minute, black, imbedded sclerotia developing in the ashy-colored plant tissue. On potato the tops of plants may be killed by the fungus that infects the seed piece and grows up the stem to the soil surface. It also enters the stolon and causes a stem-end decay of the tubers in the form of a sunken, dark-colored, shrunken area. The fungus may also enter through eyes and lenticels.

Etiology

The sclerotia form in the stolons. The parasitic, sterile stage of the fungus is *Sclerotium bataticola*.

References

4, 17, 18.

Leaf blotch, *Cercospora concors* (Casp.) Sacc.

Symptoms

Blotch appears on older leaves as a faint, yellowish green, irregular area, up to 1/2 in. in diameter. It gradually spreads to newer growth. The spots become yellow and brown; their lower surfaces become covered with the violet gray fungus growth.

Etiology

The mycelium is hyaline to faintly colored, septate, and branched. The conidiophores emerge in clusters through the stomata, and measure 20–75×7 μ. The conidia are hyaline to pale-colored, septate, mostly straight, and 20–100× 4–6 μ. The disease is most prominent toward the end of the growing season.

Reference

5.

Early blight, *Alternaria solani* (Ell. & Martin) Jones & Grout

Symptoms

Early blight first appears as dark brown to black spots on the leaves; they may be irregularly oval to angular, up to 1 cm in diameter, slightly sunken, and as they grow older they become characteristically zonate with a yellowish border. Some petiole and stem infection may develop. Sometimes the tubers become inoculated after removal from the soil, resulting in the formation of scattered, dark-colored, sunken, circular spots, often several millimeters deep, that may develop in transit and storage.

Etiology

The mycelium is brown to dark-colored, septate, and branched. The conidiophores are brown, septate, appear in the killed tissue, and measure $50–91 \times 8–9\ \mu$. The conidia are dark-colored, obclavate, muriform septate, and $120–300 \times 12–20\ \mu$. The disease is often severe in warm, dry weather.

References

7, 8.

Wilt, *Verticillium albo-atrum* Reinke & Berth.

Symptoms

Wilt is recognized in the field and is associated with a slight yellowing of the lower leaves. The plant appears stunted, some of the lower leaves fall off, and upper leaves curl upward. Plants are not often killed. The vascular system becomes colored with dark streaks.

Etiology

The mycelium is hyaline to slightly colored, septate, and branched. The conidiophores branch off from the hyphae and produce secondary whorls from which the conidia develop often in chains or in wet clumps. Conidia are hyaline, oval, 1-celled, and $5–7 \times 3\ \mu$.

References

7, 11.

Wilt, *Fusarium solani* f. *eumartii* (Carpenter) Wr.

Symptoms

The plant becomes stunted and pale green. The lower leaves become yellow, droop, and hang suspended, turning brown. The disease involves the root system, where the fungus gains entrance, and discolors the vascular system up to the ground level. New tubers often discolor at the stolon end, diverging into the tuber. Many *Fusarium* spp. have been associated with the wilt disease and cause similar symptoms.

Etiology

The mycelium is hyaline, septate, and branched; the conidia, produced in sporodochia, are hyaline, mostly 5-septate, and 50–60 × 5–7 μ.

References

11, 15, 17.

Tuber rot, *Fusarium solani* f. *radicicola* (Wr.) Snyd. & Hans.

Symptoms

Tuber rot is mostly a storage rot that may be initiated at digging time and continue thereafter. The tubers exhibit various forms of decay, soft to dry, variously colored to brown, and exterior shrinkage to interior mushy consistency. Scores of *Fusarium* spp. have been isolated from decaying tubers after removal from the soil.

Etiology

The conidia are very similar in most instances. The mycelium is hyaline, septate, and branched; it produces sporodochia and hyaline, septate, slightly curved conidia, measuring 40–65 × 4–8 μ. The ascigerous stage is considered to be *Hypomyces ipomoeae* (Halst.) Wr.

References

11, 15.

Other fungi associated with potato

Ascochyta lycopersici Brunaud
Cylindrocarpon magnusianum Wollenweber
Fusarium spp.
Gliocladium roseum (Link) Bainier
Mycosphaerella solani (Ellis & Everhart) Wollenweber
Papulaspora coprophila (Zukal) Hotson
Ramularia solani Sherbakoff
Stysanus stemonitis Corda
Trichothecium roseum Link ex Fries

References: potato

1. Abbott, E. V. 1931. Further notes on plant diseases in Peru. Phytopathology 21:1061–71.
2. Barrus, M. F. 1944. A Thecaphora smut on potatoes. Phytopathology 34:712–14.
3. Barrus, M. F., and A. S. Muller. 1943. An Andean disease of potato tubers. Phytopathology 33:1086–89.
4. Bhargava, S. N. 1965. Studies on charcoal rot of potato. Phytopathol. Z. 53: 35–44.
5. Chupp, C. 1953. A monograph of the fungus genus Cercospora. Published by the author: Ithaca, N.Y., p. 536.
6. Cox, H. E., and E. C. Large. 1960. Potato blight epidemics throughout the world. U.S. Dept. Agr., Agr. Handbook 174.
7. Dykstra, T. P. 1948. Potato diseases and their control. U.S. Dept. Agr. Farmers' Bull. (Rev.) 1881:1–53.

8. Eddins, A. H., G. D. Ruehle, and G. R. Townsend. 1946. Potato diseases in Florida. Florida Univ. Agr. Expt. Sta. Bull. 427:5–96.
9. Fischer, G. W. 1951. The smut fungi. Ronald Press: N.Y., p. 51.
10. Grove, W. B. 1935. British stem- and leaf-fungi (Coelmycetes). Cambridge University Press: London, pp. 226–27.
11. Link, G. K. K., and G. B. Ramsey. 1932. Market diseases of fruit and vegetables. U.S. Dept. Agr. Misc. Publ. 98:1–53.
12. MacNish, G. C. 1968. Pink rot of potatoes in western Australia. Plant Disease Reptr. 52:280.
13. Melhus, I. E., J. Rosenbaum, and E. S. Schultz. 1916. Spongospora subterranea and Phoma tuberosa on the Irish potato. J. Agr. Res. 7:213–53.
14. Ruehle, G. D. 1941. A Xylaria tuber rot of potato. Phytopathology 31:936–39.
15. Sherbakoff, C. D. 1915. Fusaria of potatoes. N.Y. (Cornell) State Agr. Expt. Sta. Mem. 6:87–270.
16. Tucker, C. M. 1931. Taxonomy of the genus Phytophthora de Bary. Missouri Univ. Agr. Expt. Sta. Res. Bull. 153:1–197.
17. Walker, J. C. 1952. Diseases of vegetable crops. McGraw-Hill: N.Y.
18. Watson, R. D. 1944. Charcoal rot of Irish potatoes. Phytopathology 34:433–35.
19. Weber, G. F. 1923. Potato diseases and insects. Florida Univ. Agr. Expt. Sta. Bull. 169:100–163.

Ramie, *Boehmeria nivea* Linnaeus

Leaf spot, *Cercospora boehmeriae* Peck

Leaf spot, *Cercospora boehmeriae* Peck

Symptoms

The disease is confined to the leaves. The early manifestations are yellowish brown, circular to angular, water-soaked areas up to 3 mm in diameter mostly on the upper surface. As they become older they enlarge slightly, or if numerous, coalesce and are somewhat limited by the veins. Heavy infection causes defoliation from the bottom upward. The spots become covered with conidiophores and conidia. Finally the centers of spots fade to an opaque white and may dry, crack, and fall out.

Etiology

The mycelium is hyaline becoming darker, septate, and branched. The conidiophores arise from mycelial mats of stromatic composition that are brown, subepidermal, short, fasciculate, septate, and 54×5 μ. The conidia are pale brown, cylindrical, variably obclavate, curved, septate, acute, and 84×2–4 μ. *Cercospora krugiana* may also be present.

References

1, 2, 3.

References: ramie

1. Chowdhury, S. 1957. A Cercospora leaf spot of ramie in Assam. Trans. Brit. Mycol. Soc. 40:260–62.
2. Chupp, C. 1953. A monograph of the fungus genus Cercospora. Published by the author: Ithaca, N.Y., pp. 582–83.
3. Clara, F. M., and B. S. Castillo. 1950. Leaf spot of ramie, Boehmeria nivea (Linn.) Gaudich. Philippine J. Agr. 15: 9–21.

Rice, *Oryza sativa* Linnaeus

Black rot, *Xanthomonas itoana* (Tochinai) Dowson
Bacterial blight, *Xanthomonas oryzae* (Uyeda & Ishiyama) Dowson
Seed rot, *Achyla prolifera* (Nees) De Bary
Damping-off, *Pythium* spp.
Downy mildew, *Sclerophthora macrospora* (Saccardo) Thirumalachar, Shaw, & Narasimhan
Blight, *Cochliobolus miyabeanus* (Ito & Kuribayashi) Dickson
Foot rot, *Gibberella fujikuroi* (Sawada) Wollenweber
Culm rot, *Leptosphaeria salvinii* Cattaneo
Black sheath rot, *Ophiobolus oryzinus* Saccardo
Strangle, *Balansia oryzae* (Sydow) Narasimhan & Thirumalachar
False smut, *Ustilaginoidea virens* (Cooke) Takahashi
Kernel smut, *Tilletia horrida* Takahashi
Leaf smut, *Entyloma oryzae* Hans & Paul Sydow
Rust, *Puccinia graminis* Persoon f. *oryzae* Fragoso
Seedling blight, *Botryobasidium rolfsii* (Saccardo) Venkatarayan
Sheath spot, *Thanatephorus cucumeris* (Frank) Donk
Leaf spot, *Phyllosticta oryzina* (Saccardo) Padwick
Kernel stain, *Nigrospora oryzae* (Berkeley & Broome) Petch
Black kernel, *Curvularia lunata* (Wakker) Boedijn
Blast, *Piricularia oryzae* Cavara
Stack burn, *Trichoconis caudata* (Appel & Strunk) Clements
Orange stain, *Penicillium puberulum* Bainier
Leaf scald, *Rhynchosporium oryzae* Hashioka & Yokogi
White leaf streak, *Ramularia oryzae* Dreighton & Shaw
Leaf spot, *Helminthosporium rostratum* Drechsler
Narrow leaf spot, *Cercospora oryzae* I. Miyake
Other fungi associated with rice

Black rot, *Xanthomonas itoana* (Toch.) Dows.

Symptoms

This rot is characterized by partial blackening or black spotting of the hulled grain, especially at the apical part, sometimes at the middle, and rarely at the base. The center of the black spot usually lies on the apex or groove of the grain. Morbid change is limited to the aleurone layer and upper parts of the endosperm. Affected tissues die off and turn black.

Etiology

The organism is a short, gram-negative rod, unipolar, with round ends, 1 or 2 flagella, no capsules, and measuring $0.5–5 \times 1.2–3\ \mu$. Agar colonies are yellow, glistening, convex becoming thin, granular, and viscid. This bacterium attacks rice grains through wounds during the milky ripe stage. Optimum temperature, 29C.

References

11, 13, 23.

Bacterial blight, *Xanthomonas oryzae* (Uyeda & Ishiyama) Dows.

Symptoms

Blight appears only on the leaves. In the early stage, the watery, dark green, translucent stripes run lengthwise and are confined to the areas between large veins. The stripes are 0.5–1 mm wide and 3–5 mm long. At a later stage these stripes become light brown. Amber-colored droplets of bacteria ooze from these diseased portions. With drying of infected portions these droplets harden and produce small, roundish, amber-colored beads. The disease is prevalent on succulent plants, and infection is through wounds.

Etiology

The organism is a gram-negative rod, single or in pairs, motile by 1 or 2 polar flagella, and measuring $0.5–0.8 \times 1–2\ \mu$. Agar colonies are yellow, waxy, glistening, circular, and smooth. Optimum temperature, 25–30C.

References

11, 13, 23.

Seed rot, *Achyla prolifera* (Nees) d By.

Symptoms

First evidence of this disease is an outgrowth of the whitish hyphae of the fungus on the surface of the glumes of germinating seed or on the collar of the plumules. Hyphae grow out frequently from the slit of the seed coat, which is opened or broken off by germination or threshing. Hyphae radiate in a fine halo from the affected point. In severe cases, seeds will not germinate or seedling growth is checked.

Etiology

These fungi are usually classified as water molds and their pathogenicity is variable. The disease is caused by at least 12 additional fungi such as *Achyla americana, A. flagellate, Dictyuchus sterile, Phythiogeton ramosum, Pythiomorpha miyabeana, Saprolegnia anisospora,* and other fungus saprophytes and facultative parasites attacking seedlings of low vitality.

Reference

23.

Damping-off, *Pythium* spp.

Symptoms

The disease is associated with very young plants. The embryo becomes colored, and the endosperm is darkened. The rootlets show some lack of vigor in development, and the leaves become yellowish. The seedling may survive under highly favorable conditions, but usually it is killed. Several *Pythium* species have been isolated and their pathogenicity proven, including *P. oryzae, P. monospermum, P. nagaii, P. graminicola* Subre, *P. debaryanum,* and *P. echinocarpus.* Under certain circumstances, any of them may cause infection, but inoculations have not always been successful.

Etiology

In general, these fungi produce hyaline, nonseptate mycelium and form globose to egg-shaped sporangia that may be smooth, echinulate or warty; they possess variable papillae and are 15–25 μ in diameter. The oospores are hyaline or slightly colored, spherical, and up to 21 μ in diameter. Germination is by the formation of motile zoospores, emerging from a vesicle exuded from the sporangium or oospore.

Reference

25.

Downy mildew, *Sclerophthora macrospora* (Sacc.) Thirum., Shaw, & Naras.

Symptoms

There is a malformation of the florescence on young plants at the time of flowering and formation of the panicle. Newly emerged spikes are irregular in appearance, with rachises lengthening rapidly and becoming distorted. The leaves may be spirally twisted. Panicles are reduced in size and remain green abnormally long. Occasionally the rachises are extremely reduced, bearing only a few tufts of hairs in place of rachillae and flowers. Stamens and ovaries are completely abortive.

Etiology

The mycelium is intercellular; conidiophores bear 3–4 irregularly branched, short sterigmata, each bearing small, globose sporangia that germinate by forming 8 small, ellipsoidal, biflagellate zoospores. Angular, rough oospores germinate in water by formation of biciliate protoplasmic zoospores.

References

23, 31.

Blight, *Cochliobolus miyabeanus* (Ito & Kuribay.) Dicks.

Symptoms

On the coleoptile the spots are brown, small, and circular to oval. On leaf surfaces and sheaths spots are circular or oval, scattered, variable in size, and 0.5–3 × 1–14 mm. Small spots are dark brown, larger ones are pale yellow or dirty white with brown or gray centers. Spots coalesce, causing leaves to wither and appear yellowish. Black spots appear on infected glumes; they are covered entirely with a dark brown or olivaceous velvet mat of fungus fructifications. Seeds shrivel and are discolored.

Etiology

The mycelium in the host is olive to brown, well developed, branched, septate, and up to 15 μ in diameter. Externally, the prostrate hyphae produce more or less erect conidiophores from 300 to 600 μ high with scars. Conidia are brown, 6–12-septate, slightly curved, with a distinct hilum, and 35–170×11–17 μ. Germ tubes penetrate through cuticle and epidermis. The vegetative stage of the fungus is *Helminthosporium oryzae*.

References

8, 12, 17, 23, 25.

Foot rot, *Gibberella fujikuroi* (Saw.) Wr.

Symptoms

Diseased seedlings are usually necrotic, spindly, weak, and seldom survive transplanting. In partially grown plantings there appear here and there tall, lanky tillers coming into flower earlier than the rest of the crop. They bear pale green flags shooting up conspicuously above the general level of the crop. There is an emergence of the culms of affected plants from sheaths. Leaves of infected plants dry up one after the other from below with margins first turning brown; eventually the whole plant is involved. Adventitious roots may develop from the first, second, and sometimes third node above ground level. Splitting open of culms of infected tillers reveals a distinct brown discoloration. The fungus causes a reddish, pink, or yellow discoloration of the seed and a chalky endosperm.

Etiology

The mycelium is hyaline, white, pink, or purplish, septate, and branched. Microconidia form in chains or heads and are colorless, 1- or 2-celled, and oval. The macroconidia are hyaline, awl-shaped to straight, and 3–5-septate. Chlamydospores are absent. Perithecia are dark blue, spherical to egg-shaped, outwardly somewhat rough, multicelled, with club-shaped paraphyses, and measure 120–135×15 μ. The asci are flattened above, with 4–6 spores, rarely 8. The ascospores are long-ellipsoid, 3-septate, and 15×5.2 μ. The vegetative stage of the fungus is *Fusarium moniliforme*.

References

8, 23, 25.

Culm rot, *Leptosphaeria salvinii* Catt.

Symptoms

Black, irregular lesions appear on the outer leaf sheaths near the water line. On infected plants, panicles are filled lightly, and there is a tendency to produce green shoots from the base when the rest of the crop is ripening and turning yellow. On such plants there is a discoloration of the base of the stem at the lowest distinct internode. On splitting of culms, a dark grayish weft of hyphae with black, small, round sclerotia is found scattered over the inner surface of the hollow stems. Large air spaces develop between main vascular bundles of culms and leaf sheaths. Young plants are usually killed, while old plants show partial sterility and excessive late tillering.

Etiology

The hyphae, white or hyaline and septate, develop in subepidermal layers of the culm and sheath. Sclerotia are black, spherical, smooth, shiny, and 100–300 μ in diameter. Perithecia are black, globose, and clustered in the parenchyma of the leaf sheath, with short but broad cylindrical beaks. Asci are clavate, transparent, short, and stipitate. Spores are pale yellow, oblong-fusiform, curved, 3-septate, moderately constricted at the septa, and 60×9 μ. Sclerotia remain viable several years. The sterile stage is *Sclerotium oryzae*. The conidial stage is *Helminthosporium sigmoideum*.

References

23, 25, 34, 37.

Black sheath rot, *Ophiobolus oryzinus* Sacc.

Symptoms

Affected plants are characterized by brown discoloration of sheaths from the crown to considerably above the waterline. In early stages of infection, dark, reddish brown mycelial mats are found on the inner surface of diseased sheaths. At maturity, the straw has a dull brownish cast. The leaf is killed. Tillering is reduced, producing only one head per plant, and causing premature ripening.

Etiology

The mycelium penetrates directly and forms mats between stem and sheaths; as hyphae accumulate, the perithecia primordia are initiated. Perithecia are reddish brown, globose, loosely gregarious, subcutaneous, finally vertically erumpent, and more or less ostiolate. Asci are cylindrical, with short narrow stipes, and measure 95–100×7–11 μ. Ascospores are faintly colored, greenish yellow, 3–5-septate, with many minute guttulae, and 86–100×2–3 μ.

References

8, 23, 35.

Strangle, *Balansia oryzae* (Syd.) Naras. & Thirum.

Symptoms

Upon emerging from the boot, the panicle, normally loose, is enclosed in a weft of white mycelium that confines the branches to a single stiff spike. The mycelium is dirty white, tough, and the structure becomes hard as it matures and darkens to a shiny black sclerotium.

Etiology

The mycelium produces the compact sclerotium, which covers and includes the panicle and rachis. The conidia are grouped on the tufted sporodochia that are 1–1.5 mm high. Conidia are hyaline, straight or curved, 1-celled, needle-shaped, and $22 \times 1.5\ \mu$. The perithecia are imbedded in the stroma with ostioles protruding. The ascospores are hyaline, septate, and needlelike. The vegetative stage is *Ephelis oryzae*.

References

9, 16, 23, 25.

False smut, *Ustilaginoidea virens* (Cke.) Tak.

Symptoms

The parasite is visible only in the panicles where it develops fructifications in the ovaries of individual flowers. Ovaries are transformed into large, velvety, green masses sometimes twice as large as normal grains. The green color is superficial; the inner part is orange yellow near the surface and almost white in the center. Usually only a few grains in each panicle are affected.

Etiology

The mycelium is hyaline, fine, and compact. The spores are olive brown, globulose or ovoid, united, hard, formed laterally on large filaments, and 3–5 μ in diameter. Infected rice grains are black, voluminous, bulky, and hard like sclerotia. Spores are borne on minute projections from the sides of radial hyphae. They are first smooth, becoming rough and olive green at maturity. The sexual stage is not fully described. The organism is similar to *Claviceps purpurea*.

References

3, 23, 25.

Kernel smut, *Tilletia horrida* Tak.

Symptoms

Sori appear in the ovaries, replacing the seed. Minute, black pustules or streaks appear within the glumes at time of ripening, sometimes destroying the whole grain. Upon crushing, such grains contain a powdery mass of black spores. Recognition of this powdery mass dusted on adjacent leaves before harvest is the easiest way to detect the disease. Rarely are more than 3 or 4 grains attacked in a panicle.

Etiology

Spore masses are black, pulverulent, produced within ovaries, and remain covered by glumes. Spores are globose, or irregularly rounded or sometimes broad-elliptical, and 18–23 μ in diameter. Epispores are deep olive brown, opaque, and thickly covered with conspicuous spines. The spines are hyaline or slightly colored, pointed at the apex, irregularly polygonal at the base, more or less curved, and 2–3 μ. Sporidia are filiform or needle-shaped, curved, and up to 10–12 in number.

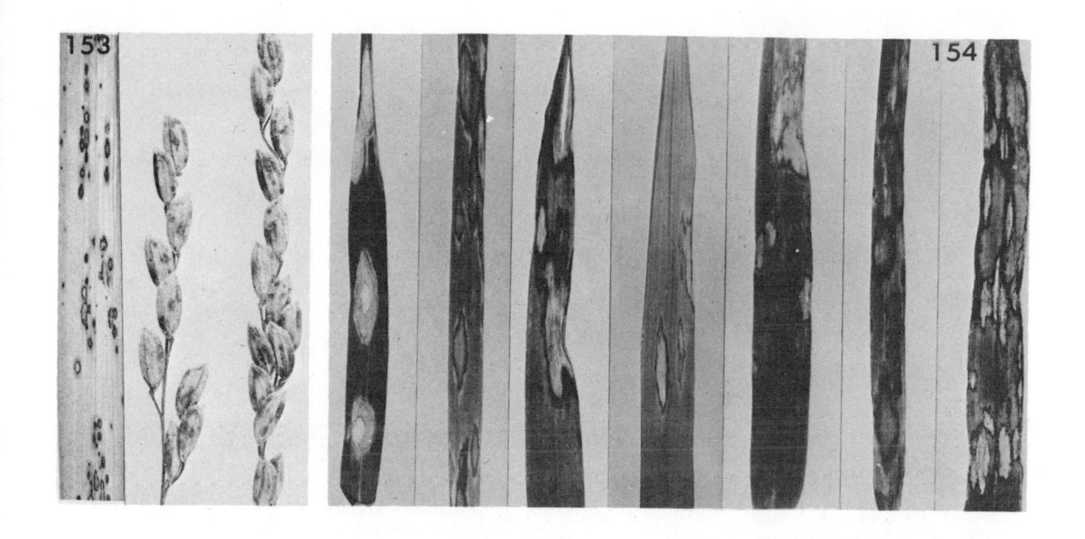

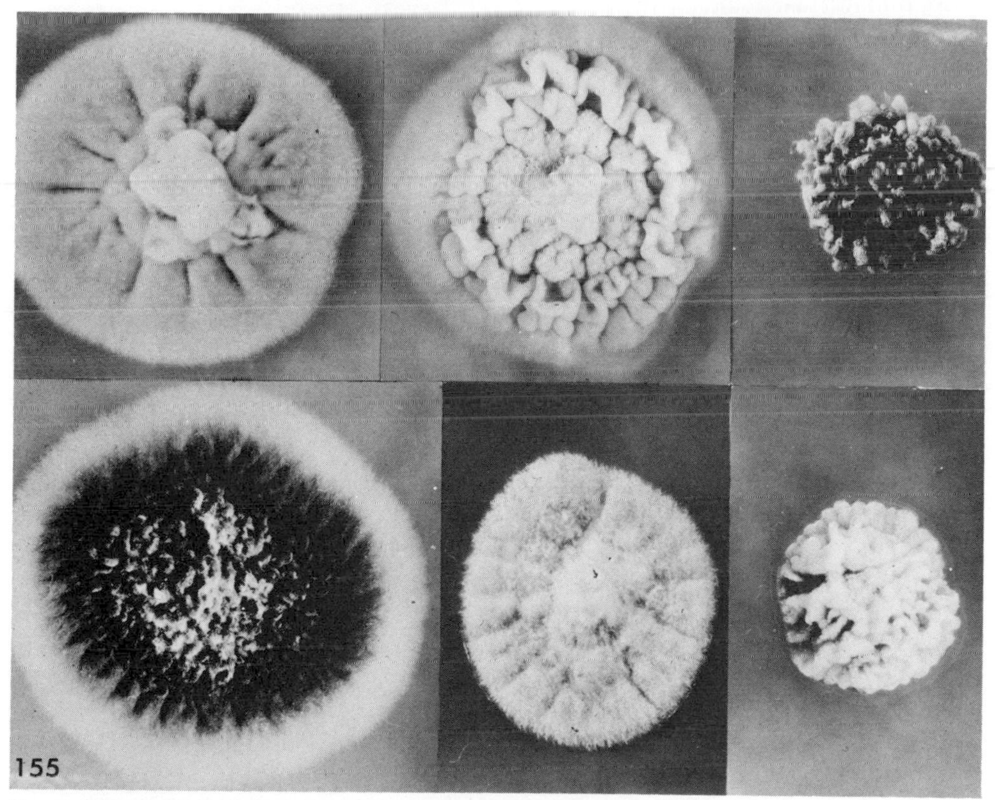

Figure 153. Blast, *Piricularia oryzae,* on rice foliage and glumes.

Figure 154. Leaf scald, *Rhynchosporium secalis,* on cereals and grasses. (Photograph by R. M. Caldwell.)

Figure 155. Month-old cultures of races of *Rhynchosporium* spp. from grasses. (Photograph by R. M. Caldwell.)

References

5, 8, 23, 24.

Leaf smut, *Entyloma oryzae* H. & P. Syd.

Symptoms

Black, irregularly shaped sori appear on leaves, sheaths, and rachises. They are linear, rectangular or angular-elliptical, and 0.5–2×0.5–1.5 mm. Spots are lead black, usually quite distinct, not confluent, and covered by the epidermis. On soaking in water the epidermis ruptures, revealing a black mass of spores.

Etiology

The sorus consists of a mass of closely agglutinated spores, pale to brown to smoky, angular-globose to angular-ovate, smooth, and 7–11×8–17 μ. The epispore is 1–1.5 μ thick. Chlamydospores germinate by formation of sporidia in situ.

References

8, 14, 23.

Rust, *Puccinia graminis* Pers. f. *oryzae* Frag.

Symptoms

This disease causes rust on numerous Gramineae worldwide and produces pustules on all aerial plant parts. Fructifications are uredia that are brown red, isolated or confluent, elongate, and 6–10 mm. Telia are black, arranged in stripes, and appear especially on glumes and sheaths.

Etiology

The uredospores are ovoid, with 3–5 equatorial pores, and measure 22–40 by 16–22 μ. The teliospores are ellipsoid or club-shaped, round at the apex, and measure 35–60×16–23 μ.

References

10, 25.

Seedling blight, *Botryobasidium rolfsii* (Sacc.) Venkat.

Symptoms

Areas of blighted seedlings are the first symptoms. Killed plants become water-soaked to dark as the fungus mycelium overgrows them. Bases of stems and roots of older diseased plants have a frosty appearance due to the presence of wefts of mycelium of the fungus. Brown, spherical sclerotia may be found on the roots at the soil surface. Diseased plants are easily pulled up.

Etiology

The white mycelium surrounds the plant stems near the soil surface. It penetrates the host tissues and forms white to brown sclerotia that are superficial, spherical, and 0.5–0.8 mm in diameter. Basidia and basidiospores are not frequently found, but when present they appear on aboveground stems and lower leaf surfaces. Basidia measure 7–9 × 4–5 μ and bear 4 sterigmata, 2–5 μ long. The spores are hyaline, obovate, pointed at the base, smooth, and 3–5 × 6–7 μ.

References

23, 39, 40, 41.

Sheath spot, *Thanatephorus cucumeris* (Frank) Donk

Symptoms

Spots are greenish gray to reddish brown, ellipsoid, and 10 × 3–4 mm. They later turn grayish white with brownish black margins. The spots develop on the sheaths near the base of the plants and may continue up to the flag leaf, sometimes affecting the leaf blades but generally confined to the leaf sheaths. The lesions may coalesce, stunting the plant, arresting the development of the panicle, and possibly causing sterility. Considerable loss results if the attack is on seedlings. There have been several descriptions of the disease since the symptoms are widely variable and often depend on the seasonal conditions and probably on the stage of development of the plant and the varieties of rice grown. Sclerotia are tan to dark brown, superficial, flattened to irregular in shape, and 5–8 mm in diameter.

Etiology

The basidia are hyaline to faintly yellow, sparingly produced, single or in clumps, and 10–15 × 6–9 μ. Each basidium produces 4 sterigmata, each measuring 5–8 × 2–3 μ. The basidiospores are hyaline, obovate, smooth, show a basal attachment, and measure 8–11 × 5–7 μ. The disease has been known as oriental sheath spot, rhizoctonia sheath spot, sheath spot, and sheath rot.

References

23, 26, 30, 32, 38, 42.

Leaf spot, *Phyllosticta oryzina* (Sacc.) Padw.

Symptoms

Leaf spot causes a whitening of the glumes that are later speckled with black pycnidia that may also appear on sheaths or old foliage. Kernels within the glumes are often darkened.

Etiology

Pycnidia are submerged becoming erumpent, flattish, with a small ostiole, and 100–120 μ in diameter. Pycnidiospores are hyaline, ellipsoid, straight with round ends, and 4.8–6 × 1.8–2 μ.

References

6, 23.

Kernel stain, *Nigrospora oryzae* (Berk. & Br.) Petch

Symptoms

Minute, black pustules, rarely exceeding 0.5 mm in diameter, appear on leaves, leaf sheaths, culms, and glumes. These pustules are easily mistaken for pycnidia of *Phoma* or *Phyllostica* but are more powdery in appearance.

Etiology

The disease is characterized by hyaline to brown mycelium in the leaf parenchyma. Short conidiophores extruded from stomata are inflated below the tip and bear, singly and apically, 1-celled, dark, opaque, globose or subspherical, smooth spores, measuring 13×15 μ. The perithecial stage of the fungus is *Khuskia oryzae*.

References

18, 23, 25.

Black kernel, *Curvularia lunata* (Wakk.) Boed.

Symptoms

Black grains appear in the polished product. On leaf sheaths, leaves, and glumes, very fine, small, punctiform spots may appear between veins. The disease development is favored by dryness accompanied by winds during the flowering period.

Etiology

The mycelium is fine and hyaline, becoming brownish later. Conidiophores are dark brown, erect, stiff, sometimes isolated or grouped in clusters of 3–10, each having 4–7 cells, dilated or curved, with a variable number of scars. Conidia are septate, boat-shaped or curved, and $75\text{–}200 \times 6\text{–}8$ μ. Central cells of the conidia are dark, and terminal cells are hyaline.

References

20, 25.

Blast, *Piricularia oryzae* Cav.

Symptoms

On the leaf, small, bluish flecks, 1–3 mm in diameter develop with a pale green or dull grayish green water-soaked appearance. The marginal outer rim is dark brown. Spots on leaf sheaths are similar to those on the leaf. Brown to black spots or rings on the rachises of maturing florescences are near joints of rachillae. On glumes, small brown or black spots appear in heavily infected panicles. Panicles emerging from sheaths are blasted prematurely and completely whitened long before time of ripening.

Etiology

The sporiferous hyphae are usually epiphyllous. In culture they are hyaline to olivaceous, septate, branched, and 1.5–6 μ wide. Conidiophores are colored, fasciculate, simple or branched, 2–4-septate, and pleurogenous. Conidia are hyaline to pale olive, obclavate, tapering at the apex, truncate or extended into a short basal appendage, 2-septate, and 20–22×10–22 μ. Chlamydospores are 5–12 μ in diameter. Infection occurs before, at, and after flowering periods. Penetration is through stomata or directly through the epidermis by germ tube, with the formation of appressoria.

References

1, 2, 8, 19, 23.

Stack burn, *Trichochonis caudata* (Appel & Strunk) Clem.

Symptoms

Symptoms appear on seedlings, leaves, and stored grains. Seedlings are attacked at both roots and coleoptile; infection on roots starts from point of attachment to the seed. On grains, black, spherical bodies develop, lying superficially on darkened areas. Coleoptiles are stained with brown patches and streaks also bearing sclerotia. On leaves, spots are circular or oval, not limited by the veins, and are outlined by a brown ring. Centers of spots are almost white with minute black dots.

Etiology

The mycelium is white and produces sclerotia and conidia. Spores are elongate and characteristically fusoid, with a long, thin appendage exceeding the spore in length. Spores are 3–4-septate, constricted, and 36–45×9–12 μ. Appendages are hyaline, 4–6-septate, and 35–45 μ long.

References

23, 33.

Orange stain, *Penicillium puberulum* Bainier

Symptoms

This disease is indicated by an orange discoloration of the endosperm of rice. It causes a reduction in the yield of plants and a lowering of grade on harvested rice. Many other species of this genus are commonly found associated with rice.

Etiology

On Czapek's agar, mycelium is greenish to olive, velvety, and produces a moldy odor. Conidiophores are up to 200 μ high and 3–4 μ wide; penicilli are asymmetric. Verticilli are branched, with 2–4 metulae and 3–5 sterigmata. Conidia are globose and 3–4 μ.

Reference

29.

Leaf scald, *Rhynchosporium oryzae* Hash. & Yokogi

Symptoms

Seedling blight, head blight, and leaf scald are the principal manifestations of the disease with leaf scald being the most characteristic and diagnostic. Lenticular-shaped spots become circled by breaks in the epidermis and later become water-soaked. A series of zones develop in which successive bands of dark brown border the light tan of the dried area. At first these zones vary in width from 1 to several millimeters around the original point of infection, but later they extend to the leaf margins. The disease most often develops at the leaf tips but may begin at any place along the blade.

Etiology

Conidia are hyaline, 1-septate, sickle-shaped to cylindrical, 12×3–$4\ \mu$, and lack the apical beak.

Reference

28.

White leaf streak, *Ramularia oryzae* Dreigh. & Shaw

Symptoms

Leaf spots are grayish white with brown margins, oblong to linear, 1–2.5 mm long and 0.5 mm wide, and appear on both surfaces.

Etiology

Hyphae are hyaline, branched, and septate. They emerge from host tissue through stomata, grow over the leaf surface, and produce short, distinct, erect conidiophores, which are continuous, straight, up to 20 μ long, and often show terminal scars. Conidia are hyaline, cylindric, catenulate, 1–3-septate, show a distinct hilum, and measure 1–2×20–$25\ \mu$.

Reference

7.

Leaf spot, *Helminthosporium rostratum* Drechs.

Symptoms

The leaf spots are brown to straw-colored, small, oval, delineated by the veins, and measure 2–3×1–1.5 mm. Lesions may coalesce, forming larger necrotic areas, and advance across the veins; some browning and yellowing of host tissue develops.

Etiology

Conidiophores, 2–4, emerge through the stomata or between epidermal cells. They are dark, multiseptate, with a swollen basal cell, measure $480 \times$ 5–7 μ, and bear conidia singly and successively at the tips. The conidia are olivaceous to dark brown, straight to slightly curved, and taper from a blunt base to the apex. The basal cell shows a protruding hilum.

References

4, 22.

Narrow leaf spot, *Cercospora oryzae* I. Miyake

Symptoms

Leaf lesions are linear, 3–5 mm long, and parallel to the long axis of the leaf. The spots are dark brown in the center, the border fading outwardly. Lesions are the same on sheath, peduncle, and glumes. On resistant varieties, lesions are red brown throughout and very narrow, while on susceptible ones the dark spots are wider, with narrow, light brown margins and gray brown centers.

Etiology

Conidiophores, emerging from stomata, are dark, solitary or in groups of 2 or 3, 3 septate or more, and 88–140 × 4.5 μ. Conidia are cylindrical to obclavate, 3–10 septate, and 20–60 × 5 μ. Penetration is through the stomata, and the disease is spread mainly longitudinally in epidermal cells.

References

8, 23, 27, 36.

Other fungi associated with rice

Alternaria oryzae Hara
Ascochyta oryzae Cattaneo
Brachysporium oryzae Ito & Ishiyama
Curvularia oryzae Bugnicourt
Diplodia oryzae Miyake
Gnomonia oryzae Miyake
Hendersonia oryzae Miyake
Metasphaeria oryzae (Cattaneo) Saccardo
Mycosphaerella oryzae (Cattaneo) Saccardo
Myrothecium oryzae Saccardo
Phyllosticta oryzae (Cooke & Massee) Miyake
Septoria oryzae Cattaneo
Sphaerulina oryzae Miyake

References: rice

1. Anderson, A. L., B. W. Henry, and E. C. Tullis. 1947. Factors affecting infectivity, spread, and persistence of Piricularia oryzae Cav. Phytopathology 37: 94–110.
2. Awoderu, V. A. 1970. Identification of races of Pyricularia oryzae in Nigeria. Plant Disease Reptr. 54:520–23.
3. Butler, E. J. 1918. Fungi and diseases in plants. Thacker: Calcutta, India.
4. Chattopadhyay, S. B., and C. Dasgupta. 1959. Helminthosporium rostratum

Drechs. on rice in India. Plant Disease Reptr. 43:1241–44.
5. Chowdhury, S. 1957. Studies in the bunt of rice (Oryza sativa L.). Indian Phytopathol. 4:25–37.
6. Copeland, E. B. 1924. Rice. Macmillan: London.
7. Deighton, F. C., and D. Shaw. 1960. White leaf-streak of rice caused by Ramularia oryzae sp. nov. Trans. Brit. Mycol. Soc. 43:516–18.
8. Dickson, J. G. 1956. Diseases of field crops. 2d ed. McGraw-Hill: N.Y.
9. Diehl, W. W. 1930. Conidial fructifications in Balansia and Dothichloe. J. Agr. Res. 41:761–66.
10. Diehl, W. W. 1944. Bibliography and nomenclature of Puccinia oryzae. Phytopathology 34:441–42.
11. Dowson, W. J. 1943. On the generic names Pseudomonas, Xanthomonas and Bacterium for certain bacterial plant pathogens. Trans. Brit. Mycol. Soc. 26: 4–14.
12. Drechsler, C. 1923. Some graminicolous species of Helminthosporium. J. Agr. Res. 24:641–739.
13. Elliott, C. 1951. Manual of bacterial plant pathogens. 2d ed. Chronica Botanica: Waltham, Mass.
14. Fischer, G. W. 1953. Manual of the North American smut fungi. Ronald Press: N.Y.
15. Flentije, H. T., H. Stretton, and E. J. Hawn. 1963. Nuclear distribution and behavior throughout the life cycles of Thanatephorus, Waitea and Ceratobasidium species. Australian J. Biol. Sci. 16:450–67.
16. Govindu, H. C. 1969. Occurrence of Ephelis on rice variety IR-8 and cotton grass in India. Plant Disease Reptr. 53: 360.
17. Grist, D. H. 1955. Rice. 2d ed. Longmans, Green Co.: N.Y.
18. Hudson, H. J. 1963. The perfect state of Nigrospora oryzae. Trans. Brit. Mycol. Soc. 46:355–60.
19. International Rice Research Institute. 1965. Rice blast disease. Johns Hopkins Press: Baltimore, Md.
20. Martin, A. L. 1939. Possible cause of black kernels in rice. Plant Disease Reptr. 23:83–84.
21. Matsumoto, T. 1952. Monograph of sugarcane diseases in Taiwan. Taiwan University: Taipei, Taiwan.
22. Mundkur, B. B. 1949. Fungi and plant disease. Macmillan: London.
23. Padwick, G. W. 1950. Manual of rice diseases. Commonwealth Mycological Institute: Kew, Surrey, England.
24. Reyes, G. M. 1933. The black smut or bunt of rice (Oryza sativa Linnaeus) in the Philippines. Philippine J. Agr. 4: 241–68.
25. Roger, L. 1953. Phytopathologie des pays chauds. Encyclopedie Mycologique 18:606–25, 884–85, 1224–59, 1360–71, 1500–1516, 1535–40, 1976–80, 2048–50.
26. Ryker, T. C., and F. S. Gooch. 1938. Rhizoctonia sheath spot of rice. Phytopathology 28:233–46.
27. Ryker, T. C., and N. E. Jodon. 1940. Inheritance of resistance to Cercospora oryzae in rice. Phytopathology 30: 1041–47.
28. Schieber, E. 1962. Rhynchosporium leaf scald of rice in Guatemala. Plant Disease Reptr. 46:202.
29. Schroeder, H. W. 1963. Orange stain, a storage disease of rice caused by Penicillium puberulum. Phytopathology 53: 843–45.
30. Singh, R. A., and M. S. Pavgi. 1969. Oriental sheath and leaf spot of rice. Plant Disease Reptr. 53:444–45.
31. Steindl, D. R. L., and R. J. Steib. 1961. Sugarcane diseases of the world. Vol. 1. Elsevier Publishing Co.: N.Y., pp. 311–25.
32. Talbot, P. H. B. 1965. Studies of 'Pellicularia' and associated genera of Hymenomycetes. Persoonia 3:371–406.
33. Tisdale, W. H. 1922. Seedling blight and stockburn rice and the hot water seed treatment. U.S. Dept. Agr. Bull. 1116.
34. Tullis, E. C. 1933. Leptosphaeria salvinii, the ascigerous stage of Helminthosporium sigmoideum and Sclerotium oryzae. J. Agr. Res. 47:675–87.
35. Tullis, E. C. 1933. Ophiobolus oryzinus, the cause of a rice disease in Arkansas. J. Agr. Res. 46:799–806.
36. Tullis, E. C. 1937. Cercospora oryzae on rice in the United States. Phytopathology 27:1005–8.
37. Tullis, E. C., and E. M. Cralley. 1933. Laboratory and field studies on the development and control of stem rot of rice. Arkansas Univ. Agr. Expt. Sta. Bull. 295:1–23.
38. Valdez, R. B. 1955. Sheath rot of rice. Philippine Agriculturist 39:317–36.
39. Venkatarayan, S. V. 1950. Notes on some species of Corticium and Pellicularia. Indian Phytopathol. 3:81–86.
40. Weerapat, P., and H. W. Schroeder. 1966. Effect of soil temperature on resistance of rice to seedling blight caused by Sclerotium rolfsii. Phyto-

pathology 56:640–44.

41. West, E. 1947. Sclerotium rolfsii Sacc. and its perfect stage in climbing fig. Phytopathology 37:67–69.

42. Witney, H. S. 1964. Sporulation of Thanatephorus cucumeris (Rhizoctonia solani) in the light and in the dark. Phytopathology 54:874–75.

Rubber, *Hevea brasiliensis* Mueller-Argov

Black thread, *Phytophthora palmivora* Butler
Patch canker, *Pythium complectens* Braun
South American leaf blight, *Microcyclus ulei* (P. Henning) Von Arx
Moldy rot, *Ceratocystis fimbriata* Ellis & Halsted
Dieback, *Botryodiplodia theobromae* Patouillard
Anthracnose, *Glomerella cingulata* (Stoneman) Spaulding & Schrenk
Stinking root disease, *Sphaerostilbe repens* Berkeley & Broome
Dry root disease, *Ustulina zonata* Léveillé
Black crust, *Catacauma huberi* (P. Henning) Theis & Sydow
Bark disease, *Kretzschmaria micropus* (Berkeley) Saccardo
Target leaf spot, *Thanatephorus cucumeris* (Frank) Donk
Pink disease, *Botryobasidium salmonicolor* (Berkeley & Broome) Venkatarayan
Red root rot, *Poria hypobrunnea* Petch
Root rot, *Ganoderma amazonensis* Weir
Wet red rot, *Ganoderma pseudoferreum* Wakefield
Brown root rot, *Fomes lamaensis* (Murrill) Saccardo & Trotter
Brown crust root disease, *Fomes noxius* (Berkeley) Corner
White root rot, *Fomes lignosus* (Klotzsch) Bresadola
Horsehair blight, *Marasmius equicrinis* Mueller
Seedling disease, *Pestalotia palmarum* Cooke
Rim blight, *Ascochyta heveae* Petch
Leaf spot, *Periconia heveae* Stevenson & Imle
Powdery mildew, *Oidium hevea* Stein
Leaf spot, *Alternaria* sp.
Bird eyespot, *Helminthosporium heveae* Petch
Algal spot, *Cephaleuros virescens* Kunze
Leaf spot, *Corynespora cassiicola* (Berkeley & Curtis) Wei
Other fungi associated with rubber

Black thread, *Phytophthora palmivora* Butl.

Symptoms

This disease occurs in three stages. (a) Black thread or black stripe is characterized by short, narrow, vertical, parallel black lines on the renewing bark above the tapping cut. They may be 1/4 in. broad and up to twice as deep. (b) Patch canker. This stage of infection at first does not produce a visible symptom. There may appear a bleeding of reddish liquid, and when decay is advanced, boring insects invade the area. The discolored, decayed area is revealed as a purplish red patch. These areas later crack along the margins and the bark peels up, forming a prominent scaling effect. (c) Leaf fall. Dieback of twigs and branches and infection of fruit is accompanied by fading of the green leaves to a yellowish white and shedding.

Etiology

The mycelium is hyaline, nonseptate, branched, coarse, and variable in diameter. The sporangia are thin-walled, spherical to pear-shaped, with a clear papillae, and measure 36–75 × 21–36 μ. The germinating sporangia produce

15–20 zoospores, measuring 7–11 μ. They are motile for a short time, round up, and produce a germ tube. Chlamydospores found in the substrata are spherical, 23–50 μ in diameter, and germinate by producing mycelium. Oospores, not common in nature, are colorless, spherical, germinate by producing mycelium, and measure 21–28 μ in diameter.

References

2, 8, 11, 12, 24, 25, 31, 35, 37.

Patch canker, *Pythium complectens* Braun

Symptoms

The infection may occur in association with the tapping panel or at the collar. There is some association of the patch canker with lightning injury. A reddish purple exudate flows from the canker borders of the diseased area. The removal of the outer cortex reveals a thin, moist, dark layer which may be purple red. The outer cortex is invaded first, followed by the inner cortex and the cambium. Under favorable conditions the stem may be girdled, and within a few weeks the tree is killed.

Etiology

Mycelium is abundant; sporangia are spherical, and upon germination develop a vesicle containing the zoospores which germinate to form mycelium. The organism is probably synonymous with *P. vexans.*

References

11, 13, 37.

South American leaf blight, *Microcyclus ulei* (P. Henn.) Arx

Symptoms

The disease, recognized in 1910, first appears on the youngest leaves which, under favorable conditions of temperature and moisture, are killed and become black. Leaves that are older show more or less translucent, olive green spots. Infection areas are 1–5 cm in diameter and develop dark specks within the lesion. The central part of the lesion may drop out, leaving an irregular, ragged hole. There is often a definite malformation of the leaflet blades. Infection spots may appear on the flowers, green seed pods, petioles, and branches. More or less complete defoliation often follows heavy infection. Foliage 12–20 days old is immune to infection.

Etiology

The mycelium is hyaline, intercellular, branched, septate, and mostly limited to the immediate lesion area. The conidia are hyaline to olivaceous, 1–2-septate, and $5–8 \times 12–30$ μ, averaging 6×20 μ. Pycnidia and pycnidiospores are usually found in abundance on older leaves. As inoculum, pycnidiospores have failed, up to the present time, to cause infection. Perithecia are found usually around the borders of fully mature leaf lesions. They are black, small, spherical, sparingly developed, and imbedded in small, unilocular stroma. The ascus is clavate, and the ascospores are hyaline, oblong, 2-celled, 13–20 × 4–5 μ, and have not been used to reproduce the disease. Thus, there are 3 spore stages: *Scolecotrichum* (conidia), *Aposphaeria* (pycnidia), and *Dothidella* (perithecia). The conidial stage is responsible for the development of inoculum and dissemination of the disease.

References

5, 22, 23, 28.

Moldy rot, *Ceratocystis fimbriata* Ell. & Halst.

Symptoms

This is a major disease of the tapping panel of rubber trees. Freshly tapped areas are susceptible to infection that results in a rot of the bark and cambium and causes discolored wood. The lesions are dark, depressed areas often parallel to the tapping cut. At the lower margin, a thick, grayish mass of mycelium is produced that becomes progressively thinner above and causes a dark brown color. The gray mycelial growth develops over the cut surface, and conidia form within 24 hours.

Etiology

Perithecia are black, more or less globose, with long necks and fimbriate ostioles. They appear in a few days, discharging ascospores at their tips where droplets accumulate. The ascospores are hyaline, ellipsoidal, often with unequally curved sides, and measure 7–10 × 2–4 μ. In addition to the hyaline endoconidia and ascospores, dark conidia are produced in typical infections.

References

9, 26, 37.

Dieback, *Botryodiplodia theobromae* Pat.

Symptoms

Dieback of branches and twigs and an ashy gray discoloration of the wood occurs with canker formations. Infection enters through tender buds or wounds and frequently is accompanied by secondary organisms. On petioles and healthy stems dieback produces hypertrophied lenticels that are covered by a conspicuous trickle of latex.

Etiology

Pycnidia are immersed, scattered or clustered, or in erumpent, glabrous or villous, globose stromata and measure 0.25–0.4 mm in diameter. Paraphyses are numerous, linear, and up to 80 μ long. Spores are fuliginous to black or black brown, oval, 1-septate, and 24–30 × 12–15 μ. The ascospore stage is *Physalospora rhodina*.

References

24, 31, 36, 37.

Anthracnose, *Glomerella cingulata* (Ston.) Spauld. & Schrenk

Symptoms

This disease causes large, irregular, discolored patches on the leaves. They are sometimes several inches long and are often marginal. At first the spots are brown; later the central parts are gray, the margin darker, and there is usually a light yellow band where the spot joins the healthy tissue. Minute, usually blackish, dots form on both surfaces, sometimes in distinct concentric circles.

Etiology

The acervuli rupture the epidermis and in humid conditions produce pink spore masses that later become dark brown. Short conidiophores are colorless, brown at the base, unbranched, nonseptate, and 7–8×3–4 μ. Sometimes rigid, sterile setae are formed. Conidia form singly at the tips of the conidiophores. The spores are colorless, nonseptate, cylindrical or a little narrowed at the ends, straight or very slightly curved, and 10–15×3.5–4 μ. The ascospore stage is *Gloeosporium albo-atrum.*

References

9, 37, 39.

Stinking root disease, *Sphaerostilbe repens* Berk. & Br.

Symptoms

The foliage becomes thin on diseased plants, and branches gradually die back. Trees are especially susceptible following flooding. When the root is dug up and the cortex is removed, black or red flattened fungus strands are seen undulating over the surface of the wood. The inside of the cortex and outer layers of the wood are usually deep blue or purple when fresh and have a foul odor. Mycelial strands are usually about 2 mm broad but may reach 5 mm. Mycelium enters the smaller roots and advances until it reaches the larger ones. There its rhizomorphs spread out, branch, and may form continuous reddish sheets.

Etiology

The vegetative stage appears as red stalks with red to pinkish heads; stalks are 2–8 mm high and 0.5–1 mm wide. The heads are globose and produce conidia measuring 10–20×5–9 μ. Fructifications are produced on diseased tissues or on the rhizomorphs. Conidia appear on short, erect, red stalks that are 2–8 mm high and 5 mm in diameter with pinkish globose heads. The perithecia are composed of small, dark red bodies, rounded below and conical above, measuring 4 mm in diameter, and crowded together on the bark.

References

31, 37.

Dry root disease, *Ustulina zonata* Lév.

Symptoms

On old trees, the disease develops as a collar rot where the bark is decayed, from ground level to a height of 1–2 feet. The fungus also attacks the taproot. The cortex of the root then turns bluish to purple and has an offensive odor. The tree often falls over before any sign of disease is noticeable in the crown. There is no external mycelium. Roots frequently bear black nodules, 2 or 3 mm in diameter, which are white internally. Between the bark and the wood there is generally a thin film of white or brownish mycelium, arranged in fans, frequently with black lines bordering the outer ends of the fans.

Etiology

Fructifications are purple gray becoming black, thin, crustlike, brittle, and generally concentrically zoned or corrugated. Conidia are produced on the white mycelium. They are borne on erect stalks, measuring 4–8×2–3 μ. Perithecia are imbedded in the stroma with only the ostioles protruding. They are oblong, about 1 mm in diameter, and contain cylindrical asci, each with 8 opaque ascospores that measure 27–38×7–13 μ.

References

31, 39.

Black crust, *Catacauma huberi* (P. Henn.) Theis & Syd.

Symptoms

The fungus in its stromatic condition, specific to rubber, is very conspicuous. The black, round, shiny incrustations, 5–10 mm in diameter and on the undersurface of leaves, produce yellowish spots on the opposite side. The crusts may appear in circles, radiating outward along the veins, or they may be more confined in one or more concentric, circular spots. Often these spots are so plentiful as to cause defoliation.

Etiology

The perithecia of the fungus are imbedded in the crusty stroma with the ostioles protruding. The asci measure 50–65×16–20 μ, each containing 8 hyaline, oval, 1-celled ascospores, measuring 14–18×8–10 μ.

References

36, 39.

Bark disease, *Kretzschmaria micropus* (Berk.) Sacc.

Symptoms

This fungus produces a root rot, collar rot, or stump rot in which the roots become decayed and purplish. The fungus occurs frequently on old canker wounds and on tree stumps.

Etiology

Fructifications appear first as a cluster of erect, branched, white stalks, brown toward the tips, and up to 1 in. long. These subsequently darken from the base upward. The apex of the stalk swells out laterally so that a continuous cushion is formed that is grayish white at first, becoming black when old. Many consider the fungus to be a variation of *Ustulina zonata* Lév., possibly distinguished by the thinness of the stroma and the cluster of hemispherical bodies on the upper surface.

References

31, 36, 37, 39.

Target leaf spot, *Thanatephorus cucumeris* (Frank) Donk

Symptoms

Target leaf spot first appears on the leaves as small, translucent, circular areas, 2–10 mm in diameter. As the disease progresses and the leaves mature, the lesions become irregular in outline, light-colored, zonate, and often have a purple brown margin. Diseased areas become dry and shatter, leaving an irregular hole with a ragged edge. In nurseries and young plantings, the disease may cause defoliation.

Etiology

The basidial layer is formed by the compact hyphae that are white or pinkish. Basidia are formed in dense masses. They are club-shaped with sterigmata 4–6 μ long. Basidiospores are hyaline, pear-shaped, and $9–12 \times 6–7$ μ.

References

7, 17, 22, 31, 37, 39.

Pink disease, *Botryobasidium salmonicolor* (Berk. & Br.) Venkat.

Symptoms

This disease is conspicuous as a pink growth of interwoven hyphae over the bark. This pink patch gradually extends and may ultimately cover the whole circumference of the stem and adjacent branches for up to several feet. Under the central part of the patch, the bark has usually been killed and is brown, but toward the margin it is alive. The advancing margin is generally superficial. The pink patches are externally thin and, when old, split in lines more or less at right angles. Old specimens lose their pink color and become ochraceous or bleached to white.

Etiology

The basidial layer is formed by the compact hyphae that are white or pinkish. Basidia are club-shaped and formed in dense masses; the sterigmata are 4–6 μ long. Basidiospores are hyaline, pear-shaped, and $9–12 \times 6–7$ μ.

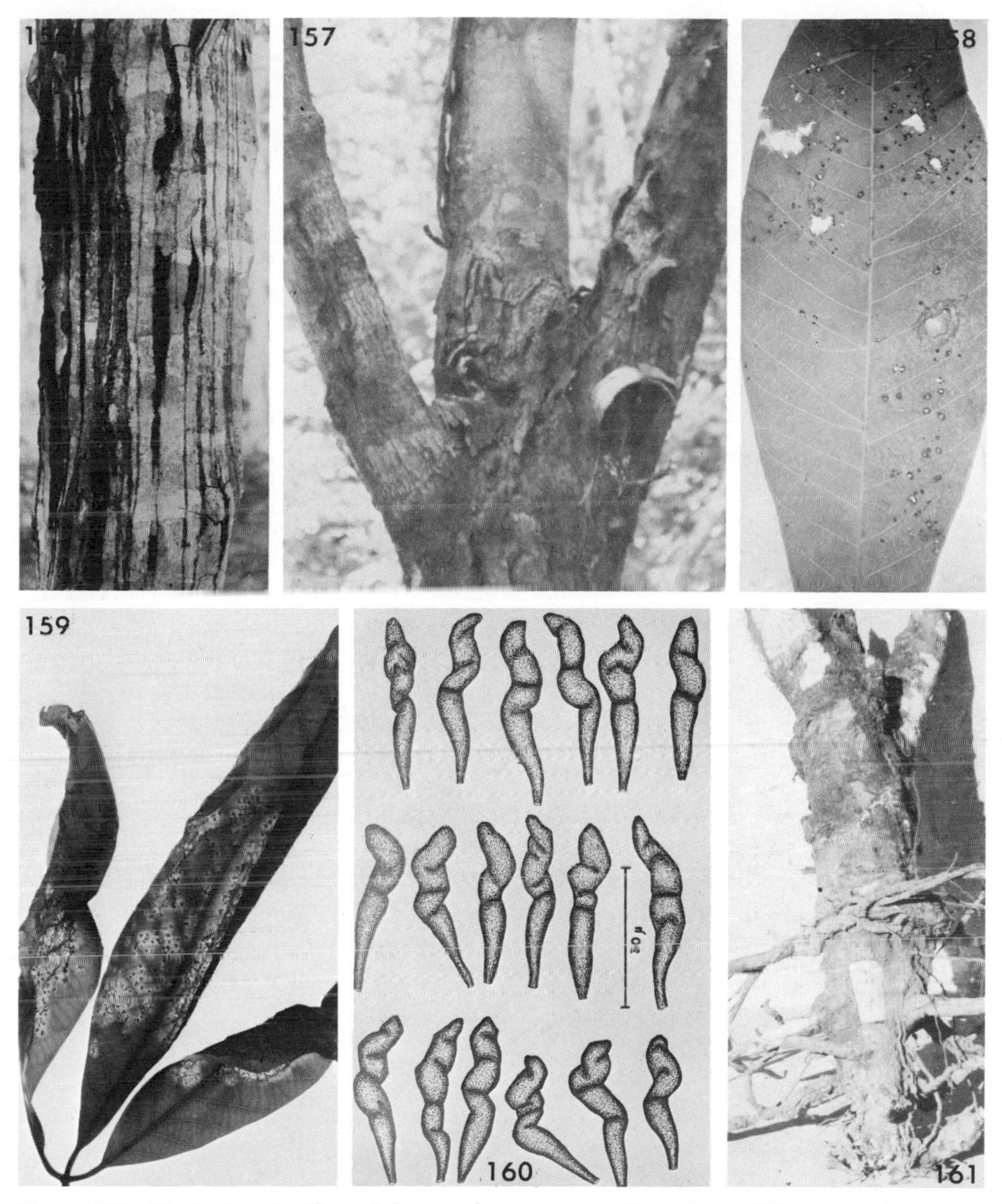

Figure 156. Black thread, *Phytophthora palmivora,* on rubber plant. (Photograph by J. Gonçalves.)

Figure 157. Black thread, *Phytophthora palmivora,* canker girdling branches of rubber tree. (Photograph by J. Gonçalves.)

Figure 158. Natural infection of South American leaf blight, *Microcyclus ulei,* on rubber foliage. (Photograph by C. H. Blazquez.)

Figure 159. Rubber foliage inoculated with *Microcyclus ulei* (South American leaf blight). (Photograph by C. H. Blazquez.)

Figure 160. Conidia of *Microcyclus ulei* (South American leaf blight). (Photograph by C. H. Blazquez.)

Figure 161. Brown crust root disease, *Fomes noxius,* girdling root of rubber plant. (Photograph by J. Gonçalves.)

References

7, 17, 31, 37, 39.

Red root rot, *Poria hypobrunnea* Petch

Symptoms

The aboveground indications are a gradual increase in a continuing unhealthy status of the tree. Root examination reveals stout, red strands of the fungus over the exterior surface that change to black with age. The roots may be more or less encrusted with soil particles. The wood is soft, pliable, and permeated with red layers or sheets.

Etiology

The sporophore is reddish brown and located at the crown of the plant aboveground. It is about 1/4 in. thick, resupinate, and may extend for several inches. The hymenium is shallow and composed of many pores, which are oriented vertically with the openings of the lower side.

References

31, 37.

Root rot, *Ganoderma amazonensis* Weir

Symptoms

The fungus is a wound parasite and is found on old trees, invading them at the root crown. The white, spongy decay on the submerged roots is wet and may extend for considerable distances along the laterals and above the soil surface. Removal of the bark exposes the brown wood of the root and reveals a white mycelium.

Etiology

The fungus produces a sporophore at the crown or on exposed roots in the form of an irregular bracket. The upper surface is brown and zonate; the margin and pores are white, and the context is white to brown.

Reference

39.

Wet red rot, *Ganoderma pseudoferreum* Wakef.

Symptoms

External deterioration, in the form of dieback, thinning of the foliage, and yellowing and withering of the leaves, develops rather slowly on old, bearing trees. They usually die as the root system is completely invaded. The disease has also been found on young trees up to 5 years of age. The woody parts of diseased trees become soft and spongy, and water can be pressed from this tissue. Red rhizomorphs develop on the external surfaces of the roots. There often develops a proliferation of adventitious roots above the diseased areas.

Etiology

The sporophore, not frequently found, is attached to the tree trunk at its base, is shelving, sessile, brown, with a white lower surface composed of shallow pores. The margin is light-colored. The conks may be 4–6×10–12 in. in width and length. The basidiospores are brown, oval, and 6–8×4–5 μ.

Reference

37.

Brown root rot, *Fomes lamaensis* (Murr.) Sacc. & Trott.

Symptoms

A diseased tree shows a slow decline, yellowing and shedding of leaves, and finally death. Roots are encrusted with a mass of sand, earth, and small stones to a thickness of 3 or 4 mm. The encrusting masses intermingled with brown threads serve to distinguish this disease. Roots become dark brown and almost black. Taproots may be in an advanced stage of decay before the fungus has spread to the laterals sufficiently to cause any marked symptoms at the crown. If the encrusting mass is cut away, the cortex on diseased roots is brown or mottled with internal white patches. Wood is also brown or brown mottled with white internal patches and is soft and pliable with a network of fine brown lines. The disease is spread almost entirely by root contact.

Etiology

Mycelium strands consist of tawny brown threads which are collected into small sheets or loose masses, either on the surface or embedded in the crust of soil. They are frequently brownish to almost white, and the masses are intermingled with the tawny brown. The sporophores are brown, perennial, zoned, hard, shelved, and 13–25 cm wide; the growing margin is creamy yellow, and the spores measure 3–5×3–4 μ.

References

31, 37.

Brown crust root disease, *Fomes noxius* (Berk.) Corner

Symptoms

The infected plants show wilting, drying, breaking, falling of branches, and roots covered with a thin crust of mycelial stroma and soil. At a later stage, the crust is black, thick, velvety, and 2 cm in diameter. In the wood, networks of black or brown lines and white mycelium grow between wood and bark. Fructifications are rare; they appear at the base of the trunk arranged in thin, hard crusts, purple brown at the surface and pale brown within.

Etiology

Sporophores are yellow to brown, dimidiate, shelved or resupinate, dry, surface glabrous, and form a crust on old stumps. Hymenial pores are deep brown and stratified; spores are hyaline, 1-celled, smooth, round or oblong, and 5 μ long. The following fungi are most important on the roots of rubber trees: Stinking root disease, *Sphaerostilbe repens*; dry root rot, *Ustulina zonata*; white root rot, *Fomes noxius*; wet red root rot, *Ganoderma pseudoferreum;* poria red root rot, *Poria hypobrunea*; and mushroom root rot, *Armillaria mellea.* The differential characteristics by which they may be correctly diagnosed are frequently not distinct or mostly lacking entirely.

References

7, 30, 31, 34, 35, 37.

White root rot, *Fomes lignosus* (Klotzsch) Bres.

Symptoms

Young trees show a typical yellowing and an inward curling of the leaf margins in one or more leaf whorls. Early flowering before the fifth year is frequently considered an indication of root infection. The fungus forms thin mats, composed of strands of mycelium, about 1 mm thick on the roots of affected plants. These rhizomorphs are white when young and become yellowish brown with age. A diseased root shows a verrucose pattern externally. When the fungus in lateral roots of a young tree reaches the taproot, it forms a mat or collar around it, and the tree soon dies.

Etiology

The sporophore is sessile, reflexed, intricate, leathery when fresh, and hard and rigid when dry. The margins are thin, entire, incurved, and the top of the pileus is reddish to cinnamon brown, glabrous, applanate to convex, and zoned. The hymenial surface is pink; pores are in one layer, often cover an older hymenium, and 45–80 μ in diameter. Basidiospores are hyaline, smooth, globose, apiculate, and 3.6×4.6 μ.

References

6, 13, 31, 34, 37, 40.

Horsehair blight, *Marasmius equicrinis* Muell.

Symptoms

The fungus is composed of mycelial strands, cords, or rhizomorphs that are black, long, horsehairlike in size and structure, extend from twig to leaf or vice versa, and adhere to the host by the formation of small brown pads. These cords do not fan out to cover the leaf surface, neither are they twining. The organism is considered nonparasitic.

Etiology

The spore-producing structures are small, stipitate mushrooms that are not frequently observed. They are white at first, then yellow, reddish, and brown. They arise directly from the mycelium and are sessile or have a short stipe. The pileus is 4–8 mm in diameter. They are membraneous, hemispheric, with 5–8 lamellae, and as many as 12. The basidia are ovoid, 12–18 × 4–6 μ, and produce 4 sterigmata, 2–3 μ long. The basidiospores are hyaline, ovoid with round and pointed ends, somewhat reniform, and 7–11 × 4–6 μ.

References

31, 35, 39.

Seedling disease, *Pestalotia palmarum* Cke.

Symptoms

The fungus frequently appears close to the soil line of seedlings that exhibit a stem lesion definitely limited in extent. The lesions are usually deep enough to cause the seedling to bend over and succumb. The fungus appears sparingly on the surface of these areas.

Etiology

Small, black pustules produce an abundance of conidia that possess 3 diverging setulae at the tip; each setula is 15–22 μ long and a single pedicel at the base of each is attached to hyaline cells with 3 brown cells in between. The conidia are slightly curved and measure 14–30 × 5–7 μ.

References

14, 31.

Rim blight, *Ascochyta heveae* Petch

Symptoms

Leaves show numerous yellowish, small, circular spots in the first stage of infection. They are crowded in groups between the principal veins and along the leaf margins. They increase in size, coalesce, and may involve large areas around the outer borders. Loss of leaf area and lower yield are the principal results of the disease.

Etiology

The pycnidia are black, submerged, ostiolate, and 60–100 μ in diameter. The pycnidiospores are hyaline, oval, 1-septate, and 15–20 × 3–4 μ.

Reference

31.

Leaf spot, *Periconia heveae* Stevenson & Imle

Symptoms

Leaf spots are circular to oval, not vein-limited, and appear the same on both surfaces. Primary lesions vary in diameter from 2 to 10 mm but frequently coalesce. Spots are brown at first, become ashen at the center, and have a brown border; on mature leaves they are often ringed by a chlorotic halo. Necrotic areas split irregularly and may partially fall away. Petiole lesions may extend to twigs, causing cankers or dieback.

Etiology

Vegetative mycelium is light brown, scanty, septate, branched, and 3–4 μ in diameter. Conidiophores are dark brown, numerous, amphigenous, scattered, erect, rigid, unbranched, 2- rarely 3-septate, and 250–400 μ high. The bulbous basal cell is 45–90 μ long, 24–30 μ in diameter at the base, and 15–18 μ in diameter above. Apical cells are light brown, short, clavate, slightly constricted at the septum, and $30–45 \times 20–25$ μ. Sporogenous cells, in a whorl at the base of the apical cell, measure $10–15 \times 18–24$ μ. Conidia are deep brown, globose, strongly verrucose, 25–45 μ in diameter, and in short, catenulate chains with a terminal conidium.

References

24, 38.

Powdery mildew, *Oidium hevea* Stein

Symptoms

This disease is most commonly found on young foliage, which appears dull in contrast to healthy leaves. The infected leaves are crinkled at the tips and are darkened, blue to black. They are rapidly shed. Often the loss of leaves results in barren twigs and branches. New leaves and flowers are similarly attacked.

Etiology

The fungus may appear white and chalky on the leaves but more often is observed with difficulty, as the superficial mycelium is rather scanty and spore production limited. The hyphae are hyaline to white, branched, septate, and produce haustoria within the epidermal cells. Erect conidiophores are hyaline, 1-celled, and cylindrical to elongate. Conidia are produced, often in chains of several spores, each spore measuring $35–82 \times 12–28$ μ.

References

3, 20, 37.

Leaf spot, *Alternaria* sp.

Symptoms

In heavily infected young leaves, partial or entire defoliation occurs early. Such leaves may have a scorched appearance. Infected leaves that develop to maturity may have brown, concentrically zonate spots along the main veins. Many of the infected leaves that reach maturity have dead distal portions.

Etiology

Mycelium is fluffy, and gray in culture. Conidiophores measure 200×4–6 μ. Conidia are single or in short chains, contain 3–9 transverse septa and 0–3 longitudinal septa, and measure 50–106×10–17 μ.

Reference

27.

Bird eyespot, *Helminthosporium heveae* Petch

Symptoms

Eyespot is characterized by scattered spots on the leaf surface. They are purple, turning white or semitransparent in the center, and, at a later stage, become concentrically zoned and surrounded by a narrow brown purple line. The spots are round or irregular and 1–5 mm in diameter. Infection produces elongate lesions on the stems, petioles, and main leaf veins. The fungus infects very young plants in the nursery, rarely older plants.

Etiology

Conidiophores arise in the center of spots as olive dots, black en masse, isolated, simple, and 80–200 μ long. Conidia are brown or dark brown, elongate, 8–11-septate, and 100–200×15–18 μ.

References

15, 21, 31, 36.

Algal spot, *Cephaleuros virescens* Kunze

Symptoms

This organism causes a purplish spotting of leaves. The spots are found mostly on the older leaves, and as they mature they become covered with yellowish hairs with swollen tips.

Etiology

The tips expand and produce the orange sporangia that measure 40–50 μ. The sporangia produce biciliate zoospores, which are motile for a short period then become stationary and germinate to form a new plant.

References

6, 18, 37.

Leaf spot, *Corynespora cassiicola* (Berk. & Curt.) Wei

Symptoms

The spots on the foliage are yellow brown, small, more or less circular, and enlarge up to 6 mm in diameter. At the maximum size, the centers become dark brown and striations are formed. Leaves turn brown and are shed.

Etiology

The mycelium is olive green in culture, septate, branched, and produces brown conidiophores that are 3–5-septate and 86–273 μ long and 3–8 μ wide. The conidia are hyaline, slender, 4–13-septate, formed in chains joined by an isthmus, and are 58–195 × 8–16 μ.

Reference

1.

Other fungi associated with rubber

Armillaria mellea Vahl ex Fries
Curvularia maculans (Bancroft) Boedijn
Diaporthe heveae Petch
Didymella oligospora Saccardo
Fusarium javanicum Koorders
Guignardia heveae Sydow
Helicobasidium compactum Boedijn
Macrophomina phaseolina (Tassi) Goidanich
Meliola heveae Vincens
Mycosphaerella heveae Vincens
Ophiobolus heveae Henning
Phoma heveae Petch
Phomopsis heveae (Petch) Boedijn
Phyllosticta heveae Zimmermann
Physalospora rhodina (Berkeley & Curtis) Cooke
Polyporus rugulosus Léveillé
Rosellinia bunodes (Berkeley & Broome) Saccardo
Scolecotrichum heveae Vincens
Septobasidium heveae Couch
Sphaerella heveana Saccardo
Xylaria thwaitesii Berkeley & Cooke

References: rubber

1. Awoderu, V. A. 1969. A new leaf spot of para rubber (Hevea brasiliensis) in Nigeria. Plant Disease Reptr. 53:406–8.
2. Barley, E. F., and S. B. Silverborg. 1952. Black thread of Hevea brasiliensis in Liberia. Phytopathology 42:547–58.
3. Beeley, F. 1939. Oidium heveae, a report on the 1939 outbreak. J. Rubber Res. Inst. Malaya 9:59–67.
4. Blazquez, C. H. 1963. Bird's eye leafspot, a severe disease of mature rubber trees. Phytopathology 53:24. (Abstr.)
5. Blazquez, C. H., and J. H. Owen. 1963. Histological studies of Dothidella ulei on susceptible and resistant Hevea clones. Phytopathology 53:58–65.
6. Bose, S. R., and B. K. Bakshi. 1957. Polyporus lignosus Klotzsch and its identity. Trans. Brit. Mycol. Soc. 40:456–60.
7. Butler, E. J. 1918. Fungi and diseases in plants. Thacker: Calcutta, India.
8. Carpenter, J. B. 1954. An epidemic of phytophthora leaf fall of hevea rubber trees in Costa Rica. Phytopathology 44:597–601.
9. Carpenter, J. B. 1954. Moldy rot of the

hevea rubbertree in Costa Rica. Plant Disease Reptr. 38:334–37.
10. Carpenter, J. B., and M. H. Langford. 1950. Target leaf spot of hevea rubber in Costa Rica. Plant Disease Reptr. 34: 56–57.
11. Chee, K. H. 1968. Patch canker of Hevea brasiliensis caused by Phytophthora palmivora. Plant Disease Reptr. 52:132–33.
12. Chee, K. H., T. M. Lim, and R. L. Wastie. 1967. An outbreak of phytophthora leaf fall and pod rot on Hevea brasiliensis in Malaysia. Plant Disease Reptr. 51: 443–46.
13. Dijkman, M. J. 1951. Hevea. University of Miami Press: Florida.
14. Guba, E. F. 1929. Monograph of the genus Pestalotia, de Notaris. Phytopathology 19:191–232.
15. Hilton, R. N. 1952. Bird's eye spot leaf disease of the hevea rubber tree caused by Helminthosporium heveae Petch. J. Rubber Res. Inst. Malaya Comm. 280, 14:40–82.
16. Hilton, R. N. 1955. South American leaf blight. J. Rubber Res. Inst. Malaya Comm. 293, 14:287–354.
17. Hilton, R. N. 1958. Pink disease of hevea caused by Corticium salmonicolor Berk. et Br. J. Rubber Inst. Malaya Comm. 322, 15:275–92.
18. Hilton, R. N. 1959. Maladies of hevea in Malaya. Rubber Research Institute: Kuala Lumpur, Malaya, pp. 1–101.
19. Holliday, P. 1970. South American leaf blight, Microcyclus ulei of Hevea brasiliensis. Commonwealth Mycological Institute, Pathological Paper 12.
20. Hubert, F. P. 1957. Diseases of some export crops in Indonesia. Plant Disease Reptr. 41:55–63.
21. Kevorkian, A. G. 1948. Bird's eye spot disease of hevea rubber in Nicaragua. Phytopathology 38:1025–27.
22. Kotila, J. E. 1945. Rhizoctonia foliage disease of Hevea brasiliensis. Phytopathology 35:739–41.
23. Langford, M. H. 1945. South American leaf blight of hevea rubbertrees. U.S. Dept. Agr., Tech. Bull. 882:1–31.
24. Langford, M. H., et al. 1954. Hevea diseases of the western hemisphere. Plant Disease Reptr. Suppl. 225:37–41.
25. Manis, W. E. 1954. Phytophthora leaf fall and dieback. Plant Disease Reptr. Suppl. 225:49–52.
26. Martin, W. J. 1947. Alternaria leaf blight of hevea rubber trees. Phytopathology 37:609–12.
27. Martin, W. J. 1949. Moldy rot of tapping panels of hevea rubbertrees. U.S. Dept. Agr. Circ. 798:1–23.
28. Miller, J. W. 1966. Differential clones of hevea for identifying races of Dothidella ulei. Plant Disease Reptr. 50:187–90.
29. Müller, E., and J. A. von Arx. 1962. Dis gattungen der didymosporen Pyrenomyceten. Beitr. kryptog. Flora Schweiz 11: 374.
30. Newsam, A., et al. 1964. Conference review. Root diseases of hevea. Rubber Res. Inst. Malaya, Planters Bull. 75: 207–62.
31. Petch, T. 1921. The diseases and pests of the rubber tree. Macmillan: London.
32. Rands, D. R. 1924. South American leaf diseases of para rubber. U.S. Dept. Agr. Bull. 1286:1–18.
33. Riggenbach, A. 1960. On Fomes lignosus (Klotzsch) Bres., the causative agent of the white root disease of the para rubber tree, Hevea brasiliensis Mull. Arg. Phytopathol Z. 40:186–212.
34. Riggenbach, A. 1966. Observations on root diseases of hevea in West Africa and the East. Trop. Agr. (London) 43:53–58.
35. Roger, L. 1951. Etude descriptive des maladies parasitaires des plantes des pays chauds. Encyclopedic Mycologique 17:648–53, 1038–39, 1091–93.
36. Roger, L. 1953. Phytopathologie des pays chauds. Encyclopedie Mycologique 18:1449–50, 1454–58, 1768–70, 2042–43.
37. Sharples, A. 1936. Diseases and pests of the rubber tree. Macmillan: London.
38. Stevenson, J. A., and E. P. Imle. 1945. Periconia blight of hevea. Mycologia 37: 576–81.
39. Weir, J. R. 1926. A pathological survey of the para rubber tree (Hevea brasiliensis) in the Amazon Valley. U.S. Dept. Agr. Bull. 1380:1–130.
40. Wijewantha, R. T. 1964. Influence of environment on incidence of Fomes lignosus in rubber replantations in Ceylon. Trop. Agr. 41:69–75.

Safflower, *Carthamus tinctorius* Linnaeus

Leaf spot, *Pseudomonas syringae* van Hall
Root rot, *Phytophthora drechsleri* Tucker

Blight, *Phytophthora palmivora* Butler
Wilt, *Sclerotinia sclerotiorum* (Libert) De Bary
Rust, *Puccinia carthami* Corda
Wilt, *Fusarium oxysporum* Schlechtendahl f. sp. *carthami* Klisiewicz & Houston
Leaf spot, *Cercospora carthami* Sundararaman & Ramakrishnan
Leaf spot, *Alternaria carthami* Chowdhury
Wilt, *Verticillium albo-atrum* Reinke and Berthold
Other fungi associated with safflower

Leaf spot, *Pseudomonas syringae* van Hall

Symptoms

Necrotic spots and streaks develop on plants in the rosette stage. The tissues in the center of the spots, which are up to 2–6 mm in diameter, become translucent with a darker margin and a dark discoloration that often extend into the stems and roots, causing the plants to become stunted and soon die.

Etiology

The organism is a gram-negative rod, single, paired, or in short chains, forming capsules, motile by 1 to several polar flagella, and measuring 1.2–1.8 × 0.6 μ. Colonies are white, transparent, round, smooth or wrinkled, and convex. A green fluorescent pigment is produced in culture. Optimum temperature, 28–30C.

References

2, 4.

Root rot, *Phytophthora drechsleri* Tucker

Symptoms

In the seedling stage a typical damping-off may be observed on plants that have escaped injury before emerging. A softening of the stem so weakens the young plants that they fall over, shrivel, and die. On older plants, black necrotic lesions appear at the soil line or just above, curtailing nutrition. The plant wilts, the leaves turn yellow, and the plant soon dies. The cortex in the lesions is brown to slate black, and the vascular tissue and pith become brown.

Etiology

The mycelium is hyaline, nonseptate, and branched. The sporangiophores are narrower than the hyphae, and the sporangia are hyaline, nonpapillate, pyriform to ovate, and 24–38 × 15–24 μ. No chlamydospores are found. Oospores are hyaline to faintly colored, spherical, smooth, granular, and 16–45 μ in diameter.

References

3, 15, 16.

Blight, *Phytophthora palmivora* Butl.

Symptoms

Seedlings are usually killed by a typical damping-off. Older plants are girdled but remain standing after death. On more mature plants, cankers develop on the lower stems; they are mostly long and narrow, causing the plant to bend over and lodge. When plants are woody, collar rot lesions develop more slowly. A dark brown decay involves the roots. Generally the disease develops rapidly, changing foliage color from dark to light green, yellowish, and finally brown.

Etiology

The mycelium is hyaline, branched, nonseptate, and appears tufted in culture. Sporangia are ovate to almost spherical, have prominent papillae, and measure 27–47×18–29 μ. They germinate by germ tube or zoospore production. Chlamydospores are hyaline, spherical, and average 21–22 μ in diameter.

References

9, 10.

Wilt, *Sclerotinia sclerotiorum* (Lib.) d By.

Symptoms

This disease usually is most evident in the field at flowering time and continues until the plants are mature. Wilting is the first symptom and increases in severity until the plant dies. A white, cottony fungus mycelium can sometimes be observed at the soil line where a cankerous lesion forms. The diseased area is somewhat sunken and a dull, water-soaked green that later becomes brown. Large, black sclerotia are produced externally and in the pith of killed plants. Often the killing of plants is rapid and sporadic. The stem becomes frayed and deteriorates into shreds.

Etiology

The mycelium is white, septate, branched, and produced abundantly. The sclerotia are black, white within, shiny, rubbery to hard, variously shaped, and 2–6 mm. The sclerotia upon germination produce 1 to several stipitate apothecia per sclerotium. The stipes may be very short or up to 2 in. long. The apothecia are pale pink to tan, saucer-shaped, and up to 1 cm in diameter. The asci are hyaline, long, and slender. The ascospores are hyaline, 1-celled, oval, uniseriately arranged, and 11–16×5–8 μ. No true conidia are formed; microconidia are small and produced in chains but apparently do not function in disease production.

References

5, 8, 12, 17.

Rust, *Puccinia carthami* Cda.

Symptoms

Cotyledons, leaves, tender stems, crowns, and roots become diseased. Chestnut brown, erumpent, scattered, pulverulent pustules appear.

Etiology

Pycnia and aecia are unknown; however, fruiting structures that resemble uredia are actually functioning aecia. These spores, which qualify as uredospores, are globoid, echinulate, with 2–3 equatorial pores, and measure 17–23×22–26 μ. Telia are chocolate brown; teliospores are ellipsoidal, slightly constricted, minutely verrucose, with short pedicels, and measure 23–29×32–42 μ.

References

11, 19.

Wilt, *Fusarium oxysporum* Schlecht. f. sp. *carthami* Klis. & Houst.

Symptoms

Plants are killed by damping-off in the seedling stage. Stem invasion results in partial yellowing and wilting. The yellowing of foliage is first evident on the lower leaves, which eventually turn brown and die; the entire plant is often killed. Frequently only one side of a plant appears affected, while the opposite side shows no indication of disease. A brown discoloration develops in the vascular bundles.

Etiology

The tufted mycelium is hyaline, white or pink en masse, branched, and septate. Microconidia are hyaline, oval to elliptical, 1-celled, slightly curved, and 5–16×2–3 μ. Macroconidia are hyaline, may be up to 5-septate but are mostly 3-septate, constricted at septa, borne in sporodochia, straight or curved, often pointed at the tip with rounded base, and measure 10–36×3–6 μ, mostly 28×4–5 μ. Chlamydospores are 1-celled, faintly colored, and 5–13×10 μ. Sclerotia are irregular in shape and up to 5 mm in diameter. The fungus form species is determined by specific pathogenicity.

References

6, 7, 13.

Leaf spot, *Cercospora carthami* Sund. & Ramak.

Symptoms

Circular, brown, slightly sunken spots appear on the lower leaves after the plants are about a month old. As the disease develops, the leaves turn brown, become distorted, and the tissue between the principal veins falls away. The bracts show reddish brown spots, and if the flower bud becomes affected it does not open. If infection occurs later, usually no seeds are formed.

Etiology

The mycelium is hyaline becoming smoky brown, septate, branched, and collects in the stomatal areas where stromata are formed. The conidiophores emerge separately or in fascicles on both leaf surfaces. They are brown, septate, and variable in length, averaging $150 \times 5\ \mu$. The conidia are hyaline, 2–20-septate, obclavate, and $140–200 \times 3–7\ \mu$.

Reference

14.

Leaf spot, *Alternaria carthami* Chowdhury

Symptoms

The leaf spots at first are yellow to brown and 1–2 mm in diameter. They increase up to 1 cm and show characteristic concentric, light and dark bands. If numerous, some coloring takes place, irregular blotches develop, and parts break away. The foliage is most severely affected, although stems and petioles may be involved. Flowers are blighted or killed.

Etiology

The mycelium is pale to dark-colored, branched, and septate with slight constrictions. Conidiophores are erect, rigid, septate, and emerge through stomata. The conidia are light brown, in chains, with long beaks, and with 3–11 cross septations and up to 6 longitudinal septations. They measure $171 \times 36\ \mu$ overall and $99 \times 36\ \mu$ without the beak.

Reference

1.

Wilt, *Verticillium albo-atrum* Reinke & Berth.

Symptoms

Although the term wilt is applied to the disease, there is very little drooping and flaccidity. The diseased plants first show a bright yellow discoloration of the lower leaves which extends to the entire plant. Seedlings may show symptoms, but generally they do not become prominent until blossoming time. Some stunting may take place, resulting in poor seed production. Plants are not rapidly killed. Occasionally there is a yellowing of one side of the leaf blade while the other half remains green, resulting in a curved or lopsided leaf. The stem shows dark, discolored vascular tissue and black pin stripes in the wood.

Etiology

The fungus mycelium is hyaline to light-colored, branched, septate, and may be soil- or seed-transmitted. The conidiophores arise in whorls at node-like points along hyphal branches, each in turn developing terminal conidia that form in chains or in wet spherical clusters. The spores are hyaline, 1-celled, oval, and $5–7 \times 3\ \mu$.

Reference

18.

Other fungi associated with safflower

Botrytis cinerea Persoon ex Fries
Cercosporella carthami Murashkinsky
Erysiphe cichoracearum De Candolle
Gloeosporium carthami (Fukui) Hori & Hemmi
Leveillula taurica (Léveillé) Arnaud
Macrosporium carthami Săvulescu
Ramularia carthami Zaprometov
Septoria carthami Murashkinsky

References: safflower

1. Chowdhury, S. 1944. An alternaria disease of safflower. J. Indian Botan. Soc. 23:59–65.
2. Elliott, C. 1951. Manual of bacterial plant pathogens. 2d ed. Chronica Botanica: Waltham, Mass., pp. 88–93.
3. Erwin, D. C. 1952. Phytophthora root rot of safflower. Phytopathology 42:32–35.
4. Erwin, D. C., M. P. Starr, and P. R. Desjardins. 1964. Bacterial leaf spot and stem blight of safflower caused by Pseudomonas syringae. Phytopathology 54: 1247–50.
5. Joshi, S. D. 1955. A wilt disease of safflower. Mem. Dept. Agr. India 13:39–46.
6. Klisiewicz, J. M., and B. R. Houston. 1962. Fusarium wilt of safflower. Plant Disease Reptr. 46:748–49.
7. Klisiewicz, J. M., and B. R. Houston. 1963. A new form of Fusarium oxysporum. Phytopathology 53:241.
8. Knowles, P. F., and M. D. Miller. 1965. Safflower. Calif. Univ. Agr. Expt. Sta. Ext. Circ. 532:1–50.
9. Larson, N. 1960. Safflower, a list of selected references. U.S. Dept. Agr. Library List 73.
10. Malaguti, G. 1950. Phytophthora blight of safflower. Phytopathology 40:1154–56.
11. Prasada, R., and H. P. Chothia. 1950. Studies on safflower rust in India. Phytopathology 40:363–67.
12. Purdy, L. H. 1955. A broader concept of the species Sclerotinia sclerotiorum based on variability. Phytopathology 45: 421–27.
13. Smith, E. F., and D. B. Swingle. 1904. The dry rot of potatoes due to Fusarium oxysporum. U.S. Dept. Agr. Bull. 55:1–64.
14. Sundararaman, S., and T. S. Ramakrishnan. 1928. A leafspot disease of safflower (Carthamus tinctorius) caused by Cercospora carthami nov. sp. Agr. J. India 23:383–89.
15. Thomas, C. A. 1951. The occurrence and pathogenicity of Phytophthora species which cause root rot of safflower. Plant Disease Reptr. 35:454–55.
16. Tucker, C. M. 1931. Taxonomy of the genus Phytophthora de Bary. Missouri Univ. Agr. Expt. Sta. Res. Bull. 153:5–197.
17. Walker, J. C. 1950. Plant pathology. McGraw-Hill: N.Y., pp. 364–69.
18. Zimmer, D. E. 1962. Verticillium wilt of safflower in the United States—a potential problem. Plant Disease Reptr. 46:665–66.
19. Zimmer, D. E. 1963. Spore stages and life cycle of Puccinia carthami. Phytopathology 53:316–19.

Salsify, *Tragopogon porrifolius* Linnaeus

White rust, *Albugo tragopogonis* Persoon ex S. F. Gray
Decay, *Sclerotinia intermedia* Ramsey
Wilt, *Verticillium albo-atrum* Reinke & Berthold
Leaf spot, *Stemphylium botryosum* Wallroth var. *tragopogoni* Linnaeus
Leaf spot, *Cercospora tragopogonis* Ellis & Everhart
Other fungi associated with salsify

White rust, *Albugo tragopogonis* Pers. ex S. F. Gray

Symptoms

Pale green to light yellow areas, irregular in outline, develop on the leaves, forming raised, blisterlike swellings that cause the epidermis to rupture and curl back, revealing a powdery mass of white spores that are readily scattered. These sori may vary in size up to 1 cm but are usually narrower. When plentiful and when infection appears in the seedling stage, foliage and sometimes entire plants are severely stunted or killed.

Etiology

The mycelium is hyaline, branched, and gives rise to short conidiophores, measuring $40–50 \times 12–15\ \mu$, that produce conidia usually in chains. They are hyaline, 1-celled, spherical to oval or angular, and $18–22 \times 12–18\ \mu$. Oospores with reticulate walls mature in the host tissue. They are brown to black and $68 \times 44\ \mu$.

References

3, 4, 8.

Decay, *Sclerotinia intermedia* Ramsey

Symptoms

This disease is characterized by a wet, soft rot that develops on the fleshy roots mostly at the soil line. After removal from the soil, diseased areas develop anywhere on the root. The affected areas are usually overgrown with white to buff mycelium that produces the sclerotia.

Etiology

The mycelium is hyaline, branched, and septate, becoming slightly colored en masse. It produces sclerotia that are usually external on the roots but associated with the mycelium and the diseased areas. They are black, hard, shiny, variable in shape, measuring up to 3 mm in diameter, mostly smaller, and often in linear strings or extensive mats. The apothecia produced upon germination are funnel- to disc-shaped, up to 3.5 mm wide, and about 9 mm high. Asci are 8-spored, uniserrate, and $121–131 \times 7–8\ \mu$. The ascospores are hyaline, oval, 1-celled, and $10–16 \times 3–6\ \mu$. Paraphyses are present. Microconidia are hyaline, globose, and $3–4\ \mu$ in diameter.

Reference

7.

Wilt, *Verticillium albo-atrum* Reinke & Berth.

Symptoms

The aboveground parts develop intermittent wilt, following an unhealthy, droopy appearance. The wilt becomes permanent, and the plant gradually declines and eventually dies. The fleshy roots usually show a brownish discoloration in streaks in the vascular regions.

Etiology

The mycelium is hyaline, delicate, septate, branched, and produces sclerotia. The conidiophores are hyaline, branch in whorls, and produce conidia at the various tips in chains, often clumped in wet heads. The conidia are hyaline, 1-celled, oval, and 4–11 × 2–5 μ.

Reference

6.

Leaf spot, *Stemphylium botryosum* Wallr. var. *tragopogoni* L.

Symptoms

The first symptoms appear as light brown lesions on the tips of the older leaves, followed by small, light brown, necrotic spots on the leaf blades. The spots enlarge to 3–4 mm and become brown with gray centers. A brown exudate is sometimes found on either surface of the spots. Heavy infection results in extensive loss of foliage, yellowing of the plant, and death.

Etiology

Hyphae are brown, septate, and branched. Conidiophores are brown, in fascicles, swollen at the apex, measure 6–8 μ, and proliferate through the apex scar, which measures 4–5.5 × 25–180 μ. Conidia are acrogenous, brown, minutely verrucose, ovoid-oblong, muriform septate, with 1 to several vertical and 1–3 cross walls, constricted, and measure 17–56 × 8–26 μ.

Reference

5.

Leaf spot, *Cercospora tragopogonis* Ell. & Ev.

Symptoms

Oval to linear leaf spots, up to 6–8 mm long, with gray centers, are surrounded by brown borders and often coalesce. Sometimes there is considerable defoliation of older leaves.

Etiology

Conidiophores are smoky to brown, in dense fascicles, septate, straight or curved, and measure 40–150 × 4–7 μ. Conidia are hyaline, acicular, multiseptate, and 30–200 × 2–4 μ.

Reference

2.

Other fungi associated with salsify

Alternaria tenuis Eisenbeck

Botryobasidium rolfsii (Saccardo) Venkatarayan
Erwinia carotovora (L. R. Jones) Holland
Erysiphe cichoracearum De Candolle
Phomopsis albicans Sydow
Phymatotrichum omnivorum (Shear) Duggar
Puccinia extensicola var. *hieraciata* (Schweinitz) Arthur
Puccinia hysterium Rohling
Sporodesmium scorzonerae Aderhold
Streptomyces scabies (Thaxter) Waksman & Henrici
Thanatephorus cucumeris (Frank) Donk
Thielaviopsis basicola (Berkeley & Broome) Ferraris
Ustilago tragopogonis-pratensis (Persoon) Rousseau

References: salsify

1. Arthur, J. C. 1934. Manual of the rusts of the United States and Canada. Purdue Research Foundation: Lafayette, Indiana, p. 199.
2. Chupp, C. 1953. A monograph of the fungus genus Cercospora. Published by the author: Ithaca, N.Y., p. 163.
3. Chupp, C., and A. F. Sherf. 1960. Vegetable diseases and their control. Ronald Press: N.Y., pp. 474–75.
4. Heald, F. D. 1926. Manual of plant diseases. McGraw-Hill: N.Y., pp. 407–14.
5. Linn, M. B. 1942. Leaf-spot disease of cultivated salsify. Phytopathology 32: 150–57.
6. Longrée, K. 1940. Wilt of salsify, caused by Verticillium sp. Phytopathology 30: 981–83.
7. Ramsey, G. B. 1924. Sclerotinia intermedia n. sp. a cause of decay of salsify and carrots. Phytopathology 14:323–27.
8. Wilson, G. W. 1907. Studies in North American Peronosporales. I. The genus Albugo. Bull. Torrey Botan. Club 34: 61–84.

Sapodilla, *Achras zapota* Linnaeus

Fruit rot, *Phytophthora palmivora* Butler
Scab, *Elsinoë lepagei* Bitancourt & Jenkins
Rust, *Scopella sapotae* Mains
Mushroom root rot, *Clitocybe tabescens* (Scopoli ex Fries) Bresadola
Leaf spot, *Pestalotia sapotae* P. Henning
Limb gall, *Pestalotia scirrofaciens* N. A. Brown
Other fungi associated with sapodilla

Fruit rot, *Phytophthora palmivora* Butl.

Symptoms

This disease has been observed only on the fruit, mainly on the lower ones; it causes a rotting and severe fruit shedding. Inoculum apparently originates in the soil. Inoculations have shown immature fruit to be susceptible as well as mature ones, with or without injury.

Etiology

Sporangia and chlamydospores are produced abundantly, the former measuring $53 \times 29\ \mu$, the latter $34\ \mu$ in diameter.

Reference

7.

Scab, *Elsinoë lepagei* Bitanc. & Jenkins

Symptoms

The first indication of the disease is the appearance of dry, small, circular to angular specks on the leaf blades. They are slightly sunken, water-soaked, less than 1 mm in diameter, and are surrounded frequently by a dark band. The spots never exceed 1 mm in diameter but may be numerous enough to cause malformed leaves and possibly some defoliation. As the spots mature, the inner parts may become light-colored and somewhat papery.

Etiology

The fruiting ascomata develop slowly below the epidermis, finally emerging. They contain irregularly placed, imbedded asci, and measure up to 250×15–18 μ. The asci are spherical to oval, 14×22 μ, and contain up to 8 ascospores, sometimes fewer. Each ascospore is hyaline, 3-septate, somewhat lanceolate in outline, and 8–14×3–8 μ. The conidia are formed in acervuli with short conidiophores. Conidiophores are hyaline, 1-celled, oval to elongate, and 5–10×2–4 μ.

References

2, 9.

Rust, *Scopella sapotae* Mains

Symptoms

The rust appears on the foliage in the form of yellow to reddish brown leaf spots. They are mostly small and slightly raised on the lower surface. The epidermis breaks open to free the spores.

Etiology

The pycnia and aecia are unknown. The uredinia are subepidermal, scattered, round or oblong, pulverulent, and up to 300 μ in diameter. The uredospores are dark cinnamon brown, finely echinulate, and 20–21×18–23 μ. Teliospores are produced in waxy telia that are subepidermal, obovoid, 32–40×16–22 μ, and 1-celled, with a hyaline pedicel.

Reference

3.

Mushroom root rot, *Clitocybe tabescens* (Scop. ex Fr.) Bres.

Symptoms

A tree in full foliage and fruit is girdled at the crown, and a complete collapse follows with a shriveling and drying of foliage. Some fruit drop occurs. Excavation of the root shows complete invasion by the mycelium of the fungus in white strands, sheets, and plates.

Etiology

The mycelium is hyaline, septate, branched, and forms white, thin plates in the cortex. When mature, honey-colored, stipitate gilled, nonannulated sporophores may develop in clusters at the soil line on the host. The caps are several inches wide, and the stipes are 5 or 6 in. tall. Basidiospores are hyaline, 1-celled, and 8–10×4–6 μ.

Reference

8.

Leaf spot, *Pestalotia sapotae* P. Henn.

Symptoms

This disease is recognized by the black spots on the upper surfaces of the leaves.

Etiology

The fruiting pustules are few and generally scattered, submerged at first becoming erumpent, raised, and breaking the epidermis into shreds. The conidia are 5-celled, clavate-fusiform, curved, and 22–27×7–10 μ. End cells are hyaline; 3 setulae develop on the terminal cell, and a short pedicel forms on the basal cell. The 3 interior cells are olive brown.

Reference

6.

Limb gall, *Pestalotia scirrofaciens* N. A. Brown

Symptoms

Woody stem swellings, almost twice the diameter of the branch, appear to be externally sterile. Sections of a swelling show enlarged wood development and a thickened cortex that is soft internally. Laboratory exploration results in the consistent isolation of a fungus that through aseptic inoculations reproduces the disease.

Etiology

The mycelium is hyaline, typically white in culture, septate, and branched. The acervuli, at first submerged, develop on the host into black fruiting structures producing multitudes of conidia in extensive masses. The conidia are slightly curved and 4-septate; the end cells are hyaline and the 3 inner cells olive brown. The terminal cell bears 3 setulae that are narrowly hyaline and up to 26 μ long; the basal cell bears a short pedicel. The conidia measure 16–24×6–10 μ.

References

1, 6.

Other fungi associated with sapodilla

Acrotelium lucumae Cummins
Armillaria mellea Vahl ex Fries
Botryobasidium salmonicolor (Berkeley & Brown) Venkatarayan
Botryodiplodia theobromae Patouillard
Fusicladium butyrospermi Griffon & Maublanc
Gloeosporium rubi Ellis & Everhart
Glomerella cingulata (Stoneman) Spaulding & Schrenk
Nigrospora oryzae (Berkeley & Broome) Petch
Pestalotia caffra Sydow
Phymatotrichum omnivorum (Shear) Duggar
Physalospora rhodina (Berkeley & Curtis) Cooke
Rosellinia bunodes (Berkeley & Broome) Saccardo

References: sapodilla

1. Brown, N. A. 1920. A pestalozzia producing a tumor on the sapodilla tree. Phytopathology 10:383–94.
2. Cohen, M., and A. S. Muller. 1957. New host records for Elsinoë lepagei Bitanc. & Jenkins. Plant Disease Reptr. 41:540.
3. Cummins, G. B. 1950. The genus Scopella of the Uredinales. Bull. Torrey Botan. Club 77:204–13.
4. Cummins, G. B. 1956. Nomenclature changes for some North American Uredinales. Mycologia 48:601–8.
5. Cummins, G. B., and P. Ramachar. 1958. The genus Physopella (Uredinales) replaces Angiospora. Mycologia 50:741–44.
6. Guba, E. F. 1961. Monograph of Monochaetia and Pestalotia. Harvard University Press: Cambridge, Mass.
7. Rao, V. G., M. K. Desai, and N. B. Kulkarni. 1962. A new phytophthora fruit rot of Achras sapota from India. Plant Disease Reptr. 46:381–82.
8. Rhoads, A. S. 1942. Notes on clitocybe root rot of bananas and other plants in Florida. Phytopathology 32:487–96.
9. Roger, L. 1953. Phytopathologie des pays chauds. Encyclopedie Mycologique 18: 1202–11.
10. Srivastava, M. P., et al. 1964. Studies on fungal diseases of some tropical fruits. Phytopathol. Z. 50:250–61.

Sesame, *Sesamum orientale* Linnaeus

Leaf spot, *Pseudomonas sesami* Malkoff

Symptoms

Lesions spread from the seedling stage to the leaf blades and the petioles. They are dark brown and more or less angular as influenced by the veination.

They become very dark, almost black toward the centers. With age, they dry, shrivel, and crack, often leaving a shot-hole effect. Under favorable conditions considerable defoliation results.

Etiology

The organism is a gram-negative rod, single or in pairs, forms no capsules, is motile by 2–5 polar flagella, and measures 0.6–0.8 × 1.2–3.8 μ. Colonies are white, opalescent, circular, entire, smooth, flat, and striate. A green fluorescent pigment is formed on agar plates. Optimum temperature, 30C.

References

4, 14, 21.

Leaf spot, *Xanthomonas sesami* Sab. & Dows.

Symptoms

The first indication of the disease is the appearance of small, dark olive green spots that rapidly enlarge up to 3 mm and become dark red brown to black and circular to angular. The involved tissue dies and may crack. Stems and pods become infected.

Etiology

The organism is a rod, produces capsules, is motile by a single polar flagellum, and measures 0.4–0.6 × 0.8–1.6 μ. On agar plates, the colonies are yellow, small, circular, convex, raised, and slimy.

Reference

16.

Blight, *Phytophthora parasitica* Dast.

Symptoms

A leaf and stem infection results in the wilting of foliage and axillary branches. Stem cankers at the collar cause a complete collapse. They develop frequently from leaf infections, forming reddish brown lesions on the stems and causing death of upper parts. Early infections prevent seed formation, and late infections sometimes include seed pods and seed.

Etiology

The mycelium is hyaline and nonseptate; the sporangiophore is branched. Sporangia measure 38 × 26 μ.

References

3, 8.

Charcoal rot, *Macrophomina phaseolina* (Tassi) Goid.

Symptoms

Wilt is caused by a girdle at, or slightly above, the soil line in young plants.

Brownish green streaks appear on the stems close to the terminal growth. The plants are killed and become brown, remaining upright in place. The pycnidia develop in a more or less stromatic accumulation of sclerotia beneath the raised, shiny epidermis.

Etiology

The pycnidiospores appear in tendrils or opaque, white masses, each oval to slightly elongate, hyaline, 1-celled, and 14–30×7–12 μ. The sclerotia are black, usually smooth, irregular in shape, shiny, hard, and 50–200 μ in diameter. Their presence in plant tissue such as cortex, epidermis, and wood is diagnostic of the sterile fungus *Sclerotium bataticola*.

References

7, 15.

Leaf spot, *Cylindrosporium sesami* Hansf.

Symptoms

Diseased spots appear first as water-soaked, darker areas on the leaf blades. They enlarge rapidly, and a dark border develops on the advancing margins. They are delineated markedly by the enlarged leaf veins, resulting in the production of a definitely angular, necrotic zone. The involved tissue dies but remains intact. The upper surface of these spots becomes speckled with dark, subepidermal acervuli.

Etiology

The acervuli measure 20–50 μ and produce an abundance of conidia. The conidia are hyaline, straight or slightly curved, and 3–8-septate. In culture the fungus produces a black, heaped up, stromatic growth.

References

12, 20.

Leaf spot, *Cercospora sesami* Zimm.

Symptoms

Small, circular spots are scattered on both leaf surfaces. They are light brown with gray centers, up to 5 mm in diameter, enlarge rapidly, coalesce into irregular blotches often 4 cm in diameter, and are concentrically zoned.

Etiology

The conidiophores are olivaceous, septate, usually single or in fascicles of up to 10, and measure 20–110×4–6 μ. Conidia are hyaline, acicular, 7–10-septate, with a broad base and acute tip, and measure 30–150×2–5 μ.

Reference

2.

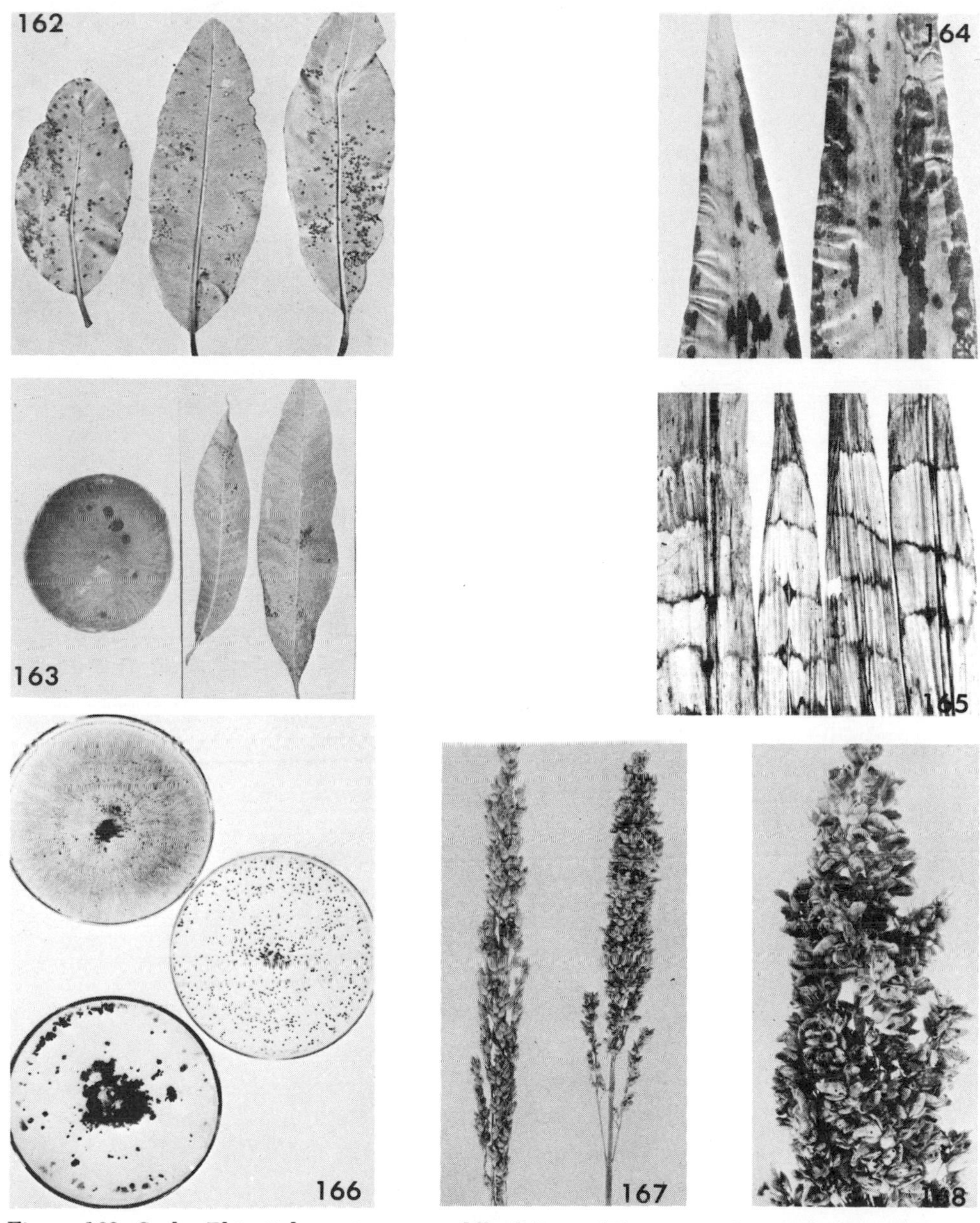

Figure 162. Scab, *Elsinoë lepagei,* on sapodilla foliage. (Photograph by H. C. Burnett.)

Figure 163. Rust, *Acrotelium lucumae,* on canistal foliage and fruit. (Photograph by H. G. McMillan.)

Figure 164. Bacterial streak, *Xanthomonas holicicola,* on sorghum.

Figure 165. Banding, *Thanatephorus sasakii,* of sorghum, millet, etc.

Figure 166. Cultures from sorghum. Top to bottom: *Botryobasidium rolfsii, Thanatephorus sasakii,* and *Thanatephorus cucumeris.*

Figure 167. Right: Covered smut, *Sphacelotheca sorghi,* on sorghum. Left: Healthy sorghum.

Figure 168. Sori showing tough peridium surrounding smutted ovaries (*Sphacelotheca sorghi*). (Photograph by D. G. Kelbert.)

Leaf spot, *Cercospora sesamicola* Moh.

Symptoms

Spots on foliage are numerous, dull brown, without a distinct border, angular, limited by principal veins, and often coalesce, killing portions of or entire leaves.

Etiology

Sporulation of the fungus is produced as a brown stromata covered with dense fascicles of pale brown conidiophores. Conidia are pale or hyaline, linear to elongate, curved, and 20–200 × 2–3 μ.

References

5, 10.

Stem rot, *Helminthosporium sesami* Miyake

Symptoms

This disease is present on leaves, pedicels, and capsules of maturing plants, where it appears as lesions, varying from 1 mm in diameter to large, dark brown, sunken spots, measuring 10 × 40 mm. Stem lesions often weaken the plant, and foliage lesions cause premature defoliation.

Etiology

The mycelium is hyaline to lightly colored, septate, and branched. The conidiophores are dark, simple, separate with swollen bases, septate, and 150–250 × 6–8 μ. The conidia are brown, obclavate, elongate, with rounded ends, slightly curved, 5–9-septate, constricted, and 46–68 × 8–11 μ. Another fungus, *Helminthosporium sesameum,* reported on this host is very similar except that the conidia measure 100–120 × 15–17 μ, and there are 18–20 septations and no constrictions.

References

13, 17, 18.

Target spot, *Corynespora cassiicola* (Berk. & Curt.) Wei

Symptoms

Early manifestations of the disease are dark, irregularly shaped spots variously scattered over the leaves and stems. They enlarge, become brown with lighter centers, and coalesce, forming a blotchy configuration. Under average conditions the fungus causes extensive defoliation and the eventual death of the plant.

Etiology

The mycelium is hyaline becoming brown, septate, and branched. Conidiophores are single or in groups of 2–6, up to 20-septate, and 6–11 × 44–380 μ.

Conidia are 10–15-septate, may be in chains of several, and measure 10 × 150–250 μ. Chlamydospores may develop.

Reference

19.

Blight, *Alternaria sesami* (Kawamura) Moh. & Behera

Symptoms

Early stages of the disease appear on the leaf blades as small, brown spots, which become larger, darker, and characteristically concentrically zonate. The dark lines are within the spot on the upper surface of the leaf. On the lower surface, the spots are a lighter brown color. Spots often coalesce and may involve large portions of the blade which become dry and are shed. Lesions appear on stems and capsules and are often observed on cotyledons of young seedlings.

Etiology

The conidiophores of the fungus are yellowish brown, simple, erect, 3-septate, not rigid, arise singly, measure 30–54 × 4–7 μ, and produce conidia at the apex. The conidia are light to dark brown, obclavate, 4–12-septate, with 0–6 longitudinal septa, and measure 120 × 9–30 μ. They are somewhat constricted at several medial septations, and the spores have a long, hyaline beak. The spores are usually borne singly but may be 1–3-catenulate.

References

1, 6, 9, 11.

Other fungi associated with sesame

Alternaria sesamicola Kawamura
Ascochyta sesami Miura
Botryobasidium rolfsii (Saccardo) Venkatarayan
Dothiorella philippinensis Petrak
Erysiphe cichoracearum De Candolle
Fusarium vasinfectum Atkinson
Gloeosporium macrophomoides Saccardo
Macrosporium sesami Kawamura
Phoma sesamina Saccardo
Phyllosticta sesami Bohovik
Pseudomonas solanacearum (E. F. Smith) Dowson
Sphaeronema sesami Sehgal & Daftari
Vermicularia sesamina Saccardo
Verticillium albo-atrum Reinke & Berthold

References: sesame

1. Berry, S. Z. 1960. Comparison of cultural variants of Alternaria sesami. Phytopathology 50:298–304.
2. Chupp, C. 1953. Monograph of the fungus genus Cercospora. Published by the author: Ithaca, N.Y., p. 436.

3. Crandall, B. S., and J. Dieguez. 1948. Phytophthora stem canker of sesame in Peru. Phytopathology 38:753–55.
4. Elliott, C. 1951. Manual of bacterial plant pathogens. 2d ed. Chronica Botanica: Waltham, Mass., p. 87.
5. Ferrer, J. B. 1960. The occurrence of angular leaf-spot of sesame in Panama. Plant Disease Reptr. 44:221.
6. Joly, P. 1964. Le genre Alternaria. Encyclopedie Mycologique 33:138–40.
7. Joshi, A. B. 1961. Sesamum. Indian Agricultural Research Institute: New Delhi, India, pp. 100–101.
8. Kale, G. B., and N. Prasad. 1957. Phytophthora blight of sesamum. Indian Phytopathol. 10:38–47.
9. Leppik, E. E., and G. Sowell. 1964. Alternaria sesami, a serious seed borne pathogen of world wide distribution. FAO Plant Protect. Bull. 12:13–16.
10. Mohanty, N. N. 1958. Cercospora leaf spot of sesame (Sesamum orientale L.). Indian Phytopathol. 11:186–87.
11. Mohanty, N. N., and B. C. Behera. 1958. Blight of sesame (Sesamum orientale L.) caused by Alternaria sesami (Kawamura) n. comb. Current Sci. (India) 27:492–93.
12. Orellana, R. G. 1961. Leaf spot of sesame caused by Cylindrosporium sesami. Phytopathology 51:89–92.
13. Poole, D. D. 1956. Aerial stem rot of sesame caused by Helminthosporium sesami in Texas. Plant Disease Reptr. 40:235.
14. Rivers, G. W., M. L. Kilman, and T. W. Culp. 1964. Inheritance of resistance to bacterial leafspot of sesame. Crop Sci. 4:455–64.
15. Roger, L. 1953. Phytopathologie des pays chauds. Encyclopedie Mycologique 18:1704–16.
16. Sabet, K. A., and W. J. Dowson. 1960. A bacterial leaf spot of sesame (Sesamum orientale L.). Phytopathol. Z. 37: 252–58.
17. Saccardo, P. A. 1931. Sylloge Fungorum 25:829.
18. Stone, W. J. 1959. Sesame blight caused by Helminthosporium sesami. Phytopathology 49:815–17.
19. Stone, W. J., and J. P. Jones. 1960. Corynespora blight of sesame. Phytopathology 50:263–66.
20. Swingle, M. K. 1945. Sesame, a list of references. U.S. Dept. Agr. Library List 20.
21. Zachos, D. G., and C. G. Panagopoulos. 1960. The bacterium Pseudomonas sesami Malkoff in Greece. Ann. Inst. Phytopathol., Benake, Greece 3:60–64.
22. Zachos, D. G., and C. G. Panagopoulos. 1961. Alternaria blight of sesame. Ann. Inst. Phytopathol., Benake, Greece 3: 163–66.

Sorghum, *Sorghum vulgare* Persoon

Bacterial stripe, *Pseudomonas andropogoni* (E. F. Smith) Stapp
Bacterial spot, *Pseudomonas syringae* van Hall
Bacterial streak, *Xanthomonas holcicola* (Elliott) Starr & Burkholder
Seedling blight, *Pythium arrhenomanes* Drechsler
Crazy top, *Sclerophthora macrospora* (Saccardo) Thirumalachar, Shaw, & Narasimhan
Downy mildew, *Sclerospora sorghi* (Ellis & Everhart) L. S. Olive & Lefebvre
Blight and stalk rot, *Gibberella zeae* (Schweinitz) Petch
Stalk rot, *Thanatephorus cucumeris* (Frank) Donk
Southern blight, *Botryobasidium rolfsii* (Saccardo) Venkatarayan
Banding, *Thanatephorus sasakii* (Shirai) Donk
Rust, *Puccinia purpurea* Cooke
Loose smut, *Sphacelotheca cruenta* (Kuehn) Potter
Covered smut, *Sphacelotheca sorghi* (Link) Clinton
Head smut, *Sphacelotheca reiliana* (Kuehn) Clinton
Long smut, *Tolyposporium ehrenbergii* (Kuehn) Patouillard
Rough spot, *Ascochyta sorghina* Saccardo
Charcoal rot, *Macrophomina phaseolina* (Tassi) Goidanich
Anthracnose, *Colletotrichum graminicola* (Cesati) G. W. Wilson
Root rot, *Periconia circinata* (Mangin) Saccardo
Sooty stripe, *Ramulispora sorghi* (Ellis & Everhart) L. S. Olive & Lefebvre
Zonate leaf spot, *Gloeocercospora sorghi* D. Bain & Edgerton
Leaf spot, *Cercospora sorghi* Ellis & Everhart
Leaf spot, *Helminthosporium rostratum* Drechsler
Target spot, *Helminthosporium sorghicola* Lefebvre & Sherwin

Leaf blight, *Helminthosporium turcicum* Passerini
Other fungi associated with sorghum

Bacterial stripe, *Pseudomonas andropogoni* (E. F. Sm.) Stapp

Symptoms

This disease is found on almost all of the various sorghums. It is manifested by smooth or roughly jagged stripes, usually reddish brown, often up to 8–10 in. long and very narrow, confined mostly between the larger veins of the leaf blade. In older more severe infections, parallel zones may show these stripes involving larger leaf portions. They may fuse laterally and extend into the sheath.

Etiology

A bacterial exudate usually accumulates in droplets under humid conditions along the stripes. It dries into thin scales on the lower leaf surface. The organism is a gram-negative rod, forms capsules, is motile by a single to several bipolar flagella, and measures $0.4–0.8 \times 1.3–2.5\ \mu$. Agar colonies are glistening, raised, viscid, and produce a fluorescent pigment in culture.

References

4, 8.

Bacterial spot, *Pseudomonas syringae* van Hall

Symptoms

Bacterial spots appear water-soaked at first and become reddish as they develop; they are circular to irregular in outline and up to 1 cm in diameter, usually smaller. They are variously scattered, but are frequently so numerous as to destroy the leaves. The spots finally dry with light-colored, sunken centers and red borders. They appear first on the lower leaves and extend upward.

Etiology

The organism is a gram-negative rod, single or in chains, forms capsules, is motile by 1 to several polar flagella, and measures $0.4–0.6 \times 1.2–2\ \mu$. Agar colonies are white, transparent, round, smooth or wrinkled, and convex. A green fluorescent pigment is produced in culture.

References

4, 9.

Bacterial streak, *Xanthomonas holcicola* (Elliott) Starr & Burkh.

Symptoms

Streak is common and widespread but limited in host range. In the early stages, the infection occurs as water-soaked areas upon which yellowish droplets appear. Gradually the narrow, reddish brown streaks enlarge up to several millimeters wide and 6 in. long. The streaks become somewhat irregular in shape, elongate-oval, with tan centers; they may coalesce, involving extensive leaf surfaces. A yellow exudate may appear and often dries into thin, creamy white scales.

Etiology

The organism is a gram-negative rod, produced singly or in short chains, forms capsules, is motile by 1 or 2 polar flagella, and measures 0.4–0.9 × 1–2.5 μ. On agar plates the colonies are yellow, circular, glistening, smooth, and butyrous. Optimum temperature, 28–30C.

References

4, 7, 9.

Seedling blight, *Pythium arrhenomanes* Drechs.

Symptoms

Blight results frequently when the early seeding conditions are cold and wet. Seeds often decay before emergence and are killed after emergence. The root system is attacked and destroyed; often the seedling is girdled at or slightly above the soil line, falls over, and dies. There is no distinct coloration. The plants appear to show some scalding effect previous to drying and dying.

Etiology

The mycelium is hyaline, scanty, nonseptate, and produces variously shaped, lobulate sporangia, measuring up to 20 μ in diameter. Oospores from culture are about 28 μ in diameter.

References

4, 6, 8.

Crazy top, *Sclerophthora macrospora* (Sacc.) Thirum., Shaw, & Naras.

Symptoms

Crazy top implies the extensive proliferation of the terminal florescence in place of the normal floral structures. The head is clothed with small leaves, and the remainder of the plant appears normal. In some instances more of the florets may be affected, and the entire head seems to become vegetative with small, short, green leaflets. There is usually some stunting, and certain leaves are narrower and thicker.

Etiology

The mycelium is systemic and nonseptate. The fungus is an obligate parasite. The sporangia are hyaline, lemon-shaped, and 66–82 × 39–55 μ. Zoospores are produced. Oospores are brown, produced in abundance, and 51 μ in diameter.

References

13, 14.

Downy mildew, *Sclerospora sorghi* (Ell. & Ev.) L. Olive & Lefebvre

Symptoms

Mildew is a systemic disease appearing on plants, from seedlings with but a few leaves to old, mature plants. The plant may show complete stunting and lack of florescence, or only axillary branches may be affected. Seedlings are often almost completely whitish, lack green areas, and usually die. Older plants may show only new growth with white streaks in the otherwise green blades; those more completely involved show all the central leaves yellow white and stunted. The fungus sporulates prolifically on the whitish streaks in the leaf blades. When past the seedling stage, plants are not often killed. A certain amount of leaf shredding occurs.

Etiology

The fungus produces broadly rotund sporangia on characteristic sporangiophores; they germinate by mycelium production and measure 18–32 × 16–23 μ. Oospore formation in the mesophyll tissue of the leaf is abundant. They are brown, spherical, and 31–39 μ in diameter.

References

1, 4, 16.

Blight and stalk rot, *Gibberella zeae* (Schw.) Petch

Symptoms

This disease often continues throughout the life of the plant. Seedlings are usually killed during cooler periods of the growing season. Those that survive show effects of the disease in the form of leaf infections and stalk lesions. The fungus invades the stalks usually through mechanical wounds and results in weakened plants, considerable breaking over, and lodging.

Etiology

Macroconidia produced in sporodochia are hyaline, elliptical, slightly curved, and 3–5-septate; no microconidia or chlamydospores are produced. The perithecia are black, globose, and smooth. The ascospores are hyaline, 3-septate, slender, curved, and pointed at the ends.

References

4, 11.

Stalk rot, *Thanatephorus cucumeris* (Frank) Donk

Symptoms

Rhizoctoniose develops on the host plants from the seedling stage to maturity and is very responsive to environmental conditions of temperature and moisture. Seedlings exhibit damping-off and girdling in cool, wet weather in a typical manner. Young plants show leaf injury when the season is wet and warm. The fungus mycelium grows up the stem, over the leaf sheaths, and onto the blade, where the tissue is invaded and killed, giving a scalded appearance. Later the leaves curl, dry, and turn brown. When plants are mature, the fungus invades the pith, producing a reddish color and causing the stalk to weaken and lodge.

Etiology

The mycelium is hyaline to brown, branched, septate, and produces no conidia. It is soil-inhabiting. The sclerotia are brown, often flat and scalelike, or almost black and loaflike. Under favorable conditions, basidiospores are formed that are hyaline, 1-celled, oval, and 5×9 μ.

References

13, 15.

Southern blight, *Botryobasidium rolfsii* (Sacc.) Venkat.

Symptoms

Seedlings crowded into low, moist areas lacking drainage and ventilation are frequently infected at the soil surface where leaf sheaths and brace roots become overgrown by the mycelium of the fungus. Tender tissues are invaded and killed, and frequently the main stalk becomes water-soaked and falls over.

Etiology

Tan to brown, spherical, hard sclerotia, less than 1 mm in diameter, develop externally on the abundant mycelium. The mycelium is hyaline, septate, branched, and produces no spores. Basidiospores are not frequently found. The imperfect stage of the fungus is *Sclerotium rolfsii.*

References

4, 13, 15.

Banding, *Thanatephorus sasakii* (Shirai) Donk

Symptoms

Banding is a striking manifestation of the effect of the fungus on the host plant. Under especially favorable, warm, humid conditions, the soil-inhabiting fungus grows up the stalk from the soil, causing little or no damage until it comes in contact with the leaf blades. There it invades the cells in a rather uniform manner, killing the tissue, except the vascular bundles, and causing it to bleach to a light tan color. The fungus produces a dark band at the margin of the advancing hyphae and later continues its growth, creating another band. The process is repeated and often leaves show as many as 6 distinctive bands, each set off by dark, horizontal bands.

Etiology

The mycelium is hyaline, becoming whitish, yellow, then brown, and abundant superficially. Sclerotia are brown, flat, irregular, and often several millimeters in diameter. Basidiospores are not frequently found. They are hyaline, obovate, 1-celled, smooth, and 8–11×5–6.5 μ.

References

4, 12, 15.

Rust, *Puccinia purpurea* Cke.

Symptoms

Rust probably occurs wherever the host plant grows and is usually not of great importance except in the seedling stage during which the leaves are killed as they develop. When the plants are more mature, the disease appears as small, reddish brown pustules, slightly raised like blisters, which break open exposing and freeing dark brown, dusty uredospores. The pustules develop on both leaf surfaces and sometimes cause the leaves to mature and 'fire' early.

Etiology

The pycnia and aecia are unknown. Uredia are dark brown. Uredospores are brown, ellipsoidal, echinulate, and 18–36×23–31 μ. Telia are small, scattered; teliospores are 2-celled, constricted, oblong, and 15–26×26–48 μ.

References

3, 4, 12.

Loose smut, *Sphacelotheca cruenta* (Kuehn) Potter

Symptoms

Smut causes a stunting of the plant as an early manifestation of the disease. Such plants are frequently of a slightly darker green color than normal.

Etiology

The fungus enters the growing tip of the seedling and remains until the florescence forms. It is present in the ovules and there produces the chlamydospores in great numbers within the fungus peridium. The peridium ruptures and frees the brown to black spores that are spherical to elliptical, with reticulated surfaces, and 5–10 μ in diameter. The distinguishing feature of the disease is the lack of a retaining membrane, thus permitting immediate spore dissemination.

References

4, 12.

Covered smut, *Sphacelotheca sorghi* (Lk.) Clint.

Symptoms

Covered smut forms as a typical disease in the florescence, causing the ovules to be replaced with dark chlamydospores of the fungus. The fungus originally enters the seedling, grows to the primordial area, and continues to the flowering head.

Etiology

The chlamydospores are brown to black, spherical, globose to angular, smooth except for a fine echinulation, and 5–8 μ in diameter. The distinguishing feature of the disease is the retaining membrane within which the spores are produced.

Reference

4.

Head smut, *Sphacelotheca reiliana* (Kuehn) Clint.

Symptoms

Infection occurs in the seedling stage. The fungus grows to the primordial area and progresses to the flowering head, where abundant mycelium is transformed into a fragile, retaining peridium. Head smut replaces the panicle or head as the fruiting structure of the plant and is not evident until the panicle emerges from the boot or flag leaf sheath.

Etiology

The chlamydospores are dusty brown to black, finely echinulate, mostly spherical, and 9–12 μ in diameter.

References

4, 12.

Long smut, *Tolyposporium ehrenbergii* (Kuehn) Pat.

Symptoms

Long smut is distinguished because of the length of the sori that replace the ovules in the fruiting panicle.

Etiology

The sori are long, cylindrical, slightly curved, and 8–13 $\times$ 3–5 mm, frequently attaining a length of 25 mm. Often only a few ovules are involved in the production of sori. The chlamydospores are attached to a central columella. The spores are spherical to oval or angular, 9–12 μ, and often in compact balls, 50–120 μ in diameter.

Reference

4.

Rough spot, *Ascochyta sorghina* Sacc.

Symptoms

Rough spot is first observed in the form of light-colored to reddish, circular to oblong spots on the leaf blades. They become purple red to brown and up to 1 in. long and about 1/4 in. wide. When numerous they coalesce.

Etiology

The pycnidia are black, gregarious, protruding, sandpapery to the touch, and develop in the diseased areas on the leaves, sheaths, and occasionally on the stalks. They are densely clustered, globose, depressed, and papillate. The pycnidiospores are hyaline, oblong-ellipsoidal, 1-septate, and $20 \times 8\ \mu$.

References

4, 12, 13.

Charcoal rot, *Macrophomina phaseolina* (Tassi) Goid.

Symptoms

This rot is recognized by the presence of small, black sclerotia imbedded in the gray host tissue. The fungus may kill seedlings and cause older plants to become dry, brittle, and form soft, discolored areas near the soil line. The pith is invaded and softens as the vascular bundles separate in a disintegration process that weakens the stalk and causes it to break over or lodge.

Etiology

The fungus develops up the stalk, where it produces the characteristic sclerotia, often for several nodes above the soil. The sclerotia are black, rounded to irregularly shaped, hard, and 50–150 μ in diameter. The pycnidia develop from a stromatic mass similar to clusters of sclerotia. The pycnidiospores are oval to elongate, 1-celled, and $14\text{–}30 \times 7\text{–}12\ \mu$.

References

4, 12.

Anthracnose, *Colletotrichum graminicola* (Ces.) G. W. Wils.

Symptoms

Anthracnose first appears on the foliage as small, circular to elliptical spots, up to 3 mm in diameter. The spots are reddish purple and later fade to a soft tan in the centers with dark borders. They are usually scattered but may cause extensive leaf loss when plentiful.

Etiology

The acervuli develop in the lighter-colored areas of the leaf spots and are characterized by the black, hairlike setae that protrude from the spot. The conidia, pink en masse, are produced in the acervulus among the setae. The conidia are hyaline, clavate to spindle-shaped or oblong-ovate, slightly curved,

1-celled, and 8–19 × 4–6 μ. The fungus is very common on many hosts and ranges from a saprophyte to a facultative parasite.

References

4, 12, 13.

Root rot, *Periconia circinata* (Mang.) Sacc.

Symptoms

Root rot affects plants past the seedling stage up to about 10 in. tall. The plants appear to be stunted, and yellow tips develop on the leaves which also show some rolling from the margins. The plant becomes distinctly yellow, and the older leaves turn brown until the entire plant becomes involved and gradually dies. If the plant survives longer, it may develop a very poor florescence of little or no value. The causal organism attacks the roots, destroying them. A reddish brown discoloration appears in the affected areas as well as at the crown. It is thought that the fungus produces a toxin that is detrimental to the growth of the host plant.

Etiology

The mycelium is hyaline and intermixed with brown conidiophores and chlamydospores that are 9–18 μ in diameter. The conidia are dark brown, 15–27 μ in diameter, spherical, coarsely verrucose or spiny, and produced on curved, septate, erect conidiophores, which terminate in several sporogenous cells.

References

4, 12.

Sooty stripe, *Ramulispora sorghi* (Ell. & Ev.) L. Olive & Lefebvre

Symptoms

The leaf lesions are elliptical, often up to 1/2 in. wide, and up to several inches long. The margins are smooth and often develop a red purple color. The centers are at first yellowish to dark gray but gradually darken as the sooty-colored sporodochia and sclerotia mature. Aggregates of hyphae form beneath the stomata into a compact stroma from which the conidiophores arise.

Etiology

The conidia are confined to a gelatinous mass which becomes compact, dry, and forms hard, black sclerotium. The conidia are long, slender, curved, 3–8-septate, 38–87 × 1–3 μ, and produce 1–3 lateral branches, which measure up to 53 × 1–3 μ and are up to 3-septate.

References

4, 12, 13.

Zonate leaf spot, *Gloeocercospora sorghi* D. Bain & Edg.

Symptoms

This leaf spot when first noticed on the foliage appears as small, circular, reddish brown, water-soaked areas that have a pale green halo. As the spots enlarge, they become dark red and may be somewhat elongate. They eventually appear as large, often circular, zonate blotches, sometimes exceedingly uniform, showing up to 5 concentric, dark bands interspaced with tan, papery tissue. Often, however, these areas are semicircular, half-moonlike, or merely irregular blotches up to several centimeters in diameter. The small, spherical to elongate sclerotia appear in these blotches.

Etiology

The fungus hyphae aggregate in the stomatal chamber and emerge through the stoma, forming sporodochia of a salmon color, which are up to 0.5 mm in diameter and rise above the epidermis. Hyphae may be in clusters, rows, or scattered. The conidiophores are hyaline, short, and densely clustered. The conidia are borne in a pink, gelatinous matrix. They are hyaline, straight or curved, up to several septate, filiform, and $20–195 \times 3\ \mu$. The black sclerotia develop in the older host tissue and often in lines parallel to the veins.

References

4, 12, 13.

Leaf spot, *Cercospora sorghi* Ell. & Ev.

Symptoms

Leaf spot develops as dark purple lesions, extending in elongate stripes 1 in. or more long and about 1/4 in. wide, usually appearing almost rectangular. They lack a distinct border, and they become dry and smooth except for tufts of brown, septate conidiophores that develop in the centers.

Etiology

The conidia are hyaline, septate, slightly curved, and $70–80 \times 3\ \mu$.

References

12, 13.

Leaf spot, *Helminthosporium rostratum* Drechs.

Symptoms

This disease normally appears as purplish leaf spots, variously scattered, oval to linear, and increasing in length as they mature but remaining mostly confined by the lateral veins. The young spots measure $1–2 \times 2–5$ mm but later may be twice as wide and several centimeters long with some irregularity of the margins where they extend across the veins.

Etiology

The conidiophores are olivaceous to brown, septate, erect, develop from swollen bases, and are up to 140 μ high. The conidia are brownish, straight or slightly curved, tapering toward the tips but widest below the middle; they are 8–15-septate and 32–184 × 14–22 μ.

References

5, 17, 18.

Target spot, *Helminthosporium sorghicola* Lefebvre & Sherwin

Symptoms

Target spot is distinguished, at least in the older lesions, by well-defined, reddish purple or brown zonate markings with light centers surrounded by a darker band and another lighter-colored band. At first the spots are small, but they enlarge up to 15 mm, change from oval to slightly elongate, become irregular, and are not limited markedly by the veins.

Etiology

The conidiophores produce conidia sparingly on both surfaces of the spots on the dead tissue. The conidia are golden yellow to olivaceous, curved, 2–8-septate, widest near the middle, tapering toward the rounded ends, and 20–105 × 8–21 μ.

Reference

13.

Leaf blight, *Helminthosporium turcicum* Pass.

Symptoms

Seedlings are often killed when infection is abundant and lesions coalesce, causing the leaves to wrinkle, wilt, and become purplish as the plant dies. On older plants, the diseased areas are conspicuously elongate and widen to form elliptical, reddish-brown-margined lesions. In the central area of the lesions, the fungus produces conidiophores and conidia plentiful enough to cause the area to darken.

Etiology

The conidiophores are olivaceous, 2–4-septate, and up to 260 μ long. The spores are straight to slightly curved, widest in the middle, tapering to the ends, and 14–132 × 15–25 μ. The basal end is somewhat pointed.

References

5, 12, 13.

Other fungi associated with sorghum

Cladosporium herbarium Persoon ex Link

Diplodia natalensis P. Evans
Gibberella fujikuroi (Sawada) Wollenweber
Heterosporium variabile Cooke
Nigrospora sphaerica (Saccardo) Mason
Penicillium oxalicum Currie & Thom
Phyllosticta sorghina Saccardo
Pseudomonas lapsa (Ark) Starr & Burkholder
Septoria pertusa Heald & Wolf
Sphacelotheca holci Jackson
Titaesspora andropogonis (Miura) Tai
Trichothecium roseum Link ex Fries
Ustilago bulgarica Bubok
Verticillium albo-atrum Reinke & Berthold

References: sorghum

1. Bain, D. C., and W. W. Alford. 1969. Evidence that downy mildew (Sclerospora sorghi) of sorghum is seedborne. Plant Disease Reptr. 53:802–3.
2. Butler, E. J. 1918. Fungi and diseases in plants. Thacker: Calcutta, India, pp. 203–18.
3. Cummins, G. B., and J. A. Stevenson. 1956. A check list of North American rust fungi (Uredinales). Plant Disease Reptr. 240:109–93. (Suppl.)
4. Dickson, J. G. 1956. Diseases of field crops. 2d ed. McGraw-Hill: N.Y., pp. 188–205, 399.
5. Drechsler, C. 1923. Some Graminicolous species of Helminthosporium. J. Agr. Res. 24:641–739.
6. Drechsler, C. 1928. Pythium arrhenomanes n. sp., a parasite causing maize root rot. Phytopathology 18:873–75.
7. Elliott, C. 1930. Bacterial streak diseases of sorghums. J. Agr. Res. 40:963–76.
8. Elliott, C. 1937. Pythium root rot of milo. J. Agr. Res. 54:797–834.
9. Elliott, C. 1951. Manual of bacterial plant pathogens. 2d ed. Chronica Botanica: Waltham, Mass.
10. Fischer, G. W. 1953. Manual of the North American smut fungi. Ronald Press: N.Y.
11. Futrell, M. C., and O. J. Webster. 1967. Fusarium scab of sorghum in Nigeria. Plant Disease Reptr. 51:174–78.
12. Leukel, R. W., J. H. Martin, and C. L. Lefebvre. 1960. Sorghum diseases and their control. U.S. Dept. Agr. Farmers' Bull. 1959:1–46.
13. Sprague, R. 1950. Diseases of cereals and grasses in North America. Ronald Press: N.Y.
14. Ullstrup, A. J. 1961. Corn diseases in the United States and their control. U.S. Dept. Agr., Agr. Handbook 199:1–29.
15. Weber, G. F. 1963. Pellicularia spp. pathogenic on sorghum and pearl millet in Florida. Plant Disease Reptr. 47:654–56.
16. Weston, W. H., and B. N. Uppal. 1932. The basis for Sclerospora sorghi as a species. Phytopathology 22:573–86.
17. Whitehead, M. D., and O. H. Calvert. 1959. Helminthosporium rostratum inciting ear rot of corn and leaf spot of thirteen grass hosts. Phytopathology 49: 817–20.
18. Young, G. Y., C. L. Lefebvre, and A. G. Johnson. 1947. Helminthosporium rostratum on corn, sorghum, and pearl millet. Phytopathology 37:180–83.

Soybean, *Glycine max* (Linnaeus) Merrill

Bacterial blight, *Pseudomonas glycinea* Coerper
Wildfire, *Pseudomonas tabaci* (Wolf & Foster) F. L. Stevens
Bacterial pustule, *Xanthomonas phaseoli* (E. F. Smith) Dowson var. *sojense* (Hedges) Starr & Burkholder
Brown rot, *Pseudomonas solanacearum* E. F. Smith
Damping-off, *Pythium debaryanum* Hesse

Yeast spot, *Nematospora coryli* Peglion
Stem rot, *Sclerotinia sclerotiorum* (Libert) De Bary
Downy mildew, *Peronospora manshurica* (Naumov) Sydow ex Gäumann
Anthracnose, *Glomerella glycines* (Hori) Lehman & Wolf
Stem canker, *Diaporthe phaseolorum* (Cooke & Ellis) Saccardo var. *caulivora* (Harter & Field) Athow & Caldwell
Pod and stem blight, *Diaporthe phaseolorum* (Cooke & Ellis) Saccardo var. *sojae* (Lehman) Wehmeyer
Powdery mildew, *Erysiphe polygoni* De Candolle
Root rot, *Thanatephorus cucumeris* (Frank) Donk
Web blight, *Botryobasidium microsclerotia* (Matz) Venkatarayan
Southern blight, *Botryobasidium rolfsii* (Saccardo) Venkatarayan
Leaf spot, *Phyllosticta glycinea* Tehon & Daniels
Brown spot, *Septoria glycines* Hemmi
Ashy stem, *Macrophomina phaseolina* (Tassi) Goidanich
Brown stem rot, *Cephalosporium gregatum* Allington & Chamberlain
Frogeye, *Cercospora sojina* Hara
Purple stain, *Cercospora kikuchii* (T. Matsumoto & Tomoyasu) Gardner
Target spot, *Corynespora cassiicola* (Berkeley & Curtis) Wei
Wilt, *Fusarium oxysporum* Schlechtendahl f. sp. *tracheiphilum* (E. F. Smith) Snyder & Hansen
Other fungi associated with soybean

Bacterial blight, *Pseudomonas glycinea* Coerper

Symptoms

Blight appears on leaves as yellow, small, angular, water-soaked spots, 1–2 mm in diameter. As they become older, a brown area appears in the centers. The centers become slightly sunken and are usually surrounded by a narrow yellow halo that is more noticeable on the upper surface. Often many spots coalesce and large areas of leaf tissue fall out, leaving ragged edges which are associated with considerable defoliation. The disease, most common on the leaves, also infects the petioles, stems, peduncles, and pods. The darkened, pepper-sized spots characterize the disease.

Etiology

The organism is a gram-negative rod, forms capsules, is motile by 1 to several polar flagella, and measures $2\text{–}3 \times 1\text{–}1.5\ \mu$. The pathogen is seed-borne. Agar colonies are creamy white to slightly brown, circular, entire, smooth, convex, and butyrous. A green fluorescent pigment is produced in culture. Optimum temperature, 14–26C.

References

6, 8, 10.

Wildfire, *Pseudomonas tabaci* (Wolf & Foster) F. L. Stev.

Symptoms

Wildfire is characterized by light brown, circular, necrotic spots with well-defined margins and prominent yellow halos. Spots measure up to 4–5 mm in diameter and may enlarge or coalesce, often causing distortion of the leaves. Small, brown, sunken spots may appear on the stalks, and some may develop on floral parts and on the pods.

Etiology

The organism is a gram-negative rod, forms in chains, produces no capsules, is motile by 1–6 polar flagella, and measures 1.4–2.8×0.5–0.75 μ. Agar colonies are white, circular, and raised with translucent margins. A green fluorescent pigment is produced in culture. Optimum temperature, 24–28C.

References

8, 10.

Bacterial pustule, *Xanthomonas phaseoli* (E. F. Sm.) Dows. var. *sojense* (Hedges) Starr & Burkh.

Symptoms

This disease appears as small, yellowish green pustules with reddish brown centers, peppery, raised, and scattered on the lower leaf surface and larger and more conspicuous on the upper leaf surface. No water-soaked margin is present. When plentiful, the individual spots may coalesce, causing brown, killed leaf blotches that may result in some defoliation. Some stem and pod infection occurs, but it is most important on the foliage.

Etiology

The organism is a gram-negative rod, forms in chains, produces no capsules, is motile by 1–2 polar flagella, and measures 1.4–2.3×0.5–0.9 μ. The pathogen is seed-borne. Agar colonies are pale to yellow, circular, small, smooth, entire, and butyrous to viscid. Optimum temperature, 30–33C.

References

8, 10, 11.

Brown rot, *Pseudomonas solanacearum* E. F. Sm.

Symptoms

Bacterial brown rot, reported as causing some stunting and wilt, is not generally important. The organism survives in the soil and invades the root tissue and the vascular system, resulting in dark, slate brown discoloration. Wilting and stunting are the aboveground symptoms that may be associated with the vascular discoloration. This disease is known to be caused by a vascular rather than a parenchymal inhabitant.

Etiology

The organism is a gram-negative rod, produces no capsules, is motile by 1 polar flagellum, and measures 1.5×0.5 μ. Agar colonies are opalescent becoming brown, wet, shiny, irregular, and smooth. Optimum temperature, 35–37C.

Reference

10.

Damping-off, *Pythium debaryanum* Hesse

Symptoms

Damping-off is usually recognized by the sudden drooping of young leaves and a stunting and collapse of the seedling. The stem is invaded at the soil line, softens, becomes water-soaked, and falls over with no definite discoloration or breaking of the epidermis. The disease usually appears under cool, wet conditions.

Etiology

The soil-inhabiting fungus produces hyaline, nonseptate, branched mycelium and oval to spherical sporangia, measuring 15–26 μ in diameter. Zoospores measure 7–12 μ in diameter. Oospores are spherical, smooth, 12–20 μ in diameter, and germinate directly.

References

8, 12.

Yeast spot, *Nematospora coryli* Pegl.

Symptoms

Yeast spot is associated with the pods and the seed. The disease on the seed is indicated by slight depressions and yellow to brown discolorations of various irregular areas. Close examination may reveal a wound, indicating a puncture of the seed coat. The seed pods are punctured by certain plant bugs in their feeding. They penetrate the pod and encounter the seed in its various stages of development. In this mechanical process the yeast is introduced by the insect, and following the inoculation the fungus continues to develop in the seed.

Etiology

The organism develops as a vegetative yeast cell. It increases by budding from hyaline, spherical, mature cells averaging 15–20 μ in diameter and by ascospores that are 2-celled, hyaline, elongate, with one whiplike end, and measuring 38–40 × 2–3 μ. Upon germination, the ascospores initiate the vegetative budding stage.

Reference

21.

Stem rot, *Sclerotinia sclerotiorum* (Lib.) d By.

Symptoms

Stem rot attacks the stems of the plant near the soil line and in the main branches. It develops cankers at the soil line that girdle the plant, causing wilt and death.

Etiology

The fungus produces white mycelium that is usually present in cottonlike bunches on the main stem or on the branches. Associated with the cottony mycelium, black, variously shaped, hard, shiny sclerotia, measuring up to 1 cm in length, often appear superficially or within the pith of the diseased stems. The disease is spread by the dissemination of sclerotia and ascospores.

References

8, 12, 21.

Downy mildew, *Peronospora manshurica* (Naum.) Syd. ex Gäum.

Symptoms

Angular to irregularly shaped, yellowish green areas on the upper leaf surface are early indications of the disease. These spots darken and become gray to dark brown surrounded by a yellowish halo. Severe infections result in defoliation. A moldlike mildew develops on the lower leaf surface opposite the spots and produces spores of the fungus. Seed pods and seed are usually infected under favorable conditions.

Etiology

The mycelium is hyaline, branched, nonseptate, and scanty. The sporangia are hyaline, globular to oval, germinate directly, and measure 20–28 × 16–22 μ. The conidiophores are pale violet, branched, slender, emerge from the stomata, and measure 6–9 × 300–500 μ. Oospores are yellowish, smooth, globular, and 20–28 μ in diameter.

References

8, 12, 13, 14, 25.

Anthracnose, *Glomerella glycines* (Hori) Lehman & Wolf

Symptoms

All parts of the plant are affected from seed destruction in pre-emergence to damping-off of seedlings, cotyledon spotting, and foliage browning. The cotyledon spots are reddish brown, shallow, and are the sources of inoculum for stems, foliage, and pods.

Etiology

The spores are produced abundantly in acervuli that are usually numerous and dark-colored because of the abundant production of black, stiff setae. The conidia are hyaline, 1-celled, curved, and 16–24 × 3–5 μ. Perithecia are black, submerged, oval, and ostiolate; the ascospores are hyaline, slightly curved, 1-celled, and 18–28 × 4–6 μ.

References

8, 12.

Stem canker, *Diaporthe phaseolorum* (Cke. & Ell.) Sacc. var. *caulivora* (Harter & Field) Athow & Caldwell

Symptoms

Canker is found on young plants before midseason. The main stem is usually girdled, killing the plant. It remains erect with the brown leaves attached and pendant. The girdling lesion is slightly sunken and brown.

Etiology

The fungus produces no vegetative spores and develops perithecia only on dead, overwintering stalks, where they develop dark stroma and cylindrical beaks. The perithecia measure 15–260 $\times$ 190–335 μ. The asci are elongate and 8-spored. The ascospores are hyaline, 2-celled, constricted, and 10–19 $\times$ 3–6 μ.

References

1, 8, 23.

Pod & stem blight, *Diaporthe phaseolorum* (Cke. & Ell.) Sacc. var. *sojae* (Lehman) Wehm.

Symptoms

Blight is mainly a disease of mature plants, affecting the stems and pods. This fungus forms numerous black, small pycnidia on the stems and pods, mostly scattered or in contoured circular rows on the pods and also definitely arranged in long rows on the stems.

Etiology

The fungus produces pycnidiospores of the alpha and beta types, the former hyaline, oval, 1-celled, measuring 6–10 $\times$ 3–4 μ and the latter, long, nonseptate, undulating or curved, narrow, and measuring 50–90 $\times$ 1–2 μ. The ascospores are hyaline, elongate, 2-celled, each biguttulate, and 8–11 $\times$ 3–6 μ. The imperfect stage of the fungus is *Phomopsis sojae*.

References

1, 8, 12.

Powdery mildew, *Erysiphe polygoni* DC.

Symptoms

Mildew covers the upper leaf surface with a thin network of mycelium, developing from centers that seldom coalesce. The fungus causes the foliage to become an off-color yellow green and finally yellow, terminating in browning and often shedding.

Etiology

The fungus produces conidia on aerial hyphae that sometimes are so numerous that they appear as white chalk dust. Stems, petioles, peduncles, and pods are subject to infection. The conidia are hyaline, 1-celled, elongate, ellipsoidal or cylindrical, produced in chains, and 26–52 × 15–23 μ. The perithecia appear later than the conidia and are light to almost black, spherical, appendaged, up to 120 μ in diameter, and contain up to 10 asci, each containing 2–4 ascospores that measure 24–28 × 11–13 μ.

References

8, 12.

Root rot, *Thanatephorus cucumeris* (Frank) Donk

Symptoms

Root rot results from attacks by the soil-inhabiting fungus during unfavorable growing conditions, causing damping-off and decay of the main and secondary roots, followed by wilting and dying of the plants. In the seedling stage, small patches of plants are frequently killed, and in drills the adjacent plants, for several inches, are killed. In average seasons the fungus is limited to the soil zone. High humidity and poor ventilation are conditions under which the fungus may ascend the plant stalk and kill some foliage.

Etiology

The parasitic stage of the fungus, *Rhizoctonia solani*, is sterile but may produce some typically large, brown sclerotia. The basidial spore stage is rarely produced, but when present, it appears on a thin network of hyphae on the lower stem.

References

2, 3.

Web blight, *Botryobasidium microsclerotia* (Matz) Venkat.

Symptoms

This blight is sporadic and seasonal in its appearance and destructiveness. The fungus is seldom associated with the soil, but is aerial in its development as it grows upon stems, petioles, foliage, peduncles, and pods. These parts are overgrown with an abundance of light brown hyphae that is particularly responsible for the killing of the leaf blades.

Etiology

The parasitic, sterile stage of the fungus, *Rhizoctonia microsclerotia*, is diagnostically characterized by the production of minute, brown sclerotia, adhering to the hyphae superficially, mostly on the main stems and petioles. The sclerotia are small enough for 60 per cent to pass through a 100-mesh screen. The basidiospore stage is produced on diseased foliage.

References

2, 21, 22.

Southern blight, *Botryobasidium rolfsii* (Sacc.) Venkat.

Symptoms

The soil-inhabiting fungus may cause some seedling damping-off but more frequently it kills plants at blossoming time and thereafter. The stems are girdled at the soil line by the penetrating hyphae, causing the plant to wilt, die, and become brown; the plant remains upright. White mycelium is produced around the stem at the soil surface and in the immediate surrounding soil.

Etiology

Brown, spherical, hard sclerotia, 1 mm in diameter, usually adhere to the loose hyphae on the soil surface or upon the lower stem. The basidiospore stage is rarely found. The parasitic, sterile stage of the fungus is *Sclerotium rolfsii.*

Reference

12.

Leaf spot, *Phyllosticta glycinea* Tehon & Daniels

Symptoms

Leaf spot develops on both leaf surfaces as round to oval spots, tan-colored at first but becoming brown with a purple border. They vary in size up to 1/4 in. but often coalesce into large blotches that are frequently marginal. The disease is also found on the stems and petioles.

Etiology

The black pycnidia of the fungus develop in the brown areas. They are small, globose, slightly imbedded, ostiolate, and 90–170 μ. The pycnidiospores are hyaline, oblong to ellipsoid, 1-celled, often exuded in tendrils, and $4\text{–}7 \times 2\text{–}3\ \mu$.

Reference

12.

Brown spot, *Septoria glycines* Hemmi

Symptoms

This spotting first appears on the cotyledons and later on the foliage, stems, and pods. The leaf lesions are reddish brown to black, angular in outline, enlarging up to 5 mm in diameter, and conspicuous on both surfaces. The surrounding tissue is pale green to yellow and defoliation may occur. The brown discolorations are often several centimeters long on the stems and may girdle them.

Etiology

The pycnidia are brown to black, imbedded, ostiolate, globose, and 60–125 μ

in diameter. The pycnidiospores are hyaline, filiform, curved, septate, and 35–40 × 1–2.5 μ.

References

8, 12, 25.

Ashy stem, *Macrophomina phaseolina* (Tassi) Goid.

Symptoms

Ashy stem is manifested by a sudden wilting and rapid dying of plants that are half grown or older. The fungus, which is often associated with the soil, contacts the plant at the soil line, where a rapid penetration and girdling take place. The cortex is killed but remains intact as the fungus continues to grow into the wood.

Etiology

The disease is readily diagnosed by the presence of black, small, imbedded, irregular, hard, shiny sclerotia, measuring 30–150 × 22–25 μ, which cause the diseased tissue to appear a slate brown color. The spores of the fungus are produced in pycnidia that are globose, ostiolate, and 100–200 μ in diameter. The pycnidiospores are hyaline, 1-celled, oval-elongate, and 18–30 × 7–11 μ. The sclerotial stage is *Sclerotium bataticola*.

References

8, 12.

Brown stem rot, *Cephalosporium gregatum* Allington & Chamberlain

Symptoms

Stem rot is demonstrated by splitting the main stem, revealing the brown to black pith in the central area. The fungus, common in the soil, enters the main root and lateral branches and continues to grow in these parts and up the main stem. External symptoms appear late in the season in the form of a blighting and rapid drying out, often particularly in evidence between the larger leaf veins where the parenchyma deteriorates most rapidly. Badly diseased fields appear brown when approaching maturity rather than light greenish yellow.

Etiology

The mycelium is hyaline, septate, and branched; the conidiophores are hyaline, septate, branched or not, and 4–15 μ long. Conidia are produced at the apex in succession, sometimes in long chains, and, when moist, in multispored balls or heads; they are hyaline, 1-celled, ovoid to elliptical, and 3–8 × 1–4 μ.

References

8, 12.

Frogeye, *Cercospora sojina* Hara

Symptoms

Frogeye is conspicuous on the foliage, where it forms a characteristic reddish brown spot that bleaches in the center as the spot enlarges. The white central area later becomes covered with the darker spore structures and spores. The disease is also found on stems, peduncles, pods, and seeds. When infection is heavy, some shedding may occur. Seeds become infected as the fungus grows through the pod, where it forms circular brown spots.

Etiology

The spores of the fungus are hyaline, cylindric to obclavate, usually slightly curved, multiseptate, and 20–30 $\times$ 4–8 μ. The ascospore stage of the fungus is *Mycosphaerella phaseolicola*.

References

5, 8, 12, 16.

Purple stain, *Cercospora kikuchii* (T. Matsu. & Tomoyasu) Gardner

Symptoms

Purple stain of the seeds consists of variable, irregular blotches of the seed coat, from pinpoint specks to almost the entire surface and from light to dark purple. Such seeds produce infected seedlings. The stem is girdled and the plant dies. When infection is less severe, plants appear stunted and weak. Infection spreads from seedling cotyledons to stems and foliage. Leaf spots are reddish purple, somewhat angular, appear on both surfaces, and expand up to 1 cm in diameter; pod infections also occur.

Etiology

The conidiophores and conidia are produced on stem lesions and foliage spots. The spores are slender to slightly curved with acute tips, up to 40 septations, and measure 210 $\times$ 3–4 μ.

References

5, 8, 12, 15, 19.

Target spot, *Corynespora cassiicola* (Berk. & Curt.) Wei

Symptoms

Target spot appears on the foliage as reddish brown, irregularly circular areas that develop from small points to spots up to 3 cm in diameter and may be surrounded by a yellow green halo. The larger spots usually show more or less zonation. Spots on the stems and petioles are dark brown and elongate, while pod spots are more or less circular and up to 3 mm in diameter. The fungus causes seed infection through the pod.

Etiology

The conidiophores appear on both leaf surfaces and are more plentiful on the darker rings of the zonate spots. The conidia are straight or curved, septate, mildly obclavate, long and slender, and 150–250 × 10 μ.

Reference

20.

Wilt, *Fusarium oxysporum* Schlecht. f. sp. *tracheiphilum* (E. F. Sm.) Snyd. & Hans.

Symptoms

This disease produces very little severe wilt, but infected plants show an earlier yellowing of the foliage that results in some defoliation. Such plants show a distinct discoloration of brown to black in the vascular tissue. The fungus is usually associated with the soil and gains entrance through the root system of the plant.

Etiology

The macroconidia are produced in abundance in salmon-colored sporodochia on white to rose to purple mycelium and are hyaline, mostly 3–4-septate, elongate to sickle-shaped, curved, showing an acute tip and a basal point of attachment, and measuring 40–50 × 3–5 μ. The microconidia are produced on verticillately branched conidiophores and are hyaline, oval, nonseptate, and 5–12 × 2–4 μ. There are few chlamydospores, and sclerotia of various sizes are common.

References

7, 8.

Other fungi associated with soybean

Alternaria atrans Gibson
Ascochyta sojae Miura
Aspergillus oxyzae (Ahlburg) Cohn
Cercospora canescens Ellis & G. Martin
Cercospora cruenta Saccardo
Cercospora glycines Cooke
Colletotrichum truncatum (Schweinitz) Andrus & W. O. Moore
Coniothyrium sojae Bouriquet
Epicoccum neglectum Desmazieres
Isariopsis griseola Saccardo
Macrophoma mame Hara
Myrothecium roridum Tode ex Saccardo
Phakopsora pachyrhizi Sydow
Phymatotrichum omnivorum (Shear) Duggar

Thielaviopsis basicola (Berkeley & Broome) Ferraris
Uromyces sojae (Henning) Sydow

References: soybean

1. Athow, K. L., and R. M. Caldwell. 1954. A comparative study of diaporthe stem canker and pod and stem blight of soybean. Phytopathology 44:319–25.
2. Atkins, J. G., Jr., and W. D. Lewis. 1954. Rhizoctonia aerial blight of soybeans in Louisiana. Phytopathology 44:215–18.
3. Boosalis, M. G. 1950. Studies on the parasitism of Rhizoctonia solani Kuehn on soybeans. Phytopathology 40:820–31.
4. Chamberlain, D., and B. Koehler. 1951. Soybean diseases in Illinois. Illinois Univ. Agr. Ext. Circ. 676:1–32.
5. Chupp, C. 1953. A monograph of the fungus genus Cercospora. Published by the author: Ithaca, N.Y.
6. Coerper, F. M. 1919. Bacterial blight of soybean. J. Agr. Res. 18:179–94.
7. Cromwell, R. O. 1917. Fusarium-blight, or wilt disease, of the soybean. J. Agr. Res. 8:421–39.
8. Dickson, J. G. 1956. Diseases of field crops. 2d ed. McGraw-Hill: N.Y.
9. Dunleavy, J. M., D. W. Chamberlain, and J. P. Ross. 1966. Soybean diseases. U.S. Dept. Agr., Agr. Handbook 302:1–38.
10. Elliott, C. 1951. Manual of bacterial plant pathogens. 2d ed. Chronica Botanica: Waltham, Mass.
11. Hedges, F. 1924. A study of bacterial pustule of soybean, and a comparison of Bact. phaseoli sojense Hedges with Bact. phaseoli EFS. J. Agr. Res. 29: 229–51.
12. Johnson, H. W., and B. Koehler. 1943. Soybean diseases and their control. U.S. Dept. Agr. Farmers' Bull. 1937:1–24.
13. Johnson, H. W., D. W. Chamberlain, and S. G. Lehman. 1954. Diseases of soybeans and methods of control. U.S. Dept. Agr. Circ. 931:1–40.
14. Johnson, H. W., D. W. Chamberlain, and S. G. Lehman. 1955. Soybean diseases. U.S. Dept. Agr. Farmers' Bull. 2077:1–16.
15. Jones, J. P. 1968. Survival of Cercospora kikuchii on soybean stems in the field. Plant Disease Reptr. 52:931–34.
16. Lehman, S. G. 1934. Frog-eye (Cercospora diazu miura) on stems, pods, and seeds of soybean, and the relation of these infections to recurrence of the disease. J. Agr. Res. 48:131–48.
17. Lehman, S. G., and F. A. Wolf. 1926. Pythium root rot of soybean. J. Agr. Res. 33:375–80.
18. Ling, L. 1951. Bibliography of soybean diseases. Plant Disease Reptr. 204:110–73. (Suppl.)
19. Murakishi, H. H. 1951. Purple seed stain of soybean. Phytopathology 41: 305–18.
20. Stone, W. J., and J. P. Jones. 1960. Corynespora blight of sesame. Phytopathology 50:263–66.
21. Walker, J. C. 1952. Diseases of vegetable crops. McGraw-Hill: N.Y.
22. Weber, G. F. 1939. Web-blight, a disease of beans caused by Corticium microsclerotia. Phytopathology 29:559–75.
23. Welch, A. W., and J. C. Gilman. 1948. Hetero- and homo-thallic types of diaporthe on soybeans. Phytopathology 38: 628–37.
24. Wolf, F. A., and S. G. Lehman. 1926. Brown-spot disease of soy bean. J. Agr. Res. 33:365–74.
25. Wolf, F. A., and S. G. Lehman. 1926. Diseases of soy beans which occur both in North Carolina and the Orient. J. Agr. Res. 33:391–96.
26. Wolf, F. A., and S. G. Lehman. 1926. Pythium root rot of soybean. J. Agr. Res. 33:375–80.

Squash, *Cucurbita pepo* Linnaeus

Blossom blight, *Choanephora cucurbitarum* (Berkeley & Ravenel) Thaxter
Other fungi associated with squash

Blossom blight, *Choanephora cucurbitarum* (Berk. & Rav.) Thaxt.

Symptoms

Blossoms of squash and other related plants show the disease on the second

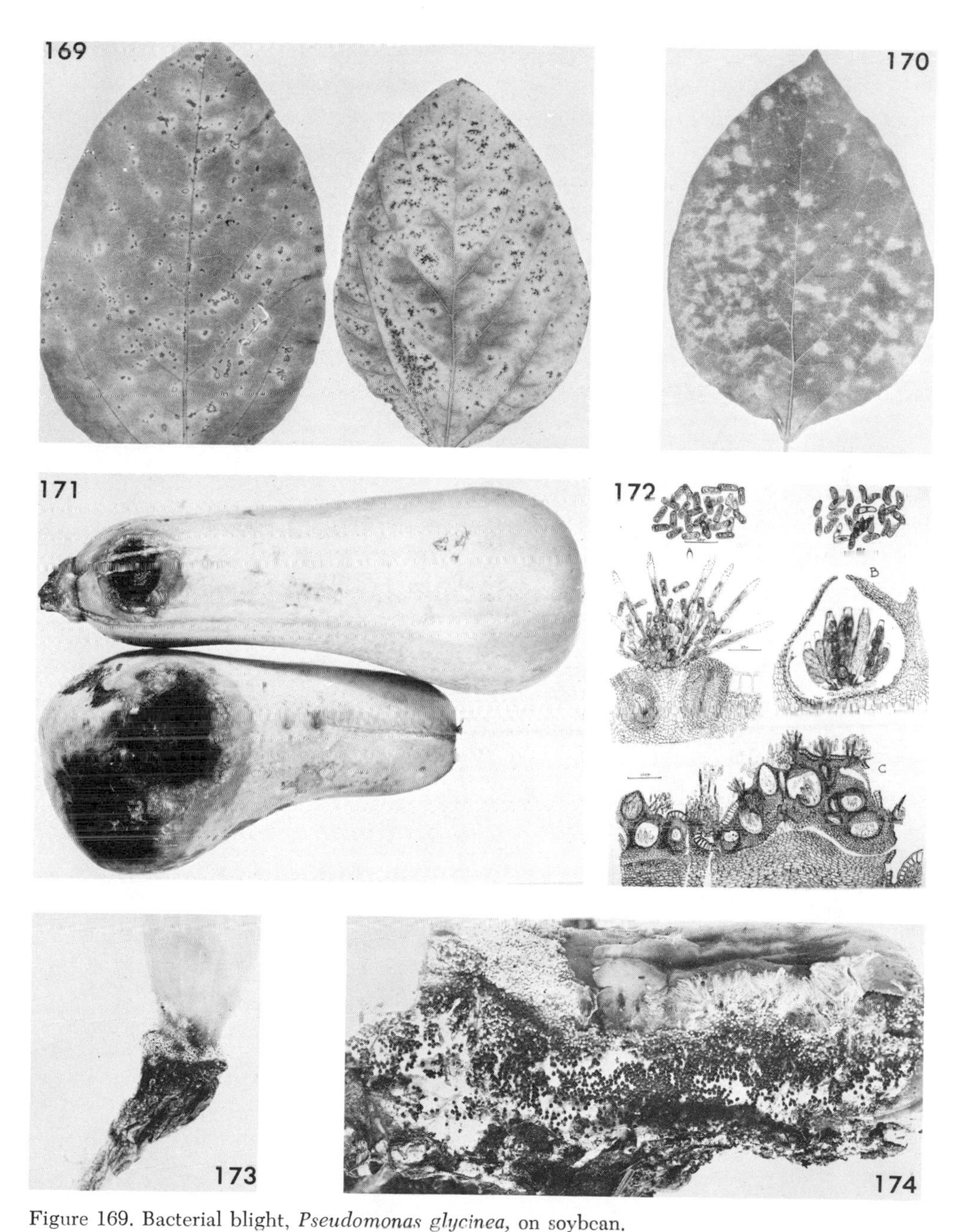

Figure 169. Bacterial blight, *Pseudomonas glycinea,* on soybean.

Figure 170. Downy mildew, *Peronospora manshurica,* on soybean.

Figure 171. Anthracnose, *Glomerella cingulata,* showing disease development stages on squash.

Figure 172. Anthracnose, *Glomerella cingulata.* A. Acervuli, conidia, setae. B. Asci, perithecia, and ascospores. C. Stages A and B in stroma. (Photograph by S. T. Doeung.)

Figure 173. Blossom blight, *Choanephora cucurbitarum,* on squash.

Figure 174. Southern blight, *Botryobasidium rolfsii,* on squash, showing sclerotia and mycelium.

day of blooming; it is usually associated with a lack of turgidity and sometimes a slight wilting. The parts are overgrown and covered with the conidiophores of the fungus, which have white conidial heads. These mature rapidly, turning brown then dark purple. The fungus mycelium penetrates the base of the corolla and on pistillate flowers enters the young squash. A water-soaked, soft decay is produced. The involved squash shrinks and may be shed. It becomes covered with the sporulating fungus in the form of a black mold.

Etiology

The conidiophores are supported by a hyaline, nonseptate, branched, rather coarse mycelium. They are white or hyaline becoming metallic, erect, and produce the capitate vesicle. From this structure many ramuli arise, each in turn producing conidia on a capitellum completely covering it. The conidia are brown, oval to elliptical, striate, and $15–25 \times 7–11\ \mu$. The sporangial stage has not appeared on squash but has been developed in artificial culture. Sporangia are black, ovoid to elongate, and $18–30 \times 10–15\ \mu$. Chlamydospores and zygospores, measuring $50–90\ \mu$ in diameter, have been produced.

Reference

9.

Other fungi associated with squash

Alternaria cucumerina (Ellis & Everhart) J. A. Elliot
Botryobasidium rolfsii (Saccardo) Venkatarayan
Botrytis cinerea Persoon ex Fries
Cercospora cucurbitae Ellis & Everhart
Cladosporium cucumerinum Ellis & Arthur
Colletotrichum lagenarium (Passerini) Ellis & Halsted
Erysiphe cichoracearum De Candolle
Fusarium oxysporum Schlechtendahl f. sp. *niveum* (E. F. Smith) Snyder & Hansen
Fusarium solani (Martius) Appel & Wollenweber f. *cucurbitae* Snyder & Hansen
Mycosphaerella citrullina (C. O. Smith) Grossenbacher
Phoma lagenariae (Thuemen) Saccardo
Phyllosticta orbicularis Ellis & Everhart
Phytophthora cactorum (Lebert & Cohn) Schroeter
Pseudomonas lachrymans (E. F. Smith & Bryan) Carsner
Pseudoperonospora cubensis (Berkeley & Curtis) Rostowzew
Pythium aphanidermatum (Edson) Fitzpatrick
Pythium ultimum Trow
Rhizopus stolonifera (Ehrenberg ex Fries) Lind
Sclerotinia sclerotiorum (Libert) De Bary
Septoria cucurbitacearum Saccardo
Stemphylium cucurbitacearum Osner
Thanatephorus cucumeris (Frank) Donk

References: squash

1. Bryan, M. K. 1930. Bacterial leaf spot of squash. J. Agr. Res. 40:385–91.
2. Chupp, C., and A. F. Sherf. 1960. Vegetable diseases and their control. Ronald Press: N.Y.
3. Guba, E. F. 1950. Spoilage of squash in storage. Massachusetts Agr. Expt. Sta. Bull. 457:9–11.
4. McCullock, L. P. 1962. Alternaria rot following chilling injury of acorn squashes. U.S. Dept. Agr., Market Res. Rept. 518:1–19.
5. Newhall, A. G., and R. E. Wilkinson. 1949. Storage rots of squash in New York State. Plant Disease Reptr. 33: 220–22.
6. Ramsey, G. B., and M. A. Smith. 1961. Market Diseases. U.S. Dept. Agr., Agr. Handbook 184.
7. Thompson, R. C., S. P. Doolittle, and D. J. Caffrey. 1955. Growing pumpkins and squashes. U.S. Dept. Agr. Farmers' Bull. 2086.
8. Walker, J. C. 1952. Diseases of vegetable crops. McGraw-Hill: N.Y.
9. Wolf, F. A. 1917. A squash disease caused by Choanephora cucurbitarum. J. Agr. Res. 8:319–28.

Sugarcane, *Saccharum officinarum* Linnaeus

Bacterial mottle, *Pectobacterium carotovorum* var. *graminarum* Dowson & Hayward
Leaf scald, *Xanthomonas albilineans* (Ashby) Dowson
Red stripe, *Xanthomonas rubrilineans* (Lee et al.) Starr & Burkholder
Mottle stripe, *Xanthomonas rubrisubalbicans* (Christopher & Edgerton) Săvulescu
Gummosis, *Xanthomonas vasculorum* (Cobb) Dowson
Dry top rot, *Sorosphaera vasculorum* (Matz) Schroeter
Seed piece rot, *Phytophthora megasperma* Drechsler
Root rot, *Pythium arrhenomanes* Drechsler
Root necrosis, *Pythium* spp.
Crazy top, *Sclerophthora macrospora* (Saccardo) Thirumalachar, Shaw, & Narasimhan
Leaf splitting, *Sclerospora miscanthi* Miyake
Downy mildew, *Sclerospora sacchari* Miyake
Sooty mold, *Capnodium* spp.
Wilt, *Ceratocystis adiposum* (Butler) Moreau
Pineapple disease, *Ceratocystis paradoxa* (Dade) Moreau
Brown stripe, *Cochliobolus stenospilum* (Drechsler) Matsumoto & Yamamoto
Leaf blast, *Didymosphaeria taiwanensis* Yen & Chi
Red leaf spot, *Dimeriella sacchari* (B. de Haan) Hansford
Spot anthracnose, *Elsinoë sacchari* (Lo) Bitancourt & Jenkins
Pokkah boeng, *Gibberella moniliformis* (Sheldon) Wineland
Iliau, *Gnomonia iliau* Lyon
Ring spot, *Leptosphaeria sacchari* B. de Haan
Tangle top, *Myriogenospora aciculisporae* Vizioli
Dry rot, *Physalospora rhodina* (Berkeley & Curtis) Cooke
Red rot, *Physalospora tucumanensis* Spegazzini
Leaf blight, *Leptosphaeria taiwanensis* (Matsumoto & Yamamoto) Yen & Chu
Rind disease, *Pleocryta sacchari* (Massee) Petrak & Sydow
Smut, *Ustilago scitaminea* Sydow
Rust, *Puccinia kuehnii* (Krueger) Butler
Root rot, *Thanatephorus cucumeris* (Frank) Donk
Banded disease, *Thanatephorus sasakii* (Shirai) Donk
Sheath rot, *Botryobasidium rolfsii* (Saccardo) Venkatarayan
Root rot, *Marasmius sacchari* Wakker
Wound rot, *Schizophyllum commune* Fries
Blob disease, *Sphaerobolus stellatus* (Tode) Persoon
Leaf spot, *Phyllosticta sorghina* Saccardo
Collar rot, *Hendersonia sacchari* Butler
Sheath rot, *Cytospora sacchari* Butler
Leaf scorch, *Stagonospora sacchari* Lo & Ling
Black stripe, *Cercospora atrofiliformis* Yen, Lo, & Chi
Yellow spot, *Cercospora koepkei* Krueger

Brown spot, *Cercospora longipes* Butler
Red sheath spot, *Cercospora vaginae* Krueger
Wilt, *Cephalosporium sacchari* Butler
Veneer blotch, *Deightomilla papuana* Shaw
Leaf spot, *Helminthosporium ocellum* Faris
Eyespot, *Helminthosporium sacchari* (B. de Haan) Butler
Other fungi associated with sugarcane

Bacterial mottle, *Pectobacterium carotovorum* var. *graminarum* Dows. & Hayward

Symptoms

Creamy white stripes, 1–2 mm wide, parallel with the vascular bundles, form from lower portions of the leaf toward the tip. Color may change to brownish red as leaves become older. The stripes may be few or many and, when plentiful, may develop into a definite mottling and a chlorosis. Often secondary shoots are entirely chlorotic. Numerous white droplets of bacterial exudate may appear on the lower leaf surfaces on the stripes and frequently guttation water is whitish and sticky because of bacteria. The disease probably originates on wild grasses. Natural spread takes place during warm, rainy seasons. Infection is probably through wounds and hydathodes.

Etiology

The organism is gram-negative, has 4–6 peritrichous flagella, and measures $0.4\text{–}0.6 \times 1.5\text{–}3\ \mu$. It is readily isolated and on nutrient agar develops white, opaque colonies that become gray; they are glistening, slightly raised, and butyrous. The pathogen is also known as *Erwinia carotovora* var. *graminarum.*

Reference

24.

Leaf scald, *Xanthomonas albilineans* (Ashby) Dows.

Symptoms

There is an acute and a chronic phase. The acute form causes plants to wilt and die. First the leaves wither, then the stalk gradually shrivels and dries up. In the chronic phase, the characteristic symptom is the white, straight, and narrow streaking of the leaves. Another feature is the production of side shoots and small suckers. Pronounced etiolation of the leaves also occurs in the more advanced stages. Still another feature is the tendency of the affected leaves to curl upward and inward in the form of a bow. The internal symptoms consist of tiny red streaks. The bacteria are disseminated by infected planting material, knife transmission, and other means. The organism invades the vascular system in which no tissue breakdown has occurred. The bacteria are confined strictly to the xylem elements. A frequently observed phenomenon is the splitting open of the leaf sheath streak and a consequent discharge of large masses of bacteria.

Etiology

The organism is a gram-negative rod, single or in chains, motile by a single polar flagellum, and 0.2–0.3×0.6–1 μ. Agar colonies are honey yellow, transparent, glistening, moist, entire, and viscid.

References

15, 16, 26, 27, 28.

Red stripe, *Xanthomonas rubrilineans* (Lee et al.) Starr & Burkh.

Symptoms

The disease is characterized by dark green to watery brown stripes which gradually change to deep red. They are elongated, water-soaked, and 1.5–20 mm wide. The stripes arise near midribs in basal parts of younger leaves and spread out later. Lesions are also found on the lower side of the midrib, rarely on the upper, spreading up and down along veins between tip and base of leaves but not on the sheath. Stripes remain uniform in width but vary in length. Broader bands are made by the coalescence of red and chlorotic streaks. Infection is severe on younger canes, young ratoons, and middle-aged leaves; it is most successful under moist and warm conditions. The disease occurs only on sugarcane, particularly POJ-2878, POJ-2725, and Badila.

Etiology

The organism is a gram-negative rod, forms no capsules, is motile by 3 polar flagella, and measures 0.7×1.7 μ. In culture, colonies are yellow, opalescent, glistening, slimy, and smooth. Optimum temperature, 34C.

References

16, 27, 28.

Mottle stripe, *Xanthomonas rubrisubalbicans* (Christopher & Edg.) Săvul.

Symptoms

Mottle stripe is a foliage disease characterized by linear, creamy white stripes with irregular margins and centers that are frequently white. Upon coalescence of streaks, mottled red and white bands are formed across the leaf blade. Top rot is not associated with the disease, and there are no exudation marks.

Etiology

The organism is a short, curved, gram-negative rod, growing singly, in pairs, or in short chains. It produces capsules and is motile by means of polar flagella. Agar colonies are grayish white to buff.

References

16, 24.

Gummosis, *Xanthomonas vasculorum* (Cobb) Dows.

Symptoms

In early stages of development, the disease is characterized by pale green to yellow stripes flecked with reddish dots, regular in outline when young but diffused later, forming along the margin and apex of the leaf blade. Longitudinal streaks enlarge and turn red to brown with age, followed by necrosis. In older canes, inner leaves of the apical whorl develop linear stripes, whereas on lower leaves red blotches and brown streaks appear. The organism in the vascular tissues of leaves moves down into the stalk. Advanced leaf symptoms are associated with dwarfing of plants and necrotic pockets in stalk tissues. The characteristic symptom is the presence of honey yellow, bacterial exudate in conductive tissues. Exudate is formed in wet weather as a slime on the infected surface.

Etiology

The organism is a gram-negative rod, motile with a single polar flagellum, and measures $0.4\text{–}0.5 \times 1\text{–}1.5\ \mu$. In culture, colonies are yellow and circular. Optimum temperature, 28C.

References

15, 16, 28.

Dry top rot, *Sorosphaera vascularum* (Matz) Schroet.

Symptoms

The central leaf whorl shows a fading of green color followed by rolling, wilting, and gradual dying. Some longitudinal stripes of the blade continue to dry, causing the death of the growing point. Severely affected stalks are frequently stunted. The dying of the top results from an apparent clogging of the vascular system. There is no wet, soft decay. Sections of the stalks show bright-colored, vascular bundles extending through nodes.

Etiology

Spherical, orange, smooth spores, measuring 0.4–0.6 mm, and fungus plasmodia fill the xylem. The fungus produces so-called spore balls that are typically spherical and hollow.

References

10, 15, 24.

Seed piece rot, *Phytophthora megasperma* Drechs.

Symptoms

At first a water-soaked condition is found upon splitting the stem. A salmon to red streak develops, later becoming reddish brown and producing a distinct etherlike odor.

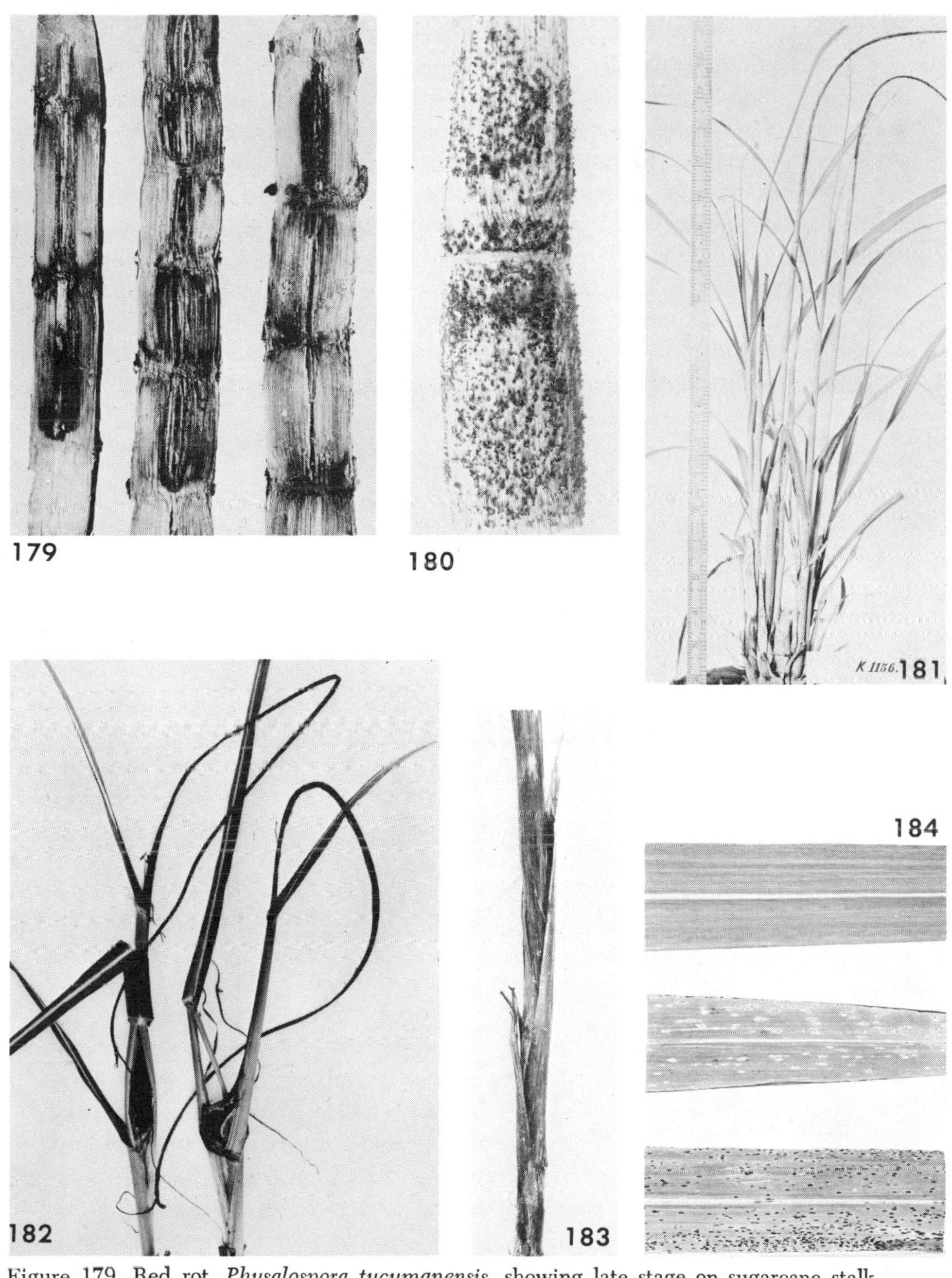

Figure 179. Red rot, *Physalospora tucumanensis,* showing late stage on sugarcane stalk.
Figure 180. Rind disease, *Pleocryta sacchari,* on sugarcane stalk.
Figure 181. Smut, *Ustilago scitaminea,* on tips of sugarcane plants. (Photograph by V. V. Chenulu.)
Figure 182. Sugarcane smut, *Ustilago scitaminea,* showing nodal sori. (Photograph by H. T. Chu.)
Figure 183. Sheath rot, *Cytospora sacchari,* on sugarcane. (Photograph by U.S.D.A.)
Figure 184. Brown spot, *Cercospora longipes,* on sugarcane. (Photograph by U.S.D.A.)

Etiology

The mycelium is usually well-developed, hyaline, branched, flat on agar surfaces or largely aerial. Sporangia are variable in form. Some are filamentous, composed of lobulated branches; others are spherical to subspherical, smooth or spiny, terminal or intercalary, averaging 20–30 μ in diameter, and germinating by germ tube or zoospore formation. Oospores are hyaline to slightly colored, usually spherical, thick-walled, and mostly smooth. Zoospores average about 8–10 μ in diameter. The following fungi have been reported causing root necrosis or have been associated with browning and decay of sugarcane roots: *Pythium aphanidermatum, P. butleri, P. debaryanum, P. dissotocum, P. graminicola, P. irregulare, P. monospermum,* and *P. splendens.* From the disease viewpoint, their identity may be difficult to discern because of possible variation under different conditions.

References

15, 38.

Crazy top, *Sclerophthora macrospora* (Sacc.) Thirum., Shaw, & Naras.

Symptoms

The consistent diagnostic symptoms are a severe stunting of the plant and the production of profuse tillers. Leaves of diseased plants are coarse and brittle, yellowish green, or show whitish streaks up to several centimeters wide that are parallel with the venation of the leaves. Chlorotic blotches, irregular in size and shape, occur on most leaves. The off-green color is traced back to the growing bud and is considered systemic in nature. All shoots from a plant do not necessarily become infected. Buds showing growth on the stalks may contract the disease and produce proliferate clusters of leaves and shoots at each node.

Etiology

Studies have shown this fungus to be slightly different from the genera *Sclerospora* and *Phytophthora* but possessing certain characteristics of each. The oospores resemble those usually encountered in plant tissues produced by *Sclerospora* spp., while the vegetative sporangia resemble *Phytophthora* spp. The sporangia are produced in abundance on the lower surface of diseased foliage and germinate by the production of zoospores. Sporangia measurements are 66–82×39–55 μ. They are thin-walled, and lemon-shaped, with apical papillae. The sporangia are produced on hyphal branches, emerging through stomata which function as sporangiophores and bear 1–5 sporangia at their tips. Oospores develop in immense numbers in the leaf tissue in the parenchyma cells between the vascular strands. When the host tissue dies, it frays out into shredded and twisted clusters, freeing the oospores. They are brown, usually spherical, and 51 μ in diameter. The disease is almost entirely spread by diseased planting stock.

References

28, 41.

Leaf splitting, *Sclerospora miscanthi* Miyake

Symptoms

Greenish yellow stripes extend the full length of the leaf, which later becomes yellow, matted, and reddish brown to dark red. In the last stages, the leaf tissue between the vascular areas disintegrates, and the leaves break apart into a tangled terminal cluster of fibers.

Etiology

The causal organism has been found on cultivated and wild species of sugarcane. The conidiophores of the fungus appear on the lower surfaces of the leaves in white, downy masses opposite the upper surface lesion. The conidiophores are rather thick and stout, twice branched at their tips, with stubby sterigmata, and measure 97–300 × 12–37 μ. The conidia are oval to elongate, measuring 37–49 × 14–23 μ. The oospores are spherical, produced abundantly, mostly in the parenchyma tissue, and 32–57 μ in diameter. Infection has been obtained by planting in inoculated soil.

Reference

24.

Downy mildew, *Sclerospora sacchari* Miyake

Symptoms

Mildew is characterized by the appearance on the leaves of a few yellowish, long stripes parallel with the veins and 3–6 mm wide. They increase in number and, by running together, occasionally cause leaves to be uniformly yellowish. Finally, they turn brown and necrotic. Leaves are easily torn along diseased areas or are shredded at the tips. Stems elongate by lengthening of the internodes, often to twice normal size.

Etiology

A white, fluffy mass of conidia-bearing conidiophores appears on the undersurface of the leaf. One or 2 conidiophores emerge from stomata; they are slightly narrowed at the base, enlarge toward the tip, and are 160–170 × 10–15 μ wide. They are branched 1–2 times at the tip, from which short branchlets arise in groups. Conidia are hyaline, elliptic, less frequently oblong or elongately ovoid, thin-walled, germinate directly by germ tubes, and measure 25–54 × 15–23 μ. The oogonia are yellowish brown, globose or irregularly elliptic, and 55–82 × 49–58 μ. Oospores are yellow, globose, and 40–50 μ in diameter. The epispore is 3.8–5 μ thick.

References

16, 28, 39.

Sooty mold, *Capnodium* spp.

Symptoms

Sooty mold is characterized by large, bright red lesions on leaf sheaths. The lesions are at first small, oval to circular spots but later become larger and irregular by coalescing into ill-defined areas with brown centers. Lesions extending through to inner, successive leaf sheaths become progressively smaller, to a red dot at the limit of invasion. In later stages, production of sooty-mold-like growth appears on infected areas, especially inside the sheath.

Etiology

Several fungi are implicated as the cause of the disease. Olivaceous sooty mold, *Fumago vagans*, appears as a fine, powdery, thin subiculum on the upper surface of the leaf. Conidiophores are yellowish brown to dark brown, simple, erect, and 3–16-septate. Conidia are catenulate in short chains, ellipsoidal or oblong, 1-septate, and $5–16 \times 3.6–5\ \mu$. Chlamydospores are irregularly ellipsoid-ovoid, 2 to several muriform septate, thick-walled, and $7–23 \times 5–9\ \mu$. Black sooty mold, *Caldariomyces fasciculatus,* develops a compact, velvety, thick mycelium on both sides of leaves. The conidiophores aggregate into synema, somewhat swollen at the end and pycnidialike. The synema are 210–476 μ long; conidia are dark olive, ellipsoid, 1–2-septate, continuous, constricted at the septum, and $7–16 \times 3.5–8\ \mu$. Brown sooty mold, *Chaetothyrium spinigerum,* forms a thin mycelial net on the surface. The perithecia are dark brown, subglobose, scattered, and ostiolate; the 1–5-septate, setae-bearing perithecia are dark brown. Asci are clavate with a thickened apex. Ascospores are hyaline, ellipsoid, obtuse at the ends, 1–3-septate, and $10–24 \times 4.3–6.5\ \mu$. The triposoid conidia are 3–4-radiate, subhyaline, 2–4-septate, and $22–44 \times 5–9\ \mu$. Stalks are short, cylindric, subhyaline, continuous, and $10–12 \times 5–7\ \mu$.

References

8, 9, 27, 29.

Wilt, *Ceratocystis adiposum* (Butl.) Moreau

Symptoms

Cut or broken ends of canes injured by animals or wind become covered with a grayish black growth, and the cane pith is blackened at the point of injury. The inner pith is dusky red in streaks. No obvious change in the exterior of cane occurs until later stages when drying up of the pith causes shrinkage. At this point the pith is muddy brown, and the redness disappears gradually. An odor of fermenting pineapple is often detected from split stalks.

Etiology

Grayish brown hyphae penetrate readily into cells of cut or over-ripe cane. Conidia are borne on short, special, lateral branches and are septate, usually forming a row of 2 or 3 cells. Endoconidia are produced unicellularly. They are hyaline, extremely variable in size and shape, elongate, smooth, formed in chains of 40 or more spores, and measure 9–25×4.5–18 μ. Beaks, 2–6 mm high, develop from perithecia. Ascospores, exuded from the ostiole in droplets, are crescent-shaped, and expelled in a translucent, fatty drop. Spores are hyaline, unicellular, and 6.5–8×3–4 μ. Both types of spores germinate rapidly in water. The fungus is not a virulent parasite.

References

8, 24, 25, 38.

Pineapple disease, *Ceratocystis paradoxa* (Dade) Moreau

Symptoms

This disease is characterized on sugarcane by blackening of the core, associated with reddening of tissues; it finally results in rotting. Infected cane has an odor somewhat like that of pineapple. Disease is usually noticed on seed cuttings. Infection takes place from cut ends of canes, extending from one or both ends, but in standing cane it occurs commonly by way of borer holes.

Etiology

The fibrovascular bundles are not disintegrated. Perithecia are hyaline, gregarious, globose, 200–350 μ in diameter, with brown, stellate appendages. Beaks are black, long, and shining, with hyaline, fimbriate tips. The asci are clavate, stipitate, 25 μ long, and 10 μ in diameter; the ascospores are hyaline, ellipsoidal or with unequally curved sides, and 7–10×2.5–4 μ. There are two types of conidia. Microconidia are hyaline, cylindric, produced endogenously in a chain by subdivision of internal contents of hyphae, and measure 10–15×3.5–5 μ. Macroconidia are dark-colored, oval or spherical, produced endogenously in chains by septation at tips of the hyphae, and measure 10–18×7–10 μ. The fungus is a wound parasite.

References

15, 25, 27, 28, 30.

Brown stripe, *Cochliobolus stenospilum* (Drechs.) Matsu. & Yamamoto

Symptoms

The earliest symptom appears on the young leaves as minute, watery spots about half the size of a pinhead. These initial infections quickly turn reddish and assume an elongated shape with their long axes parallel to the veins. As the lesions mature, they become brownish red and form stripes. Surrounding the stripes is a definite yellowish halo. No streak extends from the primary infection toward the leaf tip as is the case with eyespot disease. When the disease is severe, the lesions coalesce, causing older leaves to appear prematurely dried.

Etiology

The imperfect stage, *Helminthosporium stenospilum,* of the fungus is readily isolated from diseased cane tissue. Conidia are brown, straight or slightly curved, cylindric, and variable in size, measuring 50–130×11–18 μ. Perithecia are dark brown, flask-shaped, imbedded, ostiolate, smooth, with beaks protruding. The asci are cylindrical, round at the apex with a stipitate base, and measure 127–195×20–33 μ. Ascospores are hyaline filaments, helicoidal, with an average of 6–9 septa, and measure 130–300×6–8 μ.

References

12, 27, 38.

Leaf blast, *Didymosphaeria taiwanensis* Yen & Chi

Symptoms

Blast first appears as yellowish, elongate, narrow spots on the foliage, parallel with the vascular tissue. The color becomes purplish red as the leaf matures, and the spots appear on both surfaces. Occasionally the spots coalesce. These leaves cease to function from the tip downward and die. Small, black perithecia develop in the dead leaves.

Etiology

The perithecia are scattered, erumpent, with short papillae and spherical ostioles. The asci are oval to oblong or clavate, thin-walled and slender toward the base, and 58–98×9–14 μ. The 8 biseriate ascospores are yellow to brown, rod-shaped, straight, usually 1-septate, distinctly restricted, with a larger upper cell, and measure 16.3–24×3–5 μ. Paraphyses are hyaline, numerous, and septate. The imperfect stage is not known.

Reference

24.

Red leaf spot, *Dimeriella sacchari* (B. de Haan) Hansf.

Symptoms

The leaf spots begin as red dots with yellowish borders; later they become purple red. They are circular to elliptical, up to 2 mm in diameter, and may coalesce. This disease is distinguished from ring spot by the lack of fading in the center of spots.

Etiology

Perithecia are brown, superficial, subglobose, aparaphysate, ostiolate, and 34–88 μ in diameter. The asci are hyaline, ovoid or ellipsoidal, sessile, 8-spored, and 26–35×14–19 μ. Ascospores are pale olivaceous, elongate-elliptical, with rounded ends, 1-septate, not constricted, and 11–15×4.5–6 μ. The conidial stage is not known.

Reference

24.

Spot anthracnose, *Elsinoë sacchari* (Lo) Bitanc. & Jenkins

Symptoms

Minute yellow to purple, elongate to fusiform spots appear on the blade and midrib. They turn pale brown and finally chalky with a reddish halo, average 0.5–1 mm in size, become covered with a fine lacework, and appear in powdery streaks.

Etiology

The sphaceloma stage shows solitary or confluent acervuli that are subcuticular and erumpent; the conidiophores are hyaline, measuring $6–9 \times 3$ μ. Ascomata are pulvinate, intraepidermal, and 50×31 μ. The asci are spherical to oval and measure $10–13 \times 9–11$ μ. They are scattered in the stroma of the ascomata. The ascospores are hyaline, oblong to oval, usually straight, 1–3-septate, often constricted, and $8.6–10 \times 3–3.3$ μ.

References

24, 42.

Pokkah boeng, *Gibberella moniliformis* (Sheldon) Wineland

Symptoms

The earliest symptom is a chlorotic condition at the base of the blade, also a wrinkling, twisting, and stunting of the blade and a discoloration of the stalk. In advanced stages the top rots. The young spindle is stunted and dies, often producing a curly top. A reddish to purple discoloration is associated with diseased sugarcane stalks and roots, and a wilting is evident in the seedling stages. Discolored sets begin to decay. Few or no roots are formed, and buds may swell but fail to produce growth. Internally, the parenchyma and vascular tissues are a wet, reddish brown color. Reddish stem rot and pokkah boeng are caused by the imperfect stage of the same fungus.

Etiology

The causal fungus was first described as *Fusarium moniliforme*. The wilt phase of the disease was attributed to *Cephalosporium sacchari*. The microspores of these fungi are similar. The former produces macrospores while the latter does not, and the former is purple in culture rather than white. The mycelium is hyaline, dirty white to deep rose and violet, branched, and septate. The conidiophores are simple or branched, forming at the tips of mycelial branches, which produce microconidia in abundance in long chains or in heads or clusters. The microconidia measure 1.5×4.2 μ. The macroconidia are borne on aerial mycelium in sporodochia. They are usually curved, 3–5-septate, slightly constricted at the septa, and measure 41×4.3 μ. Perithecia in culture are superficial or imbedded, scattered, and $250–300 \times 300–350$ μ. Ascospores,

8 per ascus, in 2 rows, are almost straight, with rounded ends, 1–3-septate, constricted, and 15.3×5 μ. The fungus is distinguishable from *Gibberella saubinetti* by both microconidia and macroconidia.

References

28, 35, 38, 45.

Iliau, *Gnomonia iliau* Lyon

Symptoms

The fungus invades the leaf sheaths and firmly binds or cements them into a hard case, within which is the growing point. The outer leaves die and assume a pinkish brown color. When the stem is attacked, a bluish gray color develops. On plants moderately attacked, the growing point continues to develop within the hardened case and often breaks through the cemented leaf sheaths, thus causing a distortion of the stem at this point. The iliau fungus is regarded as a weak parasite and attacks young shoots which are not growing normally. The fungus survives in the soil and may be transmitted by infected seed stalks.

Etiology

Spores develop in small blisters on the inner leaf sheaths and are liberated when the stalk disintegrates. Pycnidiospores are brown, oval, and 7–10×15–28 μ. They ooze from the pycnidia in black tendrils. The fungus produces perithecia imbedded in the outer leaf sheaths. The ostiolate beaks are up to 550 μ high. The ascospores are discharged into the air from the perithecia. They are hyaline, 2-celled, slightly curved, constricted, and 22–30×5.7 μ. The imperfect stage is *Melanconium iliau.*

References

27, 30, 38.

Ring spot, *Leptosphaeria sacchari* B. de Haan

Symptoms

The first symptom is the appearance on the older leaves of small, light greenish purple spots which soon change to a bronze brown. The centers of the mature lesions are straw-colored and are surrounded by a thin, reddish brown margin. The spots are most numerous on the older leaves and are usually most plentiful on the tip. Individual spots are elongate, parallel to the veins, and often appear on the stalks. Frequently, older lesions may be somewhat circular with concentric rings, which are accentuated by rows or rings of black specks, which are the perithecia of the fungus.

Etiology

The perithecia are embedded in the host tissue with only the ostiole projecting. They are most conspicuous on the upper leaf surface. They are spherical, about 140 μ in diameter, and paraphyses are present. Asci are numerous, slender, 8-spored, and biscriately arranged. Ascospores are hyaline to faintly colored, slender, 3-septate, and 20–24×5 μ. Inoculation experiments failed to show infection from ascospores. Conidia of an apparent saprophyte, *Phyllosticta sorghina,* mixed with the ascospores resulted in the production of typical leaf spotting, and the development of typical perithecia resulted. The relationship of these has not been further investigated.

References

6, 15, 30.

Dry rot, *Physalospora rhodina* (Berk. & Curt.) Cke.

Symptoms

This disease appears mostly on over-ripe canes, which are entered through wounds resulting from mechanical injury and borers. The stalks shrivel and become dark-colored, and the fruiting bodies appear as black, raised structures in rows along cracks in the rind or variously scattered, producing long tendrils of spores.

Etiology

The vegetative stage of the fungus, *Diplodia natalensis,* produces pycnidia that are black, scattered, covered, later erumpent, papillate, and 150–180 μ. Pycnidiospores are dark, elliptical, 1-septate, not constricted, and measure 24×15 μ, with faintly striated markings pole to pole. The ascigerous stage frequently develops following the vegetative stage by producing black, submerged, ostiolate perithecia. Asci are oval to elongate, saclike structures, containing 8 hyaline, oval spores, measuring 24–42×7–17 μ.

References

24, 30, 35.

Red rot, *Physalospora tucumanensis* Speg.

Symptoms

Plants show a severe wilting of leaves, a loss of normal green color, and premature drying of the older leaves. Standing cane, stubble, sets, and leaf midribs are mostly infected. When diseased stalks are cut open, interior areas show a distinct reddish color. The discoloration may be confined to a few internodes, or it may extend throughout a number of joints. Discolored tissue on the stalk often becomes dry and pithy. The fungus is considered chiefly a wound parasite. The disease is transmitted by spores or by hyphae that contaminate the soil.

Etiology

The vegetative stage of the fungus, *Colletotrichum falcatum,* produces hyaline, septate mycelium. The conidia, produced in acervuli, are hyaline, 1-celled, with a rounded and a pointed end, and measure 16–48×4–8 μ. Setae appear among the pink masses of conidia. The perithecia are black, crowded, submerged at first, then erumpent; the asci are clavate, and the ascospores are hyaline, 1-celled, oval, and thicker through the middle.

Reference
1.

Leaf blight, *Leptosphaeria taiwanensis* (Matsu. & Yamamoto) Yen & Chu

Symptoms

Spots appear on the young foliage. They are small at first, narrow, elliptical, yellowish and red, and visible from both sides of the leaf. They later coalesce, become more elongate and reddish brown. Older diseased leaves die and wither, adhering to the stalk. Close examination reveals numerous, dark fruiting clusters associated with the more mature spots.

Etiology

The sporulating stage of the fungus, *Cercospora taiwanensis,* can be observed with a lens on the lower leaf surface where it arises from superficial hyphae. The subhyaline conidiophores are upright, unbranched, straight and geniculate, 2–4-septate, and taper toward the tip. Conidia are straight or slightly curved, usually subhyaline, 5–8-septate, not constricted, and measure $30–150 \times 2–4$ μ. The perithecia of the fungus are black, scattered, spherical to ovoid, glabrous, and 80–160 μ in diameter. The ascospores are hyaline, dark brown, fusiform, slightly curved, mostly 3-septate, restricted at the septa, and $39–46 \times 6–12.5$ μ. Paraphyses are hyaline, filiform, and unbranched.

References
24, 29.

Tangle top, *Myriogenospora aciculisporae* Vizioli

Symptoms

Extreme stunting and the persistence of the tips of the unfolding leaf to expand are characteristic. The new leaf is prevented from developing and separating from the preceding leaf by the black stroma of the fungus. The new shoot is often killed, and new buds may develop laterally.

Etiology

Perithecia of the fungus develop on the stomatal area. They are black, oval to flask-shaped, sunken in the stroma, short-necked, not erumpent, with a circular ostiole. Asci in fascicles are polyspored, fusiform, and $210–250 \times 12–16$ μ. Ascospores are hyaline, simple, straight or slightly curved, unicellular, acicular, and $18–25 \times 1.5$ μ.

References
1, 11, 17, 27, 30, 35, 38, 39.

Rind disease, *Pleocryta sacchari* (Mass.) Petr. & Syd.

Symptoms

The rind or epidermis becomes somewhat roughened, and numerous small eruptions or pustules appear on the internodes. The affected areas become discolored, and from the eruptions, black, hairlike tendrils of conidia develop. When the affected stalks are split, the tissue beneath the discolored areas on the stalk is red to reddish brown, and the internodes are shrunken and have lost their natural color. A sour odor is common. Leaf blades and sheaths may be affected, resulting in premature yellowing and drying.

Etiology

The vegetative stage of the fungus produces acervuli in great numbers in the rind tissue. The conidia are exuded in black, threadlike, tendrils often inches long. The spores are dark-colored, cylindric to oval, 1-celled, smooth, and $10–15 \times 3–4\ \mu$. The hyphae gain entrance through the wounds and grow within the tissue of plants. The stromata develop from a cushiony mass of mycelium and assume a brown to black color.

References

16, 27, 30.

Smut, *Ustilago scitaminea* Syd.

Symptoms

The smut disease is most easily and correctly diagnosed by the long, whiplike structures at the top of growing stalks, varying from a few inches to several feet in length, terete, and exceedingly narrow. They are unbranched, erect or bent over depending on length, and composed of tightly rolled leaves around a florescence. The outer tissue appears silvery, and as it matures the color changes to black, and the peridium breaks open and scales off, exposing the spores. After the terminal smutted whip matures and frees the spores, the secondary buds farther down on the stalk begin to grow, and each in turn develops a smutted whip. First infection takes place in the soil at the time of germination and growth of the nodal buds. There the fungus penetrates the growing point and persists as the stalk elongates, continually invading all buds at the nodes.

Etiology

The mycelium is intercellular and produces haustoria. As the parasite matures, the mycelium is broken up, and cells form into chlamydospores, which are brown, spherical, smooth, more or less uniform in size, and $4–9\ \mu$ in diameter. They germinate readily by producing a promycelium, usually 3-septate, and measuring $16 \times 3–4\ \mu$. Each of the 4 cells may produce 1 or more sporidia which are hyaline, oval, 1-celled, and $6 \times 2\ \mu$. They germinate by producing septate mycelium, infection hyphae, and secondary sporidia.

References

5, 28, 29, 30, 40.

Rust, *Puccinia kuehnii* (Krueger) Butl.

Symptoms

Minute orange uredisori appear on both leaf surfaces, but mostly on the lower surface. The spots have a pale yellow green halo, become brownish, are elongate, parallel with veins, and are up to 30 mm long and 1–3 mm wide. Severe infection results in killing of leaf tissue.

Etiology

The pycnial and aecial stages are not known; uredospores are orange to brown, oval or pear-shaped, and very variable in size, measuring 29–57 × 18–34.5 μ. They are club-shaped or cylindrical-echinulate. Paraphyses develop at the margins of the sori. Spore germination shows direct penetration of the germ tube through stomata; appressoria may also form. Teliospores are pale to brown, oblong to club-shaped, borne on short stalks, and 28–45 × 14–21 μ.

References

8, 15, 24, 29.

Root rot, *Thanatephorus cucumeris* (Frank) Donk

Symptoms

The wide range of symptoms includes damping-off, root rot, eyespots, brown patch, sheath spotting, leaf blight, and others, depending on the host plants, their stage of maturity, temperatures, humidity, air drainage, and soil reaction. These symptoms are usually accompanied by signs of the fungus such as brown mycelium and sclerotia and basidiospore formation. The host parts are usually killed, especially in the seedling stage. Later in maturity, more definite tissues are subject to invasion, and the wide range of diagnostic characteristics are manifested.

Etiology

The mycelium is tan to brown, septate, with right-angle branching, and 4–15 μ in diameter. Sclerotia are usually pale yellow, changing to dark brown or blackish at maturity, soft, spongy in early formation, oval to flat, irregular, and superficial. The basidial stage consists of delicate, separable, thin, netlike aggregations of hyphal branches, short thick cells, and a more or less definite hymenium. The hymenium contains elongate aerial basidia that support 4 sterigmata, each of which produces a hyaline, ellipsoidal, 1-celled basidiospore that measures 7–12 × 7–7 μ.

Reference

38.

Banded disease, *Thanatephorus sasakii* (Shirai) Donk

Symptoms

A superficial thin web of mycelium extends its advancing margin very rapidly under conditions of high temperatures and humidity. Growth is virtually stopped in the middle of the day when the air and leaf surfaces are dry; this accounts for the banded pattern which is repeated daily when conditions are favorable. The parenchyma tissue beneath the superficial mycelium is killed, appears scalded, and bleaches to a contrasting dirty green, yellow tan color. The areas are bordered with thin, reddish brown lines. The disease on plants less than 2 feet tall causes stunting and death.

Etiology

The basidia and basidiospores are formed on the mature mycelium 2–3 days old and on several bands back of the advancing margin. Sclerotia appear on the older dead leaves and sheaths. The sclerotia are chocolate brown, irregularly flat, and 5–8 mm in diameter. The aerial hyphae over the surface of the bands is white at first, becoming yellow to brown. The basidia are hyaline, clavate, and 15–18 × 7–9 μ. The basidiospores are borne singly or 2 at a time on respective sterigmata; they are hyaline, obovate, smooth, 1-celled, and 8–11 × 5–6.5 μ.

References

29, 32, 34, 46, 47.

Sheath rot, *Botryobasidium rolfsii* (Sacc.) Venkat.

Symptoms

This rot is characterized by an orange red discoloration on the leaf sheath and is found less frequently on young shoots. Lesions are irregular in outline with indistinct margins that spread over the sheath surface. The infected area soon becomes covered by white, radiating mycelium that passes onto the inner leaf sheaths and often binds them loosely together. In later stages of the disease, pale yellow to brown, spherical sclerotia may appear on the lesions. In severe cases, stem areas close to or adjacent to sheath lesions may be infected.

Etiology

Mycelium is tan to brown, septate, and branched usually at right angles. Sclerotia are brown, spherical, smooth, and 0.6–1.6 mm. Basidia are hyaline, clavate, and 2–4 sterigmata form at the apex. Basidiospores are hyaline, smooth, obovate or slightly cylindric, flattened on one side and apiculate at the base, and measure 5–10 × 3.5–6 μ.

References

15, 27, 29, 38, 43.

Root rot, *Marasmius sacchari* Wakk.

Symptoms

Leaves are yellowish, and the edges roll inward, dry, and wither prematurely. Sheaths are matted and cemented together by white, threadlike strands of mycelium. Underground portions of the affected stalks show a definite internal rot of reddish brown color which extends up the stalk for several internodes. At times, a soft rot develops on the roots at the surface of the soil. The rot may invade the stalk for only a few millimeters.

Etiology

The fungus is a very feeble, facultative parasite. It may live for long periods in the soil as a saprophyte on dead cane trash, causing infection only when conditions are favorable and the cane is growing abnormally. Under field conditions, the fungus spreads by its underground mycelium from diseased to healthy plants. The hyphae have cross walls, many of which show clamp connections. At the base of affected plants, small mushrooms, 5–17 mm high, are frequently found in the early morning, following rain. These delicate white to grayish structures are the fruiting bodies of the fungus. When mature, masses of basidiospores, $8–12 \times 4–6\ \mu$, are released.

References

14, 27, 29, 38.

Wound rot, *Schizophyllum commune* Fr.

Symptoms

This disease somewhat resembles red rot. The affected areas are more or less grayish brown, shrunken, and covered with numerous, small, fan-shaped, resupinate, shell-like mushrooms. The leaves dry up as the infection spreads to a few internodes, causing them to wrinkle. Upon cutting longitudinally, deep red spots, alternating with white spaces, are revealed, followed later by a carpet of thick, white mycelium.

Etiology

Sporophores are gregarious or scattered; the pileus is flabelliform, resupinate, sessile or with a very short lateral stipe, and attached by a narrow base. Sporophores are thin, coriaceous, 6–45 mm broad, tomentose, and lobed on incurved margins. Gills radiate from the base, split longitudinally along the edge, and revolute at the margin of the pileus; they are whitish or gray or purplish gray with age. Basidia are clavate-cylindrical, with 4 sterigmata, and measure $19–22 \times 4–5\ \mu$. Spores are hyaline, elliptical, and $4–6 \times 2–3\ \mu$.

References

34, 41.

Blob disease, *Sphaerobolus stellatus* (Tode) Pers.

Symptoms

Numerous sooty black spots or splashes appear on the lower surface of leaves. They may appear anywhere up to several feet above the soil surface but are most numerous on the lower leaves. There is no discoloration on the upper surface.

Etiology

These spore masses are produced in a peridium of a saprophytic fungus that measures up to 4 mm in diameter and ejects the small, 2.5 mm, global spore ball explosively for several feet as a means of dissemination.

References

4, 18.

Leaf spot, *Phyllosticta sorghina* Sacc.

Symptoms

The fungus infects the host specifically at points of attachment of sheath to the stalk, producing small, dry spots, straw-colored, with a dark red or purple margin. Necrosis occurs and infection goes deeper into internodal tissues. Pycnidia are borne on the spot surfaces. There exists considerable divergence of opinion among investigators regarding the relationship of the fungus with the ringspot disease. It is a vegetative stage of *Leptosphaeria sacchari*, and synonyms listed are *Phyllosticta sacchari*, *P. parici*, and *P. kawariensis*.

Etiology

The pycnidia are black, globose, erumpent, and 170–190 × 135–140 μ. They have 1 or several papillate ostioles containing hyaline, elliptical, smooth conidia that are 2–3-guttulate and measure 8.5–12 × 2–3.8 μ.

References

6, 27, 35, 38.

Collar rot, *Hendersonia sacchari* Butl.

Symptoms

The top leaves wither back from the tip along the edges. The loss of foliage results in less productive canes, because of pithiness. The pith may appear water-soaked and show some browning and red streaks at the nodes. At the crown, the red color is very conspicuous, and roots become brown or black.

Etiology

The mycelium is white, septate, and branching. The pycnidia are brown, with numerous ostioles, confluent, and lined with hyaline, fasciculated sterigmata. Pycnidia with parenchymatous walls, imbedded in the stroma, form numerous irregular cavities with or without separation walls. There are two kinds of spores. One is brownish, ellipsoidal or elongate, straight or curved, with round ends, unicellular or 1–2-septate, and 15–24 × 3.5–5.0 μ. The other is filiform, straight or flexuous, multiguttulate, and 20–60 × 0.6–2 μ.

References

8, 24, 35.

Sheath rot, *Cytospora sacchari* Butl.

Symptoms

This disease is primarily found on the sheaths but also infects seed pieces, cuttings, stubble, and young shoots and stalks. Spots are brick red to brown patches on the sheath near the soil line. As the spots become older, the red color fades to pale yellow. It is somewhat modified by the appearance of the black pycnidia that are mostly imbedded with a pointed, protruding ostiole that feels rough when rubbed. As the plants mature, the older leaves droop and die, bending down while still attached. Later these leaves are shed, and the sheaths disintegrate and slough off.

Etiology

The thick-walled pycnidia are produced in dark stromatic masses in the sheath tissues. They are black, usually singular but often crowded, erumpent, and protrude 1–2 mm above the cuticle. The conidia are hyaline, elliptical to ovate, $3\text{–}4 \times 1\text{–}1.5\ \mu$, and are exuded as a glistening droplet at the mouth of the ostiole. The fungus is usually considered a weak parasite.

References

16, 24, 29, 38.

Leaf scorch, *Stagonospora sacchari* Lo & Ling

Symptoms

The early symptom on young foliage is very small, red to reddish brown, densely or sparsely scattered, spindle-shaped spots, measuring $0.5\text{–}3 \times 0.3\text{–}1$ mm. These spots gradually elongate to many times this length, yet become only a few millimeters wider. The lesions become straw-colored in the centers and have a reddish margin. The leaves become streaked to their tips between the vascular bundles, and eventually all the older leaves are involved. On the older lesions, numerous pycnidia appear.

Etiology

The mycelium is mostly intercellular, white at first but gradually becoming dark. The pycnidia are immersed in the leaf tissue, being more common on the upper surface. They are dark brown, more or less spherical, 150–228 μ in diameter, with a membranous wall and a slightly protruding ostiole. The conidia are hyaline, elliptic-fusiform, with acute apex and rounded base, are straight or slightly curved, and measure $44\text{–}52 \times 10\ \mu$. They are 3-septate and constricted slightly at the septa; the conidiophores are short.

References

16, 19, 28.

Black stripe, *Cercospora atrofiliformis* Yen, Lo, & Chi

Symptoms

This disease appears on the leaf blades as narrow streaks, measuring $3\text{–}35 \times$

0.1–1.2 mm. They begin as yellow, minute, oval to circular spots, later becoming linear and brownish black. They lack a definite center and halo.

Etiology

Mycelium is readily observed on the lower surface of the lesions. The conidiophores, arising singly, are straight or slightly geniculate, 2–6-septate, and 20–78 × 33–46 μ. Conidia are hyaline or subhyaline, filiform, obclavate, slightly curved, 2–22-septate, not constricted at the septa, and measure 14–212 × 2–5 μ.

Reference

24.

Yellow spot, *Cercospora koepkei* Krueger

Symptoms

The disease is first visible on leaf blades as yellowish green, irregular spots of various sizes. They become tinged with reddish color, and evolve into irregular, large, red patches due to coalescence. Leaves turn red or reddish brown in advanced stages. The disease occurs on all leaves but is only occasionally found on younger ones near the spindle. Dirty-gray, moldlike fungus growth appears on red blotches.

Etiology

The mycelium is light brown, septate, branched, and granular. Conidiophores are fasciculate in bundles of 3–10, emerging from stomata or directly from the epidermis. They are light grayish olive, paler at the apex, 1–12-septate (mostly 3–6), and geniculate at the upper part. Conidia are hyaline or subhyaline, elongated or fusiform, straight, slightly curved, 1–5-septate, (mostly 3–4), not distinctly constricted at the septa, and 26–55 × 4.3–5.7 μ.

References

16, 28, 29.

Brown spot, *Cercospora longipes* Butl.

Symptoms

Spots on leaf blades are numerous, small, elongate, red at first, and surrounded by a white yellow zone or halo. They appear first on lower leaves and progress upward. In the advanced stage, spots are oval and pale with brown red and yellow borders. They coalesce at plant maturity with only a brown border zone to be seen against the dried background. Fruiting bodies of the fungus are black points on the spots on the lower leaf surface.

Etiology

Conidiophores appear on both leaf surfaces, arising in clusters. They are brown, long, slender, septate, very flexuous and angular, bear traces of attachment points at the apex, and measure 140–250 × 3–5 μ. The conidia are hyaline, elongate, club-shaped, narrow, attenuate, straight or slightly curved, 4–5-septate, and 40–80 × 5 μ.

References

3, 8, 11, 24, 38.

Red sheath spot, *Cercospora vaginae* Krueger

Symptoms

Large, bright red lesions that develop from small, circular spots appear on the leaf sheaths. They eventually become irregular, often coalesce and lose their typical shape, and develop a brownish center. Sooty fruiting structures develop on these spots as they become older. These conidiophores and conidia are mostly on the inner surface of the affected sheath.

Etiology

The conidiophores are dark brown, fasciculate, erect, 1–5-septate, and 31–136 × 3–4 μ. The conidia are hyaline or slightly colored, cylindric, obclavate, with a pointed tip and an abrupt base, up to 3-septate, often constricted, and 14–56 × 3–7 μ.

References

24, 29, 38.

Wilt, *Cephalosporium sacchari* Butl.

Symptoms

Half-grown canes show a wilt of leaves, which gradually turn yellow and then dry up. The stems become pithy and hollow. Internally the pith is purple to red in conspicuous streaks. The odor of the decay is offensive but not sour.

Etiology

The mycelium is white, septate, slender, and penetrates all tissue, mainly in the center of the stem. Cavities are produced that are lined with velvety white films and fertile hyphae that branch verticillately. Conidia are produced abundantly and successively, detached from the tip one by one, but generally remaining aggregated in form, and measuring 4–12 × 2–3 μ. There appears to be considerable morphological variation among the several isolates of the fungus, certain ones of which closely resemble the microconidial stage of *Fusarium moniliforme*. Infection occurs through wounds. Infested soil has proven to be a source of inoculum. The fungus is known to survive in soil for 31 months.

References

8, 15, 24, 30.

Veneer blotch, *Deightomilla papuana* Shaw

Symptoms

Lesions appear on the foliage as small, oval, light green spots with a thin, reddish brown, distinct border. As the spot enlarges, it advances parallel with the leaf veins, above and below the original infection. Secondary extension occurs in a similar manner, thus producing concentric zones one after the other, often exceeding a dozen. Each zone becomes more extensive than the one preceding. The pattern is striking and often exceeds 60 cm in length by 1–2 cm in width.

Etiology

On the lower leaf surface, dense black, hirsute, simple conidiophores form, measuring 39–70 × 6–8 μ. They are brown, cylindrical, tapering and twisted, with a smaller base. Conidia are pale brown, globose, 1-celled, finely echinulate, and 15–20 × 15–18 μ.

Reference

24.

Leaf spot, *Helminthosporium ocellum* Faris

Symptoms

This disease is characterized by yellowish, small, elliptical spots on the leaves. They are elongated rather than spindle-shaped, distinct in outline, 3–28 × 1–3 mm, and become dark brown and form characteristic brown stripes.

Etiology

Conidiophores are grayish yellow brown, paler at the apex, single or fasciculate, simple or branched, straight or subflexuous on leaves, at times geniculate at the upper part, 2–10-septate, and 145–380 μ long. Conidia are long, elliptical or fusiform, widest at the middle, slightly tapering toward the ends, straight or slightly curved, thick-walled, 3–12-septate, and 69 × 12.2 μ. *Cercospora sacchari* is listed as a synonym.

References

7, 20, 29.

Eyespot, *Helminthosporium sacchari* (B. de Haan) Butl.

Symptoms

Minute, watery spots are present on the youngest leaves. The centers of the lesions become reddish. Surrounding the reddish center is a very narrow margin of straw-colored tissue, making the lesions very conspicuous on the green leaves. Streaks caused by toxin often develop from the lesions, extending toward the leaf tips and killing large expanses of leaf tissue. This is primarily a foliage disease and only in very severe cases does it attack the stalk.

Etiology

Conidia are olive to brown, cylindrical, curved, 3–10-septate, and 22–110 × 9–21 μ. They develop singly on short, upright branches on conidiophores which push up between the collapsed epidermal cells of the leaf. In some instances, they grow out of the stomata. They are yellow brown and 70–380 μ long.

References

15, 16, 27, 31, 38.

Other fungi associated with sugarcane

Apiospora camtospora Penzig & Saccardo
Arthrobotrys superba Corda
Asterostroma cervicolor (Berkeley & Curtis) Massee
Cladosporium herbarum Persoon ex Link
Curvularia lunata (Wakker) Boedijn
Eriosphaeria sacchari (B. de Haan) Went
Erwinia ananas Serrano
Erwinia flavida (Fawcett) Magrou
Gloeocercospora sorghi D. Bain & Edgerton
Hypocrea gelatinosa Tode ex Fries
Lophodermium sacchari Lyon
Macrophoma sacchari (Cooke) Berlese & Voglino
Mycosphaerella striatiformans Cobb
Nigrospora oryzae (Berkeley & Broome) Petch
Odontia saccharicola Burt
Papularia sphaerosperma (Persoon ex Link) Hoehnel
Papularia vinosa (Berkeley & Curtis) Mason
Periconia sacchari J. R. Johnston
Plectospira gemmifera Drechsler
Pseudomonas desaiana (Burkholder) Săvulescu
Pseudomonas lapsa (Ark) Starr & Burkholder
Pyrenochaeta terrestris (Hansen) Gorenz, Walker, & Larson
Rosellinia paraguayensis Spegazzini
Trichoderma lignorum Tode ex Harz
Tubercularia saccharicola Spegazzini
Vermicularia graminicola West

References: sugarcane

1. Abbott, E. V. 1938. Red rot of sugarcane. U.S. Dept. Agr. Tech. Bull. 641:1–93.
2. Abbott, E. V., and R. L. Tippett. 1941. Myriogenospora on sugar cane in Louisiana. Phytopathology 31:564–66.
3. Bancroft, K. 1910. A handbook of the fungus diseases of West Indian plants. Pulman & Sons: London, p. 55.
4. Birchfield, W., et al. 1957. Chinese evergreen plants rejected because of glebal masses of Sphaerobolus stellatus on foliage. Plant Disease Reptr. 41:537–39.
5. Bock, K. R. 1964. Studies on sugar-cane smut (Ustilago scitaminea) in Kenya. Trans. Brit. Mycol. Soc. 47:403–17.
6. Bourne, B. A. 1934. Studies on the ring spot disease of sugarcane. Florida Univ. Agr. Expt. Sta. Tech. Bull. 267:1–76.
7. Bourne, B. A. 1941. Eye spot of lemon grass. Phytopathology 31:186–89.
8. Butler, E. J. 1918. Fungi and diseases of plants. Thacker: Calcutta, India.
9. Cobb, N. A. 1906. Fungus maladies of the sugarcane. Hawaiian Sugar Planters' Expt. Sta. Bull. 5:1–254.

10. Cook, W. R. I. 1932. On the life-history and systematic position of the organisms causing dry top rot of sugar cane. Porto Rico Dept. Agr. J. 16:409–18.
11. Delacroix, G. 1911. Maladies des plantes cultivees dans les pays chauds. Augustin Challamel: Paris, pp. 513–17.
12. Dickson, J. G. 1956. Diseases of field crops. 2d ed. McGraw-Hill: N.Y.
13. Diehl, W. W. 1934. The myriogenospora disease of grasses. Phytopathology 24:677–81.
14. Divinagracia, G. G. 1957. Marasmius stem rot of sugar cane. Philippine Agriculturist 40:469–85.
15. Earle, F. S. 1928. Sugarcane and its culture. John Wiley & Sons: N.Y.
16. Edgerton, C. W. 1955. Sugarcane and its diseases. Louisiana University Press: Baton Rouge.
17. Edgerton, C. W., and F. Carvajal. 1944. Host-parasite relations in red rot of sugar cane. Phytopathology 34:827–37.
18. Ellis, M. B. 1960. The blob disease of sugarcane, Commonwealth Phytopathol. News 6:43–44.
19. Exconde, O. R. 1963. Leaf scorch of sugar cane in the Philippines. Philippine Agriculturist 47:271–97
20. Faris, J. A. 1928. Three helminthosporium diseases of sugar cane. Phytopathology 18:753–74.
21. Hayward, A. C. 1961. Gumming disease of sugarcane. Commonwealth Phytopathol. News 7:1–2.
22. Holmes, F. O. 1939. Handbook of phytopathogenic viruses. Burgess Co.: Minneapolis, Minn.
23. Hubert, F. P. 1957. Diseases of some export crops in Indonesia. Plant Disease Reptr. 41:55–64.
24. Hughes, C. G., E. V. Abbott, and C. A. Wisner. 1964. Sugarcane diseases of the world. Elsevier Co.: N.Y.
25. Hunt, J. 1956. Taxonomy of the genus Ceratocystis. Lloydia 19:1–58.
26. Koike, H. 1968. Leaf scald of sugarcane in the continental United States—a first report. Plant Disease Reptr. 52:646–49.
27. Martin, J. P. 1938. Sugarcane diseases in Hawaii. Expt. Sta. Hawaiian Sugar Planters' Assoc.
28. Martin, J. P., E. V. Abbott, and C. G. Hughes. 1961. Sugarcane diseases of the world. Van Nostrand Co.: N.Y.
29. Matsumoto, T. 1952. Monograph of sugarcane diseases in Taiwan. University of Taiwan: Taipei, Taiwan.
30. Nowell, W. 1923. Diseases of crop-plants of the Lesser Antilles. West India Committee: London, pp. 288–330.
31. Parris, G. K. 1950. The helminthosporia that attack sugar cane. Phytopathology 40:90–103.
32. Ramakrishnan, K., and T. S. Ramakrishnan. 1948. Banded leaf blight of arrowroot, Maranta arundinacea. Indian Phytopathol. 1:129–36.
33. Rands, R. D., and E. Dopp. 1934. Variability in Pythium arrhenomanes in relation to root rot of sugarcane and corn. J. Agr. Res. 49:189–221.
34. Roger, L. 1951. Phytopathologie des pays chauds. Encyclopedie Mycologique 17:719–27, 1000–1003, 1099–1102.
35. Roger, L. 1953. Phytopathologie des pays chauds. Encyclopedie Mycologique 18:1378–95, 1495–1521, 1667, 1780–81.
36. Sartoris, G. B. 1927. A cytological study of Ceratostomella adiposum (Butl.) comb. nov., the black-rot fungus of sugar cane. J. Agr. Res. 35:577–85.
37. Smith, K. M. 1957. A textbook of plant virus diseases. 2d ed. Little Brown Co.: Boston, Mass., pp. 477–90.
38. Sprague, R. 1950. Diseases of cereals and grasses in North America. Ronald Press: N.Y.
39. Steib, R. J., and S. J. P. Chilton. 1951. Infection of sugar-cane stalks by the red-rot fungus, Physalospora tucumanensis Speg. Phytopathology 41:522–28.
40. Talballa, H. A. 1969. Smut on true seedlings of sugarcane. Plant Disease Reptr. 53:992–93.
41. Thirumalachar, M. J., C. G. Shaw, and M. J. Narasimhan. 1953. Sclerospora macrospora Sacc. Bull. Torrey Botan. Club 80:299–307.
42. Todd, E. H. 1900. Elsinoë disease of sugarcane in Florida. Plant Disease Reptr. 44:153.
43. Venkatarayan, S. V. 1950. Notes on some species of Corticium and Pellicularia. Indian Phytopathol. 3:81–86.
44. Vizioli, J. 1926. Estudo preliminar sobre um nova pyrenomyceto parasita da canna. Bol. Agr. São Paulo 27:60–69.
45. Voorhees, R. K. 1933. Gibberella moniliformis on corn. Phytopathology 23:368–78.
46. Wakker, J. H., and F. A. F. C. Went. 1898. Die Ziekten van Het Suikerriet op Jave. E. J. Brill: Leiden.
47. Weber, G. F. 1963. Pellicularia spp. pathogenic on sorghum and pearl millet in Florida. Plant Disease Reptr. 47:654–56.
48. Zwet, T. van der, and I. L. Forbes. 1961. Phytophthora megasperma, the princi-

pal cause of seed-piece rot of sugarcane in Louisiana. Phytopathology 51:634–40.
49. Zwet, T. van der, I. L. Forbes, and R. S. Steib. 1960. Studies on the phytophthora rot of sugarcane seed pieces in Louisiana. Plant Disease Reptr. 44:519–23.

Sweet potato, *Ipomoea batatas* (Linnaeus) Lamarck

Brown rot, *Pseudomonas solanacearum* E. F. Smith
Pox, *Streptomyces ipomoea* (Person & W. F. Martin) Waksman & Henrici
White rust, *Albugo ipomoeae-panduratae* (Schweinitz) Swingle
Mottle necrosis, *Pythium ultimum* Trow
Soft rot, *Rhizopus nigricans* Ehrenberg
Mucor rot, *Mucor racemosus* Fresenius
Blossom blight, *Choanephora cucurbitarum* (Berkeley & Ravenel) Thaxter
Stem rot, *Sclerotinia sclerotiorum* (Libert) De Bary
Dry rot, *Diaporthe batatis* Harter & Field
Black rot, *Ceratocystis fimbriata* Ellis & Halsted
Scab, *Elsinoë batatas* (Sawada) Viégas & Jenkins
Sooty mold, *Meliola ipomoeae* Earle
Rust, *Coleosporium ipomoeae* (Schweinitz) Burrill
Southern blight, *Botryobasidium rolfsii* (Saccardo) Venkatarayan
Soil rot, *Thanatephorus cucumeris* (Frank) Donk
Leaf blight, *Phyllosticta batatas* (Thuemen) Cooke
Foot rot, *Plenodomus destruens* Harter
Java black rot, *Diplodia natalensis* Evans
Leaf spot, *Septoria bataticola* Taubenhaus
Blotch, *Pestalotia batatae* Ellis & Everhart
Charcoal rot, *Macrophomina phaseolina* (Tassi) Goidanich
Scurf, *Monilochaetes infuscans* Ellis & Halsted ex Harter
Gray mold, *Botrytis cinerea* Persoon ex Fries
Wilt, *Fusarium oxysporum* f. *batatas* (Wollenweber) Snyder & Hansen
Leaf spot, *Cercospora bataticola* Ciferri & Bruner
Brown rot, *Trichoderma koningi* Oudemans
Root rot, *Phymatotrichum omnivorum* (Shear) Duggar
Other fungi associated with sweet potato

Brown rot, *Pseudomonas solanacearum* E. F. Sm.

Symptoms

This rot is found in the vascular system where a brown stain indicates the disease. Plants are dwarfed, and foliage wilts and shrivels. In the later stages of severe disease, wet, darkened blotches occur on stems, and a bacterial exudate appears. Yellowing of the foliage, leaf killing, and shedding are early symptoms if the disease is mild.

Etiology

The organism is a gram-negative rod, producing no capsules, and motile by a single polar flagellum. Agar colonies are white to opalescent, gradually darkening, small, irregularly circular, smooth, and shiny.

Reference

9.

Pox, *Streptomyces ipomoea* (Person & W. F. Martin) Waks. & Henrici

Symptoms

Pox is most frequently observed on the surface of fleshy roots where brown, scattered, circular spots with depressed centers appear. As these spots enlarge, they become deeper, and the surface becomes dry and eventually falls away, leaving somewhat conical depressions. The surface area enlarges very slowly, whereas the interior extension continues, often involving large internal areas. In severe infections, the potatoes become malformed. The disease is most frequently found in alkaline soils during dry seasons.

Etiology

The aerial mycelium as produced in culture is grayish green to almost green. Sporulation takes place in the form of spiral chains of oval to elliptical spores that are hyaline and 1-celled. The spores form on erect hyphae tips from the aerial mycelium. The conidia measure $1.3–1.8 \times 0.9–1.3\ \mu$.

References

10, 14, 19, 21.

White rust, *Albugo ipomoeae-panduratae* (Schw.) Swing.

Symptoms

Rust is recognized on the foliage and sometimes on the stems and is not a destructive disease, although it is widely distributed. Light green blotches on the upper leaf surface indicate the presence of small to large, circular to irregular sori of the fungus on the opposite surface of the leaf.

Etiology

These whitish areas form under the epidermis, break open, and expose clusters of white or hyaline, small, 1-celled, subspherical sporangia, borne in chains, and each measuring $12–22 \times 10–21\ \mu$. These spores germinate directly or produce a dozen or more zoospores, motile by 2 flagella, that can germinate and cause infection. Oospores are found imbedded in the host tissue.

References

9, 20, 21.

Mottle necrosis, *Pythium ultimum* Trow

Symptoms

This disease is inconspicuous except for some slightly sunken areas surrounding small roots. Occasionally larger sunken lesions are variously scattered. Internally some or all of the tissue is involved in a decay, ranging from slight sunken surface bands to mottling. Cavity laden, rough appearing decay is soft and light-colored. Field diagnosis should be verified by laboratory tests to ascertain the correct cause, as other root decays may be involved.

Etiology

The mycelium is hyaline, branched, sometimes producing few septations, and measuring up to 6.5 μ in diameter. Conidia are terminal, spherical, germinate directly, and measure 12–28 μ in diameter. Zoospores are not formed. Oogonia are usually terminal, smooth, and 19–23 μ in diameter. Oospores are spherical, heavy-walled, germinate directly, and measure 14.7–18.3 μ in diameter. The fungus also causes damping-off.

References

5, 9, 13, 20.

Soft rot, *Rhizopus nigricans* Ehr.

Symptoms

Soft rot may occur on the tubers of this plant whenever they are injured in harvesting and exposed in storage and transit, but it seldom occurs in the soil. Foliage and aboveground parts are not affected. The fungus causes infection at the open ends of the potato, where invasion is rapid, and results in a soft, wet, moldy decay with a pleasant odor. Under less humid conditions, the decay may be rather dry. The mold consists of mycelium and sporangia of the fungus that appear black because of abundant spore production. The disease spreads by contact.

Etiology

The mycelium is hyaline, branched, nonseptate, and produces long, arching, aerial stolons. The stolons extend 1–2 cm beyond the surface and, at the mycelial margin, bend down to the host and develop rhizoids. The sporangiophore rises at the place of rhizoid growth, terminating in a sporangium. The sporangium contains many spores confined until the wall breaks, freeing them to air dissemination. The spores are dark-colored, 1-celled, oval, striate, and 14×11 μ. The resting zygospore, which is black, rough, and spherical, measures 150–200 μ.

References

9, 20.

Mucor rot, *Mucor racemosus* Fres.

Symptoms

This fungus causes infection through the ends and side blemishes of the roots after removal from the soil and while in storage or transit under humid conditions at relatively low temperatures. The decay develops slowly, forming circular, sunken, somewhat zonate spots 1 in. or more in diameter. The flesh becomes grayish and mottled; it usually remains firm but is stringy and fibrous when broken or torn apart.

Etiology

The mycelium is hyaline, nonseptate, branched, usually stout, and frequently vacuolate. Sporangiophores develop sporangia at their tips. Conidia are hyaline, 1-celled, oval, and less than 10 μ long.

Reference

9.

Blossom blight, *Choanephora cucurbitarum* (Berk. & Rav.) Thaxt.

Symptoms

Mold frequently develops in humid weather on foliage that has been injured mechanically as from the sun, cold, or hail. The light-colored mycelium and sporangia of the fungus grow over the injured tissue, which usually turns black. The fungus is not too conspicuous and may be limited in its role as a parasite.

Etiology

The mycelium is hyaline, nonseptate, branched, and mostly submerged in host tissue. Conidiophores arise, producing heads containing black, 1-celled, striate, lemon-shaped spores that measure 18–30 × 10–15 μ. The spores have 12–20 terminal, hairlike structures. The zygospores measure 50–90 μ in diameter.

References

20, 21.

Stem rot, *Sclerotinia sclerotiorum* (Lib.) d By.

Symptoms

This rot occasionally causes a canker at the soil line of the main stem or infection on the vines in contact with the soil under wet and humid weather conditions. The stem or vine is girdled; a small amount of white mycelium may be present, and large, irregular, black sclerotia may be found associated with the lesions. The fungus has not been associated with diseases of the fleshy roots in the field or in storage.

Etiology

The mycelium is hyaline, appearing white and cottony when abundant. It is septate, branched, and rapid-growing. No spores are produced that are considered infectious. Sclerotia are black, hard, oval to elongate, and aid in the survival of the fungus.

Reference

20.

Dry rot, *Diaporthe batatis* Harter & Field

Symptoms

Dry rot is encountered mostly in storage; infection begins at the ends, progresses slowly, eventually involves the entire root. Bedding stock, if diseased, causes the draws to be diseased sufficiently to spread the fungus in the field. In storage, the shriveling, drying, and wrinkling begin at the ends of the tuber and continue to destroy it, forming a hard mummy.

Etiology

The pycnidia in which spores are produced are imbedded in the outer cortex of the mummy. The pycnidiospores are hyaline, 1-celled, oblong to oval, and $6–8 \times 3–5\ \mu$. Stylospores are produced in pycnidia, often with the oval spores. They are hyaline, filiform, slightly bent at one end, nonseptate, and $16–30 \times 1–2\ \mu$. The perithecia develop in the stromata usually associated with the pycnidia. They produce beaks up to several millimeters long. The asci are clavate, sessile, and contain striate ascospores that are hyaline, 2-celled, constricted, elliptical, and $8–12 \times 4–6\ \mu$.

References

9, 17, 20.

Black rot, *Ceratocystis fimbriata* Ell. & Halst.

Symptoms

Black rot in the field causes a progressive yellowing and death of foliage and vines. The fungus is not found on the aboveground parts but may cause cankers that girdle the stem just below the soil line. On the potatoes, brownish, circular, sunken spots appear with definite, dark-colored centers. These colored centers contain many black, shiny, hairlike structures, up to 1 mm in length, that are sporulating parts of the fungus. These spots may be numerous or scattered; the black discoloration of the flesh beneath them penetrates deeply and eventually involves the entire potato, resulting in a dry, shrunken, hard mummy.

Etiology

The mycelium is colored, branched, septate, and abundant. Conidia are of two kinds. The more abundant are endospores that are hyaline, 1-celled, rectangular to oval, cylindrical, and $9–50 \times 3–5\ \mu$. The others are dark-colored, oval, 1-celled, and $10–19 \times 8–10\ \mu$. The perithecia are imbedded with protruding beaks up to 1 mm long. The asci are pear-shaped and contain 8 hat-shaped ascospores that are hyaline, 1-celled, and $5–7\ \mu$ in diameter.

References

9, 20, 21.

Scab, *Elsinoë batatas* (Saw.) Viegas & Jenkins

Symptoms

This fungus causes lesions on the leaf blades, petioles, peduncles, and veins. These spots are light tan to brown, oval to elongate, with depressed centers. The spores of the fungus are produced abundantly in the centers of the spots.

Etiology

The mycelium is hyaline, scanty, septate, branched, and up to 3 μ in diameter. The acervuli are subepidermal, emerging, and 12–25 μ in diameter. Conidiophores are hyaline, 1-celled, short, and 6–8 μ long; the conidia are 1-celled, oblong, and 6–8 × 2–4 μ. The ascomatalike structure is elongate in the stromata, forming an area containing a single row of asci which are globose and 15–16 × 10–12 μ. The ascospores are hyaline, curved, septate, and 7–8 × 3–4 μ. The conidial stage is *Sphaceloma batatas.*

References

2, 11, 20.

Sooty mold, *Meliola ipomoeae* Earle

Symptoms

The fungus grows on a honeydew produced by insects such as aphids, scales, mealy bugs, and white flies that infect many plants. It does not parasitize the sweet potato but often develops a black, filmy covering of the aerial parts.

Etiology

The mycelium is dark, branched, septate, superficial, and produces a black, velvety, membranous coating over the leaf surface, continuous or in spots. The hyphal cells are 8–10 × 6–10 μ. Conidia are 1-celled and abscised from the hyphae. The perithecia are black, oval, up to 200 μ in diameter, and contain ascospores that are dark, 3-septate, and 40–60 × 21–23 μ.

References

2, 6.

Rust, *Coleosporium ipomoeae* (Schw.) Burr.

Symptoms

Rust is recognized by the orange red, minute pustules that appear on the lower leaf surfaces. They may be few and scattered or so numerous that the leaf deteriorates rapidly, dies, and turns brown. Many species of pine are the alternate hosts on which the pycnial and aecial stages develop. The aeciospores then infect the sweet potato and produce uredospores that in turn may reinfect the sweet potato foliage.

Etiology

The fungus is an obligate parasite and produces minute and inconspicuous pycnia and aecia on the alternate host, pine trees. The pycniospores are hyaline, oval, and small. The aeciospores are oval to ellipsoidal, 22–27×17–20 μ, with orange contents, and a roughly verrucose wall. The uredia are circular and are orange yellow on the lower leaf surface; the uredospores are slightly verrucose, ellipsoidal, and 18–27×13–21 μ. The telia are subepidermal and are reddish orange on the lower leaf surface. The teliospores are oblong to clavate, germinate in place, and 60–80×19–23 μ.

References

1, 9.

Southern blight, *Botryobasidium rolfsii* (Sacc.) Venkat.

Symptoms

In the nursery beds the sprouts are often killed at the soil line not from diseased stock but from mycelium of the fungus growing in surface soil. The fungus is soil-borne and survives from season to season in the form of brown, small, spherical sclerotia that are produced in abundance. Their presence is diagnostic. Vines used as planting stock in the field often are killed in large areas by this fungus. The cuttings are often killed before they produce roots. The fungus seldom causes decay of the harvested potatoes.

Etiology

The mycelium is hyaline, white en masse becoming brown, branched, septate, and usually luxuriant. Sclerotia are light brown to brown, spherical, hard, superficial, and measure less than 1 mm in diameter. The basidiospores are hyaline, 1-celled, elongate to oval, and 7–8×3–5 μ.

References

9, 17, 21.

Soil rot, *Thanatephorus cucumeris* (Frank) Donk

Symptoms

Soil rot causes stem cankers near or below the soil line and on the lower or soil side of the vines. Plants appear to be off-color or yellowish when this fungus infects the roots.

Etiology

The mycelium is light brown, septate, branched, and usually evident superficially around the stem of diseased plants at the soil line. The fungus produces brown, flat to oval sclerotia. The basidia are produced on the lower stem petioles or lower leaf surfaces in a fine network of hyphae. The basidiospores are hyaline, 1-celled, oval to elongate, and 3–6×7–10 μ.

References

2, 9.

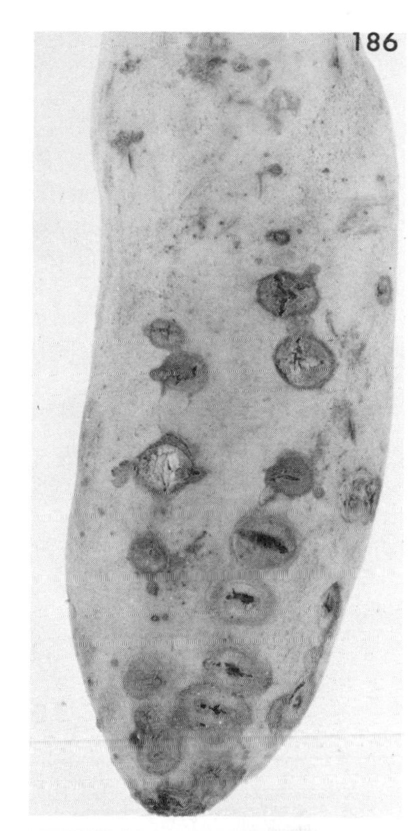

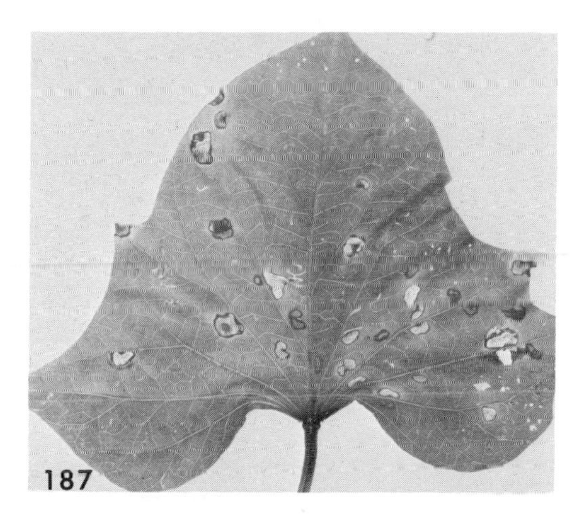

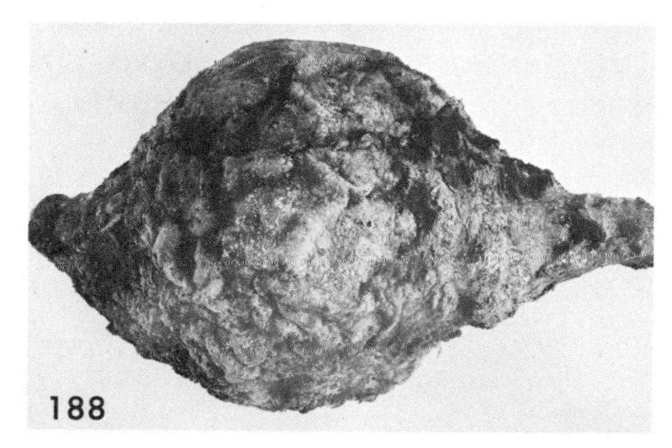

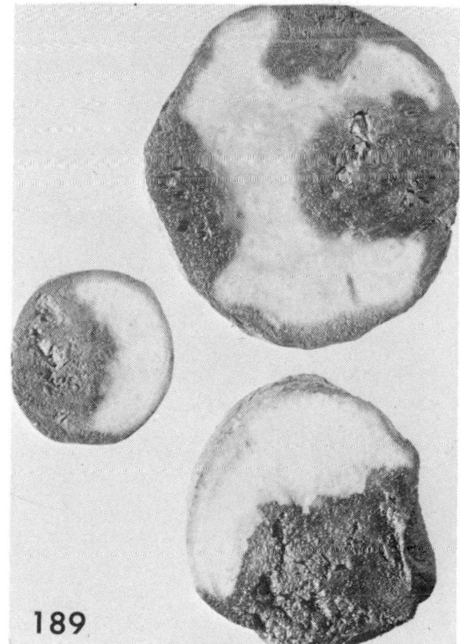

Figure 185. White rust, *Albugo ipomoeae-panduratae,* lesions on sweet potato leaf.
Figure 186. Black rot, *Ceratocystis fimbriata,* on sweet potato.
Figure 187. Leaf blight, *Phyllosticta batatas,* on sweet potato.
Figure 188. Java black rot, *Diplodia natalensis,* on sweet potato.
Figure 189. Charcoal rot, *Macrophomina phaseolina,* on sweet potato.

Leaf blight, *Phyllosticta batatas* (Thuem.) Cke.

Symptoms

Blight is common and frequently serious as a defoliator under very humid conditions. It is found only on the foliage and is first evident by the appearance of yellow spots on the leaf blades. They are usually scattered, circular at first becoming angular, and up to 1/4 in. in diameter. The central part of the spot dries and changes from light green to brown. Numerous pycnidia form on the brown tissue, which shrinks away from the margins, shrivels, and falls away.

Etiology

The pycnidia are black, ovoid, and up to 125 μ in diameter. The pycnidiospores are hyaline, oval to oblong, 1-celled, and $5–10 \times 2–4\ \mu$.

Reference

9.

Foot rot, *Plenodomus destruens* Harter

Symptoms

Foot rot most often originates in the seedbed on draws from diseased seed. The vines become infected and are transplanted into the field. There young plants develop cankers at, or just below, the soil line. They develop slowly until well established and under favorable conditions girdle the stem and quickly kill the plant. The roots are infected from the stem end and may show considerable decay at digging time. It is a root decay and not usually found aboveground.

Etiology

The mycelium is hyaline, branched, septate, and produces an abundance of pycnidia that are subepidermal. Pycnidiospores are hyaline, 1-celled, oval, $6–10 \times 3–5\ \mu$, and emerge in long, silvery tendrils. Stylospores also develop in pycnidia, but, as far as is known, they are functionless. They are hyaline, 1-celled, curved or straight, with rounded ends, and up to 15 μ long.

References

2, 9, 20.

Java black rot, *Diplodia natalensis* Evans

Symptoms

This fungus is a wound parasite that causes infection at the open ends of the potato and also through surface abrasions. The decay develops slowly and is indicated by the slate color produced in the cortex. As the decay progresses, the internal areas blacken, and the potato begins to shrivel and dry.

Etiology

The fungus produces pycnidia that are black, globose, ostiolate, and 250–350 μ in diameter. They are imbedded in the outer wall of the potato, in the cortical areas beneath the cuticle or skin. The pycnidiospores are dark-colored, oval, striate, 2-celled, and 26×15 μ. They are produced in subepidermal cortical tissue and become almost superficial. Black masses of conidia spread over part or all of the surface. The ascospore stage of the fungus is *Physalospora rhodina.*

References

2, 9, 21.

Leaf spot, *Septoria bataticola* Taub.

Symptoms

Leaf spot is characterized by small, circular spots, 2–5 mm in diameter, scattered over the upper surface of the leaf blade.

Etiology

The pycnidia, measuring 70–150 μ in diameter, appear as black specks over the whitish center area of the leaf spots and produce pycnidiospores that are hyaline, filiform, 3–7-septate, straight or curved, and 8–60×1–2 μ.

Reference

9.

Blotch, *Pestalotia batatae* Ell. & Ev.

Symptoms

Blotch appears on the cortex and epidermis of the potato as a black, thick, roughened, raised, cushionlike crust, variously scattered but often coalescing.

Etiology

The fruiting pustules develop in the cortical area, at first imbedded, and cause the cuticle to become rough and dark, later breaking open. The conidia are dark-colored, 5-celled, constricted, oblong, clavate, straight, and 23–28 μ. The terminal cells are hyaline and small, and the 3 interior cells are dark-colored. There are 3 terminal setulae and a short basal pedicel.

Reference

7.

Charcoal rot, *Macrophomina phaseolina* (Tassi) Goid.

Symptoms

This disease, usually considered a storage rot, has been observed in the field on newly set vines and mature plants. Infection takes place through wounds and at the open ends of the potato. The disease develops slowly and goes almost unnoticed except for a slight darkening of the cortical and epidermal layers from chocolate brown to a dark, reddish brown. The sclerotia of the fungus are formed imbedded in the tissue, and, because of their dark color and abundance, the decay appears almost black. The darkened tissue often involves most of the potato, which shrivels and becomes mummified.

Etiology

The sclerotia are diagnostic of the disease; they measure 32–150 × 22–25 μ. The pycnidia are globose, ostiolate, and 100–200 μ in diameter. The conidia are hyaline, oval to elliptical, and 16–32 × 9–10 μ. The sterile stage of the fungus is *Sclerotium bataticola.*

References

9, 20, 21.

Scurf, *Monilochaetes infuscans* Ell. & Halst. ex Harter

Symptoms

Scurf causes a superficial series of brown patches on the underground parts of the plant. The diseased potato shrinks because of loss of moisture. The disease is spread in the field by seedbed draws grown from scurfy seed.

Etiology

The mycelium is septate, branched, and hyaline becoming dark. The conidiophores are septate and up to 175 μ long. The conidia are produced on the surface of the scurf areas. They are hyaline or colored, oval to elongate, 1-celled, and 12–20 × 4–7 μ.

References

8, 20, 21.

Gray mold, *Botrytis cinerea* Pers. ex Fr.

Symptoms

Mold causes a slow, soft, watery decay of potatoes after harvest and in storage. The tissues are easily pulled apart and are of a stringy consistency. The fungus is common and widespread, but infrequently causes potato decay.

Etiology

The mycelium is branched, septate, and hyaline becoming dark. The conidiophores are hyaline to dark, erect, and branched. The conidia of the fungus are hyaline, 1-celled, oval, and 11–14 × 8–10 μ. They are produced abundantly on successive conidiophores. The sclerotia are black, hard, irregularly shaped, and up to several millimeters in diameter.

References

2, 20.

Wilt, *Fusarium oxysporum* f. *batatas* (Wr.) Snyd. & Hans.

Symptoms

Wilt develops slowly in the field and is characterized by the off-color of the foliage. The older leaves often die and are shed, resulting in long, leafless vines with terminal foliage. The plants often remain alive for extensive periods but are not productive. The parasite is soil-inhabiting and remains viable therein for years.

Etiology

The mycelium is hyaline, septate, branched, and produces hyaline, elongate, slightly curved macroconidia that are up to 5-septate and measure 37–50 × 3–4 μ. Microconidia are hyaline, 1-celled, oval to elongate, and 5–12 × 2–4 μ. Chlamydospores are formed.

References

9, 20.

Leaf spot, *Cercospora bataticola* Cif. & Bruner

Symptoms

This disease appears as small, circular, tan spots with a narrow brown margin. They are mostly scattered and up to 1/3 in. in diameter.

Etiology

Conidiophores are produced on the lower leaf surface in dense olivaceous clusters. They are septate, 80–140 × 2–5 μ, and bear the hyaline to faintly colored conidia. The conidia are filiform to obclavate, septate, usually straight, taper toward the tip, and measure 50–160 × 3–6 μ.

References

2, 20.

Brown rot, *Trichoderma koningi* Oud.

Symptoms

This is a storage rot and probably a wound parasite that appears as circular, light brown spots. The invaded flesh is water-soaked and hard, and the entire potato tends to shrivel. A black zone separates invaded brown tissue from tan healthy tissue. Eventually the whole potato rots.

Etiology

The fungus is soil-inhabiting and goes into storage as a saprophyte on the potato. The conidia are produced on the mycelium. They are hyaline to olivaceous, elliptical, 1-celled, and 3–4 × 2–3 μ.

References

2, 4, 18.

Root rot, *Phymatotrichum omnivorum* (Shear) Dug.

Symptoms

This rot is characterized by root depressions covered with fine, brown, mycelial threads. The fungus is usually disposed on the soil surface. It usually infects the roots at the ends and gradually grows through the entire potato, leaving it a brown, decayed mummy. The vines may be diseased for several inches from the main stem, but terminal tips continue to grow from the nodal roots.

Etiology

The fungus survives from season to season as sclerotia which appear at the soil surface. The spores of the fungus, although rarely functional, are borne on erect conidiophores up to 28 μ high; they are hyaline, circular to oval, and 3–10 $\times$ 4–8 μ. This is referred to as the spore mat stage. The vegetative stage of the fungus is *Ozonium omnivorum*.

References

2, 16, 20.

Other fungi associated with sweet potato

Aspergillus niger Tieghem
Blakeslea trispora Thaxter
Cercospora ipomoeae Winter
Leptosphaerulina bataticola Kohkhryakoff & Dyurinski
Phoma batatae Ellis & Halsted
Phytophthora areacae (Condit) Pethybridge
Pythium scleroteichum Drechsler
Rhizopus oryzae Went & Pringle
Thielaviopsis basicola Berkeley & Broome
Verticillium albo-atrum Reinke & Berthold

References: sweet potato

1. Arthur, J. C., and G. B. Cummins. 1962. Manual of the rusts in United States and Canada. Hafner: N.Y., p. 438.
2. Chupp, C., and A. F. Sherf. 1960. Vegetable diseases and their control. Ronald Press: N.Y., pp. 489–524.
3. Cook, H. T. 1955. Sweet potato diseases. U.S. Dept. Agr. Farmers' Bull. 1059: 1–26.
4. Cook, M. T., and J. J. Taubenhaus. 1911. Trichoderma köningi: the cause of a disease of sweet potatoes. Phytopathology 1:184–89.
5. Drechsler, C. 1934. Pythium scleroteichum n. sp. causing mottle necrosis of sweet potatoes. J. Agr. Res. 49:881–90.
6. Fawcett, H. S. 1936. Citrus diseases and their control. McGraw-Hill: N.Y.
7. Guba, E. F. 1961. Monograph of Monochaetia and Pestalotia. Harvard University Press: Cambridge, Mass., p. 342.
8. Harter, L. L. 1916. Sweet-potato scurf. J. Agr. Res. 5:787–91.
9. Harter, L. L., and J. L. Weimer. 1929. A monographic study of sweet potato diseases and their control. U.S. Dept. Agr. Tech. Bull. 99:1–117.
10. Hooker, W. J., and L. E. Peterson. 1952. Sulfur soil treatment for control of sweet potato soil rot incited by Streptomyces ipomoea. Phytopathology 42: 583–91.
11. Jenkins, A. E., and A. P. Viégas. 1943. Stem and foliage scab of sweet potato (Ipomoea batatas). J. Wash. Acad. Sci.

33:244.

12. Lauritzen, J. I. 1935. Factors affecting infection and decay of sweet potatoes by certain storage rot fungi. J. Agr. Res. 50:285–329.
13. Matthews, V. D. 1931. Studies on the genus Pythium. University of North Carolina Press: Chapel Hill.
14. Person, L. H., and W. J. Martin. 1940. Soil rot of sweet potatoes in Louisiana. Phytopathology 30:913–26.
15. Ramsey, G. B., B. A. Friedman, and M. A. Smith. 1959. Market diseases, sweet potatoes. U.S. Dept. Agr., Agr. Handbook 155:22–35.
16. Streets, R. B. 1937. Phymatotrichum (cotton or Texas) root rot in Arizona. Ariz. Univ. Agr. Expt. Sta. Tech. Bull. 71:299–410.
17. Taubenhaus, J. J. 1918. Diseases of truck crops and their control. Dutton & Co.: N.Y., pp. 151–84.
18. Taubenhaus, J. J., and T. F. Manns. 1915. The diseases of the sweet potato and their control. Delaware Univ. Agr. Expt. Sta. Bull. 109:23–25.
19. Waksman, S. A., and A. T. Henrici. 1943. The nomenclature and classification of the Actinomycetes. J. Bacteriol. 46:337–41.
20. Walker, J. C. 1952. Diseases of vegetable crops. McGraw-Hill: N.Y.
21. Weber, G. F., and E. O. West. 1930. Diseases of sweet potatoes in Florida. Florida Univ. Agr. Expt. Sta. Bull. 212: 1–40.

Tea, *Thea sinensis* Linnaeus

Leaf spot, *Bacillus theae* Hori & Bokura
Brown blight, *Glomerella cingulata* (Stoneman) Spaulding & Schrenk
Leaf spot, *Phaeosphaerella theae* Petch
Leaf spot, *Calonectria theae* (Petch) Loos
Copper blight, *Guignardia camelliae* (Cooke) Butler
Black stem, *Massaria theicola* Petch
Red root disease, *Sphaerostilbe repens* Berkeley & Broome
Black blight, *Asterina camelliae* Sydow & Butler
Sooty mold, *Limacinula theae* Sydow & Butler
Dieback, *Nectria cinnabarina* (Tode) Fries
Root decay, *Rosellinia arcuata* Petch
Root rot, *Ustulina zonata* Léveillé
Thorny stem blight, *Tunstallia aculeata* (Petch) Agnihothrudu
Thorny stem blight, *Tunstallia aculeata* (Petch) Agnihothrudu var. *kesabii* Agnihothrudu
Blister, *Exobasidium vexans* Massee
Thread blight, *Corticium theae* Bernard
Thread blight, *Ceratobasidium stevensii* (Burt) Venkatarayan
Pink disease, *Botryobasidium salmonicolor* (Berkeley & Broome) Venkatarayan
Red root, *Poria hypolateritia* Berkeley
Brown root, *Fomes lamaensis* (Murrill) Saccardo & Trotter
White root rot, *Fomes lignosus* (Klotzsch) Bresadola
Shoestring rot, *Armillaria mellea* Vahl ex Fries
Horse hair blight, *Marasmius equicrinis* Mueller
Dieback, *Stilbella theae* Bernard
Leaf spot, *Phoma theicola* Petch
Black spot, *Discosia thea* Cavara
Canker, *Macrophoma theicola* Petch
Internal decline, *Diplodia natalensis* Evans
Gray blight, *Pestalotia theae* Sawada
Leaf spot, *Pestalotia guipini* Desmazieres
Eyespot, *Cercospora theae* Petch
Leaf spot, *Cercosporella theae* Petch
Collar rot, *Rhizoctonia solani* Kuehn
Red rust, *Cephaleuros virescens* Kunze
Other fungi associated with tea

Leaf spot, *Bacillus theae* Hori & Bokura

Symptoms

Small, pale brown, circular spots, 2–3 mm in diameter, appear on fully developed leaves. They enlarge and coalesce, forming irregular, reddish brown, well defined, concentrically zoned patches. Injured leaves are often shed, leaving bare twigs. Twig infection results in a blackening and withering, and buds in leaf axils are usually blackened. The disease is favored by wet weather and appears most often on plants fully exposed to the sun.

Etiology

The organism is a gram-negative rod, forms in short chains, is motile by 5–8 peritrichous flagella, and measures 0.8–1.8 μ. Colonies on gelatin are grayish white becoming dark brown. The organism is probably also included in *Erwinia.*

References

2, 4, 7, 13.

Brown blight, *Glomerella cingulata* (Ston.) Spauld. & Schrenk

Symptoms

The brown blight of the leaf is characterized by the formation of round or oval spots that are diffuse at first, nonzonated, and yellow green, later becoming brown. They are somewhat reddish and finally chocolate brown, surrounded by a yellow margin. Fruiting bodies appear in concentric circles on the spots, and severely infected leaves are often shed. Dieback of young branches occurs on weak plants; the bark becomes dry, brown or dark, and dies.

Etiology

Fruiting bodies are readily produced in culture. Perithecia are brown, membranous, spherical, ostiolate, isolated or in groups on the black stroma, and have a conical rostrum. The ascus is oblong or claviform, measuring 55–70 $\times$ 9 μ. The ascospores are hyaline, curved, 1-celled, and 12–22 $\times$ 3.5–5 μ. The conidial stage is variable, *Colletotrichum* spp. and *Gloeosporium* spp.

References

5, 13, 16, 17.

Leaf spot, *Phaeosphaerella theae* Petch

Symptoms

This disease is characterized by the formation of numerous irregular holes in the leaf. They occur on older leaves and on recently plucked shoots. The leaves are covered with small, yellow spots that are gray in the center, somewhat circular, but later becoming angular, and frequently extending in narrow prolongations. The spots dry and fall out, leaving an irregular hole bordered by a narrow, brown zone of dead tissue.

Etiology

Perithecia are black, immersed, epiphyllous, minute, and 80–100 μ in diameter. Asci are few, clavate, and 50×12 μ. Ascospores are brown, fusoid, 1-septate, constricted, and 9–14×4–5 μ. The upper cell is larger than the lower and rounded at the apex, and the lower cell is somewhat oblong and obtuse at the apex. Fruiting bodies are produced as minute black points in dead brown tissue at wound margins.

References

13, 16.

Leaf spot, *Calonectria theae* (Petch) Loos

Symptoms

Minute, circular, brown or black spots appear on young leaves. When numerous, they cause the leaf to wrinkle or become distorted. In wet weather, the leaves become black and decay. On older leaves the spots enlarge, often up to 5 mm; they are black at first, then gray, and finally whitish with a well-defined light to dark purple, slightly raised margin on the upper surface.

Etiology

The hyphae form a thin film on those spots on the lower leaf surfaces. They are hyaline and sometimes unite into strands or threads. Conidiophores show a main axis from which branches arise in irregular fashion. The conidia, which appear in clusters, are hyaline, variable in width and length, 62–100×3–7 μ, multiseptate, and cylindrical with rounded ends. The perithecia are orange red, superficial, globose to obovate, and 1/3–1/2 mm in diameter. They are surrounded by orange mycelium, the walls are rough, and the ostiole is raised and smooth. The asci are hyaline, and the ascospores are hyaline, elongate-fusoid, 1–3-septate with distinct medial restriction, and measure 42–70×6–8 μ. The vegetative stage of the fungus is *Cercosporella theae.*

References

11, 13, 16.

Copper blight, *Guignardia camelliae* (Cke.) Butl.

Symptoms

Spots are usually large and irregular, often extending inward from the leaf margins. They are dark brown on the upper leaf surface, mottled with yellow brown or gray brown without zoning, and grayish brown on the undersurface of the leaf. Lesions are covered with radiating mycelium, clearly evident toward the margins of the spots. The diseased area includes minute, black points that are arranged in irregular, concentric circles.

Etiology

Perithecia are globose, 200–250 μ in diameter, imbedded in tissue, and contain elliptical asci that are elongate and slender. Ascospores are hyaline, 1-celled, ovoid, and 8–18 × 3.5–6 μ. Perithecial features are variable, depending on the host area involved. The imperfect stage is *Phoma camelliae*.

References

5, 13, 16.

Black stem, *Massaria theicola* Petch

Symptoms

Black stem is characterized by gradual death of the upper parts of the bush, branch by branch, and is often accompanied by the production of new shoots. Leaves wither and branches die. Death of the whole bush is rare. The internal cortex is black, underlaid by dark brown wood.

Etiology

The mycelium grows in vessels, and black perithecia are borne in groups, each 250–300 μ in diameter. The asci among the paraphyses are cylindrical, very long and slender, and 120–160 × 20 μ. Eight ascospores are arranged in an oblique row; they are light olive, oval, slender, 3-celled, and 17–22 × 7 μ.

References

5, 16.

Red root disease, *Sphaerostilbe repens* Berk. & Br.

Symptoms

Diseased plants show typical root-infection symptoms such as stunted growth, yellow, droopy leaves that dry and fall, cortical tissues that become dry, and shrunken bark at the collar. The mycelium is located between the outer cortex and cambium. It is common in water-soaked soil where it is ramified, digitated, first white, later brown red, and black when older.

Etiology

Conidiophores are 2–8 mm high; the stalk is red brown and tomentose with spinulose hairs. The conidia are hyaline, oval, continuous, and 9–22 × 6–10 μ. Perithecia are dark red, aparaphysate, clustered, flask-shaped, and 0.6 mm high and 0.4 mm in diameter. The asci are cylindric with a short stalk and measure 190–220 × 10–20 μ. The ascospores are pale brown or reddish brown, uniseriate, oval, 1-septate, slightly constricted, and 19–21 × 8 μ.

References

12, 13, 16.

Black blight, *Asterina camelliae* Syd. & Butl.

Symptoms

This disease is characterized by black crusts on the leaves. The stromata are epiphyllous, very hard, and in the form of shields composed of radiating hyphae.

Etiology

The thallus is often in groups, forming large spots irregular in shape, and 1 cm in diameter. Perithecia are epiphyllous, membranous, composed of radiating hyphae, dehisce stellately, and measure 200–300 μ in diameter. They are clustered and confluent in groups, form black crusts up to 1 cm wide, and produce asci that are elliptical, obovate, and $70–100 \times 25–35$ μ. Ascospores are brown, ellipsoid, with ends rounded, centrally 1-septate, constricted at the septum, and $30–33 \times 16$ μ.

References

5, 13, 16.

Sooty mold, *Limacinula theae* Syd. & Butl.

Symptoms

Leaves and twigs are covered with a black film, usually with a powdery appearance. This black covering is an easily detached, thin layer of mycelium, bearing fructifications and often splitting away in flakes if dried. The fungus lives on insect secretions and is not a plant parasite. No direct injuries to the plants are caused by the organism, except that it acts indirectly by hampering photosynthesis.

Etiology

Mycelium is black, effused, membranous, covers the upper surface of leaves, and is composed of septate, fuscous hyphae with black, simple, acute, setae. Pycnidia are erect, cylindric, inflated in the middle, attenuated above or broadly cylindric, 80 μ high, and 30 μ in diameter. Pycnidiospores are hyaline, cylindric, straight, continuous, and $2.5–3 \times 1.5$ μ. Perithecia are superficial, globose then depressed, and ostiolate. The asci are subsessile, ovoid, and bear ascospores that are hyaline, oblong with rounded ends, muriform with 5 transverse and 1 longitudinal septa, and measure $25–33 \times 9–11$ μ. Also see *Meliola* sp. and *Capnodium* sp.

References

5, 13, 16.

Dieback, *Nectria cinnabarina* (Tode) Fr.

Symptoms

Infected bushes gradually decline but rarely die completely. Partial dieback is usually followed by the growth of new, weak, thin shoots. Small, coral to pink-colored cushions appear subsequently. The fungus is a wound parasite.

Etiology

Perithecia are cinnamon-colored, spherical, and rough, with a papillate ostiole; they cluster on pulvinate, fleshy stroma. The asci are cylindric-clavate, attenuated above, and measure 60–90 × 8–12 μ. Ascospores are hyaline, oblong, obtuse, straight or slightly curved, uniseptate, and 14–16 × 5–7 μ. The vegetative stage of the fungus is *Tubercularia vulgaris*.

References

10, 13, 16.

Root decay, *Rosellinia arcuata* Petch

Symptoms

This disease is characterized by roots covered with black strands of woolly mycelium that grow longitudinally and unite to form a network. Exterior strands of the fungus bear pear-shaped swellings at one end. There is a rapid spread of mycelium within the top 2–3 in. of soil and on aerial parts of the plant. Mycelium grows along the stem up to 6 in. above the soil.

Etiology

Perithecia are gregarious, first imbedded in purple brown mycelium, black, superficial, globose, slightly depressed, and smooth. The ostiole is conical, 0.1 mm high, and 0.4 mm wide at the base. The asci are about 300 × 8 μ, cylindric, and contain obliquely uniseriate spores. Paraphyses are filiform and 2 μ in diameter. The ascospores are black, cymbiform, pointed, and 40–47 × 5–7 μ. Conidiophores are black, erect, up to 2 mm high, and 0.1 mm in diameter. Conidia are hyaline, narrow, oval, and 4–6 × 2 μ. The vegetative stage of the fungus is *Graphium* sp.

References

5, 13, 16, 19.

Root rot, *Ustulina zonata* Lév.

Symptoms

Tea bushes gradually lose their leaves. They occasionally wilt and die suddenly. There is no other external indication of the disease aboveground. Root examination shows a whitish mycelium in fan-shaped patches in the cambium area and frequently occupying the inner cortex.

Etiology

The mycelium is mostly brownish white. The sporophore is white when young, becomes gray and finally black, and shows many dots which are the ostioles of the perithecia. Conidia are hyaline and 6–8 × 2–3 μ. Perithecia are about 1 mm in diameter. Asci are cylindric, elongate, and 250×10 μ; ascospores are black and 30–38×9–13 μ.

References

5, 13, 16.

Thorny stem blight, *Tunstallia aculeata* (Petch) Agniho.

Symptoms

The disease is characterized by the presence of so-called black thorns produced on dead branches. These thorns are the tops of the fruiting bodies of the fungus. Each thorn is surrounded by an area of raised and cracked bark on twigs and branches that have been killed by the fungus. Dead branches are brittle and easily broken off.

Etiology

The perithecia are immersed becoming erumpent, in a thick black stroma, 3–5 mm in diameter, and in clusters of up to 5 or 6. Ostioles of each opening converge into one that protrudes above the stroma as a black point, 0.75–2.0 mm high with paraphyses. Asci are cylindrical, slender, with curved pedicel and truncated tip, and measure 160–210×30–35 μ. Ascospores are hyaline, cymbiform, 1-celled, guttulate, and 80–110×6–13 μ. The conidial stage is not known.

References

1, 13, 16.

Thorny stem blight, *Tunstallia aculeata* (Petch) Agniho. var. *kesabii* Agniho.

Symptoms

Same as above.

Etiology

Same as above except the asci measure 148–266×22–37 μ. Ascospores are fasciculate, parallel, twisted, anguilliform, guttulate, and 116–152×6–12 μ.

References

1, 13, 16.

Blister, *Exobasidium vexans* Mass.

Symptoms

This disease first appears on leaves as small, pale yellow spots, 6–15 mm in diameter, with a red point in the center. As the spots enlarge, the center turns light brown. Swellings appear on the spots, causing the leaf blade to become convex and pale green on the upper surface. The spots generally elongate along blade axes and have a powdery texture at maturity. Spots, 1–30 per blade, are more extensive on young leaves, and the fructifications are on the lower surface. Mature spots and blisters dry out.

Etiology

The hymenium is white, chiefly hypophyllous, and tomentose. Conidia are hyaline, 1-septate, narrow-oval, solitary, and $12–21 \times 4.5–6\ \mu$. Conidiophores are simple. Basidia are clavate and $30–90 \times 3.7–6\ \mu$. Basidiospores are hyaline, elliptical, continuous, and $9–12 \times 3–3.5\ \mu$.

References

5, 13, 15, 19.

Thread blight, *Corticium theae* Bern.

Symptoms

This disease is characterized by white or pinkish hyphae growing along the branches, ramifying, and uniting. It is found only on young branches, 2–3 years old. Upon reaching a petiole, it grows to the undersurface of the leaf blade, covering it with a thin, iridescent, silky film.

Etiology

The fungus is pulverulent and produces basidia that are ovoid to elongate and $20–25 \times 6–8\ \mu$. Basidiospores are hyaline, elliptic, and $7–9 \times 5–7\ \mu$. The vegetative parasitic fungus is *Hypochnus thea.*

References

5, 13, 15.

Thread blight, *Ceratobasidium stevensii* (Burt) Venkat.

Symptoms

This disease is first noticed because of the brown, dead foliage hanging pendent from the oldest part of a flush of growth. The fungus survives on old wood. It grows up the stems of new growth and proceeds up the leaf petiole and flower peduncles in the form of brown, undulating cords. On the leaf blades, the brown thread fans out, covering the underside with a fine, silky network of hyphae, which invade the host tissues.

Etiology

The cords or threads formed by fascicles of hyphae grow along the lower surface of twigs and survive in that fashion from the past to the current season. Basidia are produced on the leaf surface scattered in the hyphae. They are short, round to oval, and $8–12\ \mu$. They develop long sterigmata that support the hyaline, oval basidiospores that measure $8–9 \times 3–5\ \mu$. The vegetative stage is *Hypochnus stevensii.*

Reference

15.

Pink disease, *Botryobasidium salmonicolor* (Berk. & Br.) Venkat.

Symptoms

A withering and dying of the foliage on specific twigs and branches are the first indications of the disease. As the branches are killed, a pink, resupinate fungus growth, composed of compact, septate mycelium, spreads over the cortex. It surrounds the limb and extends in both directions. The terminal part dies, the pinkish hymenium dries, and the surface cracks. The interwoven hyphae make up a pink, compact layer.

Etiology

The hyaline, internal, septate hyphae, developing from the external mycelium, produce white to pink, small, protruding pustules followed by erumpent sporodochia, bearing oblong, catenulate, 1-celled conidia that measure $14–18 \times 9–13$ μ. The basidia are short, cylindrical, $16–34 \times 5–8$ μ, and form hyaline, pyriform basidiospores that measure $9–12 \times 6–7$ μ.

References

4, 5, 13.

Red root, *Poria hypolateritia* Berk.

Symptoms

The only aboveground indication of disease is the decline of the bushes, culminating in death. Examination of the roots of diseased plants reveals mycelial strands of the fungus on the outer surfaces, forming a sort of network, at first white and soft but later compact, tough, and reddish. By the time older bushes succumb, these strands become black.

Etiology

The sporophore of the fungus is a reddish to pink, thin, flat structure, resupinate with the stem, up to 4 in. in diameter, and less than 1 in. thick. The tubes are in a layer and, upon drying, curl. The fruiting structure is usually at the collar of the plant.

References

13, 14.

Brown root, *Fomes lamaensis* (Murr.) Sacc. & Trott.

Symptoms

Plants affected by the disease are first recognized by a general unthrifty appearance, loss of leaves, some dieback, wilting and browning of leaves, and finally death. Examination of the roots shows a layer of earth, several millimeters thick, encrusting them, held there by the brown mycelium of the fungus. Between the cortex and the wood, there is a layer of brownish mycelium. As the disease develops, the mycelium and crust become blackish. The disease develops slowly but usually kills the plant.

Etiology

The fruiting structure is a bracket, sessile, and up to several inches thick. It is purple brown to black with a yellow margin, pored, and smooth or slightly furrowed. These sporophores are not frequently observed.

References

13, 15.

White root rot, *Fomes lignosus* (Klotzsch) Bres.

Symptoms

Aboveground indication of the disease is a general unthriftiness of the plant, which gradually sheds its leaves and deteriorates until dead. Roots are covered with whitish, stout threads of mycelium in a netlike pattern. Sources of the fungus are old, dead stumps left in the planting from previous clearing.

Etiology

The conks are shelving, sessile, perennial, and formed on the stem of the plant. They are reddish brown with a yellow margin, up to several inches in diameter, and 1 in. thick.

References

3, 13, 15.

Shoestring rot, *Armillaria mellea* Vahl ex Fr.

Symptoms

There are no aboveground indications that a plant is diseased until it begins to show a decline, becoming off-color green, usually yellow or tan; there is a drooping of the leaves and dieback of the twigs. Often the entire plant may collapse before any other visual character appears. Root examination reveals a girdling effect taking place at the crown or on the main roots. The affected roots are enlarged, cracked, and wet. Cortex removal shows the whitish fungus sheets in the cambium and inner cortex areas. Honey-colored sporophores may be observed seasonally on diseased plants at the soil line.

Etiology

The fungus usually spreads by means of rhizomorphs, which grow in the soil and contact healthy plants as they radiate out from a diseased plant. The fungus can be identified by the presence of black, shiny rhizomorphs, the white layer of mycelium in the cortex, the pale tan, stipitate sporophores with an annulus, and the production of white basidiospores.

References

8, 13, 14.

Horse hair blight, *Marasmius equicrinis* Muell.

Symptoms

Blight is characterized by black, shiny, horsehairlike mycelium attached at intervals by small, brown discs to upper branches, leaves, and twigs. Small mushrooms may appear on the aerial mycelium.

Etiology

The pileus of the sporophore is up to 8 mm in diameter, hemispherical, umbilicate, deeply radially sulcate, somewhat membranous, and yellow brown, red brown, or ochraceous. The 5–8 gills are white then cream-colored, distant, broad, attenuated, and united into a collar around the stalk. The stipe is a black, shining, rhizomorphic strand of mycelium. The basidiospores are white, oval, and 10–14×4 μ.

Reference

13.

Dieback, *Stilbella theae* Bern.

Symptoms

This disease is recognized by the withering, drying, and death of leaves and stems. On diseased branches, there appear a large number of fructifications in the form of minute, orange red stalks, expanded above into small heads; these structures may also appear on leaves. In severe cases, the infected branch appears to be covered by a velvety layer.

Etiology

Spore-producing stalks are red or dark, red brown at the base, clear orange red above, papillate, almost smooth, and 20–800 μ high. The head is rose-colored, more or less globose, 150–300 μ in diameter, and bears conidia that are hyaline and 5–7×2.5 μ. This fungus is probably identical with *Stilbum nanum*.

References

13, 16, 19.

Leaf spot, *Phoma theicola* Petch

Symptoms

This disease appears on leaves between the lateral veins, especially on young plants or on the lower leaves of mature plants. Small spots are brown red or red, at first circular enlarging to oval, and becoming angular or irregular.

Etiology

Pycnidia are black, parenchymatous, in groups, imbedded in host tissue, with a definite ostiole, and about 100 μ in diameter. Pycnidiospores are hyaline, 1-celled, oval, and 8–10×5–6 μ.

References

13, 16.

Black spot, *Discosia thea* Cav.

Symptoms

This disease is recognized by the presence of black, circular, definite spots of varying sizes, often confluent, on the upper surface of the leaf. The spots, formed beneath the cuticle, are elevated in the center and ultimately split.

Etiology

Pycnidia are black, scattered, superficial, flattened, rugose, with a prominent ostiole. Spores are cylindric, slightly curved, obtuse, obliquely uniciliate at each end, 3-septate, and 18–20×2–3 μ. Terminal cells are hyaline, and the median cell is greenish. Setae are hyaline, narrow, and 6–8 μ long.

References

13, 16.

Canker, *Macrophoma theicola* Petch

Symptoms

The disease is indicated by black spots on the cortex of twigs; the spots are round to elongate and 2–3×1 cm. Lesions delimited by raised tissue surrounding them are thick, irregular, and form a callus by girdling. Leaves turn yellow and fall. The disease is occasionally found on leaves, producing large spots on the blade that are irregular or granular, brown, then whitish, with a definite margin.

Etiology

The pycnidia, imbedded in the cortical tissue, are black, 250 μ in diameter, and 16–25 μ. The pycnidiospores are hyaline, oval, elongate with rounded ends, and 27–32×5–7 μ.

References

13, 16.

Internal decline, *Diplodia natalensis* Evans

Symptoms

The leaf veins become very prominent and a yellowing develops as though from malnutrition. Defoliation and dieback occur. The whole plant or any part of it may show these conditions. Occasionally an entire plant will, without outward indicators, collapse. Often the disease begins following pruning. Recently diseased plants show no recognizable root symptoms. Older specimens, however, show a black, cankered area in the cortex.

Etiology

The mycelium exists in the cortex, cambium, and xylem. The pycnidia imbedded in the outer cortex are black, flask-shaped with frequently protruding ostioles through which tendrils of hyaline, 1-celled, oval conidia exude. As the conidia mature, they become dark-colored, 2-celled, and somewhat striated. The ascospore stage of the fungus is *Physalospora rhodina.*

References

5, 8, 13.

Gray blight, *Pestalotia theae* Saw.

Symptoms

Irregularly circular or oval spots appear in the middle of leaves or extend inward from the tip or margin; the upper leaf surface is concentrically zoned and ridged with pale spots covered with minute black points. Spots are first surrounded by a narrow, greenish yellow or purple zone.

Etiology

Acervuli are submerged becoming erumpent, and conidiophores measure 4–9 × 1 μ. Conidia are fusiform, 4-septate, and slightly constricted. The 3 inner cells are dark brown, measuring 16–21 μ; the basal and apical cells are hyaline, measuring 4–6 μ. The 3–4 setulae are terminal, slightly swollen at the apices, and measure 28–36 × 1–2 μ. The stipe is up to 6 μ long.

References

5, 9, 13, 16, 19.

Leaf spot, *Pestalotia guipini* Desm.

Symptoms

Leaf spots are ash gray, usually variable with a reddish brown border, papery, and up to 1.5 cm in diameter. They appear on blades, petioles, and twigs. Pycnidia are scattered or in groups on the lower surface.

Etiology

Pycnidia are black, globose, prominent, ostiolate, scattered or subgregarious, and amphigenous. The pycnidiospores are pale brown, cylindric, with rounded ends, 3-septate, and 14 × 21 μ. Terminal cells are hyaline, 10–12.5 μ long or longer, and have a hyaline pedicel. The 1–3 setulae are hyaline, divergent, and 10–24 μ long.

References

9, 13, 16.

Eyespot, *Cercospora theae* Petch

Symptoms

This disease generally occurs on older leaves, but occasionally infects young leaves and flowers. Spots are circular, first purple red, with indefinite, yellow green borders. Centers of spots become sunken below the leaf surface. The entire spot is white when fully developed, circular, not over 2–3 mm in diameter, and surrounded by a narrow, purple red border. Fruiting bodies develop on the spots as minute black points.

Etiology

Stromata are black, minute, on either side of the leaf, and produce scattered conidiophores. Conidia are hyaline, elongate, irregularly curved, multiseptate, and up to 140 μ long and 3–4 μ in diameter.

References

5, 13, 16.

Leaf spot, *Cercosporella theae* Petch

Symptoms

Small, circular, water-soaked, brownish black spots appear on young leaves. When the spots are numerous, the leaf becomes distorted. On older leaves, the spots, up to 5 mm, are dark but become grayish white, with a well defined purple, slightly raised border on the upper surfaces. Plants become defoliated and often show a purple lesion on the green stems. The mycelium of the fungus appears in white clumps on the surface of the older spots.

Etiology

Mycelium is hyaline, up to 5 μ in diameter, and sometimes accumulates into strands. The conidia are cylindric, elongate with obtuse ends, up to 6-septate, not restricted, and 64–130 $\times$ 5–14 μ. The ascospore stage is *Calonectria theae.*

References

13, 16.

Collar rot, *Rhizoctonia solani* Kuehn

Symptoms

Various stages of yellowing and withering seedlings are observed. Associated with stem discolorations are splitting of the bark, stem girdling, and lesions showing bark shedding and callus formation.

Etiology

Occasionally the fungus mycelium grows up and invades the succulent stems and foliage. This is not a common disease but develops under high humidity and air stagnation. The perfect stage is *Thanatephorus cucumeris*.

Reference

18.

Red rust, *Cephaleuros virescens* Kunze

Symptoms

The disease is a most destructive pathogen of tea. It occurs on leaves as circular or elongate spots, which are up to 15 mm in diameter, slightly raised, velvety, orange or red purple, and become green or gray when old. On young twigs, there is a browning of the bark and a fissuring of the cortex. In severe cases, an etiolation, weakening, and occasional dieback of branches occur.

Etiology

A green alga, the pathogen causes circular spots, 1–15 mm in diameter, with a radiated or lobed, fimbriated border. The thallus is thin, isolated or confluent, and very adherent. The alga produces a few very slender filamentous rhizoids in host parenchyma, passing through cuticle and becoming external where sterile and fertile hairs develop. The fertile hair is the sporophore, terminated at the tip by a round vesicle with a lateral septum, with 3–6 or 8 ovoid sporangia. Upon germination, sporangia produce zoospores with 2 cilia. The parasite develops between the cuticle and the epidermis, mostly on the upper surface of the leaf.

References

5, 12, 13, 19.

Other fungi associated with tea

Ascochyta theae Hara
Cercoseptoria theae (Cavara) Curzi
Coniothyrium theae Petch
Fomes applanatus Persoon
Fomes lucidus (Leysser) Fries
Fomes noxia (Berkeley) Corner
Hendersonia theicola Cook
Irpex subrinosus (Berkeley & Broome) Petch
Leptosphaeria camelliae Cooke & Massee
Monochaetia camelliae Miles
Polyporus interruptus Berkeley & Broome
Polyporus mesotalpae Lloyd
Ramularia theicola Curzi
Sclerotinia camelliae Hara
Septobasidium acaciae Sawada
Stagonospora theicola Petch

Trametes theae Zimmermann
Venturia speschnervii Saccardo & Saccardo
Vermicularia microchaeta Passerini

References: tea

1. Agnihothrudu, V. 1960. Tunstallia gen. nov. causing the thorny stem blight of tea (Camellia sinensis L.) O. Kunze. Phytopathol. Z. 40:277–82.
2. Bergey, D. H., et al. 1939. Manual of determinative bacteriology. 5th ed. Williams & Wilkins: Baltimore, Md., p. 471.
3. Bose, S. R., and B. K. Bakshi. 1957. Polyporus lignosus Klotzsch, and its identity. Trans. Brit. Mycol. Soc. 40: 456–60.
4. Bouriquet, G. 1946. Les maladies des plantes cultivees a Madagascar. Encyclopedie Mycologique 12:184–86.
5. Butler, E. J. 1918. Fungi and diseases in plants. Thacker: Calcutta, India.
6. Delacroix, G. 1911. Maladies des plantes cultivees dans les pays chauds. Augustin Challamal: Paris, pp. 415–41.
7. Elliott, C. 1951. Manual of bacterial plant pathogens. 2d ed. Chronica Botanica: Waltham, Mass., p. 159.
8. Gadd, C. H. 1930. What is the diplodia root disease of tea? Tea Quart. 3:44–48.
9. Guba, E. F. 1961. Monograph of Monochaetia and Pestalotia. Harvard University Press: Cambridge, Mass., pp. 93–96, 108–13.
10. Hubert, E. E. 1931. An outline of forest pathology. John Wiley & Sons: N.Y., pp. 194–96.
11. Loos, C. A. 1950. Calonectria theae n. sp. the perfect stage of Cercosporella theae Petch. Trans. Brit. Mycol. Soc. 33:13–18.
12. Mann, H. H., and C. M. Hutchinson. 1907. Cephaleuros virescens Kunze, the red rust of tea. Mem. Dept. Agr. India, Botan. Ser. 1:1–35.
13. Petch, T. 1923. The diseases of the tea bush. Macmillan: London.
14. Petch, T. 1928. Tropical root disease fungi. Trans. Brit. Mycol. Soc. 13:238–53.
15. Roger, L. 1953. Phytopathologie des pays chauds. Encyclopedie Mycologique 17:932–38, 948–55, 1049–56.
16. Roger, L. 1954. Phytopathologie des pays chauds. Encyclopedie Mycologique 18:1273, 1293–94, 1352–55, 1357–59, 1404–9, 1440, 1454–57, 1475–78, 1491–95, 1635, 1673–74, 1683–84, 1793, 1871–74, 1960–61, 2114–15.
17. Tunstall, A. C. 1934. A new species of Glomerella on Camellia theae. Trans. Brit. Mycol. Soc. 19:331–36.
18. Venkataramani, K. S., and C. S. Venkata Ram. 1959. Rhizoctonia solani inciting a collar rot of tea. Phytopathology 49:527.
19. Watt, G. 1898. The pests and blights of the tea plant. Government Printing Office: Calcutta, India, pp. 412–28, 433–67.

Teff, *Eragrostis abyssinica* Link

Smut, *Tilletia baldratii* Mentem
Rust, *Uromyces eragrostidis* Tracy
Leaf spot, *Phoma depressitheca* Bubák
Leaf spot, *Septoria eragrostidis* Castellani & Ciccorone
Smudge, *Helminthosporium miyakei* Nisikado

Smut, *Tilletia baldratii* Mentem

Symptoms

The systemic diseased plants are more erect than normal, and spikelets are less dense, with dry swollen glumes. The infected ovules are dark violet to gray, swollen, and irregular in shape. The entire florescence is involved, and at maturity the dark brown to black chlamydospores are released.

Etiology

The sori are olive brown, involve the entire embryo, give off a fetid odor when mature or broken, and measure 1–1.5×1 mm. Chlamydospores are brown, mostly globose, 16–22 μ, and have a reticulate wall marking, 4–7 μ wide and 1–1.5 μ high, which is externally gelatinous.

References

2, 3.

Rust, *Uromyces eragrostidis* Tracy

Symptoms

Leaf spots are tan to brown, scattered, circular, small to 1 mm, and show powdery pustules on the lower leaf surface.

Etiology

The pycnial and aecial stages are not known. The uredia on the lower leaf surface are light brown and oblong, scattered, 1 mm in diameter, and produce uredospores that are brown, ellipsoidal, echinulate, and 20–26×15–19 μ. The telia on the upper leaf surface are dark brown, elongate, and produce teliospores that are oval to spherical or clavate, 19–27×16–23 μ, with a medium-sized stipe about 25 μ long.

References

1, 5, 6.

Leaf spot, *Phoma depressitheca* Bub.

Symptoms

The leaf spots are not conspicuous; the black dots on the spots are few, striately located between veins, and often laterally depressed.

Etiology

The mycelium is hyaline, septate, and branched. The pycnidia are black, ostiolate, subcuticular, and produce pycnidiospores that are hyaline, elliptical to oval, and 4×1.8 μ.

Reference

2.

Leaf spot, *Septoria eragrostidis* Castell. & Ciccor.

Symptoms

Small, circular to elongate, colored areas appear on the apical parts of the lower leaves. In the center of these spots on the withering leaves, black pycnidia develop.

Etiology

The subspherical, subepidermal pycnidia measure 90×100 μ, are ostiolate, and emit tendrils of spores. The pycnidiospores are hyaline, straight to curved, and 20–30×2–2.7 μ. Associated with these pycnidia were perithecia of *Mycosphaerella eragrostidis,* assumed to be related but unproven.

Reference

2.

Smudge, *Helminthosporium miyakei* Nisikado

Symptoms

An olivaceous, velvety accumulation develops around the glume margins or completely covers the outer surface. It may cause shrinkage and deformed ovaries.

Etiology

The mycelium is hyaline to olivaceous, branched, septate, frequently fasciculate or tangled, and produces conidiophores that are simple to branched and tortuous with inflated tips. The conidia are olivaceous, 3–4-septate, fusiform to sickle-shaped, restricted at the septa, have rounded bases, and taper toward the apices.

References

2, 4.

References: teff

1. Arthur, J. C., and G. B. Cummins. 1962. Manual of the rusts in United States and Canada. Hafner Co.: N.Y., p. 162.
2. Castellani, E., and A. Ciccorone. 1961. Cryptoyamic diseases of teff. Agr. and Economics Working Papers No. 8. Imperial Ethiopian Government: Addis Ababa, pp. 43–59.
3. Castellani, E., and R. Ciferri. 1937. Prodromus mycoflorae africae orientalis talicae. Firenze, 1st Agric. Colonial Garden. Reviewed in No. 3.
4. Drechsler, E. 1923. Some graminicolous Helminthosporium. J. Agr. Res. 24: 641–740.
5. Rouk, H. F., and H. Mengesha. 1963. An introduction to teff, Eragrostis abyssinica Schrod. Haile Selassie 1. Univ. Dire Dawa. Ethiopia Expt. Sta. Bull. 26.
6. Tracy, S. M. 1893. Descriptions of new species of Puccina and Uromyces. J. Mycol. 7:281.

Tobacco, *Nicotiana tabacum* Linnaeus

False broom rape, *Corynebacterium fascians* (Tilford) Dowson
Black leg, *Erwinia aroideae* (Townsend) Holland
Wild fire, *Pseudomonas tabaci* (Wolf & Foster) Stevens
Angular leaf spot, *Pseudomonas angulata* (Fromme & Murray) Holland
Granville wilt, *Pseudomonas solanacearum* E. F. Smith
Damping-off, *Pythium debaryanum* Hesse
Black shank, *Phytophthora parasitica* Dastur var. *nicotianae* (B. de Haan) Tucker
Downy mildew, *Peronospora tabacina* Adam
Argentine mildew, *Peronospora nicotianae* Spegazzini
Metallic mold, *Blakeslea trispora* Thaxter
Seedling blight, *Olpidium brassicae* (Woronin) Dangeard
Cottony rot, *Sclerotinia sclerotiorum* (Libert) De Bary
Powdery mildew, *Erysiphe cichoracearum* De Candolle

Rust, *Uredo nicotianae* Anastasia, Saccardo, & Splendore
Soil rot, *Thanatephorus cucumeris* (Frank) Donk
Southern blight, *Botryobasidium rolfsii* (Saccardo) Venkatarayan
Leaf spot, *Phyllosticta nicotiana* Ellis & Everhart
Leaf spot, *Ascochyta phaseolorum* Saccardo
Anthracnose, *Colletotrichum tabacum* Böning
Gray mold, *Botrytis cinerea* Persoon ex Fries
Black root rot, *Thielaviopsis basicola* (Berkeley & Broome) Ferraris
Frog eye, *Cercospora nicotianae* Ellis & Everhart
Brown spot, *Alternaria longipes* (Ellis & Everhart) Mason
Sooty mold, *Fumago vagans* Persoon
Blotch, *Septomyxa affinis* (Sherbakoff) Wollenweber
Wilt, *Fusarium oxysporum* Schlechtendahl f. *nicotianae* (J. Johnson) Snyder & Hansen
Other fungi associated with tobacco

False broom rape, *Corynebacterium fascians* (Tilford) Dows.

Symptoms

Many white, short, succulent, irregular, fleshy, thick, and abortive outgrowths appear below ground. These growths may emerge and in the light develop chlorophyll. Affected plants may show some stunting or inferior blossoming, or may remain symptomless.

Etiology

The organism is a gram-positive rod, in short chains, single or in pairs, non-motile, and $0.5 \times 1.5–4\ \mu$. Agar culture colonies, appearing after 72 hours, are light cream to orange, glistening, smooth, opaque, raised, and circular.

References

4, 5, 12, 23.

Black leg, *Erwinia aroideae* (Towns.) Holland

Symptoms

Stem bases and the roots of infected plants become blackened. Leaves that are in contact with the soil become involved in a wet rot that may be spread throughout the petioles into the stems. The disease on field-grown plants usually appears when the crop has reached the topping and suckering stage. Leaves wilt on isolated plants or on a few consecutive plants in rows. Usually upper leaves are first affected; these collapse but remain attached. Brown, sunken lesions are formed in the veins of hanging leaves. Parts of the midribs decay. The pith of the entire stem becomes rotten and slimy, and dries, leaving the stem hollow. The bacterium is considered a soft-rotting organism and is associated with plants diseased by *Erwinia carotovora*.

Etiology

The organism is a gram-negative rod, usually single but sometimes in pairs, produces no capsules, is motile by 2–8 peritrichous flagella, and measures $0.5 \times 2–3\ \mu$. Agar colonies are white to slightly opalescent, glistening, and circular. Optimum temperature, 35C.

References

5, 27.

Wild fire, *Pseudomonas tabaci* (Wolf & Foster) Stev.

Symptoms

This disease may appear on all of the aboveground parts, particularly on the foliage of seedlings where it is recognized by the wide, circular, halo which surrounds the killed tissue in the center of each spot. In crowded seedbeds, leaf margins and tips often show a wet rot that dries and falls out. Diseased leaves become distorted, torn, and ragged. Lesions on capsules are brown and slightly elongate. The bacteria enter the leaf through the stromata and wounds.

Etiology

The organism is a gram-negative, aerobic rod produced in chains, motile by 1–6 polar flagella, and measuring $0.5–0.7 \times 1.4–2.8\ \mu$. A green fluorescent pigment is produced in culture. Nutrient agar colonies are white, circular, raised, and have opaque centers with translucent edges. The organism secretes an exotoxin that destroys chlorophyll and is the cause of the extensive yellow halo surrounding infection spots.

References

5, 12, 27.

Angular leaf spot, *Pseudomonas angulata* (Fromme & Murray) Holland

Symptoms

This spot may appear on a host in any stage of its growth. Angular to irregular, very dark brown to black necrotic areas, up to 8 mm in diameter, appear on any leaf of seedlings. Young spots are very dark, almost black, and become brown as the dead tissues become desiccated. The tissue bordering the lesions is yellowish, but there is never a yellowish halo. When spots are numerous, the leaves become distorted from the growth tension. During wet weather, dead tissues fall away and the veins may remain intact. Flowers may become infected, and calyx and corolla tubes become blackened and distorted. Lesions on capsules are dark brown.

Etiology

The organism is a gram-negative rod, motile by 3–6 polar flagella, and measures $0.5 \times 2–2.5\ \mu$. Agar colonies are opalescent to dull white with opaque centers, glistening, smooth, convex, and circular with undulating margins. A green fluorescent pigment is produced in culture. The pathogen is seed-borne.

References

5, 8, 27.

Granville wilt, *Pseudomonas solanacearum* E. F. Sm.

Symptoms

Plants are affected in all stages of development, from seedbed to harvest. One or more leaves may droop during the day and recover at night. Often one-half of the blade becomes flaccid, and the other half appears normal. As the disease progresses, the leaves become pale green and gradually turn yellow. By this time the midrib and secondary veins have become limp, and the leaves droop in an umbrellalike fashion. This shape persists as the leaves gradually become brown and dry. This is a vascular disease, causing dwarfing, wilting, and bundle staining. Cross sections of the diseased stalk show yellowish streaks, involving the xylem, that soon darken to brown or black. The pressure of darkened streaks of the xylem extending upward into the leaves is an important diagnostic feature. The organism survives in the soil from season to season.

Etiology

The organism is a gram-negative rod, motile by a single polar flagellum, and measures 0.5×1.5 μ. Agar colonies are opalescent becoming brown, small, irregular, smooth, wet, and shiny. The organism enters the host through root injuries. Optimum temperature, 35–37C.

References

5, 8, 27.

Damping-off, *Pythium debaryanum* Hesse

Symptoms

This disease may occur in seedbeds at any time and often appears on the main stem and roots of older plants. A brown, wet soft-rot develops at the soil line, and seedlings are so weakened that they topple over and become overgrown by the mycelium. Older plants are often girdled and die, but remain standing. Some stem lesions develop later and cause very little damage to the plant other than a brown spot.

Etiology

The fungus is soil-inhabiting and produces hyaline, branched, nonseptate mycelium. Sporangia are oval to spherical, 15–26 μ in diameter, and may germinate by the production of germ tubes or zoospores. The zoospores average 7–12 μ, and a score or more may emerge from a single sporangium. The oospores are smooth, 12–20 μ in diameter, and form germ tubes upon germination.

References

8, 12, 15, 27.

Black shank, *Phytophthora parasitica* Dast. var. *nicotianae* (B. de Haan) Tucker

Symptoms

Damping-off of seedlings and transplants occurs, and a wilt is produced in older plants. Stunted in growth, the oldest leaves become yellow and brown, gradually dying. A soil-line lesion develops, partially or completely girdling the main stem. The margins of the diseased area are sunken and become slate to black in contrast to the green, healthy cortical tissue. The decaying area remains dry and may show vertical cracks. A cross section of the stem through the lesion reveals an extensive dark color through the woody tissues into the pith. A longitudinal section of the stem shows dark-colored cambium, wood, and pith. The pith is separated into platelike discs, producing a ladderlike impression. Plants often die prematurely but may slowly decline. Dead plants lose their leaves, and the stalks may remain standing. The fungus survives from season to season in the soil.

Etiology

The fungus produces hyaline, nonseptate mycelium that branches profusely; the advance is rapid in the host tissue. The sporangia are produced on long, hyaline sporangiophores; development is continued successively at the tip. Sporangia are hyaline to light yellow, oval or lemon-shaped, 25×35 μ, papillate, and produce 5–30 zoospores upon germination, each measuring 7–11 μ and biflagellate. Chlamydospores are spherical to oval, average 25 μ in diameter, and are rarely found. Oospores average 28 μ in diameter.

References

8, 12, 18, 19, 21, 27.

Downy mildew, *Peronospora tabacina* Adam

Symptoms

Seedling leaves up to 1 in. in diameter and older leaves show yellow green, circular patches, and secondary leaves often stand upright. The tips of upper, erect leaves droop and are flaccid. The undersurface of the lower leaves may be partly or entirely covered with a superficial, purplish, downy mildew, which may also appear on the upper leaf surface. This is of common occurrence on leaves in thickly sown beds or on shaded leaves. Large, dry, irregular areas replace the yellow patches, and the general aspect of the seedbed resembles plants that have been scalded and produce a characteristic odor. The disease in the field may destroy several of the older leaves. The main stem and petioles usually escape infection.

Etiology

The mycelium is hyaline, branched, nonseptate, and confined to the host tissue. The fungus is an obligate parasite. The conidiophores arise from the submerged mycelium through the stomata in fascicles. They are hyaline, erect, several times dichotomously branched at the apex, and vary in height from 400 to 750 μ. The hyaline, lemon-shaped to oval conidia measure 17–28×13–17 μ and develop at the subulate conidiophore tips. They develop at night and by daylight are mature and are disseminated. They germinate by producing mycelium that penetrates the host directly. Oospores are produced in the host tissue; they are reddish brown, spherical, with rough bumps and ridges, and measure 20–60 μ in diameter.

References

11, 12, 14, 27.

Argentine mildew, *Peronospora nicotianae* Speg.

Symptoms

This disease of seedlings is often severe enough to cause their death. Older plants show yellowish green, small, circular to irregular spots, particularly on the older leaves. The sporangiophores appear as a tan, downy fungus growth on the lower surfaces of leaves. These areas become pale, and the tissue becomes necrotic and dies.

Etiology

The sporangiophores are many times divided, producing sporangia on each of the many slender, subulate tips. These spores measure 15–25×8–14 μ. The oospores are 50–80 μ in diameter. This fungus has not been reported outside Argentina.

Reference

27.

Metallic mold, *Blakeslea trispora* Thaxt.

Symptoms

The fungus is often observed in early morning dew on the declining corollas. It appears on harvested tobacco that is not properly cured.

Etiology

The fungus mycelium is hyaline, nonseptate, branched, and produces sporangia with a single columella and sporangia that are branched several times at the tips, supporting sporangi and spores. Chlamydospores are hyaline, intercalary, and oval. The fungus produces black spherical zygospores that measure 45–63×38–61 μ and show the 2 suspensors on the lower side.

References

9, 24, 27.

Seedling blight, *Olpidium brassicae* (Wor.) Dang.

Symptoms

A yellowing and decline of the aboveground parts of seedlings are associated with this brown decay of the roots, which become heavily infested. The star-shaped, resting sporangia are characteristic of the disease.

Etiology

The fungus is an obligate parasite that produces no mycelium. The thallus consists of a single globular cell that evolves into a zoosporangium, measuring 12–20 μ in diameter. The uniciliate zoospores measure up to 3 μ in diameter. They germinate and penetrate the host roots directly.

References

2, 12.

Cottony rot, *Sclerotinia sclerotiorum* (Lib.) d By.

Symptoms

Damping-off in the seedbeds is common, and decayed stems become covered with a white weft of mycelium in which sclerotia are imbedded. Lesions near the base of the stem may completely girdle it, resulting in wilting, yellowing, and death. Affected stems which have been split open show that the pith has been replaced by a copious mycelial web. Black sclerotia may be present internally as well as on the exposed areas of the stems. The fungus survives in the soil by means of these sclerotia. Soft-rot or wet-rot areas form within the blade tissue. The stalks become slimy and wet, and the cortex may slip away easily from the woody cylinder.

Etiology

The mycelium is hyaline, septate, branched, and plentiful. The sclerotia are produced on the mycelial wefts; they are white and spongy at first later becoming black, oval to irregular in shape, hard, with a white interior. When the sclerotial growth is renewed, several stalked, projecting structures are formed. On the soil surface they produce small apothecia connected to the imbedded sclerotia by a slender stalk. The sclerotial stage can remain dormant for extended periods.

References

12, 27.

Powdery mildew, *Erysiphe cichoracearum* DC.

Symptoms

The presence of fungus is indicated by the white, powdery film over the surface of leaves and on stems. On lower leaf surfaces, the patches enlarge rapidly, run together to form a complete white coating, and may soon occupy the upper surface of all lower leaves, which often become pale to yellowish. The numerous spores are capable of initiating the disease on other plants.

Etiology

The mycelium is white or hyaline, branched, septate, and superficial on the

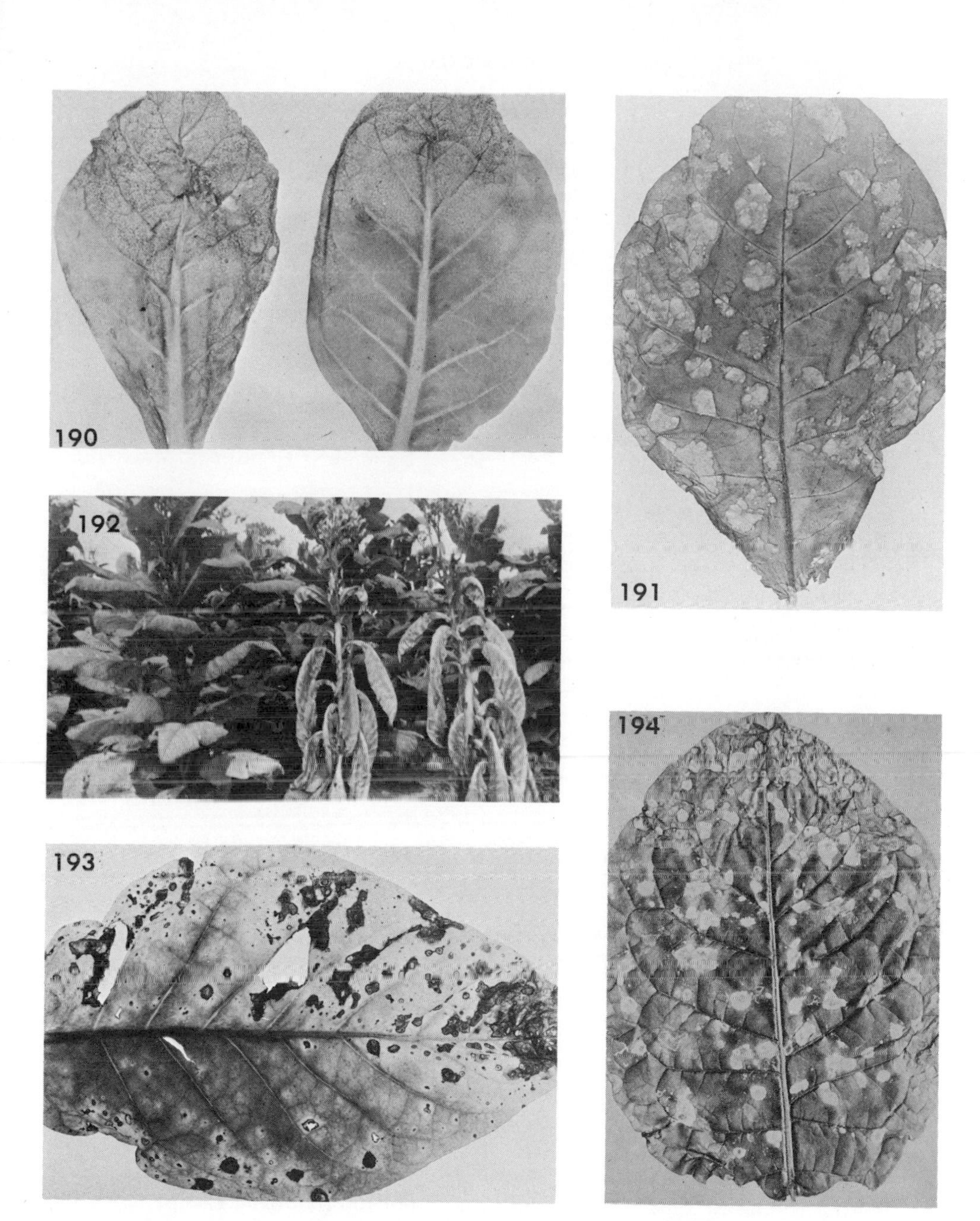

Figure 190. Downy mildew, *Peronospora tabacina,* on tobacco, seedling stage.
Figure 191. Downy mildew, *Peronospora tabacina,* on tobacco leaf, late stage. Enlarged
Figure 192. Black shank, *Phytophthora parasitica* var. *nicotianae,* on tobacco.
6.4X.
Figure 193. Brown spot, *Alternaria longipes,* on tobacco. (Photograph by W. B. Tisdale.)
Figure 194. Frog eye, *Cercospora nicotianae,* on tobacco leaf.

leaf surface. It is anchored to the host by the appressoria from which haustoria develop. The conidiophores arise from the hyphae vertically, measure 108–168 × 11–12 μ, and produce hyaline, ovoid, 1-celled conidia, one after another, often in chains, and each measuring 29–31 × 14–18 μ. The cleistothecia (perithecia) are black, spherical, with flexuous, indeterminate appendages, and measure 80–140 μ. They contain 10–25 asci, each enclosing usually 2 hyaline, 1-celled ascospores that measure 20–28 × 12–20 μ.

References

8, 12.

Rust, *Uredo nicotianae* Anas., Sacc., & Splendore

Symptoms

Brown sori produced on the leaves are chiefly hypophyllous.

Etiology

They produce globoid uredospores, closely and coarsely verrucose, with obscure pores, and measuring 23–26 × 24–32 μ. The disease is important on tobacco.

Reference

1.

Soil rot, *Thanatephorus cucumeris* (Frank) Donk

Symptoms

Damping-off of seedlings forms watery lesions at the soil level, which extend around the stems until they are girdled. The seedlings wilt and topple over, and stem tissue turns brown or is completely decayed. In the field, plants previously affected and transplanted are often broken off by winds, and the leaves of diseased plants may wither and dry. Stem lesions near the soil line appear as dry cankers and tend to girdle field plants.

Etiology

This disease is caused by the sterile vegetative stage of the soil-inhabiting fungus. The mycelium is brown and usually not too plentiful, except when the humidity is high. It can be distinguished by brown, branched, coarse, septate hyphae. The brown sclerotia consist of densely aggregated, thick-walled cells. The fruiting stage consists of short, club-shaped basidia, bearing 4 pointed sterigmata each of which supports a single, 1-celled, hyaline basidiospore and measures 5 × 9 μ. The imperfect stage of the fungus is *Rhizoctonia solani.*

References

12, 27.

Southern blight, *Botryobasidium rolfsii* (Sacc.) Venkat.

Symptoms

The fungus is soil-inhabiting and causes infections on roots and the main stem at the soil line. The first symptom is a flaccid condition or wilting of the foliage and a yellowing of the older leaves. The fungus causes considerable decay of the roots or a stem canker or lesion at the soil line. This stem infection may be from 1 to several inches in extent and usually girdles the stem, killing the cortex and cambium down to the woody tissue. An abundance of white mycelium usually surrounds the stem above the soil line. Sclerotia are produced on the mycelium on the host tissue.

Etiology

The fungus mycelium is white, thin-walled, branched, septate, and produces clamp connections. The sclerotia are white when immature, becoming brown; they are superficial, spherical or nearly so, and measure 1–2 mm. They survive in the soil from season to season. The basidia, not often observed, appear on the thin, lacy mycelium and are hyaline, measuring 7–9 × 4–5 μ. Hyaline, 1-celled, obovate, apiculate, smooth basidiospores are formed. The sterile vegetative stage is *Sclerotium rolfsii*.

References

9, 12, 27.

Leaf spot, *Phyllosticta nicotiana* Ell. & Ev.

Symptoms

This disease is recognized by brown, circular to irregularly zonate spots, measuring up to 10 mm in diameter. The spots usually become light brown in the centers, while the outer areas and margins are much darker.

Etiology

The pycnidia are black, imbedded in the dark areas of the leaf spot, and 75–150 μ in diameter. The pycnidiospores are hyaline, oval, 1-celled, and 6–10 × 3–4 μ.

References

12, 27.

Leaf spot, *Ascochyta phaseolorum* Sacc.

Symptoms

Leaf spots are gray to brown, circular to irregular, up to 1 in. or more in diameter, and may appear on seedlings or mature plants. Pycnidia are scattered in the necrotic tissue and on the stems.

Etiology

The pycnidiospores are hyaline, oblong to oval, 2-celled, and 8–10 × 3–4 μ.

Reference

12.

Anthracnose, *Colletotrichum tabacum* Böning

Symptoms

Infection appears on the seedling leaves in the form of dull, green, water-soaked spots, 2–3 mm in diameter. These spots may enlarge to 10 mm and become brown with a darker margin. Coalescence of spots kills the plant. Secondary infection results in the appearance of many pinpoint spots and may cause irregular growth and distorted and ragged foliage. Stem infection is often severe, causing cracks and stem girdling.

Etiology

The fungus produces an abundance of acervuli with stout, dark brown setae up to 90 μ in length, among which hyaline, fusoid, cylindrical, 1-celled spores, measuring $15–22 \times 4–5\ \mu$, are produced. The fungus is primarily seed-borne but may survive on vegetative debris.

References

8, 12, 16, 27.

Gray mold, *Botrytis cinerea* Pers. ex Fr.

Symptoms

Mold on tobacco appears principally on the seedlings and on the harvested crop during air curing. It develops on the lowermost leaves, appearing as large, brown, dry spots bordered by a yellowish zone. During rainy periods, wet-rot lesions are formed and are covered with a mouse gray, fruiting portion of the fungus. Petioles may show sunken, brown to black lesions, and leaves may dry and remain attached. If wet weather prevails, stem lesions increase, girdling the stem, and all distal parts die.

Etiology

The mycelium advances through the middle lamella. It is septate, branched, hyaline, and later darkens. Conidiophores are gray, erect, and branched several times at the tips, upon which small sterigmata develop, each supporting a hyaline, ovoid, 1-celled conidium, which measures $6–9 \times 8–11\ \mu$. The spores cluster around the conidiophore, producing a gray color. Black, hard, small, rough sclerotia form in and on the host tissue. The fungus persists in the soil. The ascospore stage is *Botrytinia fuckeliana*.

References

12, 26, 27.

Black root rot, *Thielaviopsis basicola* (Berk. & Br.) Ferr.

Symptoms

Seasonal developments are damping-off of seedlings, stunting of older plants, and pale green foliage with yellow tips and dry edges. There is root infection, indicated by a blackening and decay of the root system with lesions at the soil level, extending upward and downward. In initial infections on larger seedlings, the roots are blackened and die, and the stems remain stunted.

Etiology

The mycelium is hyaline, branched, septate, 3–7 μ in diameter, and gradually turns brown. The conidia are hyaline, thin-walled, and 8–30×3–5 μ. They are produced by endoconidiophores, which are short and taper slightly to their tips. The conidia develop within the terminal cell which measures 80–170×3–6 μ. Mature conidia are pushed out of the basal cell in chains; they are hyaline, changing to brownish black, up to 8-celled, and measure 8–30×10–12 μ. The 2 or 3 hyaline basal cells are capable of breaking up when mature.

References

8, 10, 12, 13, 27.

Frog eye, *Cercospora nicotianae* Ell. & Ev.

Symptoms

On mature leaves, spots are ashen gray with narrow, brown to very dark, raised borders, circular, and 1–10 mm in diameter. Young spots are tan to brown-colored and become grayish as soon as the fructifications of the causal fungus appear.

Etiology

The mycelium is limited in the host tissue by a natural walling-off in the mesophyll, thus limiting the infection to the typical spots. The conidiophores are brown, septate, geniculate, sparingly branched, and 75–100×4–5 μ. The conidia are hyaline, slender, straight or curved, multiseptate, and 38–135×3–4 μ.

References

8, 12, 27.

Brown spot, *Alternaria longipes* (Ell. & Ev.) Mason

Symptoms

This disease appears first on the lower and older leaves. It originates as small, circular, water-soaked areas, enlarging gradually to form circular, light brown, zonate leaf spots about 1/2 in. in diameter. The spots become pale with age, sunken, elongated, and often surrounded by a yellow halo. Many spots may fuse, involving large portions of leaves, which become brown. Dark brown lesions occur on the stems and produce circular patches on the seed capsules.

Etiology

The mycelium is dark-colored, branched, septate, 2–4 μ in diameter, and produces conidiophores that are effused, amphigenous, geniculate, septate, constricted, 25–65 × 5–6 μ, and emerge through the stomata in groups of 3–10. The conidia are ovate to obclavate, showing 3–7 cross septations, 1–3 longitudinal septations, and measuring 35–90 × 8–15 μ, with a distinct beak up to 50 μ long.

References

8, 12, 20.

Sooty mold, *Fumago vagans* Pers.

Symptoms

The disease is manifested in the form of a black, sooty, superficial film of dark hyphae on the surfaces of mature leaves. The fungus grows on honeydew produced by insects, principally aphids. Control of the fungus necessitates the control of the insects.

Etiology

The fungus develops simple, erect conidiophores, producing 1–2-celled, ovoid conidia, measuring 5–6 × 3–7 μ. Dark brown chlamydospores, measuring 7–23 × 5–19 μ, are often present. The sexual stage of the fungus is *Capnodium* sp.

References

8, 12.

Blotch, *Septomyxa affinis* (Sherb.) Wr.

Symptoms

The seedlings become yellow and stunted. Blotches on the upper surfaces of leaves are olivaceous and generally irregular in outline. Similar colored areas may appear on the stem and leaf petioles. A decay often affects the stems.

Etiology

The conidia are hyaline, delicate, straight, 1-septate, and 10 × 3 μ.

References

12, 17.

Wilt, *Fusarium oxysporum* Schlecht. f. *nicotianae* (J. Johnson) Snyd. & Hans.

Symptoms

A yellowing and wilting of the leaves is usually first evident on one side of maturing plants in the field. Some plants may develop the yellow color and not show the wilting condition immediately, or the wilt may involve the entire plant at one time. If some outer bark of stem lesions is removed, the wood beneath is brown or black in contrast to the white of nondiseased plants. The fungus enters the host through the roots and may kill some or most of them. It is a soil-borne fungus.

Etiology

In culture the mycelium is white or pink or bluish. Microconidia, formed abundantly, are hyaline, oval to elongate, 1-celled, or 1- or 2-septate, and formed on short, branched conidiophores that measure 5–12×2–3 μ. The macroconidia develop in sporodochia and are hyaline, 3–5-septate, sickle-shaped, and 41–50×2–5 μ. Chlamydospores are usually intercalary, averaging about 8–10 μ.

References

8, 12, 27.

Other fungi associated with tobacco

Agrobacterium tumefaciens (E. F. Smith & Townsend) Conn
Alternaria tabacina (Ellis & Everhart) Hori
Alternaria tenuis Nees
Aspergillus niger van Tieghem
Botryosporium pulchrum Corda
Cladosporium herbarum Link ex Fries
Fusarium roseum Link
Macrophomina phaseolina (Tassi) Coidanich
Monila macrospora Kingman
Mucor mucedo (Linnaeus) Brefeld
Oospora nicotianae Penzig & Saccardo
Penicillium citrinum Thom
Phymatotrichum omnivorum (Shear) Duggar
Pseudomonas mellea Johnson
Pseudomonas polycolor Clara
Pseudomonas pseudozooglosae (Honing) Stapp
Rhizopus nigricans Ehrenberg
Sclerotinia minor Jagger
Septoria nicotianae Patouillard
Verticillium albo-atrum Reinke & Berthold

References: tobacco

1. Arthur, J. C., and G. B. Cummins. 1962. Manual of the rusts in United States and Canada. 2d ed. Hafner Co.: N.Y.
2. Bensaude, M. 1923. A species of Olpidium parasitica in the roots of tomato, tobacco and cabbage. Phytopathology 13:451–54.
3. Clayton, E. E., and J. E. McMurtrey, Jr. 1950. Tobacco diseases and their control. U.S. Dept. Agr. Farmers' Bull. 2023:1–70.
4. Dukes, P. D., S. F. Jenkins, Jr., and R. W. Toler. 1963. An improved inoculation technique for transmission of false broomrape to flue-cured tobacco. Plant Disease Reptr. 47:895–97.
5. Elliott, C. 1951. Manual of bacterial plant pathogens. 2d ed. Chronica Botanica: Waltham, Mass.
6. Hahn, P. M. 1958. Flue-cured tobacco.

American Tobacco Co.: Richmond, Va., pp. 1–23.
7. Higgins, B. B. 1927. Physiology and parasitism of Sclerotium rolfsii Sacc. Phytopathology 17:417–48.
8. Hopkins, J. C. F. 1956. Tobacco diseases. Commonwealth Mycological Institute: Kew, Surrey, England.
9. Jochems, S. C. J. 1927. The occurrence of Blakeslea trispora Thaxter in the Dutch East Indies. Phytopathology 17: 181–84.
10. Johnson, J. 1916. Resistance in tobacco to the root-rot disease. Phytopathology 6:167–81.
11. Kröber, H., and W. Weinmann. 1964. Ein beitrag zur morphologie und taxonomie der Peronospora tabacina Adam. Phytopathol. Z. 51:241–51.
12. Lucas, G. B. 1965. Diseases of tobacco. Scarecrow Press: N.Y.
13. McCormick, F. A. 1925. Perithecia of Thielavia basicola Zopf, in culture. Connecticut Agr. Expt. Sta. Bull. 269: 539–54.
14. McGrath, H., and P. R. Miller. 1958. Blue mold of tobacco. Plant Disease Reptr. Suppl. 250:1–35.
15. Middleton, J. T. 1943. The taxonomy, host range and geographic distribution of the genus Pythium. Mem. Torrey Botan. Club 20:98–106.
16. Riley, E. A. 1953. Anthracnose: a serious disease of tobacco nurseries in northern Rhodesia caused by Colletotrichum tabacum Böning. Trop. Agr. (Trinidad) 31:307–11.
17. Tisdale, W. B. 1929. A disease of tobacco seedlings caused by Septomyxa affinis (Sherb.) Wr. Phytopathology 19:90. (Abstr.)
18. Tisdale, W. B. 1931. Development of strains of cigar wrapper tobacco resistant to blackshank (Phytophthora nicotianae Breda de Haan). Florida Univ. Agr. Expt. Sta. Tech. Bull. 226:1–45.
19. Tisdale, W. B., and J. G. Kelley. 1926. A phytophthora disease of tobacco. Florida Univ. Agr. Expt. Sta. Tech. Bull. 179:159–218.
20. Tisdale, W. B., and R. F. Wadkins. 1931. Brown spot of tobacco caused by Alternaria longipes (E. & E.), n. comb. Phytopathology 21:641–60.
21. Tucker, C. M. 1931. Taxonomy of the genus Phytophthora De Bary. Missouri Univ. Agr. Expt. Sta. Res. Bull. 153:1–197.
22. Valleau, W. D., and E. M. Johnson. 1936. Tobacco diseases. Kentucky Agr. Expt. Sta. Bull. 362:7–60.
23. Weber, G. F. 1954. False broomrape of tobacco in Florida. Plant Disease Reptr. 38:121–22.
24. Weber, G. F., and F. A. Wolf. 1927. Heterothallism in Blakeslea trispora. Mycologia 19:302–7.
25. Wolf, F. A. 1922. A leafspot disease of tobacco caused by Phyllosticta nicotiana E. & E. Phytopathology 12:99–101.
26. Wolf, F. A. 1931. Gray mold of tobacco. J. Agr. Res. 43:165–75.
27. Wolf, F. A. 1956. Tobacco diseases and decays. 2d ed. Duke University Press: Durham, N.C.

Tomato, *Lycopersicon esculentum* Miller

Canker, *Corynebacterium michiganense* (E. F. Smith) H. L. Jensen
Soft rot, *Erwinia aroideae* (Townsend) Holland
Brown rot, *Pseudomonas solanacearum* E. F. Smith
Speck, *Pseudomonas tomato* (Okabe) Altstatt
Bacterial spot, *Xanthomonas vesicatoria* (Doidge) Dowson
Crown gall, *Agrobacterium tumefaciens* (E. F. Smith & Townsend) Conn
Damping-off, *Pythium aphanidermatum* (Edson) Fitzpatrick
Late blight, *Phytophthora infestans* (Montagne) De Bary
Buckeye, *Phytophthora parasitica* Dastur
Soft rot, *Rhizopus nigricans* Ehrenberg
Yeast spot, *Nematospora coryli* Peglion
Canker, *Didymella lycopersici* Klebahn
Fruit rot, *Pleospora lycopersici* Elie & Emile Marchal
Stem rot, *Sclerotinia sclerotiorum* (Libert) De Bary
Powdery mildew, *Leveillula taurica* (Léveillé) Arnaud
Southern blight, *Botryobasidium rolfsii* (Saccardo) Venkatarayan
Soil rot, *Thanatephorus cucumeris* (Frank) Donk
Black spot, *Phoma destructiva* Plowright
Leaf spot, *Septoria lycopersici* Spegazzini

Anthracnose, *Colletotrichum phomoides* (Saccardo) Chester
Fruit rot, *Myrothecium roridum* Tode ex Fries
Gray mold, *Botrytis cinerea* Persoon ex Fries
Leaf mold, *Cladosporium fulvum* Cooke
Leaf spot, *Cercospora fuligena* Roldan
Watery rot, *Oospora lactis parasitica* Pritchard & Porte
Wilt, *Fusarium oxysporum* f. sp. *lycopersici* (Saccardo) Snyder & Hansen
Wilt, *Verticillium albo-atrum* Reinke & Berthold
Early blight, *Alternaria solani* (Ellis & G. Martin) Jones & Grout
Nailhead, *Alternaria tomato* (Cooke) Weber
Gray leaf spot, *Stemphylium solani* Weber
Leaf spot, *Helminthosporium lycopersici* Maublanc & Rogers
Fruit rot, *Helminthosporium carposaprum* Pollock
Other fungi associated with tomato

Canker, *Corynebacterium michiganense* (E. F. Sm.) H. L. Jens.

Symptoms

Canker is first detected by a wilting of the growing extremities and is further characterized by the streaking of the stem between nodes, often developing a break in the stem in the nature of a canker. The phloem tissue is invaded and becomes darkened. The wilting and browning may develop on one side of the plant, or a stunting may follow severe infection. The fruits at first show water-soaked spots surrounded by a white band or halo; they are more or less circular, slightly raised, and reach a size of 3 or 4 mm. As the green fruit matures, the white band surrounding each spot becomes brown. The bacteria survive in the soil for more than a year. Secondary infection may appear on all aboveground parts.

Etiology

The organism is a gram-positive, nonmotile rod, forming capsules, and measuring $0.7–1.2 \times 0.6–0.7\ \mu$. In culture it develops yellow, glistening, smooth, circular colonies. Optimum temperature, 25–27C.

References

3, 4, 8, 21.

Soft rot, *Erwinia aroideae* (Towns.) Holland

Symptoms

Soft rot is encountered in the field and in the market. The decay develops often as a secondary invader through injuries or insect wounds. Fruit are softened by an internal decay in any stage of development, in storage, or in the market. In the field, diseased fruits frequently remain attached to the peduncle and become a bag holding the watery contents of the fruit. Sometimes stem decay occurs in the form of soft, wet lesions.

Etiology

The organism is a gram-negative rod, forms no capsules, is motile by 2–8 peritrichous flagella, and measures $2–3 \times 0.5\ \mu$. In culture the colonies are white, opalescent, glistening, and more or less circular. Optimum temperature, 35C.

References

8, 14, 25.

Brown rot, *Pseudomonas solanacearum* E. F. Sm.

Symptoms

Brown rot first causes a wilt of the growing tips of otherwise healthy appearing plants. Wilting is often temporary, depending on growing conditions, but usually becomes permanent, and eventually the plant dies. Cross sections and broken leaf petioles or lateral branches reveal the slate brown discoloration of the vascular tissue. Leaf blades and fruit show no symptoms. The organism is soil-inhabiting and invades plants through wounds.

Etiology

The organism is a gram-negative rod, produces no capsules, and is motile by a single polar flagellum. Agar colonies are whitish or opalescent, turn brown, are smooth, shiny, irregularly circular, and usually small. Optimum temperature, 35–37C.

References

8, 21, 25.

Speck, *Pseudomonas tomato* (Okabe) Altstatt

Symptoms

Speck, observed mostly on the fruit, is characterized by numerous black, pimplelike, slightly raised dots, usually less than 1 mm in diameter. They are mostly scattered but frequently very numerous. All parts of the host are susceptible, and the fruit are infected only when green and immature. The lesions are mostly superficial. Leaf spots are yellowish brown, later darkening.

Etiology

The organism is a gram-negative rod, often in chains, forms capsules, is motile by 1–7 polar flagella, and measures $1.3–2.5 \times 0.6\ \mu$. Agar colonies are white, glistening, circular, and flat. A green fluorescent pigment is produced in culture. Optimum temperature, 23–25C.

References

4, 7, 8, 21.

Bacterial spot, *Xanthomonas vesicatoria* (Doidge) Dows.

Symptoms

This disease has been reported from most tomato-growing areas. All parts of the plant aboveground are susceptible. Small, circular to irregularly shaped, dark, watery spots appear on leaflets and on stems, often causing some defoliation or stem lesions. Leaf spots become dark-colored with a narrow, yellow border. The disease is not systemic. On the green fruit the disease is particularly distinctive by the small, water-soaked, scattered spots on the surface. They are surrounded by a light-colored band which disappears as the spot becomes light brown, sunken, and rough. Often they are numerous and coalesce, destroying the fruit. The organism is seed-borne, may survive in the soil for limited periods, and gains entrance through wounds caused by insects or blowing soil particles.

Etiology

The organism is a gram-negative rod, often in chains, forms capsules, is motile by a single polar flagellum, and measures 1–1.5$\times$0.6–0.7 μ. In culture it produces yellow, semitranslucent, circular to spreading, rather viscid colonies. Optimum temperature, 30C.

References

8, 24.

Crown gall, *Agrobacterium tumefaciens* (E. F. Sm. & Towns.) Conn

Symptoms

Crown gall is not a pathogen of tomatoes, although artificially inoculated plants develop stem galls in a very prolific and characteristic manner.

Etiology

The organism is a gram-negative rod, forms capsules, is motile by polar flagella, and measures 1–3$\times$0.4–0.8 μ. Agar colonies are white, glistening, translucent, small, and circular. Optimum temperature, 25–30C.

References

8, 21.

Damping-off, *Pythium aphanidermatum* (Edson) Fitzp.

Symptoms

Damping-off usually appears on seedlings at the time of emergence or shortly thereafter. The soil-inhabiting organism becomes destructive under cool, moist conditions. The germinating seed may be invaded, the young plant may be prevented from emerging, the seedling may show wilting tendencies, in the cotyledon stage, or the plant may be girdled at the soil line and killed. The stem is softened, and the plant falls over. The invaded area or canker may involve 1 in. or more of the main stem, which becomes soft, shrunken, tender, and shows no color change. White, scanty, branched, nonseptate mycelium may be observed on the diseased parts.

Etiology

The sporangia are lobulate, filamentous, and discharge 15–40 reniform, biciliate zoospores, measuring 12–14×6 μ. Oogonia are spherical and 16–34 μ in diameter. Oospores are spherical, smooth, and 12–28 μ in diameter.

References

7, 13, 25.

Late blight, *Phytophthora infestans* (Mont.) d By.

Symptoms

This disease on tomato is similar to late blight on potato and is caused by the same fungus. All parts of the plant aboveground may show the disease at any time. The leaf spots appear water-soaked at first and gradually darken. A soft, downy mold may be seen on the lower surface of these spots, along the advancing margin. The spots enlarge rapidly and may involve large portions of the leaflets, which rapidly wilt, blacken, droop, and remain attached. The fungus also causes petiole, peduncle, fruit, and stem lesions that often result in a collapse of the plant. Leaf spots, stem lesions, and fruit discoloration characterize the disease. The fungus is air disseminated.

Etiology

The mycelium is hyaline, branched, and usually scanty. It is nonseptate and supports sporangiophores that emerge through the stomata and lenticels and are hyaline, branched, and septate. Sporangia are produced successively at the tips. Each sporangium forms a septation and a characteristic swelling, measures 21–38×12–23 μ, and has an apical papilla. Sporangia form infective hyphae or zoospores upon germination. Oospores are rarely found. Oogonia are hyaline, spherical, and 31–50 μ in diameter; oospores are 25–35 μ in diameter.

References

4, 7, 21.

Buckeye, *Phytophthora parasitica* Dast.

Symptoms

Buckeye is limited to the growing fruit, although cankers on the stems have been observed. During relatively dry weather, the fungus, which is soil-inhabiting, attacks the fruit that is in contact with the soil. Initially the diseased area is a slightly lighter green than normal. As it enlarges, irregular, brown bands are formed with lighter green interspaces of a target design. In wet weather the disease spreads to fruit above the soil, up to a foot on staked plants. The interior of the diseased green fruit usually remains firm rather than watery.

Etiology

The mycelium is hyaline often becoming brown, branched, nonseptate, and often forming intercalary chlamydospores. The sporangiophores are hyaline, erect, and form sporangia at their tips that are hyaline to faintly colored, ovate, 30–40 × 25–30 μ, and germinate directly or by zoospore formation. The oogonia measure 22 μ in diameter and the oospores 20 μ in diameter.

References

4, 20, 21, 24, 25.

Soft rot, *Rhizopus nigricans* Ehr.

Symptoms

The rot develops as a wound parasite, invading the fruit through injury usually inflicted during handling, storage, and marketing activities. The fungus causes a soft, watery decay, and invaded fruit melt away rapidly. In storage the fungus often develops a black mold as it produces spores in large numbers on the many minute heads attached to the diseased areas of the fruit. The fungus is air disseminated and does not occur on uninjured fruit except possibly through contact with decaying fruit.

Etiology

The mycelium is hyaline and nonseptate; the spores are small, dark, and 1-celled.

References

14, 19, 21, 25.

Yeast spot, *Nematospora coryli* Pegl.

Symptoms

This yeast is considered a wound parasite, gaining entrance to attached fruit through mechanical injury caused by feeding bugs that are also the carriers of the fungus. The original tissue around the insect puncture is mostly white, caused by the loss of moisture and inclusion of air. When invaded by the yeast, these light, cloudy areas show light brown discoloration. The introduced organism causes the parenchyma tissue to become coarse and dry.

Etiology

Hyaline, spherical to elliptical, free-floating vegetative cells develop by budding to form colonies with scanty mycelium. Cell fusion initiates the sexual stage. The asci are unattached, produced abundantly, and free floating. They are hyaline, broadly cylindrical with rounded ends, thin-walled, and 60–70 μ long. The ascospores are in 2 clusters of 4 in the opposite ends of the ascus. They are hyaline, 2-celled, fusiform-obclavate, with unequal cells, and measure 50 × 4 μ.

References

4, 9, 16, 21.

Canker, *Didymella lycopersici* Kleb.

Symptoms

At the soil line or slightly above, on the main stem of the plant a canker is produced that often causes a wilting of the parts above or girdles the stem entirely, killing it. Leaf spots are not plentiful. They are brown and sometimes zonate.

Etiology

The mycelium is hyaline, septate, branched, and confined to the host tissue which cankers kill; many black pycnidia develop in these cankers. They are subglobose, ostiolate, and 100–270 μ. The perithecia are black, oval, ostiolate, and contain 8 hyaline, spiral-shaped, 1-septate ascospores, each measuring 70–95 × 5–7 μ.

References

4, 21.

Fruit rot, *Pleospora lycopersici* El. & Em. Marchal

Symptoms

This rot occurs on tomatoes after removal from the vine, mostly in storage and transit where wounds become infected. Normally it is a weak wound parasite that produces a slate black area usually around the peduncle scar and extending outward.

Etiology

The mycelium is dark, branched, and septate, and the black, muriform septate conidia measure 20–37 × 11–15 μ. The black, globose, ostiolate perithecia produce dark, muriform, finely echinulate ascospores that measure 30–41 × 15–35 μ.

References

4, 14, 21.

Stem rot, *Sclerotinia sclerotiorum* (Lib.) d By.

Symptoms

Stem rot is frequently recognized by wilting caused by stem girdling, a sunken, brown decay at or near the soil line. Lesions are often partially covered with a white, fluffy, cottony mycelial growth that may be limited in dry weather. The organism is often a wound parasite, but under humid conditions it causes widespread damage to the stem and other aerial parts of the plant.

Etiology

The disease is diagnosed by the presence of black, irregularly shaped, shiny, hard sclerotia mostly produced in the pith but sometimes attached to the outer

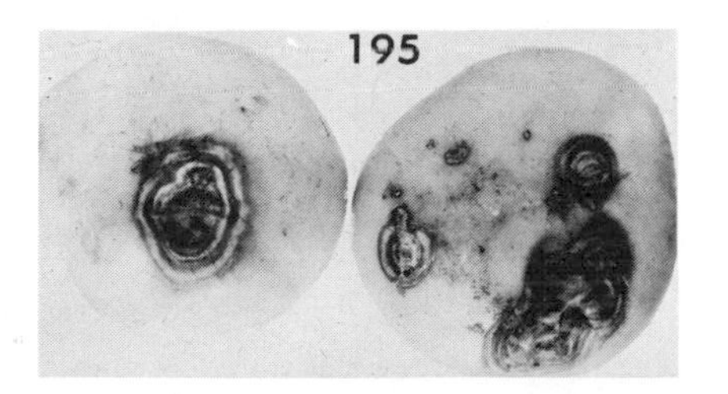
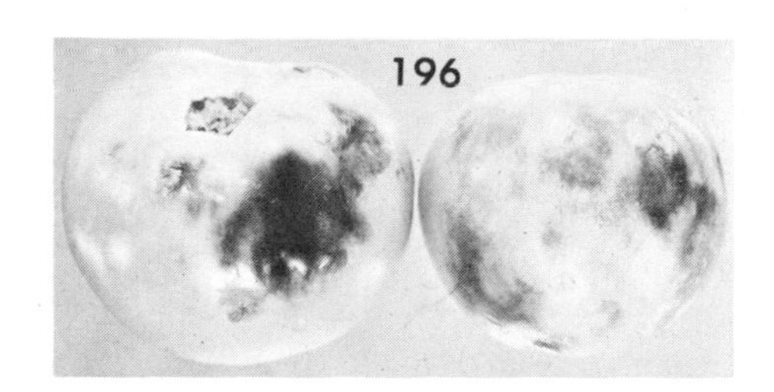
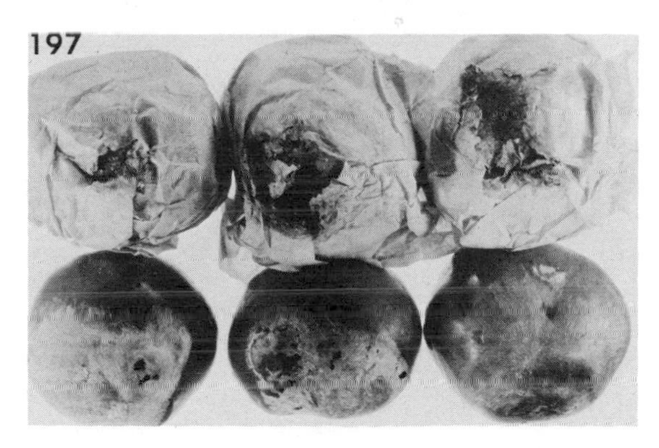

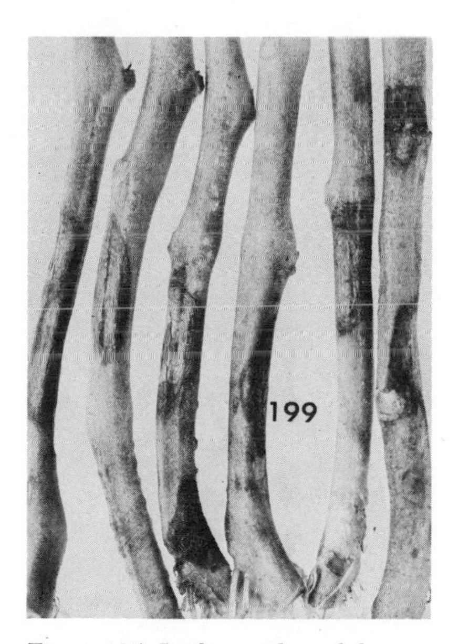
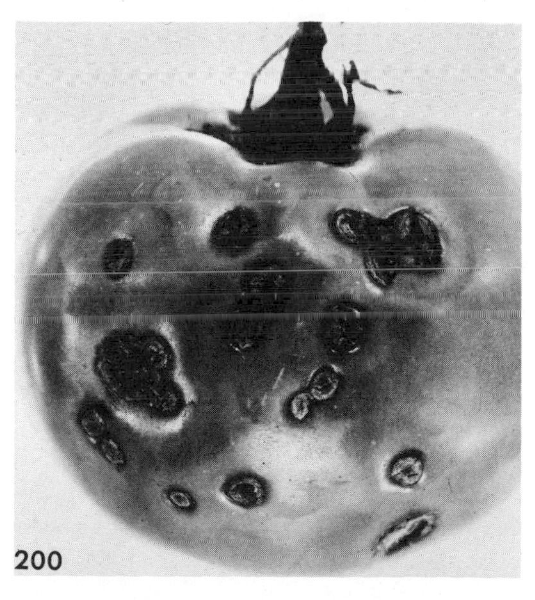

Figure 195. Buckeye, *Phytophthora parasitica,* on tomato.
Figure 196. Yeast spot, *Nematospora coryli,* on tomato.
Figure 197. Soil rot, *Thanatephorus cucumeris,* of "green wrap" tomatoes occurs in transit.
Figure 198. Black spot, *Phoma destructiva,* field infection on tomato leaves.
Figure 199. Early blight, *Alternaria solani,* lesions on stems of seedbed tomato plants. Enlarged 1.3X.
Figure 200. Nailhead, *Alternaria tomato,* lesions on tomato fruit.

cortex, which becomes dry, papery, cracked, and light brown in color. Conidia are not found, and the ascospores are produced in apothecia.

References

4, 21, 25.

Powdery mildew, *Leveillula taurica* (Lév.) Arn.

Symptoms

This mildew is recognized by the white, chalky, powderlike dust that accumulates on the lower leaf surfaces and causes the corresponding upper leaf areas to become yellowish. The leaves lose moisture and may wither and drop.

Etiology

The mycelium is hyaline, septate, branched, superficial, and produces upright conidiophores, terminated by a single, hyaline, 1-celled, oval to long, cylindric, slightly rough conidium, measuring 40–60 × 12–16 μ. There appear black, appendaged, very small perithecia scattered among the conidiophores and conidia, producing 2–4-spored asci and ascospores that measure 19–25 × 13–15 μ.

Reference

4.

Southern blight, *Botryobasidium rolfsii* (Sacc.) Venkat.

Symptoms

Blight is definitely distinguished by the production of an abundance of white mycelium and brown, spherical, hard, superficial sclerotia, which measure 1/2–1 mm in diameter. The sclerotia appear at the soil line associated with a stem canker that usually girdles the main stem. Stem girdles cause a wilting of the tops, continued decline, and eventual death. The fungus is a soil-inhabiting organism. The brown cortical tissues crack, shrink, curl, and become dry.

Etiology

Basidiospores are produced, but the fungus is disseminated largely by the movement of the sclerotia. The hyphae are hyaline, and masses of mycelium are white, branched, septate, and produce an abundance of creamy to brown, spherical, hard, superficial sclerotia, measuring less than 1 mm in diameter.

References

21, 25.

Soil rot, *Thanatephorus cucumeris* (Frank) Donk

Symptoms

Soil rot results from infection of the plant parts at all stages of their development. Pre- and post-emergence damping-off of seedlings occurs. The stem is girdled, and the plant falls over and dies. The stem becomes slightly brownish, a symptom distinguishing it from *Pythium* damping-off. The main stem of bearing plants may be girdled at the soil line. Foliage escapes infections except when in contact with the soil. The fruits in all stages of maturity in contact with the soil are rapidly involved in a semisoft decay. Affected fruit show little or no zonate browning.

Etiology

Brown mycelium is abundantly produced on the outside of the fruit. The fungus produces brown, septate, branched mycelium and brown sclerotia, measuring up to 3 mm in diameter.

References

5, 21, 25.

Black spot, *Phoma destructiva* Plowr.

Symptoms

Phoma spot develops rapidly on the foliage; the spots are black, variously scattered, small, irregularly shaped, and sunken. As the spots enlarge, they become brown, circular, and variously zoned, target fashion. The darker bands are thickly speckled with black pycnidia. The spots may enlarge to several centimeters in diameter, and frequently the killed center areas fall out. Stem and peduncle lesions are black, elongate, and often girdling. The fruit are particularly susceptible to infection around the shoulders during handling, packing, and transit.

Etiology

The mycelium is branched, septate, and hyaline becoming colored. It becomes concentrated in zones, and rows of black, globose, ostiolate, subepidermal pycnidia develop. A second circle of pycnidia may appear outside the first row. Pycnidia are 50–350 μ in diameter. The pycnidiospores are hyaline, ovate to subglobose, 1-celled, and $3.4–9 \times 1.7–3\ \mu$. These spores often exude from the pycnidium in long, continuous tendrils.

References

1, 4, 21, 25.

Leaf spot, *Septoria lycopersici* Speg.

Symptoms

This spotting is one of the several important leaf diseases that has been destructive to tomato. The infection is confined almost entirely to the foliage, sometimes appears on the stems, and is rarely found on the fruit. Seedling infection in the seedbed is often most disastrous, resulting in the loss of transplants and the spread of the disease. The leaf spots are generally small, mostly less than 5 mm in diameter, circular, depressed, water-soaked at first, bleaching to almost white, and surrounded by a dark border.

Etiology

The central area becomes speckled with the black pycnidia, 1 to several correlated with the size of the spot. The mycelium is hyaline becoming colored, branched, septate, intercellular, and confined in host tissue. The pycnidia, at first subepidermal, are black, globose, small, and usually few. The pycnidiospores are hyaline, long, thin, 3–9-septate, 60–120×2–4 μ, and often emerge from the pycnidium in long, undulating tendrils.

References

4, 21, 25.

Anthracnose, *Colletotrichum phomoides* (Sacc.) Chester

Symptoms

Anthracnose is found on fruit, seldom on the stems or foliage. Mature fruit at the stage of changing color may become affected. Scattered, circular, depressed spots appear; as the fruit softens they enlarge and appear water-soaked. The central portion of those spots becomes slightly roughened, and the margins may be extended until they are 1 in. in diameter. When numerous, the spots coalesce, and the entire fruit becomes soft and watery.

Etiology

In the centers of the spots, numerous acervuli become prominent. The mycelium is hyaline, septate, branched, and forms subepidermal stromata from which the black, septate setae develop. The acervuli produce an abundance of hyaline, 1-celled, oval to elongate conidia that measure 20–28×4–16 μ and en masse appear to be pink or salmon-colored. Setae are found in these fruiting areas. The perfect stage is *Glomerella cingulata.*

References

4, 21, 25.

Fruit rot, *Myrothecium roridum* Tode ex Fr.

Symptoms

Fruit rot is sometimes encountered in the form of brownish specks on foliage and peduncles. It is most frequently associated with the fruit, where it produces a brown to black, oval to circular, depressed but firm ring rot. The somewhat concentrically zoned area supports bands of sporodochialike fruiting structures.

Etiology

The mycelium is hyaline to colored, branched, septate, and internal; it forms stromata from which hyaline, branched conidiophores emerge. Each branch bears masses of dark olivaceous, oval, 1-celled conidia that measure 5–10×2–4 μ. The conidia are produced terminally on the tips of the conidiophores, which give rise to 4–6 phialides.

References

14, 21.

Gray mold, *Botrytis cinerea* Pers. ex Fr.

Symptoms

Mold is destructive to foliage, stems, and fruit. The fruit spots in the beginning are small, black specks that do not enlarge, but there develops around them a whitish subepidermal area often called ghost spots. Foliage and stem lesions soon become overgrown by the mycelium and conidiophores of the fungus, which produce gray spores that detach easily and spread. Fruit cracks are invaded, and a soft, watery decay is produced. The disease is common on acidic soils and not on calcic soils.

Etiology

The conidia are hyaline, oval, 1-celled, and 11–15 × 8–11 μ. Black sclerotia are produced.

References

4, 18, 21.

Leaf mold, *Cladosporium fulvum* Cke.

Symptoms

Leaf mold is mostly confined to the foliage where yellowish, circular to angular blotches appear. On the lower surface of these yellow areas, an olive green mold is produced, which comprises the fruiting structures of the fungus. The spots usually enlarge until most of the leaf is involved, droops, dies, and is shed. The disease begins on the lower leaves and gradually advances upward, causing defoliation as it progresses.

Etiology

The mycelium is hyaline becoming colored, branched, septate, and subepidermal. The branched conidiophores are dark, septate, and produce terminal conidia. The spores of the fungus are dark-colored, mostly 1-celled, sometimes 1-septate, oval to oblong, and 14–16 × 5–8 μ.

References

4, 21, 25.

Leaf spot, *Cercospora fuligena* Roldan

Symptoms

This spot develops on the leaf surface as a yellow area that gradually becomes brown. On the lower surface, opposite a brown mold, it expands into colonies that are indefinite in extent and shape and may be limited by the veins. As the disease develops, the leaf turns brown, withers, and dies. The fungus may cause brown zones on the stems but not on the fruit.

Etiology

The mycelium is septate, branched, and hyaline to faintly colored. The conidia are produced in the velvety mold on fascicles of brown conidiophores. Conidia are slightly colored, semiclavate, curved, septate, and $15–118 \times 3–5\ \mu$.

Reference

4.

Watery rot, *Oospora lactis parasitica* Pritchard & Porte

Symptoms

Watery rot results from wound infection by the fungus on injured fruit. The infection develops rapidly, rendering the internal tissue liquid but not affecting the epidermis. Fruit in any stage of development are susceptible. The decay is accompanied by a characteristic acidlike odor.

Etiology

The fungus develops a hyaline, scanty, branched, septate mycelium over the softened areas. The mycelium produces endogenous conidia and breaks up into individual oidia that are hyaline, 1-celled, short, square to slightly rounded to elongate, and $3–40 \times 2–8\ \mu$.

References

19, 21, 25.

Wilt, *Fusarium oxysporum* f. sp. *lycopersici* (Sacc.) Snyd. & Hans.

Symptoms

Wilt is a systemic disease originating in the root system and causing the plant to deteriorate slowly. Diseased plants first appear somewhat stunted followed by a slight yellowing of the older leaves, which continue to be nonfunctional, as those farther up the stem become involved. A badly diseased field of plants shows extensive browning and loss of foliage, poor yield, and undersized fruit. Fruit and foliage are not directly affected. The vascular tissue from the roots to the foliage, petioles, peduncles, and side axillary branches is a dull slate brown color. Stem lesions may develop, usually at the soil line, showing some girdling and cankers.

Etiology

The mycelium is septate, branched, and hyaline, becoming creamy yellow as it ages. It produces in abundance hyaline, 1-celled, elliptical to oval, round ended microconidia, measuring $6–15 \times 2–4\ \mu$. Macroconidia are hyaline, sickle-shaped to fusiform with obtuse ends, mostly 3-septate, and $25–33 \times 3–6\ \mu$. Chlamydospores are formed, but there are no pionnotes, sporodochia, or sclerotia.

References

4, 7, 17, 21, 25.

Wilt, *Verticillium albo-atrum* Reinke & Berth.

Symptoms

Wilt is a destructive disease gaining entrance through the root system. The affected plant may show some yellowish lower leaves and a general unthrifty appearance. The disease develops slowly from soil line to aerial extremities, producing dark vascular tissues and characteristic black streaks in the woody parts of the stem. Plants are not inclined to die under normal conditions and may survive through the season, although the fruit is reduced in number and size.

Etiology

The mycelium is hyaline to frequently colored, branched, septate, and produces conidiophores in clusters with delicate terminals. The fungus produces hyaline, small, oval, 1-celled spores cut off from the tips of the branches of the conidiophore. Spores may be in chains or in wet, spherical balls as heads. The fungus is soil-inhabiting.

References

2, 4, 21.

Early blight, *Alternaria solani* (Ell. & G. Martin) Jones & Grout

Symptoms

Seedbed infection consists of cotyledon or leaflet spotting. The spots are black, sunken, small, and variously shaped. Elongate lesions develop on the main stems, often completely girdling the small stem. On older plants the leaf spots are brown to black, more or less circular to angular, and usually concentrically zoned, particularly on the upper surface. The spots vary in size from a few millimeters to 2–3 cm. Often the centers crack and partially fall away. Lesions appear as black, elongate, sunken areas on the main stem and branches as well as on petioles and peduncles. The fungus invades the fruit from calyx lesions and produces spores in abundance on the lower leaf surfaces.

Etiology

The mycelium is dark, septate, branched, and produces short, dark, septate conidiophores, mostly on the foliage. The conidia are dark brown, obclavate, long-beaked, muriform septate, borne singly or in chains of 2–3, and measure $40–200 \times 10–19\ \mu$.

References

4, 21, 25.

Nailhead, *Alternaria tomato* (Cke.) Weber

Symptoms

Nailhead develops on all parts of the plant aboveground but is most severe on the fruit. The leaf spotting may appear in the seedbed as dark, sunken spots, and on older foliage as tan to brown, more or less circular spots, usually less than 1 cm in diameter. Similar spots develop on the stems, petioles, and peduncles. On the fruits the infections are tan, at first small, shallow, and variously scattered. The mycelium penetrates the cortex. The conidia are formed in the centers of the spots. Spores are produced on killed foliage that has fallen onto the soil as well as on the fruit spots. Often early infection causes malformed fruit and severe spotting.

Etiology

The mycelium is colored, septate, branched, 4–10 μ in diameter, and produces brown, erect, short, smooth, septate conidiophores. The conidia are brown, obclavate, with beaks that are septate, colored, uniformly narrow, and 2–4 μ wide and 60–80 μ long. The conidia are very characteristic, muriform septate, and 39–65 × 13–22 μ.

References

4, 21, 23.

Gray leaf spot, *Stemphylium solani* Weber

Symptoms

Leaf spot appears on the foliage and stems of plants but does not infect the fruit. In the seedling stage the cotyledons and leaflets show dark, irregular, depressed spots, mostly less than 2–3 mm in size. As the plants develop, the infection is spread to the upper leaves; they in turn become yellowed, droop, die, turn brown, and are shed. The fungus continues to produce conidia on the shed leaves in contact with damp soil. The small, grayish brown leaf spots are found characteristically throughout the season at the growing end of the new leaves.

Etiology

The mycelium is brown, branched, and septate, and the conidiophores are dark, short, and septate. The conidia are produced abundantly on the leaf lesions. They are dark brown, muriform septate, oblong to oval or slightly elongate, with a rounded base and obtuse terminal tip; they lack an elongate beak, are restricted at the median septation, and measure 40–65 × 13–22 μ.

References

4, 10, 22, 24.

Leaf spot, *Helminthosporium lycopersici* Maubl. & Rogers

Symptoms

Leaf spot attacks the foliage. Spots are light-colored with brown margins, circular, and up to 3 mm in diameter; they are scattered but may coalesce when numerous, causing some defoliation.

Etiology

The conidiophores are olive brown, septate, and 70–145×7–9 μ. The conidia are produced on the lower surfaces on these scattered conidiophores. The spores are olive brown, 4–12-septate, rounded at the base and generally obclavate, curved, and 50–107×10–18 μ.

References

4, 15.

Fruit rot, *Helminthosporium carposaprum* Pollock

Symptoms

Fruit show circular to irregular spots, covered by a dense gray to black mycelium, and 10–30 mm in diameter. The lesions are flat and sunken, and the tomato flesh becomes black and spongy.

Etiology

The dark mycelium that permeates the spongy tissue produces erect, septate conidiophores, which measure 140–500×6–10 μ, and are on the surface of the diseased area. The conidia are hyaline to pale olivaceous, cylindrical, taper, with rounded ends, are straight or curved in chains, with up to 15 septations, and measure 28–220×6–12 μ.

References

4, 12, 14.

Other fungi associated with tomato

Aphanomyces cladogamus Drechsler
Ascochyta lycopersici (Plowright) Brunaud
Aspergillus niger Tieghem
Basisporium gallarum Molliard
Botryosporium pulchrum Corda
Brachysporium tomato (Ellis & Bartholomew) Hiroë & Watanabe
Chaetomium bostrychodes Zopf
Colletotrichum atramentarium (Berkeley & Broome) Taubenhaus
Dendrodochium lycopersici Em. Marchal
Diaporthe phaseolorum (Cooke & Ellis) Saccardo
Fusarium oxysporum f. sp. *lycopersici* (Saccardo) Snyder & Hansen (race 2)
Isaria clonostachoides Pritchard & Porte
Macrophomina phaseolina (Tassi) Goidanich

Melanospora interna Tehon & Stout
Nigrospora oryzae (Berkeley & Broome) Petch
Olpidium brassicae (Woronin) Dangeard
Phyllosticta hortorum Spegazzini
Phymatotrichum omnivorum (Shear) Duggar
Phytophthora cactorum (Lebert & Cohn) Schroeter
Phytophthora mexicana Hotson & Hartge
Plectospira myriandra Drechsler
Pyrenochaetia lycopersici Schneider & Gerlack
Spongospora subterranea (Wallroth) Lagerheim
Thielaviopsis basicola (Berkeley & Broome) Ferraris
Trichothecium roseum Link ex Fries

References: tomato

1. Aulakh, K. S., et al. 1969. Phoma destructiva, its variability, host range and varietal reaction on tomatoes. Plant Disease Reptr. 53:219–22.
2. Bewley, W. F. 1922. "Sleepy disease" of tomato. Ann. Appl. Biol. 9:116–34.
3. Bryan, M. K. 1930. Studies on bacterial canker of tomato. J. Agr. Res. 41:825–51.
4. Chupp, C., and A. F. Sherf. 1960. Vegetable diseases and their control. Ronald Press: N.Y., pp. 525–76.
5. Conover, R. A. 1949. Rhizoctonia canker of tomato. Phytopathology 39:950–51.
6. Doolittle, S. P. 1948. Tomato diseases. U.S. Dept. Agr. Farmers' Bull. 1934:1–82.
7. Doolittle, S. P., A. L. Taylor, and L. L. Danielson. 1961. Tomato diseases and their control. U.S. Dept. Agr., Agr. Handbook 203:1–86.
8. Elliott, C. 1951. Manual of bacterial plant pathogens. 2d ed. Chronica Botanica: Waltham, Mass.
9. Fawcett, H. S. 1929. Nematospora on pomegranates, citrus, and cotton in California. Phytopathology 19:479–82.
10. Hannon, C. I., and G. F. Weber. 1955. A leaf spot of tomato caused by Stemphylium floridanum sp. nov. Phytopathology 45:11–16.
11. Hotson, J. W., and L. Hartge. 1923. A disease of tomato caused by Phytophthora mexicana, sp. nov. Phytopathology 13:520–31.
12. McColloch, L. P., and F. Pollack. 1946. Helminthosporium rot of tomato fruits. Phytopathology 36:988–98.
13. Matthews, V. D. 1931. Studies on the genus Pythium. University of North Carolina Press: Chapel Hill.
14. Ramsey, G. B., J. S. Wiant, and L. P. McColloch. 1952. Market diseases of tomatoes, peppers, and eggplants. U.S. Dept. Agr., Agr. Handbook 28:1–92.
15. Roldan, E. F. 1936. New or noteworthy lower fungi of the Philippine Islands. Philippine J. Sci. 60:119–23.
16. Schneider, A. 1917. Further note on a parasitic saccharomycete of the tomato. Phytopathology 7:52–53.
17. Stall, R. E. 1961. Development of fusarium wilt on resistant varieties of tomato caused by a strain different from race 1 isolates of Fusarium oxysporum f. lycopersici. Plant Disease Reptr. 45: 12–15.
18. Stall, R. E. 1963. Effects of lime on incidence of Botrytis gray mold of tomato. Phytopathology 53:149–51.
19. Timmer, C. P. 1963. The harmful fungi of tomatoes and their problem to the processor. Econ. Botany 17:86–96.
20. Tompkins, C. M., and C. M. Tucker. 1941. Buckeye rot of tomato in California. J. Agr. Res. 62:467–74.
21. Walker, J. C. 1952. Diseases of vegetable crops. McGraw-Hill: N.Y., pp. 431–514.
22. Weber, G. F. 1930. Gray leafspot of tomato caused by Stemphylium solani sp. nov. Phytopathology 20:513–18.
23. Weber, G. F. 1939. Nailhead spot of tomato caused by Alternaria tomato (Cke.) n. comb. Florida Univ. Agr. Expt. Sta. Tech. Bull. 332:5–51.
24. Weber, G. F., and D. G. A. Kelbert. 1940. Seasonal occurrence of tomato diseases in Florida. Florida Univ. Agr. Expt. Sta. Bull. 345:5–36.
25. Weber, G. F., and G. B. Ramsey. 1926. Tomato diseases in Florida. Florida Univ. Agr. Expt. Sta. Bull. 185:61–138.
26. Weber, G. F., S. Hawkins, and D. G. A.

Kelbert. 1932. Gray leafspot, a new disease of tomatoes. Florida Univ. Agr. Expt. Sta. Tech. Bull. 249:1–34.

Tung, *Aleurites fordii* Hemsley

Leaf spot, *Pseudomonas aleuritidis* (McCulloch & Demaree) Stapp
Leaf spot, *Phytophthora palmivora* Butler
Dieback, *Botryosphaeria ribis* Grossenbacher & Duggar
Canker, *Physalospora rhodina* (Berkeley & Curtis) Cooke
Thread blight, *Ceratobasidium stevensii* (Burt) Venkatarayan
Web blight, *Botryobasidium microsclerotia* (Matz) Venkatarayan
Shoestring root rot, *Armillaria mellea* Vahl ex Fries
Mushroom root rot, *Clitocybe tabescens* (Scopoli ex Fries) Bresadola
Leaf spot, *Cercospora aleuritidis* Miyake
Leaf spot, *Pestalotia clavispora* Atkinson
Other fungi associated with tung

Leaf spot, *Pseudomonas aleuritidis* (McCull. & Demaree) Stapp

Symptoms

Early infections appear as greenish to brown, small, translucent spots that later darken and become opaque and angular. Lesions may show a narrow, pale yellow border, and are often limited by the veins. When infections are numerous and conditions favorable, the lesions spread and become irregular. The diseased tissues, in spots up to 20 mm in extent, are dry, dull brown to almost black, opaque, shriveled, and usually cracked. Defoliation is often severe. Vascular tissues are not invaded, and no exudate is evident.

Etiology

The organism is a gram-negative rod with rounded ends; it appears singly or in chains, produces capsules, is motile by 1–5 polar or bipolar flagella, and measures $1.1–3 \times 0.6–0.7\ \mu$. On agar the colonies are circular to lobed, and white and translucent, becoming greenish and transparent. A green fluorescent pigment is produced in the culture medium. Optimum temperature, 27–28C.

References

2, 10.

Leaf spot, *Phytophthora palmivora* Butl.

Symptoms

The disease is manifested by the formation of leaf spots, particularly during humid, wet, rainy weather. They appear along the margins and at the same time are somewhat limited by the principal veins. Later these wet, frosty appearing spots on foliage and branches become brown and cause a bending of the branches; young plants become defoliated when the attack is severe.

Etiology

The mycelium is hyaline, branched, nonseptate, with haustoria and chlamydospores. Oval to lemon-shaped sporangia, measuring $32 \times 58\ \mu$, develop on undulating sporangiophores, each of which produce 15–30 biciliate zoospores that germinate by forming penetrating hyphae. Oospores are hyaline, smooth, spherical, and 18×41 μ.

Reference

14.

Dieback, *Botryosphaeria ribis* Gross. & Dug.

Symptoms

The first noticeable symptom is a blight or wilt of the new flush of growth as the tree emerges from dormancy. The primary source of the infection can usually be traced to a dead fruit peduncle still attached from the preceding season. This early symptom may expand to include a similar condition over larger branches and extended portions of the tree. Close examination may reveal bark cankers, which develop over several growing seasons and result in the death of all or part of the tree because of complete girdling.

Etiology

The mycelium invading the primary cortex results in the formation of marked protuberances. The ostiolate, papillate pycnidia, which measure 175–250 μ, produce pycnidiospores that are hyaline, fusoid, continuous, and 13–25× 6–10 μ. This stage is followed by the perithecial stage which is black, oval, and somewhat pulvinate with a rough outer surface. The perithecia appear in more or less longitudinal rows in the host tissue with about 1/3 of their walls protruding. The asci are clavate, 80–120×17–20 μ, and intermixed with paraphyses. The ascospores are hyaline, fusoid, continuous, and 19×7 μ. The vegetative stage of the fungus is *Dothiorella ribis.*

References

5, 11.

Canker, *Physalospora rhodina* (Berk. & Curt.) Cke.

Symptoms

Cankers appear as dark, sunken areas of the cortex, trunks, branches, limbs, and twigs. The girdling effect of the disease causes these parts to wilt and die. The fungus develops on the canker surfaces of dead parts, showing black pycnidia emerging from the outer cortex.

Etiology

The spores at first appear hyaline, oval, and very numerous. Later they become dark and 1-septate, typical of *Diplodia natalensis.* The perithecia found on these cankers, usually in older wood, appear very similar to the pycnidia. They are black, ostiolate, and protrude from the outer cortical areas. The asci are 63–70×20–22 μ, and the ascospores are hyaline, 1-celled, oval, and 15–24× 8–9 μ.

References

8, 16.

Thread blight, *Ceratobasidium stevensii* (Burt) Venkat.

Symptoms

The fungus can be identified by the presence of the brown, loaflike sclerotia, located on the past season's flush of growth, on the outer cortex usually above a lenticel or closely clustered around the base of terminal buds. A brown, superficial, spiderweblike mycelium grows from these sclerotia toward the new growth. It proceeds up petioles onto the leaf blades, where it invades the parenchyma tissue that becomes necrotic, brown, and dies. The silky mycelium spreads out fanlike over the entire lower surface of the leaves. Eventually the dead leaves are shed, but most often they remain suspended from the host by the fungus threads that collect in strands on the surface of the cortex. These are the threads to which the common name refers. Peduncles and fruit are likewise involved.

Etiology

As the fungus matures, it produces a superficial, thin hymenium from which basidia and basidiospores develop. As the threads become older, sclerotia are formed on the host tissue and are the principal source of the fungus, which survives from season to season particularly on deciduous perennial trees and shrubs.

Reference

9.

Web blight, *Botryobasidium microsclerotia* (Matz) Venkat.

Symptoms

Young infections are light tan, measure up to 2 cm, and appear anywhere on the leaves. A fine, silky, brownish hyphal mat radiates outward, often involving large portions of the leaf. The advancing margin is white, thin, and about 1/2 in. wide. The leaf parenchyma, when attacked earlier, turns lighter green, fades to yellow, becomes brown, and dies. The fungus spreads by contact and grows over petioles and stems.

Etiology

On all parts covered by mycelium, there develop many brown, oval to elongate, loosely compact sclerotia, which are attached to the mycelium and measure 80×600 μ, averaging about 100×250 μ. About 60 per cent of them will pass through a 100-mesh screen.

References

7, 15.

Shoestring root rot, *Armillaria mellea* Vahl ex Fr.

Symptoms

The aboveground symptom is a general decline caused by curtailment of an adequate nutrient supply due to root impairment. A crown canker is usually evident near the soil line, resulting in a stunted flush of growth, undersized leaves, a lack of set fruit, and a yellowish rather than green color. Sometimes there is a white accumulation of hyphae in the cortex, but this is not always true. Rhizomorphs develop underground on dead roots and under the bark of killed areas. Their presence is diagnostic of the disease.

Etiology

Further identification is the recognition of the honey-colored, stipitate, gill-containing mushrooms that are usually scattered but sometimes in limited clumps. The sporophore, 2–5 in. wide and 3–6 in. high, consists of a central stipe, an umbrella-shaped cap, and a more or less prominent annulus. The basidiospores are hyaline, 1-celled, elongate, with rounded ends, straighter on one side than the other, and $8–9 \times 5–6$ μ.

References

4, 12, 18.

Mushroom root rot, *Clitocybe tabescens* (Scop. ex Fr.) Bres.

Symptoms

This disease is revealed on the aboveground portions of the tree. Two manifestations are apparent. A sudden wilting of the foliage and death of the tree may occur within a month or less, or there may be a long, continuous lack of vigor, stunting of new growth, yellowing, shedding of foliage, and eventual death a year or more after the first symptoms are observed. An examination of the crown usually reveals a canker just above the soil line, extending 1 or 2 feet up the trunk. The line of demarcation is very plain in that diseased cortical tissues shrink away from healthy parts. The inner cortex, phloem, and cambium are killed, and a diagnostic white mycelium in the form of mats or plates occurs imbedded in the tissues. At certain times of the year the fungus produces tan to honey-colored clusters of mushrooms.

Etiology

The mycelium is hyaline or white, septate, branched, and produces sporophores that are tan above and lighter-colored beneath, usually in clusters of up to 50 or more, each with caps measuring 2–4 in. wide and nonannulated stipes up to 4–8 in. high. The basidiospores are hyaline or white, 1-celled, elongate to oval, and $8–10 \times 4–6$ μ.

References

11, 13.

Leaf spot, *Cercospora aleuritidis* Miyake

Symptoms

Leaf spots are light brown to dark reddish brown to dark brown, irregular to angular, and up to 10 mm in diameter. The disease is often severe by the middle of summer, causing extensive defoliation by the time the nuts are mature.

Etiology

Conidiophores, usually fasciculate, arise from dark brown stromata. They are mostly brown, septate, branched, curved, exhibit mild spore scars, and measure 10–65 × 2–5 μ. The ascospore stage is *Mycosphaerella aleuritidis.*

References

1, 3.

Leaf spot, *Pestalotia clavispora* Atk.

Symptoms

Spots are pale to dark brown, mostly irregular in shape, circular to angular, and up to 1 cm in diameter. Black pustules are variously scattered and often exude black spores, resulting in a black sooty area around the leaf spot.

Etiology

The pustules are immersed becoming erumpent and measure 150–275 μ in diameter. They are ostiolate, exuding tendrils of black, 5-celled, constricted, clavate, fusiform conidia, the apical cell of which is hyaline. Three setulae extend from the apical cell and measure 14–31 × 1–2 μ. The basal cell is hyaline with a pedicel 4–7 μ long. The 3 intervening cells are dark brown or fuliginous, the more basal of them slightly lighter-colored. The conidia measure 18–26 × 6–9 μ.

Reference

6.

Other fungi associated with tung

Botryobasidium rolfsii (Saccardo) Venkatarayan
Botryobasidium salmonicolor (Berkeley & Broome) Venkatarayan
Fomes hawaiensis Lloyd
Fomes lamaensis (Murrill) Saccardo & Trotter
Fomes lignosus (Klotzsch) Bresadola
Fusarium heterosporium var. *aleuritidis* Saccas & Drowillon
Ganoderma sessile Murrill
Gloeosporium aleuriticum Saccardo
Glomerella cingulata var. *aleuritidis* Saccas
Lophodermium aleuritidis Rehm
Melampsora aleuritidis Cummins

Phyllosticta aleuritidis Saccas
Phymatotrichum omnivorum (Shear) Duggar
Phytophthora cinnamomi Rands
Pythium aphanidermatum (Edson) Fitzpatrick
Septobasidium aleuritidis Heim & Bouriquet
Sphaerostilbe repens Berkeley & Broome
Uncinula migabei var. *aleuritidis* Wei
Ustulina vulgaris Tulasne

References: tung

1. Bain, D. C. 1960. Cercospora leaf spot of tung in Mississippi. Plant Disease Reptr. 44:190–91.
2. Boyd, O. C. 1930. A bacterial disease of tung-oil tree. Phytopathology 20:756–58.
3. Chupp, C. 1953. A monograph of the fungus genus Cercospora. Published by the author: Ithaca, N.Y., p. 212.
4. Gibson, I. A. S., and B. C. M. Corbett. 1964. Variation in isolates from Armillaria root disease in Nyasaland. Phytopathology 54:122–23.
5. Grossenbacher, J. G., and B. M. Duggar. 1911. A contribution to the life history, parasitism and biology of Botryosphaeria ribis. N.Y. State Expt. Sta. Tech. Bull. 18:114–88.
6. Guba, E. F. 1961. Monograph of Monochaetia and Pestalotia. Harvard University Press: Cambridge, Mass., pp. 218–19.
7. Large, J. R. 1944. Web blight of seedling tung trees tentatively identified as the rhizoctonia stage of Corticium microsclerotia. Phytopathology 34:648–49.
8. Large, J. R. 1948. Canker of tung trees caused by Physalospora rhodina. Phytopathology 38:359–63.
9. Large, J. R., J. H. Painter, and W. A. Lewis. 1950. Thread blight in tung orchards and its control. Phytopathology 40:453–59.
10. McCulloch, L., and J. B. Demaree. 1932. A bacterial disease of the tung-oil tree. J. Agr. Res. 45:339–46.
11. Plakidas, A. G. 1937. Diseases of tung trees in Louisiana. Louisiana Agr. Expt. Sta. Bull. 282:2–11.
12. Rhoads, A. S. 1945. A comparative study of two closely related root-rot fungi, Clitocybe tabescens and Armillaria mellea. Mycologia 37:741–66.
13. Rhoads, A. S. 1956. The occurrence and destructiveness of Clitocybe root rot on woody plants in Florida. Lloydia 19: 193–239.
14. Venkatarayan, S. V. 1932. Phytophthora on Aleurites fordi. Phytopathology 22: 222–27.
15. Weber, G. F. 1939. Web-blight, a disease of beans caused by Corticium microsclerotia. Phytopathology 29:559–75.
16. Wiehe, P. O. 1952. Bibliography of the fungi and bacteria associated with tung (Aleurites spp.). Plant Disease Reptr. Suppl. 216:189–99.
17. Wiehe, P. O. 1952. Life cycle of Botryosphaeria ribis on Aleurites montana. Phytopathology 42:521–26.
18. Wiehe, P. O. 1952. The spread of Armillaria mellea (Fr.) Quel. in tung orchards. E. African Agr. J. 18:67–72.

Turmeric, *Curcuma longa* Linnaeus

Soft rot, *Xanthomonas zingiberi* (Uyeda) Săvulescu
Rhizome rot, *Pythium graminicolum* Subramaniam
Leaf spot, *Taphrina maculans* Butler
Leaf spot, *Colletotrichum capsici* (Sydow) Butler & Bisby
Leaf spot, *Phyllosticta zingiberi* Ramakrishnan
Other fungi associated with turmeric

Soft rot, *Xanthomonas zingiberi* (Uyeda) Săvul.

Symptoms

The general early wilting of foliage and young stems is followed by a yellow-

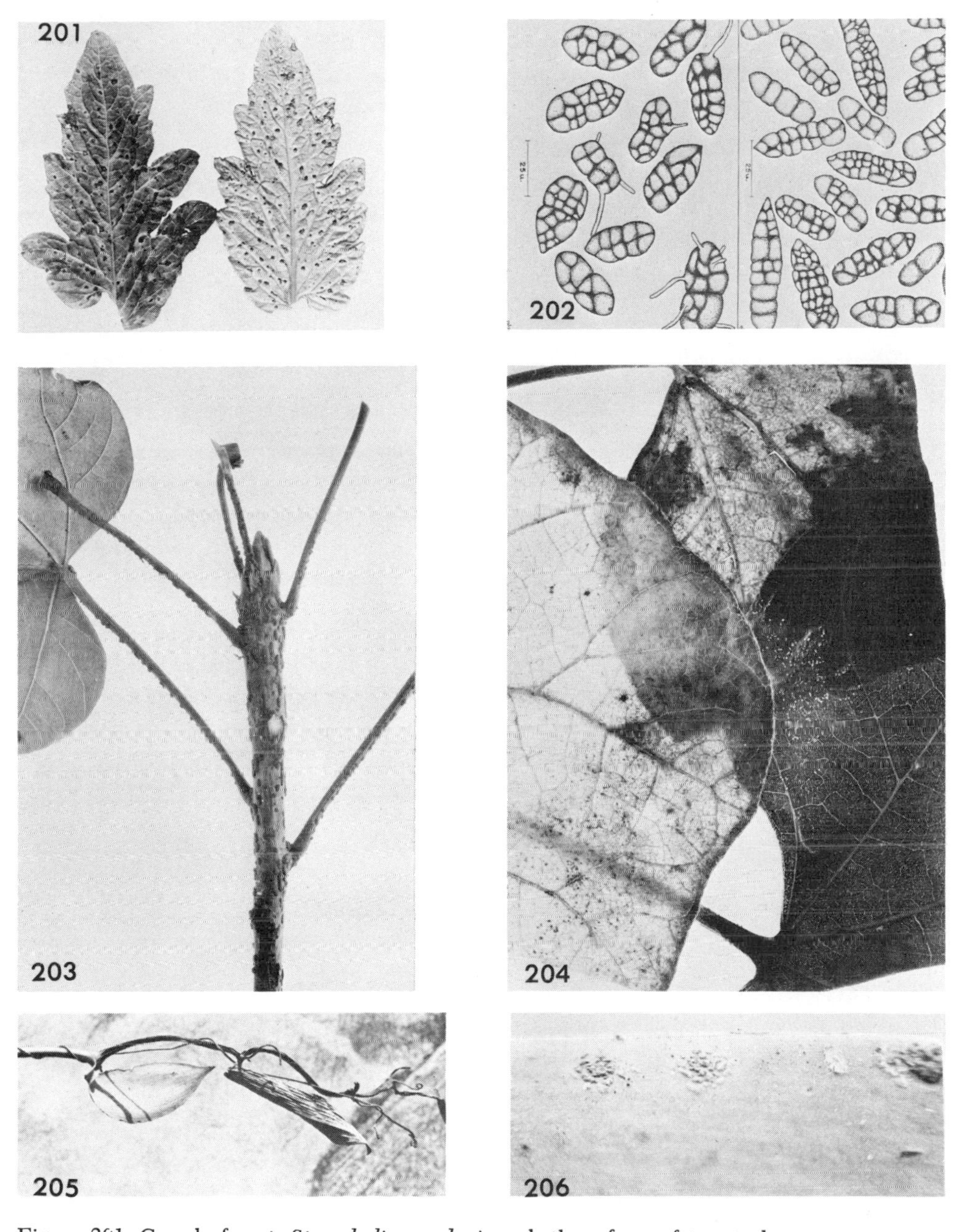

Figure 201. Gray leaf spot, *Stemphylium solani*, on both surfaces of tomato leaves.

Figure 202. Conidia of *Stemphylium solani*, left, and *S. floridana*. (Photograph by C. I. Hannon.)

Figure 203. Thread blight, *Ceratobasidium stevensii*, on twig tip, petioles, and foliage of tung.

Figure 204. Web blight, *Botryobasidium microsclerotia*, showing sclerotia and mycelium extension by contact on tung foliage.

Figure 205. Root rot, *Fusarium oxysporum* f. sp. *vanillae*, on vanilla vine, early stages. (Photograph by M. A. Porres.)

Figure 206. Rust, *Uromyces joffrini*, on vanilla foliage. (Photograph by M. A. Porres.)

ing of these parts, their collapse, and browning. Rhizome infection at the parts away from the bud end results in a slow decline, whereas bud infection causes a rapid killing of the plant. The rhizome is usually completely destroyed.

Etiology

The organism is a gram-negative rod, motile by 1–3 polar flagella, and measuring 0.75–1.8 × 0.5–1.1 μ. Agar colonies are white. Optimum temperature, 28C.

Reference

2.

Rhizome rot, *Pythium graminicolum* Subr.

Symptoms

Beginning with a gradual dying along the margins of the leaves and ending with all the leaves dry and dead, the disease as observed aboveground is one of decline. The root system of such plants is very much reduced or destroyed, and the disease often leads to the production of a soft, rotten condition of the rhizome and total loss.

Etiology

Isolated from the roots, the fungus produces hyaline, nonseptate, branched mycelium, 3–8 μ in diameter. Sporangia are produced abundantly and germinate by germ tube. The oogonia are spherical, smooth, and 28 μ in diameter; the oospores are about 23 μ in diameter.

Reference

5.

Leaf spot, *Taphrina maculans* Butl.

Symptoms

There is an excessive spotting of the foliage, although severely diseased plants are not killed. The spots are small, less than 3 mm in diameter, appear on both leaf surfaces, and cause the leaf to become reddish brown after an early yellowing. The fungus is subcuticular and may invade the epidermis and other cells.

Etiology

The hyaline, septate, branched mycelium is generally considered scanty. Haustoria penetrate the parenchyma. The accumulation of fungus tissue causes the rupture of the epidermis, and the asci develop in the exposed condition. The asci measure 20–30 × 6–10 μ and contain 8 ascospores that are hyaline, oval, 1-celled, and 4–7 × 2–3 μ. The ascospores soon produce a number of spores by budding.

References

1, 3, 7.

Leaf spot, *Colletotrichum capsici* (Syd.) Butl. & Bisby

Symptoms

Small leaf spots soon enlarge to elongate areas or patches on the foliage, measuring 1–3 in. in length and half as wide. When several coalesce, the leaf is killed. Individually the center of the spots may be grayish white and dotted with many black acervuli, which are arranged in concentric rings on both surfaces. The spots usually are surrounded by a yellow halo.

Etiology

The hyphae are hyaline, subepidermal, and septate; they later darken, and produce a stroma from which the setae and conidiophores develop. Hyaline, crescent-shaped, 1-celled conidia develop in the acervulus. The conidia measure 25×3 μ.

Reference

4.

Leaf spot, *Phyllosticta zingiberi* Ramak.

Symptoms

Small, round to irregular leaf spots appear, light-colored at first but changing to dark shades of brown. The spots enlarge rapidly and often coalesce, involving extensive leaf area and causing the leaves to curl and dry. The brown spots on dead leaves are usually speckled with black, small, spherical, imbedded fruiting structures, which become erumpent.

Etiology

The mycelium is hyaline, branched, septate, and about 2–5 mm in diameter. The pycnidia are dark brown, ostiolate, and $23–94 \times 24–96$ μ. The pycnidiospores are hyaline, 1-celled, oval, and $3–6 \times 1.5–4.5$ μ.

Reference

6.

Other fungi associated with turmeric

Colletotrichum curcumae (Sydow) Butler & Bisby
Corniothyrium zingiberi Stevens & Atienza
Nectriella zingiberi Stevens & Atienza
Piricularia zingiberi Nisikado
Rosellinia zingiberi Stevens & Atienza
Sphaceloma curcumae Thirumalachar
Xanthomonas zingiberae (Uyeda) Burkholder

References: turmeric

1. Butler, E. J. 1918. Fungi and diseases in plants. Thacker: Calcutta, India, pp. 346–48.
2. Elliott, C. 1951. Manual of bacterial plant pathogens. 2d ed. Chronica Botanica: Waltham, Mass., p. 152.

3. Mundkur, B. B. 1949. Fungi and plant diseases. Macmillan: London, pp. 110–12.
4. Ramakrishnan, T. S. 1954. Leaf spot disease of turmeric (Curcuma longa L.) caused by Colletotrichum capsici (Syd.) Butl. Bisby. Indian Phytopathol. 7:111–17.
5. Ramakrishnan, T. S., and C. K. Sowmini. 1954. Rhizome and root rot of turmeric caused by Pythium graminicolum Sub. Indian Phytopathol. 7:152–59.
6. Summanwar, A. S., and V. P. Bhide. 1962. Leaf spot of turmeric (Curcuma longa L.) caused by Phyllosticta zingiberi Ramakrishan. J. Indian Botan. Soc. 41:313–16.
7. Upadhyay, R., and M. S. Pavgi. 1967. Perpetuation of Taphrina maculans Butler, the incitant of turmeric leaf spot disease. Phytopathol. Z. 59:136–40.

Vanilla, *Vanilla planifolia* Andrews

Downy mildew, *Phytophthora jatrophae* Jensen
Anthracnose, *Glomerella vanillae* (Zimmermann) Petch & Ragunathan
Brown spot, *Nectria tjibodensis* Penzig & Saccardo
Root rot, *Fusarium oxysporum* f. sp. *vanillae* (Tucker) Gordon
Rust, *Uromyces joffrini* Delacroix
Algal spot, *Cephaleuros virescens* Kunze
Other fungi associated with vanilla

Downy mildew, *Phytophthora jatrophae* Jens.

Symptoms

The pods are attacked before maturity at either extremity during periods of wet weather, resulting in the formation of chocolate brown spots. They are widely variable in color and size. Sometimes a white florescence is produced, which is composed of small tufts emerging through the cuticle. The pods lose their turgidity and later become detached from the vine.

Etiology

The mycelium is hyaline, branched, nonseptate, about 6 μ in diameter, and produces absorption organs in the host cells. The conidiophores are external, bearing clusters of sporangia that are hyaline, 1-celled, oval with a prominent papillae, and measuring 20–73 $\times$ 15–35 μ. Sporangia germinate in water, forming zoospores or frequently mycelium directly.

Reference

1.

Anthracnose, *Glomerella vanillae* (Zimm.) Petch & Ragun.

Symptoms

The infections that occur on the stems and leaves are usually initiated through wounds. The diseased areas are distinctly sunken, oval to elongate, and the parenchyma appears to be invaded to a limited extent. These lesions or spots may elongate on the stems more so than on the foliage. They are water-soaked at first, then become yellowish and dry considerably, becoming darker as the dark-colored setae increase in number. The general symptoms are highly variable, depending on the age of the host, ecological conditions, and methods of pest control.

Etiology

The mycelium is hyaline to faintly colored, septate, branched, and mostly confined within the host tissue. The stromata develop in the epidermal cells, forming a characteristic acervulus. The conidiophores are hyaline, short, nonseptate, pointed, and 25–50×3–4 μ. The setae are brown, nonseptate, mostly curved, numerous, and 55–105×4–5 μ. The conidia are 1-celled, elongate, hyaline, and pink en masse. The perithecia are black, conical to pear-shaped, single or in groups, subepidermal becoming erumpent, and 175–200 μ in diameter. There are no paraphyses. The asci measure 60–70×9–12 μ, and the ascospores are hyaline, allantoid, 1-celled, and 14–20×3–6 μ. This fungus has been reported as *Botryosphaeria, Calospora, Colletotrichum, Gloeosporium, Vermicularia,* and *Volutella.*

References

6, 8.

Brown spot, *Nectria tjibodensis* Penz. & Sacc.

Symptoms

This disease is found mostly on the main stems of the plant, above the crown and root system, but infrequently on the leaves. The parenchyma tissues are brownish and shriveled, sometimes extensively. The fungus also causes a decay of the end shoots and fruit, which become brown and decayed from main stem infections.

Etiology

The sporodochia are plentiful, superficial, and produce conidia that are hyaline, 2-celled, cylindrical, and 16–20×3–4 μ. The perithecia develop in the same place on their stromata, which are reddish, spherical, and have projecting beaks. The asci are elongate, 50–60 μ long, and contain 8 hyaline, 2-celled ascospores that measure 9×2 μ.

Reference

8.

Root rot, *Fusarium oxysporum* f. sp. *vanillae* (Tucker) Gordon

Symptoms

Browning and dying of roots are the first symptoms of the disease visible at the surface of the soil. The aerial portion may remain viable for several months. Whenever new roots contact the soil they become diseased, yet the disease does not always become systemic. Plant growth ceases when the roots are killed, and its stored energy is dissipated in the formation of a new proliferation of roots. The growing tip dies, and the stem and leaves become yellow. The stem withers, and finally the plant dies.

Etiology

Diseased roots reveal the hyphae of the fungus and spores in the older decaying portions. The macroconidia are hyaline, septate, curved, pedicellate, slightly attenuate at the apex with no restrictions at the septa, and measure 23–45 × 2.6–4 μ. Chlamydospores occur singly or in chains up to 4. Sporodochia are borne on the vanilla roots. Microconidia are hyaline, oval, not in chains, and measure 4–7 × 2–4 μ.

References

1, 2, 4, 7.

Rust, *Uromyces joffrini* Del.

Symptoms

The rust develops on the lower surface of the leaves. It begins to show in small, oval, raised, subepidermal, pale spots that erupt, producing a characteristic scabby appearance. The spots are 2.5 mm in diameter. Some small sori develop on the upper leaf surface. The disease is very destructive to the foliage.

Etiology

The uredia are brown, circular to elliptical, and less than 1 mm in diameter. They are generally scattered but occasionally become gregarious. The uredospores are pale yellow becoming brown, oval to pyriform, with short pedicels and 2 equatorial germ pores, and measure 24–36 × 17–26 μ.

References

5, 9.

Algal spot, *Cephaleuros virescens* Kunze

Symptoms

This organism is found on the foliage and rarely on the stems. The small, rough, more or less hairy spots are irregularly circular and 5–10 mm in diameter. The alga grows under the cuticle over the epidermis.

Etiology

It is multiseptate, and its many branches extend in all directions on the same plane. The sterile hairs are upright and stiff. The fertile hairs are the sporophores and contain sporangia at their tips that produce zoospores. There are 3–9 sterigmata at the tips of the hairs; they are equal in length and bear the yellowish sporangia that are oval to spherical, thick-walled, and 40–50 μ in diameter. Upon germination, the zoospores exude from the ostiole and swim away in moisture by means of 2 cilia.

Reference

8.

Other fungi associated with vanilla

Amerosporium vanillae Hennings
Ascospora vanillae Rehm
Aspergillus niger Tieghem
Botryobasidium rolfsii (Saccardo) Venkatarayan
Botryobasidium stevensii (Burt) Venkatarayan
Botryosphaeria vanillae (Stoneman) Petch & Ragunathan
Calospora vanillae Massee
Chaetodiplodia vanillae Zimmermann
Fusicladium vanillae Zimmermann
Guignardia traversi Cavara
Lembosia rolfsii Horne
Penicillium vanillae Bouriquet
Phyllosticta vanillae Hennings
Trullula vanillae Hennings
Volutella vanillae (Delacroix) Petch & Ragunathan

References: vanilla

1. Alconero, R. 1968. Infection and development of Fusarium oxysporum f. sp. vanillae in vanilla roots. Phytopathology 58:1281–83.
2. Alconero, R., and A. G. Santiago. 1969. Mycorrhizal infections of mature portions of vanilla roots by Rhizoctonia solani as a predisposing factor to infection by Fusarium oxysporum f. sp. vanillae. Phytopathology 59:1521–24.
3. Bouriquet, G. 1946. Les maladies des plantes cultivees a Madagascar. Encyclopedie Mycologique 12:265–87.
4. Burnett, H. C. 1938. Orchid diseases. State Plant Board of Florida Bull. 12: 29–30.
5. Cook, M. C. 1886. Exotic fungi. Grevillea 15:18.
6. Feldkamp, C. L. 1945. Vanilla: culture, processing and economics. U.S. Dept. Agr. Library List 13:9–10.
7. Irvine, J., et al. 1964. Techniques for evaluating reactions of vanilla to fusarium root rot. Phytopathology 54: 827–31.
8. Roger, L. 1953. Phytopathologie des pays chauds. Encyclopedie Mycologique 18: 1423–25, 1481–82, 2260–66.
9. Sydow, P., and H. Sydow. 1924. Monographia Uredinearum 4:507.
10. Tucker, C. M. 1927. Vanilla root rot. J. Agr. Res. 35:1121–36.

Velvet bean, *Mucuna deeringianum* (Bort) Merrill

Wilt, *Phytophthora drechsleri* Tucker
Leaf spot, *Chaetoseptoria wellmanii* Stevenson
Leaf spot, *Cercospora stizolobii* Sydow
Other fungi associated with velvet bean

Wilt, *Phytophthora drechsleri* Tucker

Symptoms

Seedlings frequently are killed before emergence and also later in the 2-leaf stage. Woodier plants become infected at the soil line and may become completely girdled. Leaves become yellow when infections are limited, but the entire plant wilts and dies following stem girdling. The leaves often are shed, although some top leaves may remain attached, brown, and dead.

Etiology

The mycelium is hyaline, branched, and produces sporangia that are hyaline, oval, 1-celled, and 50–73×36–46 μ. Zoospores are hyaline, 1-celled, oval to spherical, and 17×11 μ; oospores are 36×21 μ.

Reference

3.

Leaf spot, *Chaetoseptoria wellmanii* Stevenson

Symptoms

The leaf spots are brown and scattered, small at first, later enlarging up to 6 mm in diameter with a halo. They develop several alternate light and dark brown bands, and the centers often fall out.

Etiology

The usually scattered pycnidia are sunken becoming erumpent, ostiolate, membranous, and 120–170 μ in diameter. The setae are brown, erect, straight, 3–6-septate, taper upward, and 90–225 μ. The conidia are hyaline, straight or curved, septate, and 75–160×2–4 μ.

Reference

2.

Leaf spot, *Cercospora stizolobii* Syd.

Symptoms

This spot, common in tropical countries, is tan or brown with a reddish brown border, circular to irregular, and up to 5 mm in diameter. The spots often coalesce, and parts may fall away.

Etiology

Dark brown stromata appear on either surface of the spots. Conidiophores are produced in dense fascicles, are olivaceous to brown, septate, branched with small scars, and measure 10–40×3–4 μ. Conidia are subhyaline, cylindrical to obclavate, curved, multiseptate, mostly 3, and 35–80×3–5 μ.

Reference

1.

Other fungi associated with velvet bean

Ascochyta imperfecta Peck
Botryobasidium rolfsii (Saccardo) Venkatarayan
Cercospora cruenta Saccardo
Cercospora mucunae Sydow
Macrophomina phaseolina (Tassi) Goidanich
Mycosphaerella cruenta (Saccardo) Latham
Phyllosticta mucunae Ellis & Everhart

Phymatotrichum omnivorum (Shear) Duggar
Phytophthora parasitica Dastur
Pseudomonas stizolobii (Wolf) Stapp
Pseudomonas syringae van Hall
Thanatephorus cucumeris (Frank) Donk

References: velvet bean

1. Chupp, C. 1953. A monograph of the fungus genus Cercospora. Published by the author: Ithaca, N.Y., p. 335.
2. Stevenson, J. A. 1946. Fungi novi denominati-II. Mycologia 38:524–33.
3. Sturgess, C. W., and B. T. Egan. 1960. A wilt disease of velvet bean caused by Phytophthora drechsleri Tucker. Sugar Expt. Sta. Tech. Com. Quart. pp. 9–13. Brisbane, Queensland.

Wheat, *Triticum aestivum* Linnaeus

Basal glume blotch, *Pseudomonas atrofaciens* (McCulloch) F. L. Stevens
Bacterial spike blight, *Corynebacterium tritici* (Hutchinson) Burkholder
Black chaff, *Xanthomonas translucens* f. *undulosa* (E. F. Smith, L. R. Jones, & Reddy) Hagborg
Root rot, *Pythium arrhenomanes* Drechsler
Root rot, *Pythium graminicola* Subramaniam
Downy mildew, *Sclerophthora macrospora* (Saccardo) Thirumalachar, Shaw, & Narasimhan
Powdery mildew, *Erysiphe graminis* f. sp. *tritici* Em. Marchal
Ergot, *Claviceps purpurea* (Fries) Tulasne
Take-all, *Ophiobolus graminis* Saccardo
Scab, *Gibberella zeae* (Schweinitz) Petch
Black spot, *Phyllachora graminis* (Persoon ex Fries) Fuckel
Speckled blotch, *Leptosphaeria tritici* (Garovaglio) Passerini
Glume blotch, *Leptosphaeria nodorum* (Berkeley) Mueller
Loose smut, *Ustilago nuda* (Jensen) Rostrup
Bunt, *Tilletia foetida* (Wallroth) Liro and *T. caries* (De Candolle) Tulasne
Flag smut, *Urocystis agropyri* (Preuss) Schroeter
Stem rust, *Puccinia graminis tritici* Eriksson & E. Henning
Stripe rust, *Puccinia striiformis* Westendorp
Leaf rust, *Puccinia recondita tritici* (Eriksson) Carleton
Soil rot, *Thanatephorus cucumeris* (Frank) Donk
Southern blight, *Botryobasidium rolfsii* (Saccardo) Venkatarayan
Eyespot, *Selenophoma donacis* (Passerini) Sprague & A. G. Johnson
Leaf spot, *Ascochyta sorghi* Saccardo
Anthracnose, *Colletotrichum graminicola* (Cesati) G. W. Wilson
Root rot, *Gloeosporium bolleyi* Sprague
Culm rot, *Cercosporella herpotrichoides* Fron
Blue mold, *Penicillium expansum* Link ex Thom
Spot blotch, *Helminthosporium sativum* Pammel, King, & Bakke
Leaf blight, *Helminthosporium tritici-repentis* Diedicke
Yellow spot, *Helminthosporium tritici-vulgaris* Nisikado
Leaf blight, *Alternaria triticina* Prasada & Prabhu
Other fungi associated with wheat

Basal glume blotch, *Pseudomonas atrofaciens* (McCull.) F. L. Stev.

Symptoms

Glume rot is distinguished by the appearance of pale to almost black areas at the base of the glumes. In severe infections most of the outer glume surface becomes discolored, and a bacterial rot may be found at the base of the spikelet. From there the invasion extends to the rachis and into the base of the kernel. An inconspicuous exudate is sometimes present.

Etiology

The organism is a gram-negative rod, forming chains and capsules, motile by 1–4 polar or bipolar flagella, and measuring $1\text{–}2.7 \times 0.6\ \mu$. Agar colonies are white becoming greenish, glistening, round with concentric markings, and smooth. A green fluorescent pigment is produced in culture. Optimum temperature, 25–28C.

References

6, 10.

Bacterial spike blight, *Corynebacterium tritici* (Hutch.) Burkh.

Symptoms

Spike blight appears on plants approaching maturity. The leaves become wrinkled and twisted, and a bacterial exudate forms a yellow, sticky liquid that, as it dries, seals the leaves and sheaths to the stems, preventing normal elongation. The glumes become stuck together and to adjoining parts; and the plant becomes distorted and abnormal.

Etiology

The organism is a gram-positive rod, motile by a single polar flagellum, and measures $2.4\text{–}3.2 \times 0.8\ \mu$. Agar colonies are bright yellow, deepening to orange, glistening, entire, round, and convex with opalescent margins and opaque centers.

References

6, 10.

Black chaff, *Xanthomonas translucens* f. *undulosa* (E. F. Sm., L. R. Jones, & Reddy) Hagb.

Symptoms

Black chaff occurs on the aboveground plant parts, forming yellow, translucent, irregularly linear stripes that may coalesce into blotches. On the outer glumes and awns, the lesions are brown to almost black, linear to striated, often coalescing and blackening the outer surface. On the rachis and culm, the lesions are narrowly linear. Scales of dried exudate may appear on the surfaces of lesions. In severe attacks, the kernels are shriveled and light in weight and some stunting may occur.

Etiology

The organism is a gram-negative rod, single or in pairs, motile by a single polar flagellum, and measuring 1–2.5×0.5–0.8 μ. Agar colonies are yellow to tan, round, smooth, shining, entire, and have some concentric striations. Optimum temperature, 26C.

References

3, 6, 10.

Root rot, *Pythium arrhenomanes* Drechs.

Symptoms

Root rot associated with wheat plants reveals large brown patches in the fields in early spring. The lower leaves become pale green and turn brown. The roots show reddish brown, water-soaked lesions at the tips of short crown roots. The general macroscopic symptom is a deterioration that might be expected from drought.

Etiology

The mycelium is hyaline, branched, nonseptate, and usually 2–6 μ in diameter. It supports lobulate sporangia that produce 20–50 zoospores, averaging 28–30 μ in diameter, and antheridia with 2 cilia. Oogonia are subspherical, averaging 28–30 μ in diameter. There are 15–20 crook-necked antheridia with contributing hyphae, several of which are distinct from the oogonium hyphae. Oospores are yellowish, subspherical, and 27–28 μ in diameter; they do not fill the oogonium.

References

6, 8, 15.

Root rot, *Pythium graminicola* Subr.

Symptoms

Root rot is of variable importance, depending on seasonal conditions. A general browning of the older foliage continues to develop, involving most of the leaves and eventually resulting in stunted plants and a low yield of shrunken kernels. Light tan to brown lesions develop on the fine roots and often on the primary roots, causing them to die back.

Etiology

The mycelium is hyaline, branched, nonseptate, 3–7 μ in diameter, and produces inflated, filamentous sporangia which germinate by initiating 15–48 zoospores that average 8–11 μ in diameter when encysted and are reniform when motile with 2 laterally attached cilia. Oogonia are spherical, smooth, and average about 30 μ in diameter. Antheridia, 1–6 to an oogonium, arise from the hyphae bearing the oogonium. Oospores are yellowish, spherical, average about 38 μ in diameter, and usually fill the oogonium.

References

9, 18, 24.

Downy mildew, *Sclerophthora macrospora* (Sacc.) Thirum., Shaw, & Naras.

Symptoms

Mildew produces a distinct distortion and malformation of the floral parts, causing considerable elongation of the rachis and a wide separation of the spikelets. On other plants where the infection varies only separate parts are diseased, resulting in a curling of the awns and bending and doubling over of the rachis. The heads are mostly sterile; frequently a proliferation of the head occurs with vegetative abortive foliage replacing floral parts.

Etiology

The fungus mycelium is hyaline, branched, nonseptate, and produces short sporangiophores that branch sympodially. The sporangia are lemon-shaped, papillate, 57–91 $\times$ 41–57 μ, and germinate by producing 24–32 biciliate zoospores. The oogonia and antheridia are produced on the same hyphae, and the single oospore completely occupies the oogonial wall. Oospores average 53–63 μ in diameter, and it is generally supposed that they germinate by germ tube development although zoospore production has been reported.

References

6, 24, 25, 27.

Powdery mildew, *Erysiphe graminis* f. sp. *tritici* Em. Marchal

Symptoms

Powdery mildew on wheat affects the aboveground parts of the host. It produces superficial mycelium and spores as a white, weblike growth that becomes more chalky as conidia accumulate. Diseased leaves become pale green, eventually yellow, and finally brown, shrivel and die. The florescence, rachis, glumes, and culms are often overgrown. Plants are frequently stunted, tillers limited, and grain poorly filled. The fungus is an obligate parasite.

Etiology

The mycelium is hyaline, septate, branched, grows over the epidermal surface, and gradually darkens to light brown. It is anchored by appressoria and derives nutriment from the haustoria. The conidiophores are erect, become septate, and form terminal conidia that are hyaline, 1-celled, barrel-shaped, 25–30 $\times$ 8–10 μ, and borne in chains. The ascocarp is a cleistothecium, which is brown to black, spherical, up to 220 μ in diameter, with undulating appendages. Asci are numerous, 9–30, measure 70–108 $\times$ 25–40 μ, and contain 8 ascospores, each measuring 20–23 $\times$ 10–13 μ.

References

6, 24.

Ergot, *Claviceps purpurea* (Fr.) Tul.

Symptoms

Ergot is a common disease of many grasses and cereals. It is usually first evident by the production in the flowering heads of a sweetish exudate which attracts insects. The ovule is invaded by the fungus and destroyed, causing a partial reduction in yield. The ovule is replaced by a sclerotium of the fungus, which enlarges and is conspicuous because of its color and size.

Etiology

The sclerotia are violet to black, white within, elongate, often up to 3 cm long, and curved to crescent-shaped. After a dormant period they germinate, forming capitate perithecial stromata which are pink to brown, spherical to elongate, stipitate, and up to 2 cm high. Perithecia, sunken in a knoblike head are flask-shaped, ostiolate, paraphysate, and contain the long, narrow asci, each with 8 filamentous, hyaline, septate ascospores that measure $50–75 \times 0.5–1.0\ \mu$. The ascospores infect the floral ovaries, producing conidia. The conidia are hyaline, 1-celled, very small, $10–15 \times 0.5\ \mu$, and disseminated by insects and other means.

References

6, 24.

Take-all, *Ophiobolus graminis* Sacc.

Symptoms

Take-all has been found mostly at the critical stage of culm elongation and at heading time. The foliage becomes pale green, yellow, and bleaches; tillering is reduced, and the plants become stunted. The principal roots, the stem base, and the crown show a brown decay, accompanied by mycelium of the fungus.

Etiology

The mycelium is brown to black, branched, and septate. The perithecia, formed among the hyphal strands beneath the sheaths, are hyaline and spherical, with ascospores measuring $70–80 \times 2–3\ \mu$.

References

6, 24.

Scab, *Gibberella zeae* (Schw.) Petch

Symptoms

Scab is known to cause seedling blight, foot rot, and head infection. Germinating seed may become infected, and a reddish brown, water-soaked decay develops before emergence, reducing the stand. Surviving seedlings often develop a crown rot involving the main roots and the base of the stem, developing mostly by the time the head appears. The plants continue to survive but in many instances appear off-color and weak. The head infection is very conspicuous since the diseased spikelets lose their chlorophyll and often bleach out to a pale straw yellow. Whole heads may be involved or only portions of the spikelets. During wet periods pinkish clusters of conidia develop within the glumes and around the kernels. Perithecia develop in scattered places on the glumes, rachises, and awns.

Etiology

The mycelium is hyaline, pinkish, branched, septate, and produces sporodochia containing sickle-shaped, tapering, usually 5-septate, nonconstricted conidia, measuring $40–60 \times 4–6\ \mu$; no chlamydospores are present. The perithecia, blue, purple to black and ovoid to conical, are almost superficial but imbedded in mycelium on the surface. The asci are hyaline, cylindrical, oblong to clavate, acuminate, and $60–76 \times 10–12\ \mu$. The ascospores are fusiform, straight or curved, acute, 3-septate, and $18–23 \times 2–5\ \mu$. The imperfect stages include various *Fusarium* spp.

References

6, 24.

Black spot, *Phyllachora graminis* (Pers. ex Fr.) Fckl.

Symptoms

Black spot, commonly found on many grasses, is not important on wheat. Infections are unnoticed until a discoloration develops on the leaf, causing the epidermis to darken; black stromatalike thickenings appear that are pronounced on the upper surface. Usually little damage results.

Etiology

The mycelium is dark, septate, branched, and occupies the host cells between the upper and lower epidermis. The perithecia are imbedded with ostioles protruding. They occupy loculi rather than separate fruiting bodies. The paraphyses are long and slender, and the asci are cylindric, measuring $70–80 \times 7–8\ \mu$. The ascospores are hyaline, ovoid, 1-celled, and uniseriate. No vegetative conidia are known.

Reference

24.

Speckled blotch, *Leptosphaeria tritici* (Garov.) Pass.

Symptoms

Blotch forms irregularly circular to elongate, linear spots on the foliage. The rounder spots are on the leaves of the plant in the juvenile stage, and the linear brown areas are on the culm and flag leaves at blossom time. The killed areas are light brown on the older leaves with no line of demarcation, while the spots may show a darker border on young foliage. These spots are often speckled with black pycnidia that are imbedded and in broken lines associated with the stomata.

Etiology

The ostioles coincide with the stomatal opening. The pycnidia are black, spherical to oval, thin-walled, small, and 80–150 μ in diameter. The pycnidiospores are hyaline, septate, seldom straight, and 50–90×3–4 μ. The perithecia are black, imbedded, globose, ostiolate, paraphysate, and papillate. The asci are hyaline, clavate, and stipitate, and the ascospores are hyaline to pale, biseriate, elongate, 3-septate, constricted, and 18–20×2–6 μ. The vegetative stage of the fungus is *Septoria tritici.*

References

4, 6, 13, 26.

Glume blotch, *Leptosphaeria nodorum* (Berk.) Muell.

Symptoms

Glume blotch forms tan to brownish, small, irregular spots on the outer glumes of the spikelets. These spots enlarge, often including most of the external glume area, and darken to a chocolate brown. The black, imbedded pycnidia appear in the discolored areas sometimes in lines between the prominent veins. The rachis and nodes are usually darkened, and pycnidia develop sparingly on these parts. Leaf spots are at first yellowish, and later the centers dry out and bleach, leaving a darker border. Severe infections often kill a large majority of the leaves. Pycnidia are less plentiful on the leaves than on the glumes.

Etiology

The pycnidia are black, spherical, ostiolate, and 160–210 μ in diameter; the pycnidiospores are hyaline, 3-septate, and 18–32×2–4 μ. The perithecia are black, spherical, ostiolate, and imbedded; the asci are hyaline, cylindrical, and 8-spored. The ascospores are hyaline and slightly colored, 3-septate, constricted, and 20–30×5–6 μ. The vegetative stage of the fungus is *Septoria nodorum.*

References

6, 13, 23, 26.

Loose smut, *Ustilago nuda* (Jens.) Rostr.

Symptoms

Loose smut is recognized as a disease of the spike in which the fragile, covered sori replace the grain in the spikelets. The diseased heads appear earlier than healthy heads. The sorus wall ruptures, freeing the black, powdery spores, and soon only the naked rachis remains. Fungus spores collect in glumes, included seed often germinate simultaneously, and seedling infection takes place. Flower infection may occur. The fungus continues to develop as the plant grows and finally produces spores in the head. These spores mature early enough to infect the flowers of the healthy plants.

Etiology

The mycelium is hyaline to brown, septate, and matures into spherical, finely echinulate, brown chlamydospores, which germinate to form promycelium which eventually infects the host flowers. The chlamydospores are 5–9 μ in diameter.

References

4, 6, 11, 12, 19.

Bunt, *Tilletia foetida* (Wallr.) Liro and *T. caries* (DC.) Tul.

Symptoms

Bunt cannot be confused with other smuts of the grain heads since the covering of the sorus is tough and permanent, and the black dusty chlamydospores are contained within. The symptoms usually are not visible until heading time, although some stunting may be detected along with a darker green color of plants before heading.

Etiology

The mycelium is septate and mostly hyaline. The cells round up into chlamydospores that become brownish black and are globose, smooth or with reticulate, meshlike wall markings, and 15–23 μ in diameter. They germinate by forming a promycelium or basidium that produces an apical crown of 8–16 filiform sporidia that fuse in pairs before being shed. Seedling infection takes place, and the fungus develops in the primordial tissue, culminating in ovule destruction and chlamydospore production.

References

4, 6, 11, 12, 19.

Flag smut, *Urocystis agropyri* (Preuss) Schroet.

Symptoms

Flag smut is evident throughout the life of the plant and is recognized by the black, long, narrowly linear sori that are conspicuous on the leaves and sheaths occupying the cells between the principal veins. The epidermis ruptures when the spores are mature, and they escape as dusty material. Seedling infection takes place following chlamydospore germination and continues as a systemic disease.

Etiology

The spore balls are mostly globose and composed of 1–4 chlamydospores surrounded by numerous smaller, sterile cells. The chlamydospores are 14–20 μ

in diameter and germinate by forming a basidium which produces 4 hyaline, cylindrical sporidia.

References

4, 6, 11.

Stem rust, *Puccinia graminis tritici* Eriks. & E. Henn.

Symptoms

Stem rust produces sori, containing reddish brown uredospores, on the aboveground parts of the plant. The uredia are small, blisterlike sori that may be elongate; they rupture the covering epidermis and free the dusty, orange yellow, oblong to circular, echinulate uredospores. The telia form in the same tissues, producing brown to black, oblong to linear sori that contain brown, fusiform to clavate, 2-celled, constricted pedicellate, smooth teliospores. The teliospores over-winter and germinate by forming a basidium for each cell. Four basidiospores are produced by each basidium. These basidiospores infect the alternate host, the barberry, where plus or minus pycnia containing hyaline, 1-celled pycniospores emerge, fuse, and initiate the aecia, which contain pale orange, 1-celled, echinulate, oval aeciospores in cuplike structures. These spores infect the wheat plant.

Etiology

The pycniospores are 2–4 μ in diameter. The aeciospores are 15–25 μ. The uredospores are 25–38 $\times$ 15–20 μ. The teliospores are 35–65 $\times$ 15–20 μ. The basidiospores are hyaline, linear, 1-celled, and 6–8 $\times$ 2–4 μ.

References

1, 5, 6, 22.

Stripe rust, *Puccinia striiformis* West.

Symptoms

Stripe rust is of particular interest because the pycnial and aecial stages are not known. The uredia produce pale yellow, linear, narrow stripes between the principal veins in the leaf blades and sheaths and some similar yellowing on the heads. The sori usually rupture when the uredospores are mature. The telia are mostly black, linear, and subepidermal.

Etiology

The uredospores are spherical to oval, echinulate, sometimes with spatulate paraphyses, and measure 19–30 $\times$ 16–25 μ. The telia appear on the lower leaf surfaces in long, fine, black lines with paraphyses. The teliospores are chestnut brown, 2-celled, oblong-clavate, rounded above, constricted, pedicellate, and 32–56 $\times$ 13–24 μ.

References

1, 6, 17.

Leaf rust, *Puccinia recondita tritici* (Eriks.) Carl.

Symptoms

Leaf rust may appear on the foliage from the seedling stage to maturity and usually survives from fall infections to the spring revival of growth. Heavy infection results in the deterioration of the foliage, which turns pale green, then yellow; browning begins at the tips and progresses downward. Leaves are killed successively but seldom so extensively as to kill plants. Severe infection results in reduction in yield and light-weight grain.

Etiology

The uredia are orange yellow to brown, circular sori, usually less than 1 mm in diameter. They may appear on both surfaces of the leaf, and at maturity they rupture the epidermis. The uredospores are echinulate, mostly globoid, and 16–32 × 13–24 μ. The blackish, scattered telia usually appear on the lower surface, mostly under the epidermis. Teliospores are chestnut brown, 2-celled, pedicellate, clavate to cylindrical, and 32–56 × 13–24 μ. The pycnia and aecia appear on opposite sides of the foliage of *Thalictrum polygamum*.

References

1, 5, 6.

Soil rot, *Thanatephorus cucumeris* (Frank) Donk

Symptoms

Soil rot results in the plants appearing stunted and off-color; they are weakened and sometimes killed. A brown decay develops on the roots, the basal portions of the culms, and surrounding leaf sheaths. The decay is found on plants in varying stages but is usually not severe on this host.

Etiology

The mycelium is brown, septate, branched, and frequently accumulates in the form of irregular, flat to oval sclerotia, adhering superficially to dead plant parts. The basidial stage may develop on the lower plant parts, producing basidiospores that are hyaline, 1-celled, elongate, oval, and 7–13 × 4–7 μ. The imperfect stage is *Rhizoctonia solani*.

References

6, 24.

Southern blight, *Botryobasidium rolfsii* (Sacc.) Venkat.

Symptoms

Southern blight may cause some killing of plants but is considered of little importance. The plants are girdled at or below the soil line, where inconspicuous, pale brown lesions develop in association with whitish mycelium.

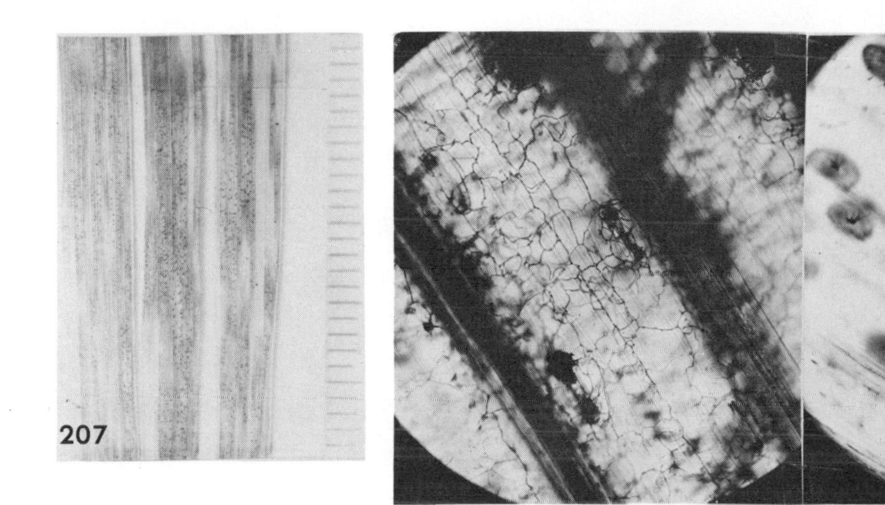

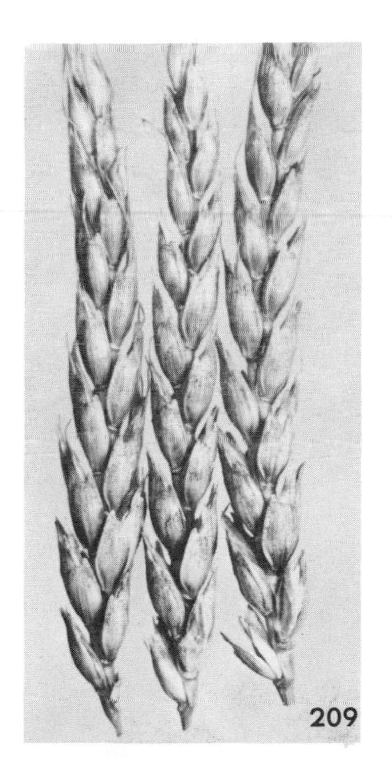

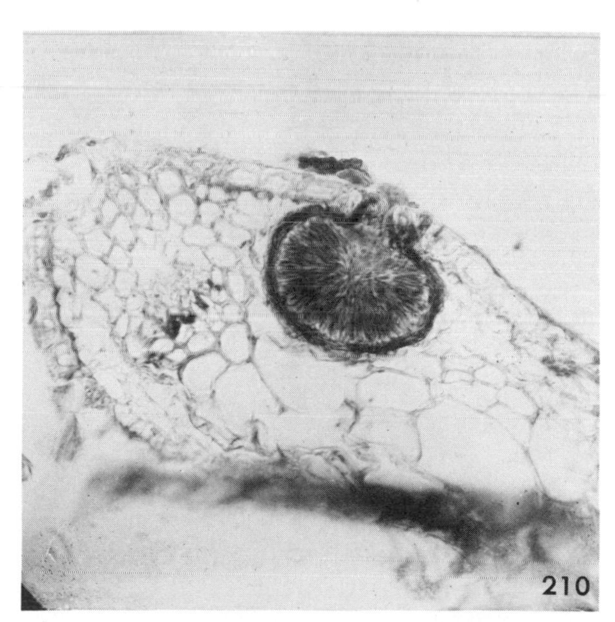

Figure 207. Speckled blotch, *Leptosphaeria tritici*, on wheat foliage. Enlarged 3X.
Figure 208. Speckled blotch, *Septoria tritici*, on wheat, showing pycnidia. Enlarged 7.6X.
Figure 209. Glume blotch, *Leptosphaeria nodorum*, on wheat.
Figure 210. Glume blotch, *Leptosphaeria nodorum*, showing pycnidium in section. Enlarged 10X.

Etiology

The sclerotia of the fungus are often present as brown, spherical, hard structures, loosely attached to the hyphae on the surface of the diseased plant parts, and usually measuring less than 1 mm in diameter. The sclerotia germinate by producing mycelium. The basidial stage of the fungus is rarely found. The basidiospores are hyaline, 1-celled, oval to elongate, and $3–10 \times 3–6$ μ. The imperfect stage is *Sclerotium rolfsii.*

References

16, 24.

Eyespot, *Selenophoma donacis* (Pass.) Sprague & A. G. Johnson

Symptoms

Eyespot is characterized by circular to elliptical, straw-colored spots with tan to lavender borders. They may enlarge considerably and coalesce, forming blotches mostly on the older half of the leaves.

Etiology

The pycnidia are brown, erumpent, globose, ostiolate, and 40–150 μ in diameter. The pycnidiospores are hyaline, falcate, 1-celled, and $18–35 \times 2–5$ μ.

References

6, 24.

Leaf spot, *Ascochyta sorghi* Sacc.

Symptoms

Leaf spot is more commonly found on the many grasses and less frequently on wheat. It causes brown, circular to elliptical foliage spots mostly on the lower leaf surface near the leaf tips. Heavily infected leaves become yellow and gradually deteriorate.

Etiology

The pycnidia are subglobose, erumpent, ostiolate, and 90–140 μ in diameter. The pycnidiospores are elongate to oval, fusoid, and $11–21 \times 1.6–4$ μ.

References

6, 24.

Anthracnose, *Colletotrichum graminicola* (Ces.) G. W. Wils.

Symptoms

Anthracnose is found extensively on most cereals and grasses following blossoming time. Dark stains develop on the lower portion of the culm and in the blades and sheaths of the older foliage that becomes brown and dies. The culm discoloration is not extensive until the sporulation of the fungus begins.

Etiology

The mycelium is hyaline, septate, branched, and usually scanty. The acervuli are largely superficial, 50–150 μ long, in clusters mixed with dark stromatic mycelium and black, septate setae that are most conspicuous and plentiful. The conidia, produced on short conidiophores, are hyaline, 1-celled, curved, broadly sickle-shaped, and $20–23 \times 3–6$ μ.

References

6, 24.

Root rot, *Gloeosporium bolleyi* Sprague

Symptoms

Root rot is known to be present on many gramineous hosts and is often considered mostly saprophytic. It has been proven to cause extensive root decay on plants growing in cold, wet soils. The root systems are reduced, and consequently nutrition is subnormal.

Etiology

The fungus mycelium is hyaline, branched, septate, and in culture is a reddish pink. There are no setae; the conidia are hyaline, oval to elongate, 1-celled, multiguttulate, and $4–9 \times 2–4$ μ.

Reference

24.

Culm rot *Cercosporella herpotrichoides* Fron

Symptoms

Culm rot lesions on the leaf sheaths are tan with brown margins of variable size, elliptical to ovate, and elongate to indeterminate. The culms immediately beneath the sheath spots contract similar spots. Decay occurs around the crown of the plants and on the uppermost roots. Often the black stroma materializes over the diseased area. The culms become brittle and lodge.

Etiology

The mycelium is septate, branched, and sometimes darkened. The conidiophores are short, erect, and protrude from the stomata. The conidia are curved, 5–7-septate, obclavate, and $30–80 \times 2–4$ μ.

References

6, 24.

Blue mold, *Penicillium expansum* Lk. ex Thom

Symptoms

Blue mold occurs when grain has gotten wet or has not been properly cured. Grain is subject to attacks by this fungus and probably several other closely related species. Some degree of decay may result when seed are planted and cold, wet conditions prevent rapid germination. The starchy endosperm is reduced by the mold, and consequently injury to germination results.

Etiology

The mycelium is hyaline, and the spores en masse are bluish green. The conidiophores are 1–2 mm high. Penicilli and verticils branch sufficiently to support whorls of sterigmata, supporting conidia in long chains, measuring 130–200 × 50–60 μ. The conidia are greenish, globose to elliptical, 1-celled, and 2–3 × 3–4 μ.

References

21, 24.

Spot blotch, *Helminthosporium sativum* Pam., King, & Bakke

Symptoms

Spot blotch not only forms spots on the foliage but also infects the spikelets and seeds. It causes root and crown rot and seedling decay in early stages of growth. It is very cosmopolitan, causing disease on a multitude of cereals and grasses. There are many races of the fungus.

Etiology

The mycelium is light brown, septate, branched, and produces conidiophores that emerge singly or in fascicles of 2 or 3, measure 11–150 × 6–8 μ, and are up to 8-septate. Conidia are brown, curved, taper toward the ends or bend to almost angular, up to 10-septate, and 60–120 × 15–20 μ. The ascospore stage is *Cochliobolus sativus.*

References

6, 24.

Leaf blight, *Helminthosporium tritici-repentis* Died.

Symptoms

Leaf blight in most instances is more or less indefinite in form and extent. Tan to brownish spots appear to be associated with a gradual deterioration of the foliage and final withering to a gray color.

Etiology

The leaves become covered with conidiophores that are dark olivaceous and 80–220 × 7–9 μ. The conidia are subhyaline to light tan-colored, mostly straight, up to 9-septate, somewhat constricted, and 45–175 × 12–21 μ. The ascospore stage is *Pyrenophora tritici-repentis.*

Reference

24.

Yellow spot, *Helminthosporium tritici-vulgaris* Nisikado

Symptoms

Yellow spot is mostly a foliage disease in which elongated, yellowish to brown, frequently zonate spots appear on the leaf blades and sheaths, causing considerable yellowing.

Etiology

The conidiophores appear singly or in clusters of 2 or 3 and are rigid, dark, up to 12-septate, and $90–400 \times 6–8\ \mu$. The conidia are yellow to brown, cylindrical, up to 10-septate, usually curved with some constrictions, and $28–180 \times 9–22\ \mu$. The ascospore stage is considered to be *Pyrenophora tritici-vulgaris*.

References

6, 24.

Leaf blight, *Alternaria triticina* Prasada & Prabhu

Symptoms

Brown, oval spots on leaves become irregular and surrounded with a bright yellow marginal zone. They coalesce when numerous and cause a severe blighting of the foliage, which becomes brown, producing a fired condition. The disease becomes conspicuous after the seedling stage.

Etiology

The mycelium is hyaline to yellow and olivaceous, septate, branched, and $3–7\ \mu$ wide. The conidiophores are olivaceous, septate, erect, emerge through the stomata, are single or fasciculate, straight, and geniculate. Conidia are brown to olivaceous, irregularly oval, ellipsoidal to conical, tapering to a beak, borne singly or in chains of up to 4, and $3–37 \times 3–7\ \mu$. Overall measurements of conidia, including beaks, are $15–89 \times 7–30\ \mu$. There may be up to 10 septations and up to 5 longitudinal setae.

References

14, 20.

Other fungi associated with wheat

Alternaria tenuis Nees ex Corda
Aspergillus glaucus Link
Botrytis cinerea Persoon ex Fries
Brachycladium spiciferum Bainier
Cephalosporium acremonium Corda
Curvularia geniculata (Tracy & Earle) Boedijn
Fusarium poae (Peck) Wollenweber
Helminthosporium cyclops Drechsler
Helminthosporium tetramera McKinney
Hendersonia crastophila Saccardo
Heterosporium avenae Oudemans

Leptosphaeria herpotrichoides De Notaris
Marasmius tritici P. A. Young
Mycosphaerella tulasnei (Janczewski) Lindau
Nigrospora sphaerica (Saccardo) Mason
Periconia circinata (Mangin) Saccardo
Pleospora gramineum Diedicke
Pseudomonas coronafaciens (Elliott) Stevens
Rhizopus stolonifera (Ehrenberg ex Fries) Lind
Stemphylium consortiole (Theumen) Groves & Skolko

References: wheat

1. Arthur, J. C., and G. B. Cummins. 1962. Manual of the rusts in United States and Canada. Hafner: N.Y.
2. Arya, H. C., and M. S. Ghemawat. 1963. Occurrence of powdery mildew of wheat in the neighborhood of Jodhpur. Indian Phytopathol. 6:123–30.
3. Bamberg, R. H. 1936. Black chaff disease of wheat. J. Agr. Res. 52:397–417.
4. Butler, E. J. 1918. Fungi and diseases in plants. Thacker: Calcutta, India.
5. Cummins, G. B. 1956. Host index and morphological characterization of the grass rusts of the world. Plant Disease Reptr. Suppl. 337:1–52.
6. Dickson, J. G. 1956. Diseases of field crops. 2d ed. McGraw-Hill: N.Y., pp. 225–92.
7. Dillon-Weston, W. A. R. 1948. Diseases of cereals. Longmans, Green & Co.: N.Y., pp. 1–64.
8. Drechsler, C. 1928. Pythium arrhenomanes n. sp., a parasite causing maize root rot. Phytopathology 18:873–75.
9. Drechsler, C. 1936. Pythium graminicolum and P. arrhenomanes. Phytopathology 26:676–84.
10. Elliott, C. 1951. Manual of bacterial plant pathogens. 2d ed. Chronica Botanica: Waltham, Mass.
11. Fischer, G. W. 1953. Manual of the North American smut fungi. Ronald Press: N.Y.
12. Fischer, G. W., and C. G. Shaw. 1953. A proposed species concept in the smut fungi, with application to North American species. Phytopathology 43:181–88.
13. Hosford, R. M., Jr., et al. 1969. Studies of Leptosphaeria avenaria f. sp. triticea on wheat in North Dakota. Plant Disease Reptr. 53:378–81.
14. Joshi, L. M., et al. 1970. Diseases of wheat in India other than rusts and smuts. Plant Disease Reptr. 54:594–97.
15. Kilpatrick, R. A. 1968. Seedling reaction of barley, oats and wheat to Pythium species. Plant Disease Reptr. 52:209–12.
16. Kilpatrick, R. A., and O. G. Merkle. 1967. Seedling disease of wheat caused by Sclerotium rolfsii. Phytopathology 57:538–40.
17. Loegering, W. Q., J. W. Hendrix, and L. E. Browder. 1967. The rust diseases of wheat. U.S. Dept. Agr., Agr. Handbook 334:1–22.
18. Matthews, V. D. 1931. Studies on the genus Pythium. University of North Carolina Press: Chapel Hill, pp. 58–61.
19. Moseman, J. G. 1968. Fungicidal control of smut diseases of cereals. U.S. Dept. Agr. Crops Res. Div., Compiled Crops Rept. 42:1–36.
20. Prasada, R., and A. S. Prabhu. 1962. Leaf blight of wheat caused by a new species of Alternaria. Indian Phytopathol. 15:292–93.
21. Raper, K. B., and C. Thom. 1949. A manual of the penicillia. Williams & Wilkins: Baltimore, Md.
22. Romig, R. W., B. J. Roberts, and A. P. Roelfs. 1968. Physiologic races of Puccinia graminis in the United States in 1965. Plant Disease Reptr. 52:512–15.
23. Scharen, A. L. 1964. Environmental influences on development of glume blotch in wheat. Phytopathology 54: 300–303.
24. Sprague, R. 1950. Diseases of cereals and grasses in North America. Ronald Press: N.Y.
25. Tyagi, P. D., and S. C. Anand. 1968. Downy mildew of wheat in India. Plant Disease Reptr. 52:569.
26. Weber, G. F. 1922. Septoria diseases of wheat. Phytopathology 12:537–85.
27. Whitehead, M. D. 1958. Pathology and pathological histology of downy mildew, Sclerophthora macrospora, on six graminicolous hosts. Phytopathology 48:485–93.

Yam, *Dioscorea alata* Linnaeus

Rust, *Goplana dioscoreae* (Berkeley & Broome) Cummins
Anthracnose, *Colletotrichum gloeosporioides* Penzig
Leaf spot, *Cercospora pachyderma* H. & P. Sydow
Leaf spot, *Cercospora uli* Raciborski
Other fungi associated with yam

Rust, *Goplana dioscoreae* (Berk. & Br.) Cumm.

Symptoms

The sori form on the foliage, causing yellow to orange spots. At first they are subepidermal and later result in a limited amount of yellowing in the small areas where sori are formed.

Etiology

The fungus is an obligate parasite, forming its various spore stages on the foliage. The pycnia and aecia are unknown. The uredinia are small, circular, and subepidermal, becoming erumpent. The yellowish uredospores are stipitate, borne singly, echinulate, subglobose, and 12–18 × 10–17 μ. The telia are produced subepidermally; as they mature, a small, knoblike protuberance evolves and later erupts, exposing the gelatinous telial matrix in which the teliospores are imbedded. The golden brown teliospores are usually 4-celled and are in a single layer. They are cylindrical, 46–60 × 8–10 μ, and germinate in place. The sterigmata are long and slender, one to each cell of the teliospores. The basidiospores are hyaline, subglobose, and 1-celled. The imperfect stage is *Uredo dioscoreae*.

Reference

3.

Anthracnose, *Colletotrichum gloeosporioides* Penz.

Symptoms

Small, barely visible, brown spots appear on the leaves and stems. Leaf spots enlarge rapidly and often coalesce, involving much area and causing leaves to yellow, wilt, die, and shed. The stem spots coalesce and become dark and shiny. Under continued humidity, brown acervuli appear on both leaf surfaces.

Etiology

The fungus grows well in culture, producing mycelium that is hyaline to olive green. Acervuli are formed early, producing light pink to brown spore masses without setae. Conidia are borne singly on the conidiophores and are hyaline, oblong to cylindrical, 11–18.5 × 3.7 μ, and guttulate.

References

1, 4, 5.

Leaf spot, *Cercospora pachyderma* H. & P. Syd.

Symptoms

Leaf spots are often indistinct in outline; however, they are indicated by a pale green or yellow on the upper surface of the leaves. On the lower surface there develops a dark smudgy area, corresponding to the lighter areas on the upper side. This dark olivaceous to almost black area extends from barely detectable origins to most or all of the lower surface. Very little host tissue is killed except when the disease is of long duration.

Etiology

The submerged, hyaline, septate, branched mycelium supports conidiophores that are pale olive brown, mostly single, septate, restricted, and 75–400 × 3–6 μ. The conidia are 1–4-septate, measuring 50–100 × 417 μ.

Reference

6.

Leaf spot, *Cercospora uli* Rac.

Symptoms

Leaf spots are slightly yellowish, circular to irregular in outline, and small, soon enlarging up to 1–2 cm. The leaves become yellowish brown to completely brown and are usually shed.

Etiology

The conidiophores are mostly brown, formed in fascicles of up to a dozen stalks, septate, straight or curved, some branched, and measure 20–150 × 3–5 μ. The conidia are hyaline to faintly colored, obclavate, sometimes slightly curved, 3–9-septate, and 20–120 × 4–8 μ.

Reference

2.

Other fungi associated with yam

Bagnisiopsis dioscoreae Wakefield
Botryobasidium cucumeris (Frank) Donk
Botryobasidium rolfsii (Saccardo) Venkatarayan
Botryodiplodia theobromae Patouillard
Cercospora carbonacea Miles
Cercospora dioscoreae Ellis & Martin
Colletotrichum dioscoreae Tehon
Gloeosporium pestis Massee
Hemiliae dioscoreae-aculatae Raciborski
Phoma oleraceae Saccardo
Phyllachora dioscoreae (Schweinitz) Saccardo
Phyllosticta dioscoreae Cooke

Pythium ultimum Trow
Ramularia dioscoreae Ellis & Everhart
Urocystis dioscoreae H. & P. Sydow

References: yam

1. Adeniji, M. O. 1970. Fungi associated with storage decay of yam in Nigeria. Phytopathology 60:590–92.
2. Chupp, C. 1953. A monograph of the fungus genus Cercospora. Published by the author: Ithaca, N.Y., pp. 195–99.
3. Cummins, G. B. 1935. Notes on some species of the Uredinales. Mycologia 27: 605–14.
4. Prasad, N., and R. D. Singh. 1959. Anthracnose disease of Dioscorea alata L. (Yam). Current Sci. (India) 29:66–67.
5. Singh, R. D., N. Prasad, and R. L. Mathur. 1966. On the taxonomy of the fungus causing anthracnose of Dioscorea alata L. Indian Phytopathol. 19: 65–71.
6. Sydow, H., and P. Sydow. 1914. Novae fungonum species. Ann. Mycol. 12: 195–204.

Bibliography

Ainsworth, G. C., and G. R. Bisby. 1961. Dictionary of the fungi. 5th ed. Commonwealth Mycological Institute: Kew, Surrey, England.
Alexopoulos, C. J. 1962. Introductory mycology. 2d ed. John Wiley: N.Y.
Anderson, H. W. 1956. Diseases of fruit crops. McGraw-Hill: N.Y.
Anonymous. 1968. Plant disease development and control. Publication 1596. National Academy of Sciences: Washington, D.C.
Arthur, J. C., and G. B. Cummins. 1962. Manual of the rusts in United States and Canada. Hafner Co.: N.Y.
Bancroft, K. 1910. A handbook of fungus diseases of West Indian plants. B. Pulman: London.
Barnes, H. V., and J. M. Allen. 1951. A bibliography of plant pathology in the tropics and in Latin America. U.S. Dept. Agr. Bibliog. Bull. 14:1–78.
Barnett, H. L. 1960. Illustrated genera of imperfect fungi. 2d ed. Burgess Pub. Co.: Minneapolis, Minn.
Bergey, D. H., et al. 1957. Manual of determinative bacteriology. 7th ed. Williams and Watkins: Baltimore, Md.
Bewley, W. F. 1923. Diseases of greenhouse plants. Benn Bros.: London.
Boyce, J. S. 1961. Forest pathology. 3d ed. McGraw-Hill: N.Y.
Briton-Jones, H. P. 1940. Diseases of the coconut palm. Bail, Tin, & Cox: London.
Brooks, F. T. 1953. Plant diseases. 2d ed. Oxford University Press: London.
Bunting, R. H., and H. A. Dade. 1924. Plant pathology. Waterlow: London.
Butler, E. J. 1918. Fungi and disease in plants. Thacker: Calcutta, India.
Butler, E. J., and S. G. Jones. 1949. Plant pathology. Macmillan: London.
Chapot, H., et al. 1968. Les agrumes au Moroc. Institut National de la Recherche Agronomique: Rabat, Morocco.
Chester, K. S. 1947. Nature and prevention of plant diseases. 2d ed. Blakiston: Philadelphia, Pa.
Chupp, C. 1953. A monograph of the fungus genus Cercospora. Published by the author: Ithaca, N.Y.
Chupp, C., and A. F. Sherf. 1960. Vegetable diseases and their control. Ronald Press: N.Y.
Cook, M. T. 1913. The diseases of tropical plants. Macmillan: London.
Copeland, E. B. 1924. Rice. Macmillan: London.
Couch, H. B. 1962. Diseases of turfgrasses. Reinhold: N.Y.
Couch, J. N. 1938. The genus Septobasidium. University of North Carolina Press: Chapel Hill.
Cramer, P. J. S. 1957. A review of literature in coffee research in Indonesia. Inter-American Institute of Agricultural Sciences: Turrialba, Costa Rica.
Crist, D. H. 1955. Rice. Longmans: London.
Cunningham, G. H. 1925. Fungous diseases of fruit trees in New Zealand and their remedial treatment. Brett Publishing and Printing Co.: Auckland, New Zealand.
Dearborn, R. C. 1925. Plant diseases of importance in the transportation of fruits and vegetables. Circular 30. A reprint of circular No. 473-A. Ann. R. R. Perishable Freight Assoc. Feb. 1918.
Fischer, G. W., and C. G. Shaw. 1953. A proposed species concept in the smut fungi with application to North American species. Phytopathology 43:181–88.
Hubert, E. E. 1931. An outline of forest pathology. John Wiley: N.Y.
Hughes, C. G., E. V. Abbott, and C. A. Wismer. 1964. Sugar-cane diseases of the world. Vol. 2. Elsevier Co.: N.Y.
Jackson, B. D. 1953. A glossary of botanic terms. Lippincott Co.: Philadelphia, Pa.
Joly, P. 1964. Le genre Alternaria. Encyclopedie Mycologique. Vol. 33. Paul LeChevalier: Paris.
Joshi, A. B. 1961. Sesamum. Indian Agricultural Research Institute: New Delhi, India.
Kamat, M. N. 1958. Handbook of tropical plant diseases. Prakash: Poona 2, India.
Kamat, M. N. 1963. Introductory plant pathology. 2d ed. Prakash: Poona 2, India.
Klotz, L. J., and H. S. Fawcett. 1948. Color handbook of citrus diseases. 2d ed. University of California Press: Berkeley.

Loesche, W. H. von. 1940. Bananas. Interscience: N.Y.
Lucas, G. B. 1965. Diseases of tobacco. Scarecrow Press: N.Y.
McAlpine, D. 1899. Fungus diseases of citrus trees in Australia and their treatment. R. S. Brain Co.: Melbourne, Australia.
McAlpine, D. 1902. Fungus diseases of stone fruit trees in Australia. R. S. Brain Co.: Melbourne, Australia.
Martin, J. P. 1938. Sugar-cane diseases in Hawaii. Advertiser Publishing Co.: Hawaii.
Martin, J. P., E. V. Abbott, and C. G. Hughes. 1961. Sugar-cane diseases of the world. Vol. 2. Elsevier Co.: N.Y.
Matsumoto, T. 1953. Monograph of sugar cane diseases in Taiwan. University of Taipei: Taipei, Taiwan.
Matthews, V. D. 1931. Studies on the genus Pythium. University of North Carolina Press: Chapel Hill.
Melhus, I. E., and G. C. Kent. 1939. Elements of plant pathology. Macmillan: N.Y.
Middleton, J. T. 1943. The taxonomy, host range and geographic distribution of the genus Pythium. Mem. Torrey Botan. Club 20:1–171.
Mortensen, E., and E. T. Bullard. 1964. Handbook of tropical and sub-tropical horticulture. Department of State, Agency for International Development: Washington, D.C.
Nowell, W. 1923. Diseases of crop-plants in the Lesser Antilles. West India Committee: London.
Owens, C. E. 1928. Principles of plant pathology. John Wiley: N.Y.
Padwick, G. W. 1950. Manual of rice diseases. Commonwealth Mycological Institute: Kew, Surrey, England.
Petch, T. 1921. The diseases and pests of the rubber tree. Macmillan: London.
Petch, T. 1923. The diseases of the tea bush. Macmillan: London.
Pirone, P. P., B. O. Dodge, and H. W. Rickett. 1960. Diseases and pests of ornamental plants. 3d ed. Ronald Press: N.Y.
Pound, G. S., et al. 1960. Symposium on training for the future in plant pathology. Phytopathology 50:500–516.
Rankin, W. H. 1929. Manual of tree diseases. Macmillan: N.Y.
Reynolds, P. K. 1927. The banana. Houghton Mifflin: Boston, Mass.
Rhoads, A. S., and E. F. DeBusk. 1931. Diseases of citrus in Florida. Florida Agr. Expt. Sta. Bull. 229.
Salmon, E. S. 1900. A monograph of the Erysiphaceae. Mem. Torrey Botan. Club 9:1–292.
Segura, C. 1965. Enfermedades de cultivos tropicales y subtropicales. José D. Segura Montoya: Lima, Peru.
Sharples, A. 1936. Diseases and pests of the rubber tree. Macmillan: London.
Shurtleff, M. C. 1962. How to control plant diseases in home and garden. 2d ed. Iowa State University Press: Ames.
Simmonds, J. H. 1938. Plant diseases and their control. D. Whyte: Brisbane, Australia.
Simmonds, N. W. 1959. Bananas. Longmans: London.
Smith, E. F. 1920. An introduction to bacterial diseases of plants. W. B. Saunders: Philadelphia, Pa.
Snell, W. H., and E. A. Dick. 1957. A glossary of mycology. Harvard University Press: Cambridge, Mass.
Sprague, R. 1950. Diseases of cereals and grasses in North America. Ronald Press: N.Y.
Stapp, C. 1961. Bacterial plant pathogens. Oxford University Press: London.
Stevens, F. L. 1913. The fungi which cause plant disease. Macmillan: N.Y.
Stevens, F. L., and J. C. Hall. 1910. Diseases of economic plants. Macmillan: N.Y.
Tidbury, G. E. 1949. The clove tree. C. Lockwood: London.
Toussoun, T. A., and P. E. Nelson. 1968. A pictorial guide to the identification of *Fusarium* species. Pennsylvania State University Press: University Park. Pp. 1–51.
Vestal, E. F. 1950. A textbook of plant pathology. Kitabistan: Allahabad, India.
Walker, J. C. 1952. Diseases of vegetable crops. McGraw-Hill: N.Y.
Walker, J. C. 1957. Plant pathology. 2d ed. McGraw-Hill: N.Y.
Wardlaw, C. W. 1961. Diseases of the banana. 2d ed. John Wiley: N.Y.
Watt, G. 1898. The pests and blights of the tea plant. Calcutta, India.
Webber, H. J., and L. D. Bachelor. 1948. The citrus industry. Vol. 1. University of California Press: Berkeley.
Weiss, F., et al. 1960. Index of plant diseases in the United States. U.S. Dept. Agr. Handbook 165.

Wellman, F. L. 1961. Coffee. Leonard Hill: London.
Westcott, C. 1960. Plant disease handbook. 2d ed. Van Nostrand: Princeton, N.J.
Whetzel, H. H. 1912. History of plant pathology. Saunders: Philadelphia, Pa.
Wolf, F. A. 1935. Tobacco diseases and decay. Duke University Press: Durham, N.C.
Wormold, W. 1946. Diseases of fruits and nuts. C. Lockwood: London.
Zundel, G. L. 1953. The Ustilaginales of the world. Penn. State Univ. Contribution 176:1–410.

Useful Periodicals

Agricultural Gazette of New South Wales
American Journal of Botany
Annales Mycologici
Annals of Applied Biology
Annals of Botany
Australian Journal of Agricultural Research
Botanical Gazette
Bulletin de la Societe Mycologique de France
Bulletin of the Torry Botanical Club
Cacao
Canadian Journal of Research
Citrograph
Citrus Industry
Comptes Rendus
Current Science, India
East African Journal of Agriculture
Empire Journal of Experimental Agriculture
Encyclopedic Mycologique
Farlowia
Florida Agricultural Experiment Station Bulletin
Florida Department of Agriculture Bulletin
Hilgardia
Indian Agricultural Journal
Indian Journal of Agricultural Science
Indian Journal of Horticulture
Indian Phytopathology
Journal of Agricultural Research
Journal of the Department of Agriculture of Puerto Rico
Journal of the Elisha Mitchell Scientific Society
Journal of the Indian Botanical Society
Journal of Mycology
Journal of the Rubber Research Institute of Malaya
Lloydia
Malaysian Agricultural Journal
Memoirs of the Department of Agriculture, India
Memoirs of the Torrey Botanical Club
Mycologia
Mycopathologia et Mycologia Applicata
Philippine Agriculturist
Philippine Journal of Agriculture
Philippine Journal of Science
Phytopathologische Zeitschrift
Phytopathology
Plant Disease Reporter
Plant Pathology
Proceedings of the Florida State Horticultural Society
Queensland Agricultural Journal
Queensland Journal of Agricultural Science
Review of Applied Mycology
Revue de Pathologie Vegetale
Sitzungsberichte, Wein.
Tea Research Institute of Ceylon, Bulletin
Transactions of the British Mycological Society
Tropical Agriculture, Ceylon
Tropical Agriculture, Trinidad
Tropical Agriculturist

United States Department of Agriculture: Agricultural Handbooks, Agricultural Monographs, Agricultural Technical Bulletins, Agricultural Bulletins, Agricultural Yearbook 1953 (Plant Diseases), Miscellaneous Bulletins
West Indian Bulletin
Zeitschrift fuer Pflanzenkrankheiten

Index of Hosts by Common Name

Index of Hosts by Genus

Index of Diseases

Index of Bacteria and Fungi*

*Italic page numbers refer to illustrations.

WINTER PARK PUBLIC LIBRARY
460 E. NEW ENGLAND AVENUE
WINTER PARK, FLORIDA 32789

DISCARD